Lecture Notes in Control and Information Sciences

Volume 474

Series editors

Frank Allgöwer, Stuttgart, Germany
Manfred Morari, Zürich, Switzerland

Series Advisory Boards

P. Fleming, University of Sheffield, UK
P. Kokotovic, University of California, Santa Barbara, CA, USA
A.B. Kurzhanski, Moscow State University, Russia
H. Kwakernaak, University of Twente, Enschede, The Netherlands
A. Rantzer, Lund Institute of Technology, Sweden
J.N. Tsitsiklis, MIT, Cambridge, MA, USA

About this Series

This series aims to report new developments in the fields of control and information sciences—quickly, informally and at a high level. The type of material considered for publication includes:

1. Preliminary drafts of monographs and advanced textbooks
2. Lectures on a new field, or presenting a new angle on a classical field
3. Research reports
4. Reports of meetings, provided they are
   (a) of exceptional interest and
   (b) devoted to a specific topic. The timeliness of subject material is very important.

More information about this series at http://www.springer.com/series/642

Thor I. Fossen · Kristin Y. Pettersen
Henk Nijmeijer
Editors

Sensing and Control for Autonomous Vehicles

Applications to Land, Water and Air Vehicles

Editors
Thor I. Fossen
Norwegian University of Science and
 Technology (NTNU)
Trondheim
Norway

Henk Nijmeijer
Eindhoven University of Technology (TuE)
Eindhoven
The Netherlands

Kristin Y. Pettersen
Norwegian University of Science and
 Technology (NTNU)
Trondheim
Norway

ISSN 0170-8643   ISSN 1610-7411 (electronic)
Lecture Notes in Control and Information Sciences
ISBN 978-3-319-55371-9   ISBN 978-3-319-55372-6 (eBook)
DOI 10.1007/978-3-319-55372-6

Library of Congress Control Number: 2017934064

© Springer International Publishing AG 2017
This work is subject to copyright. All rights are reserved by the Publisher, whether the whole or part of the material is concerned, specifically the rights of translation, reprinting, reuse of illustrations, recitation, broadcasting, reproduction on microfilms or in any other physical way, and transmission or information storage and retrieval, electronic adaptation, computer software, or by similar or dissimilar methodology now known or hereafter developed.
The use of general descriptive names, registered names, trademarks, service marks, etc. in this publication does not imply, even in the absence of a specific statement, that such names are exempt from the relevant protective laws and regulations and therefore free for general use.
The publisher, the authors and the editors are safe to assume that the advice and information in this book are believed to be true and accurate at the date of publication. Neither the publisher nor the authors or the editors give a warranty, express or implied, with respect to the material contained herein or for any errors or omissions that may have been made. The publisher remains neutral with regard to jurisdictional claims in published maps and institutional affiliations.

Printed on acid-free paper

This Springer imprint is published by Springer Nature
The registered company is Springer International Publishing AG
The registered company address is: Gewerbestrasse 11, 6330 Cham, Switzerland

Preface

Sensing and Control for Autonomous Vehicles: Applications to Land, Water and Air Vehicles contains a collection of contributions presented at an invited workshop with the same name held June 20–22, 2017 in Ålesund, Norway.

The subject of the book is sensing and control with applications to autonomous vehicles. Guidance, navigation and motion control systems for autonomous vehicles are increasingly important in land-based, marine and aerial operations. Autonomous underwater vehicles may be used for pipeline inspection, light-intervention work, underwater survey, and collection of oceanographic/biological data. Autonomous aerial and ground vehicles can be used in a large number of applications such as inspection, monitoring, data collection, surveillance, etc. At present, vehicles operate with limited autonomy and intelligence. There is a growing interest for cooperative and coordinated multi-vehicle systems, localization and mapping, path planning, robust autonomous navigation systems, and robust autonomous control of vehicles. Unmanned vehicles with high levels of autonomy may be used for safe and efficient collection of environmental data, for assimilation of climate and environmental models and to complement global satellite systems.

With an appropriate balance between mathematical theory and practical applications, academic and industrial researchers working on sensing and control engineering aspects of autonomous vehicles will benefit from this comprehensive book. It is also suitable for final year undergraduates and postgraduates, lecturers, development officers, and practitioners in the areas of guidance, navigation and control of autonomous vehicles.

Acknowledgements

We are grateful to our sponsors:

- Center for Autonomous Marine Operations and Systems (NTNU AMOS) at the Norwegian University of Science and Technology, Trondheim, Norway (Norwegian Research Council grant no. 223254).
- Rolls-Royce Marine, Alesund, Norway

Trondheim, Norway   Thor I. Fossen
Trondheim, Norway   Kristin Y. Pettersen
Eindhoven, The Netherlands   Henk Nijmeijer

Contents

Part I Vehicle Navigation Systems

Observers for Position Estimation Using Bearing and Biased Velocity Information... 3
Florent Le Bras, Tarek Hamel, Robert Mahony and Claude Samson

Nonlinear Camera- and GNSS-Aided INS for Fixed-Wing UAV Using the eXogenous Kalman Filter 25
Lorenzo Fusini, Thor I. Fossen and Tor Arne Johansen

Motion Control of ROVs for Mapping of Steep Underwater Walls.. 51
Stein M. Nornes, Asgeir J. Sørensen and Martin Ludvigsen

Part II Localization and Mapping

Underwater 3D Laser Scanners: The Deformation of the Plane 73
Albert Palomer, Pere Ridao, David Ribas and Josep Forest

Advances in Platforms and Algorithms for High Resolution Mapping in the Marine Environment 89
R. Thomas Sayre-McCord, Chris Murphy, Jeffrey Kaeli, Clayton Kunz, Peter Kimball and Hanumant Singh

New Design Techniques for Globally Convergent Simultaneous Localization and Mapping: Analysis and Implementation............ 121
Pedro Lourenço, Bruno Guerreiro, Pedro Batista, Paulo Oliveira and Carlos Silvestre

Pose-Graph SLAM for Underwater Navigation 143
Stephen M. Chaves, Enric Galceran, Paul Ozog, Jeffrey M. Walls and Ryan M. Eustice

Exploring New Localization Applications Using a Smartphone......... 161
Fredrik Gustafsson and Gustaf Hendeby

Part III Path Planning

Model-Based Path Planning....................................... 183
Artur Wolek and Craig A. Woolsey

**Constrained Optimal Motion Planning for Autonomous
Vehicles Using PRONTO** 207
A. Pedro Aguiar, Florian A. Bayer, John Hauser, Andreas J. Häusler,
Giuseppe Notarstefano, Antonio M. Pascoal, Alessandro Rucco
and Alessandro Saccon

Part IV Sensing and Tracking Systems

Observability-Based Sensor Sampling............................. 229
Kristi A. Morgansen and Natalie Brace

**Tracking Multiple Ground Objects Using a Team of Unmanned
Air Vehicles**.. 249
Joshua Y. Sakamaki, Randal W. Beard and Michael Rice

**A Target Tracking System for ASV Collision Avoidance Based
on the PDAF** 269
Erik F. Wilthil, Andreas L. Flåten and Edmund F. Brekke

**Detection and Tracking of Floating Objects Using a UAV
with Thermal Camera** 289
Håkon Hagen Helgesen, Frederik Stendahl Leira, Tor Arne Johansen
and Thor I. Fossen

Part V Identification and Motion Control of Robotic Vehicles

**Experimental Identification of Three Degree-of-Freedom Coupled
Dynamic Plant Models for Underwater Vehicles** 319
Stephen C. Martin and Louis L. Whitcomb

**Model-Based LOS Path-Following Control of Planar Underwater
Snake Robots.**... 343
Anna M. Kohl, Eleni Kelasidi, Kristin Y. Pettersen
and Jan Tommy Gravdahl

Robotized Underwater Interventions.............................. 365
Giuseppe Casalino, Enrico Simetti and Francesco Wanderlingh

Adaptive Training of Neural Networks for Control of Autonomous Mobile Robots 387
Erik Steur, Thijs Vromen and Henk Nijmeijer

Modeling, Identification and Control of High-Speed ASVs: Theory and Experiments 407
Bjørn-Olav Holtung Eriksen and Morten Breivik

Part VI Coordinated and Cooperative Control of Multi-vehicle Systems

From Cooperative to Autonomous Vehicles 435
Tom van der Sande and Henk Nijmeijer

Coordination of Multi-agent Systems with Intermittent Access to a Cloud Repository 453
Antonio Adaldo, Davide Liuzza, Dimos V. Dimarogonas and Karl H. Johansson

A Sampled-Data Model Predictive Framework for Cooperative Path Following of Multiple Robotic Vehicles 473
A. Pedro Aguiar, Alessandro Rucco and Andrea Alessandretti

Coordinated Control of Mobile Robots with Delay Compensation Based on Synchronization 495
Yiran Cao and Toshiki Oguchi

Index 515

Part I
Vehicle Navigation Systems

Observers for Position Estimation Using Bearing and Biased Velocity Information

Florent Le Bras, Tarek Hamel, Robert Mahony and Claude Samson

Abstract This chapter considers the questions of observability and design of observers for position estimates of an object moving in $\mathbb{R}^n$. The scenario considered assumes a possibly biased measurement of velocity along with *bearing* or *direction* measurements to one or multiple fixed points in the environment. The motivating example is a robot moving in either two- or three-dimensional space with a sensor, such as a camera, capable of providing bearing but not distance to observed fixed points in the environment. We provide a comprehensive observability analysis, and discuss stability and convergence of the observer design and observer error dynamics, under persistence of excitation conditions on the vehicle motion. Interestingly, we can show uniform observability even for a single direction measure and with unknown velocity bias as long as the vehicle trajectory is persistently exciting. Some extensions to the case of multiple directions with known- or unknown-observed point location are addressed and observers endowed with exponential stability of error dynamics are derived. Simulation results are presented that demonstrate the performance of the proposed approach.

F. Le Bras (✉)
French Direction Générale de l'Armement (Technical Directorate), Paris, France
e-mail: florent.le-bras@polytechnique.org

T. Hamel
I3S UNS-CNRS, Université Côte d'Azur, Nice, France
e-mail: thamel@i3s.unice.fr

R. Mahony
Australian Centre for Robotic Vision, Research School of Engineering, Australian National University, Canberra, Australia
e-mail: Robert.Mahony@anu.edu.au

C. Samson
INRIA and I3S UNS-CNRS, Université Côte d'Azur, Nice, France
e-mail: csamson@i3s.unice.fr

© Springer International Publishing AG 2017
T.I. Fossen et al. (eds.), *Sensing and Control for Autonomous Vehicles*,
Lecture Notes in Control and Information Sciences 474,
DOI 10.1007/978-3-319-55372-6_1

1 Introduction

There is a rich literature in vision-based pose estimation driven by advances in the structure from motion in the field of computer vision [1]. Most of the recent structure from motion algorithms are formulated as an optimisation problem over a set of selected images [2], however, recent work has emphasised the importance of considering motion models and filtering techniques [3] for certain classes of important problems, especially those with a smaller number of environment features but many measurements such as is common in robotics applications. Recursive filtering methods for vision-based structure from motion and pose estimation themselves have a rich history primarily associated with stochastic filter design such as EKF, unscented filters, and particle filters [4–7]. A comparison of EKF and particle filter algorithms for vision-based SLAM is available in [8]. Although nonlinear observer design does not provide a stochastic interpretation of the state estimate they hold the promise to handle the nonlinearity of the vision pose estimation problem in a robust and natural manner [9]. Ghosh and subsequent authors consider nonlinear observers on the class of perspective systems [10–14], that is systems with output in a projective space obtained as a quotient of the state space. Defining x as the position of the vehicle, the perspective outputs $y(x)$ are of the form

$$y^P = (x_1/x_n, \ldots x_{n-1}/x_n, 1)$$

and can be used to model the nonlinear projection along rays through the origin onto an affine image plane perpendicular to the focal axis of a classical perspective camera geometry. There are a number of works that consider filtering for y^P directly, rather than estimating the camera position [15–17], corresponding to image tracking. Although significant work has been based on this output representation, it tends to lead to algebraically complex observer and filter design and difficult analysis [10–14]. The observability problem from a single-direction measurement (expressed directly in S^2) in the case where the measured velocity is biased by an unknown constant perturbation has been studied recently in [18]. The authors use an augmented state to linearize the problem. The same problem was also studied in [19] using nonlinear techniques without requiring an additional state. The present chapter offers an extension and generalization of previous studies by the authors and deals in particular with single or multiple points with known or unknown location as well as velocity bias.

Rather than using the perspective outputs favoured in previous papers we use direction output representation of the measurement

$$y = x/|x| = y^P/|y^P|$$

corresponding to projection onto a virtual spherical image plane and differing from perspective outputs only in the scaling. The two formulations are essentially equivalent from a systems perspective in the region where perspective outputs are defined, however, we believe that the direction output representation contributes to the algebraic simplicity of the observer proposed in the present chapter. Much of the material presented in this chapter is already available in the literature, however, the focus on estimating the position as well as the bias in the measured velocity of an object moving in n (≥ 2)-dimensional Euclidean space from direction measurements changes the perspective and it is well worth covering the material again in a variety of situations while providing new solutions and new results in closely related cases. The structural question of observability is addressed and explicit and simple observability conditions are derived. The observers are designed first for the case of strong observability, corresponding to the situation in which the position can be algebraically obtained from at least 2 uniformly non-collinear directions to target-points with known locations. We then describe the proposed observer design methodology when a single observed point is considered. We characterize the observer convergence in terms of the persistence of excitation property and we provide some insights on the observer's performance via a pseudo-Riccati equation introduced to improve the rate of convergence of the observer error to zero of the estimation error but also in order to estimate the eventual bias on the velocity measurement. The approach is then generalized to the case of multiple target-points with known location.

The SLAM problem is also considered as a natural extension of the proposed approach to estimate simultaneously the position of the vehicle and the relative location of the target-points involved in the measurements. We show in particular that the solutions we present in the chapter exploits a reduced dimension of the state and hence ensures a low computational complexity. To the authors knowledge, the solutions we proposed in the chapter have not been considered in the nonlinear observer literature. We provide rigorous proofs of the main results of the chapter as well as simulations to demonstrate the performance of the observers.

The chapter is organized in five sections. Following the present introduction, Sect. 2 introduces the system under consideration and points out observability properties attached to it. Section 3 describes the observer design methodology in the case of known location of the target-points. The case of unbiased velocity is first considered in Sect. 3.1 and then a specific state augmentation is proposed when the velocity measurements are biased by a constant vector in Sect. 3.2. We show in Sect. 4 how to modify the observers via state augmentation when the location of the target-points are unknown. This augmentation addresses the SLAM problem, where both the position of the vehicle as well as the relative target-point location must be determined. Section 5 presents a few illustrative simulations. Concluding remarks are provided in Sect. 6.

2 Problem Statement

The motivating systems considered are the kinematics of a vehicle moving in two dimensions (the planar case) or in three dimensions (the spatial case). For the sake of generality, we will assume simply that the vehicle is moving in $\mathbb{R}^n$ with $n \geq 2$ and develop a general theory. In this work, we consider the case where the orientation of the vehicle is known a-priori or separately estimated from a attitude observer. It follows that we can derotate all measurements and consider only inertial representations of position, velocity and direction. One has

$$\dot{x} = v + a \tag{1}$$

$$y_i = \frac{x - p_i}{|x - p_i|} \in S^{n-1}, \ i = 0, \ldots, l-1 \tag{2}$$

where $v \in \mathbb{R}^n$ is the velocity of the object, $a \in \mathbb{R}^n$ represents an unknown bias and $p_i \in \mathbb{R}^n$ is the position of the target-point i. The unit sphere S^{n-1} is the space of measurements $y_i \in \mathbb{R}^n$ such that[1] $|y_i| = 1$. An example of such a measurement is the bearing in S^2 obtained from a moving camera in 3D space with known orientation looking at a fixed target. In most applications the unknown velocity bias $a \in \mathbb{R}^n$ represents the velocity of the fluid in which the moving object is evolving, as well as any measurement bias.

We assume that the position x, as well as the velocity v, the bias a and the coordinates of the target-points are bounded. We emphasize that the measurements $y_i \in S^{n-1}$ and $v \in \mathbb{R}^n$ must be known at all times. In case where $l = 1$, we assume without loss of generality that the target-point is located at the origin of the inertial frame such $y = y_0 = x/|x|$.

2.1 Observability Analysis

We first give general observability criteria using well-known concepts regarding observability of nonlinear systems.

Recall that two different points $x_1^0, x_2^0 \in \mathbb{R}^n$ are said distinguishable, if there exists an input $v(t) \in \mathbb{R}^n$ and a time t_1 such that for solutions $x_1(t), x_2(t)$ of (1) with $x_1(0) = x_1^0, x_2(0) = x_2^0$ we have $y(x_1(t_1)) \neq y(x_2(t_1))$. Equivalently, in this case one says that the admissible input distinguishes the two initial states, and also that two initial states of system (1)–(2) are indistinguishable if they are not distinguished by any admissible input. This description leads to the following definition:

Definition 1 A system is called *strongly observable* if all pairs of distinct initial states are distinguished by all admissible inputs. It is called *weakly observable* if every pair of distinct initial states is distinguished by at least one admissible input.

Reasons to differentiate between strong observability and weak observability are well explained in the nonlinear control literature. For complementary details on this subject we refer the reader to Sussmann [20].

The following persistence of excitation condition yields an observability result for system (1)–(2).

---

[1] Here $|.|$ stands for the Euclidean norm of vectors and $||.||$ is the induced matrix norm.

Definition 2 A direction $y(t) \in S^{n-1}$, is called *persistently exciting* if there exist $\delta > 0$ and $0 < \mu < \delta$ such that for all t

$$\int_t^{t+\delta} \pi_{y(\tau)} d\tau \geq \mu I, \text{ with } \pi_y = (I - yy^\top) \tag{3}$$

For future use, note that (3) is equivalent to

$$\text{For all } b \in S^{n-1} \text{ then } \int_t^{t+\delta} |\pi_{y(\tau)} b|^2 d\tau \geq \mu, \tag{4}$$

Another characterization of persistence of excitation, in terms of the property that the time-derivative of $\dot{y}$ must satisfy, is provided in the following lemma.

Lemma 1 *Assume that $\dot{y}(t)$ is uniformly continuous, then relation (3) (respectively (4)) is equivalent to:*

$$\text{There exists } (\delta, \varepsilon) > 0 \text{ and } \tau \in [t, t+\delta] \text{ such that } |\dot{y}(\tau)| \geq \varepsilon \tag{5}$$

Proof The proof of this lemma is given in Appendix.

Lemma 2 *The system (1)–(2), complemented with the equation $\dot{a} = 0$, with $X = (x^\top \ a^\top)^\top$ as the system state vector, v as the system input, and a single direction y as the system output, is weakly observable but not strongly observable.*

Proof Choose, for instance, the input $v(t) = (\cos(t), \sin(t), 0 \ldots 0)^T$. The solutions to the system are then given by $x(t) = x(0) + (\sin(t) + at, -\cos(t) + at, 0 \ldots 0)^T$ and one easily verifies that $y_1(t) = y_2(t), \forall t$, implies that $x_1(0) = x_2(0)$. This establishes the weak observability property of the system. Note also that the chosen input renders both outputs $y_1(t)$ and $y_2(t)$ persistently exciting in the sense of the definition (2). On the other hand, one verifies that, if the input v is constant, then initial states $x_1(0) = k_1 v$ and $x_2(0) = k_2 v$, with k_1 and k_2 denoting arbitrary positive numbers, cannot be distinguished because y_1 and y_2 are constant and equal in this case. This proves that the system is not strongly observable.

The weak observability property of the system justifies the introduction of the persistence condition (3) to characterize "good" outputs (produced by "good" inputs) yielding a property of "uniform" observability that renders the state-observation problem (addressed in the next section) well posed.

For the case for more than 1 known target-points location, the criterion of strong observability is given by items 1 and 2 of the following lemma.

Lemma 3 *Let define $Q := \sum_{i=0}^{l-1} \pi_{y_i(t)}$. Then the persistency of excitation (Definition 2):*

$$\int_t^{t+\delta} Q(\tau) d\tau \geq \mu I,$$

is satisfied if Lemma 1 apply at least for one of the measures y_i. It is automatically satisfied independently from the input, if:

1. $l \geq 2$ provided that at least two directions y_i and y_j ($i \neq j$) are uniformly non-collinear. That is, for all $t \geq 0$ there exists an $\varepsilon_1 > 0$ such that $|y_i(t)^\top y_j(t)| \leq 1 - \varepsilon_1$,
2. $l \geq 3$ provided that at least three target-points are not aligned.

Proof The proof of the first claim of this lemma is a direct application of Lemma 1. As for the proof of items 2 and 3, it suffices to verify that Q is uniformly positive definite (i.e. there exists an $\varepsilon > 0$, such that $Q \geq \varepsilon$) and hence by choosing $\varepsilon = \frac{\mu}{\delta}$ one can ensure that the above persistent excitation condition is automatically satisfied independently of the time variation of x.

3 Observer Design Methodology

The problem of state observation refers to the design of an algorithm that allows one to recover actual state values from the observation of previous outputs. Let $\mathcal{M}_n$ denote the set of $n \times n$ real matrices and let $\mathcal{M}_n^+$ denote the set of positive-definite symmetric matrices. We start by the observer design for the problem of estimating the state $X = \begin{pmatrix} x^\top & a^\top \end{pmatrix}^\top$ from $l \geq 1$ measurements $y_i = (x - p_i)/|x - p_i|$, in case where p_i ($i \in \{0, \ldots, l-1\}$) are known vectors of coordinates of fixed target-points. Each measure y_i is the unit vector measuring the direction between the vehicle and the target-point i.

3.1 The Case of Unbiased Velocity

This subsection is devoted to the situation when the unknown constant velocity bias is zero. It is important to distinguish first between the strong ($l \geq 2$) and the weak ($l = 1$) observability cases in the observer design. Using the fact that $\pi_{y_i}(x - p_i) = 0$, for all $i = 0, \ldots, l-1$, one can remark that if the matrix $Q(t)$, defined in Lemma 3, is uniformly positive definite for $l \geq 2$, then the collected measurements y_i can be used to construct an instantaneous algebraic measurement x^m of x:

$$x^m(t) = Q(t)^{-1} \sum_{i=1}^{l-1} \pi_{y_i(t)} p_i,$$

and hence one can verify that a linear observer of the form:

$$\dot{\hat{x}} = v + K_x(x^m - \hat{x}) = v + K_x \tilde{x}, \quad \text{with } K_x \in \mathcal{M}_n^+, \tag{6}$$

ensures the global exponential convergence of the error $\tilde{x}$ to zero.

The main drawback of this observer is that it is not well defined at times when $Q(t)$ is singular and that it is ill conditioned when $Q(t)$ tends to become singular corresponding, for instance, to the weak observability case. A solution to this problem consists in choosing $K_x = K_x(t)$ such that the problematic term $Q(t)^{-1}$ (or the direct measure x^m of x) is not needed anymore. To this aim, it suffices to set $K_x = KQ(t)$, with $K \in \mathcal{M}_n^+$.

The following lemma summarizes the effect of the above choice of K_x leading to an observer that encompasses the strong and weak observability cases with any number of target-points with known location.

Lemma 4 *Consider the system (1)–(2) and redefine observer (6) with $K_x = KQ(t)$ as follows:*

$$\dot{\hat{x}} = v - K \sum_{i=0}^{l-1} \pi_{y_i(t)}(\hat{x} - p_i), \quad \hat{x}(0) = \hat{x}^0 \in \mathbb{R}^n \text{ and } K \in \mathcal{M}_n^+ \tag{7}$$

Assume that $a \equiv 0$, x is bounded and $x - p_i$ never crosses zero, so that the output $y_i(t)$ is always well defined for $i = 0, \ldots, l - 1$. Assume that the persistence condition on $Q(t)$, Lemma 3, is satisfied. Then the estimation error $\tilde{x} = x - \hat{x} = 0$ is uniformly globally exponentially stable (UGES).

Proof Using the fact that $\pi_{y_i(t)}(x - p_i) = 0$, $\forall i = 1, \ldots, l - 1$, on can remark that

$$\dot{\hat{x}} = v - KQ(t)\tilde{x},$$

and hence $\dot{\tilde{x}} = -KQ(t)\tilde{x}$. Defining $w = K^{-\frac{1}{2}}\tilde{x}$ and using the above dynamics of $\tilde{x}$ one gets the following linear time-varying system:

$$\dot{w} = -K^{\frac{1}{2}}Q(t)K^{\frac{1}{2}}w, \tag{8}$$

Using the assumption of persistent excitation characterized by relation (3) a direct application of [21, Lemma 5] proves that $\tilde{x}$ is UGES. More explicitly, one verifies that the transition matrix Φ_K associated with the above system (8) satisfies

$$\|\Phi_K(t, \tau)\| \leq \exp^{-\gamma(t-\tau)}, \tag{9}$$

where $\gamma = \mu k_m / \left(\delta(1 + lk_M\delta)^2\right)$ and k_m and k_M represent respectively the lowest and the largest eigenvalues of K.

3.2 The Case of Biased Velocity

For the sake of completeness, we extend results of the previous section to the case of biased velocity. Thus, we consider now the problem of estimating the state $X = \left(x^\top\ a^\top\right)^\top$.

Lemma 5 *Consider the system (1)–(2) and the above filter (7). Assume that $a \in \mathbb{R}^n$ is constant, v and $x \in \mathbb{R}^n$ is bounded, $x - p_i \neq 0$ such that $y_i(t)$ is always well defined for $i = 0, \ldots, l-1$. If the persistence of excitation of $Q(t)$ is fulfilled, then $|\tilde{x}|$ (and hence $|\hat{x}|$) is uniformly bounded with respect to initial conditions and ultimately bounded by $(k_M/k_m)^{\frac{1}{2}} |a|/\gamma$.*

Proof It is straightforward to verify that, in this case, the error-system equation is

$$\dot{\tilde{x}} = a - KQ\tilde{x}. \tag{10}$$

Using the previous change of variable $w = K^{-\frac{1}{2}} \tilde{x}$ with K a constant positive-definite matrix, one gets

$$\dot{w} = K^{-\frac{1}{2}} a - K^{\frac{1}{2}} Q K^{\frac{1}{2}} w,$$

whose general solution, leads to

$$\tilde{x}(t, 0) = K^{\frac{1}{2}} \Phi_K(t, 0) K^{-\frac{1}{2}} \tilde{x}^0 + \int_0^t K^{\frac{1}{2}} \Phi_K(\tau, t) K^{-\frac{1}{2}} a\, d\tau$$

Using (9) it follows that $|\tilde{x}(t)| \leq (k_M/k_m)^{\frac{1}{2}} (|\tilde{x}^0| + |a|/\gamma)$ and $\overline{\lim}_{t \to +\infty} |\tilde{x}(t)| \leq (k_M/k_m)^{\frac{1}{2}} |a|/\gamma$. Since x is bounded by definition, it follows that $\hat{x}$ is also bounded.

Now, for the design of an exponentially stable observer in the case where $a \neq 0$ the following two technical lemmas are instrumental.

Lemma 6 *Assume that the persistence condition on $Q(t)$ is satisfied. Choose a matrix-valued function $M(t) \in \mathcal{M}_n$ as the solution to*

$$\dot{M}(t) = I - KQ(t)M(t), \quad K \in \mathcal{M}_n^+, \quad M(0) = M(0)^\top = M^0 > 0 \tag{11}$$

Then $M(t)$ is bounded and always invertible, and its condition number is bounded.

Proof See Appendix 6.2.

Lemma 7 *If the persistence condition on $Q(t)$ is satisfied, then there exists $(\delta', \mu') > 0$ such that:*

$$\forall t > 0, \quad \int_t^{t+\delta'} M(\tau)^\top Q(\tau) M(\tau) d\tau > \mu' I \tag{12}$$

Proof The proof of this lemma is given in Appendix 6.3.

The following Theorem specifies the observer design and its convergence properties with the time variable t being omitted for the sake of legibility.

Theorem 1 *Consider the system (1)–(2) along with observer (7) and time-varying matrix-valued function (11). Define the estimates $(\hat{x}^a, \hat{a})$ as follows:*

$$\hat{x}^a = \hat{x} + M\hat{a}$$

$$\dot{\hat{a}} = -M^\top Q M \hat{a} - \sum_{i=0}^{l-1} M^\top \pi_{y_i}(\hat{x} - p_i) \qquad (13)$$

If the persistent condition on $Q(t)$ in the sense of Lemma 3 holds then $(\tilde{x} = x - \hat{x}^a, \tilde{a} = a - \hat{a}) = (0, 0)$, is uniformly globally exponentially stable.

Proof Let us first define a virtual state

$$z = \hat{x} + Ma$$

One easily verifies, using (7) and (11), that:

$$\dot{z} = v + a - KQ(z - x) \qquad (14)$$

Defining $\tilde{x}_z = x - z$, and using (1) and (14), one obtains:

$$\dot{\tilde{x}}_z = -KQ\tilde{x}_z$$

This equation being the same as the one for $\tilde{x}$ in the case where $a = 0$, one concludes as in Lemma 4 that $\tilde{x}_z = 0$ is uniformly globally exponentially stable, provided that the persistent condition on Q is satisfied.

Using the fact that $x = \hat{x} + Ma + \tilde{x}_z$, one gets

$$x - p_i = \hat{x} - p_i + \tilde{x}_z + Ma,$$

By multiplying the result by $M^\top \pi_{y_i}$ and taking the sum for $i = 0, \ldots, l-1$, one gets:

$$M^\top Q M a = -M^\top Q(\hat{x} + \tilde{x}_z) + \sum_{i=0}^{l-1} M^\top \pi_{y_i} p_i$$

Considering this result and the dynamics of $\hat{a}$ given by (13), it follows that:

$$\dot{\tilde{a}} = -M^\top Q M \tilde{a} - M^\top Q \tilde{x}_z$$

Knowing that $\tilde{x}_z = 0$ is uniformly globally exponentially stable and that $M^\top Q M$ verifies the persistence condition (12), one concludes that $(\tilde{a} = 0, \tilde{x} = 0)$ is also uniformly globally exponentially stable.

Remark 1 Note that the system of interest (1)–(2) is of dimension $2n$, while the observer as it is defined by (7), along with (11) and (13) is of dimension $2n + n^2$ leading to an excessive implementation cost. This is due the definition of M (11) that requires n^2 elements. A solution to this problem consists in the symmetrization of M by choosing the gain matrix K to be time varying $K = K(t) = kM(t)$. This leads to the fact that $M(t)$ becomes the solution to the so-called Continuous Riccati Equation

$$\dot{M} = I - kMQ(t)M, \; M(0) = M^0 > 0, \; k > 0. \tag{15}$$

This solution reduces the computational cost by reducing the state of M (that becomes an element of $\mathscr{M}_n^+$) from n^2 to $\frac{1}{2}n(n+1)$. It also does not affect the stability issue of the observer. To substantiate this claim it is necessary only to show that Lemma 4 is still valid for this modification. Consider the candidate Lyapunov function $V(t) = \frac{1}{2}\tilde{x}^\top(t)K^{-1}(t)\tilde{x}(t)$, one can easily verify that

$$\dot{V} = -\frac{k}{2}\tilde{x}^\top(t)K^{-2}(t)\tilde{x}(t) - \frac{1}{2}\tilde{x}^\top(t)Q(t)\tilde{x}(t) \leq -\frac{k}{2}\left(\frac{k_m}{k_M^2}\right)V,$$

which implies that the estimation error $\tilde{x} = 0$ is uniformly globally exponentially stable.

4 Extension to the SLAM Problem

We consider now the fundamental problem of the Simultaneous Localisation and Mapping (SLAM) which consists in the estimation of the vehicle's position, the bias on the velocity, and of p_i, $i = \{0, \ldots, l-1\}$ (vector of coordinates of fixed target-points) in the inertial frame from the the measurements of the directions y_i between the vehicle and the target-points in an unknown environment. For the sake of simplifying forthcoming relations, we consider first the unbiased velocity case.

Proposition 1 *Assume that p_0 ($p_0 = 0$) is the origin of the inertial frame. Assume also that $a \equiv 0$, x is bounded and $x - p_i \neq 0$ along trajectories of the system so that the output $y_i(t)$ is always well defined for $i = 0, \ldots, l-1$. Define $\tilde{x} = \hat{x} - x$ and $\tilde{p}_i = \hat{p}_i - p_i$. If the persistence of excitation condition (3) is fulfilled for each measurement y_i, $\forall i = \{0, \ldots, l-1\}$ then, the observer*

$$\dot{\hat{x}} = v - K\pi_{y_0}\hat{x}, \; \hat{x}(0) = \hat{x}^0 \in \mathbb{R}^n \text{ and } K \in \mathscr{M}_n^+, \tag{16}$$

$$\dot{\hat{p}}_i = K_p \pi_{y_i}\hat{x}_i - K\pi_{y_0}\hat{x}, \; K_p \in \mathscr{M}_n^+, \; \hat{x}_i = \hat{x} - \hat{p}_i, \; \forall i = \{1, \ldots, l-1\}, \tag{17}$$

ensures that $(\tilde{x} = 0, \tilde{p}_i = 0)$ is uniformly globally exponentially stable.

Proof The proof of the uniform global exponential stability of the error $\tilde{x} = 0$ is a direct application of Lemma 4. By defining $x_i = x - p_i$ and $\tilde{x}_i = x_i - \hat{x}_i = \tilde{x} - \tilde{p}_i$, for $i = 1, \ldots, l-1$ one verifies that

$$\dot{\tilde{x}}_i = K_p \pi_{y_i} \hat{x}_i = -K_p \pi_{y_i} \tilde{x}_i,$$

Lemma 4 again ensures that $\tilde{x}_i = 0$ is uniformly globally exponentially stable. The uniform exponential stability of $\tilde{x}$ and of $\tilde{x} - \tilde{p}_i$, in turn implies the uniform global exponential stability of $(\tilde{x} = 0, \tilde{p}_i = 0)$.

Remark 2 Note that the following expression of $\dot{\tilde{p}}_i$

$$\dot{\tilde{p}}_i = K_p \pi_{y_i} \hat{x}_i, \tag{18}$$

also ensures the uniform global exponential convergence of $\tilde{p}_i$ to zero for $i = 1, \ldots, l-1$. Indeed, in this case one has

$$\dot{\tilde{x}}_i = K \pi_{y_0} \hat{x} - K_p \pi_{y_i} \hat{x}_i = -K \pi_{y_0} \tilde{x} - K_p \pi_{y_i} \tilde{x}_i$$

Since $\tilde{x} = 0$ is globally exponentially stable, one deduces that $(\tilde{x}, \tilde{x}_i) = (0, 0)$ is globally exponentially stable. Therefore, $\tilde{p}_i = \tilde{x} - \tilde{x}_i$ globally exponentially converges to zero.

Note that Proposition 1 describes a solution to the position estimation and environment identification in a cascaded observer. It differs from the classical SLAM solutions in which the position and the environment are estimated in a consensus manner. The following proposition describes a more generic representation of the SLAM problem in which all measurements y_i are simultaneously involved in the estimation problem. This allows to improve the estimation by a spatial filtering on the space of measures.

Proposition 2 *Assume that p_0 is the origin of the inertial frame. Assume also that $a \equiv 0$, x is bounded and $x - p_i \neq 0$ so that the output $y_i(t)$ is always well defined for $i = 0, \ldots, l-1$. Define $X = (x^\top, p_0^\top, \ldots, p_{l-1}^\top)^\top$. If the persistence of excitation condition (3) is fulfilled for each measurement y_i, $\forall i = \{0, \ldots, l-1\}$ then, the following observer,*

$$\dot{\hat{X}} = \begin{pmatrix} v \\ 0 \\ \vdots \\ 0 \end{pmatrix} - K\Pi\hat{X}, \text{ with } \hat{X} = \begin{pmatrix} \hat{x} \\ \hat{p}_1 \\ \vdots \\ \hat{p}_{l-1} \end{pmatrix}, \Pi = \begin{pmatrix} \sum_{i=0}^{l-1} \pi_{y_i} & -\pi_{y_1} & -\pi_{y_2} & \cdots & -\pi_{y_{l-1}} \\ -\pi_{y_1} & \pi_{y_1} & 0 & \cdots & 0 \\ -\pi_{y_2} & 0 & \pi_{y_2} & \cdots & 0 \\ \vdots & \vdots & \vdots & \vdots & \vdots \\ -\pi_{y_{l-1}} & 0 & 0 & 0 & \pi_{y_{l-1}} \end{pmatrix}, \tag{19}$$

for $K \in \mathcal{M}_{nl}^+$ any positive-definite matrix, ensures that $\tilde{X} = X - \hat{X} = 0$ is uniformly globally exponentially stable.

Proof Using the fact that $\pi_{y_0} x = \pi_{y_i}(x - p_i) \equiv 0$, one deduces that $\Pi X \equiv 0$, and hence

$$\dot{\tilde{X}} = -K\Pi\tilde{X}, \quad \text{with } \tilde{X} = X - \hat{X}$$

Now using the fact that the persistent excitation condition (3) is fulfilled for each measurement y_i, $\forall i = \{0, \ldots, l-1\}$, it follows that there exist $\delta > 0$ and $0 < \mu < \delta$ such that for all t, all non-zero components $\sigma_i \in \mathbb{R}^{n \times n}$ of

$$\Sigma = \int_t^{t+\delta} \Pi d\tau = \begin{pmatrix} \sum_{i=0}^{l-1} \sigma_i & -\sigma_1 & -\sigma_2 & \cdots & -\sigma_{l-1} \\ -\sigma_1 & \sigma_1 & 0 & \cdots & 0 \\ -\sigma_2 & 0 & \sigma_2 & \cdots & 0 \\ \vdots & \vdots & \vdots & \vdots & \vdots \\ -\sigma_{l-1} & 0 & 0 & 0 & \sigma_{l-1} \end{pmatrix}$$

are definite positive matrices ($\sigma_i \geq \mu I$, $\forall i = 0 \ldots l-1$). It follows that for any non-zero column vector $W \in \mathbb{R}^{nl}$ with real entries $w_i \in S^{n-1}$, $i = 0 \ldots l-1$, one has

$$W^\top \Sigma W = w_0^\top \sigma_0 w_0 + \sum_{i=1}^{l-1} (w_0 - w_i)^\top \sigma_i (w_0 - w_i) > \mu > 0$$

This implies that Σ is uniformly positive-definite matrix and hence direct application of Lemma 4 ensures the uniform global exponential stability of $\tilde{X}$ to zero.

Note that the matrix gain K may be chosen time-varying. For instance, one may choose $K = kM(t)$, with $M(t) \in \mathcal{M}_{nl}^+$ solution to the Riccati equation:

$$\dot{M} = I - kM\Pi M, \quad M(0) = M^0 > 0, \quad k > 0, \tag{20}$$

4.1 The Case of Biased Velocity

The following two propositions extend Theorem 1 to the SLAM problem:

Proposition 3 *Assume that p_0 is the origin of the inertial frame. Assume also that x is bounded and $x - p_i \neq 0$ so that the output $y_i(t)$ is always well defined for $i = 0, \ldots, l-1$. Let $\hat{x}$ denote the solution of (16). Consider the following particular choice of the observer (13) of Theorem 1:*

$$\hat{x}^a = \hat{x} + M\hat{a}, \quad \dot{\hat{a}} = -M^\top \pi_{y_0} M \hat{a} - M^\top \pi_{y_0} \hat{x}$$

If the persistence of excitation condition (3) is fulfilled for each measurement y_i, $\forall i = \{0 \ldots l-1\}$, then the following dynamics of $\hat{p}_i$:

$$\dot{\hat{p}}_i = K_p \pi_{y_i} \hat{x}_i^a, \quad \text{with } \hat{x}_i^a = \hat{x}^a - \hat{p}_i \tag{21}$$

ensures the uniform global exponential convergence of $\tilde{p}_i = \hat{p}_i - p_i$ to zero, for $i = 1, \ldots, l-1$.

Proof The proof of this lemma is straightforward and left to the motivated reader.

Proposition 4 *Consider the system (1)–(2) along with (19). Define $A = (a^\top, 0, \ldots, 0)^\top$ and the matrix-valued function $M \in \mathscr{M}_{nl}^+$ as the solution to the equation:*

$$\dot{M} = I - kM\Pi M, \quad M(0) = M^0 \in \mathscr{M}_{nl}^+, \tag{22}$$

Use the following definition of the state estimate:

$$\hat{X}^a = \hat{X} + M\hat{A} \tag{23}$$

$$\dot{\hat{A}} = -M\Pi M\hat{A} - M\Pi\hat{X}, \tag{24}$$

Then (1) $M(t)$ is a positive-definite matrix, with bounded condition numbers, (2) $M\Pi M$ is persistently exciting in the sense of Lemma 7 and (3) $(\tilde{X}^a = X - \hat{X}^a, \tilde{A} = A - \hat{A}) = (0, 0)$ is uniformly globally exponentially stable.

Proof The proof proceeds item by item:

1. The proof of positiveness, boundedness and good conditioning of $M(t) \in \mathscr{M}_{nl}$ is a direct application of [22, Lemma 2.6].
2. $M\Pi M$ is persistently exciting by a direct application of Lemma 7.
3. Analogously to the proof of Theorem 1, we define a virtual state

$$Z = \hat{X} + MA$$

Differentiating Z and using (19) and (22), one verifies that:

$$\dot{\tilde{X}}_z = -kM\Pi\tilde{X}_z, \quad \text{with } \tilde{X}_z = X - Z$$

Therefore, by direct application of Lemma 4, one concludes that $\tilde{X}_z = 0$ is uniformly globally exponentially stable. Using the fact that $X = \hat{X} + \tilde{X}_z + MA$ and $\Pi X = 0$, one gets

$$M\Pi MA = -M\Pi(\hat{X} + \tilde{X}_z)$$

Introducing the above equality in the dynamics of $\hat{A}$, one obtains

$$\dot{\tilde{A}} = -M\Pi M\tilde{A} - M\Pi\tilde{X}_z$$

The remaining part of the proof is identical to the proof of Theorem 1.

5 Simulation

We consider the example of estimating the 3D position of a vehicle equipped with a camera by using one up to three direction measurements of fixed points in the environment with unknown locations. The vehicle moves along a circular trajectory at a fixed altitude ($z = 3$) above the ground. The camera's optical axis is aligned with the z-axis of the inertial frame and looks down at the target-points. The measure $y_i \in S^2$, $i = \{0, 1, 2\}$ corresponds to the spherical projection of the observed point i, given by the algebraic transformation $y_i = y_i^P/|y_i^P|$, where y_i^P is the projective measure provided by the camera. The measurement of the velocity v is biased by $a = (0.33,\ 0.66,\ 0.99)^\top$, and v is chosen so that $v + a = (-1.5\sin 0.5t,\ 1.5\cos 0.5t,\ 0)^\top$.

Two scenarios are considered. For the first one, the position and the velocity bias are estimated from a single direction measurement. The point of interest p_0 is located at the origin of the inertial frame. In the second scenario the SLAM problem is addressed from the measurements of directions of three fixed points. In this case p_0 is chosen as the origin of the inertial frame, and the other two points p_1 and p_2, whose

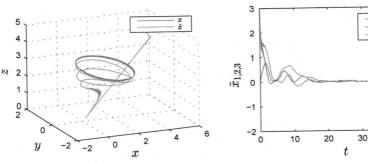

(a) Evolution of the system/observer pair in 3D space

(b) Evolution of the position error

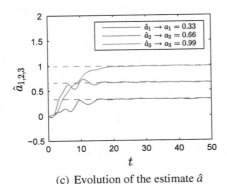

(c) Evolution of the estimate $\hat{a}$

Fig. 1 Scenario 2: the case of noisy measurements

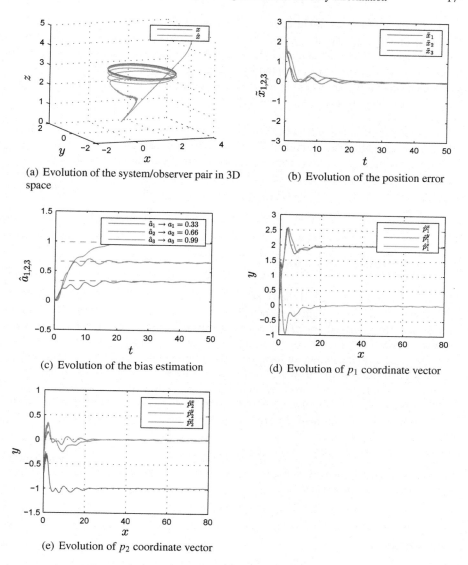

Fig. 2 Scenario 2: the case of noisy measurements

locations have to be estimated along with the camera position, have coordinates $(2, 2, 0)^\top$ and $(-1, 0, 0)^\top$ respectively.

The matrix gain K is chosen as the solution to Eq. (15) for Scenario 1 and as the solution to (22) with $k = 1$ for Scenario 2. Initial conditions are $x(0) = (3, 0, 3)^\top$, $\hat{x}(0) = (4, 0, 5)^\top$ and $\hat{a} = (0, 0, 0)^\top$, and for the SLAM problem $\hat{p}_1 = \hat{p}_2 = (0, 0, 0)^\top$.

For each scenario, simulations are carried out with measurements corrupted by noise to illustrate the resulting (and inevitable) slight degradation of the observers performance after the transient phase when the estimation errors become small. The body velocity measure v is corrupted by a Gaussian zero mean additive noise with a standard deviation equal to 0.1 m/s. As for the directions, they are calculated from a body position corrupted by a Gaussian zero mean noise with standard deviation equal to 0.03 m/s.

Figure 1a–c show the performance of the observer of Scenario 1. Figure 2a–c show the performance of the observer of Scenario 2. By comparison with Scenario 1 the camera's position estimation error is less noisy. This is coherent with the increased number of measurements involved in the estimation process and spatial filtering on the space of measures. Finally, Fig. 2a–e show the time evolution of the target-points estimates.

6 Concluding Remarks

In this chapter, we discussed the issue of observability of a vehicle's position from bearing measurements along with (possibly biased) velocity measurement. The nonlinear observers derived are effective when the bearing measurement satisfies a condition of persistent excitation. There are an increasing number of applications for mobile vehicles using vision for location and environment estimation, and the present chapter contributes to the growing body of work that supports this developing application domain. Extending the observer design methodology described in the chapter to the estimation of the relative pose between mobile objects evolving in $SE(n)$, with applications in $SE(3)$, is also possible. This is part of our future work plans.

Acknowledgements This work was supported by the ANR-ASTRID project SCAR "Sensory Control of Unmanned Aerial Vehicles", the ANR-Equipex project "Robotex" and the Australian Research Council through the ARC Discovery Project DP120100316 "Geometric Observer Theory for Mechanical Control Systems".

Appendix

6.1 Proof of Lemma 1

Let us first show that (4) implies (5). For $\tau \in [t, t+\delta]$ one has

$$|b^\top y(\tau)|^2 = |b^\top (y(t) + \begin{pmatrix} \dot{y}_1(s_1) \\ \vdots \\ \dot{y}_n(s_n) \end{pmatrix} (\tau - t))|^2$$

$$\geq |b^\top y(t)|^2 - nd(\delta)(\tau - t) - 2n^2 d(\delta)^2 (\tau - t)^2$$

for some $s_i \in [t, \tau]$ ($i = 1, \ldots, n$) and $d(\delta) = \sup_{\tau \in [t,t+\delta]} |\dot{y}(\tau)|$. Choose $b = y(t)$ so that $b^\top y(t) = 1$, then

$$|b^\top y(\tau)|^2 \geq 1 - nd(\delta)(\tau - t) - 2n^2 d(\delta)^2 (\tau - t)^2$$

and

$$\int_t^{t+\delta} |b^\top y(\tau)|^2 d\tau \geq \delta - nd(\delta)\delta^2 - \frac{1}{3} n^2 d(\delta)^2 \delta^3$$

Clearly there exists $\varepsilon > 0$ (independent of t) such that

$$d(\tau) \leq \varepsilon \implies \int_t^{t+\delta} |b^\top y(\tau)|^2 d\tau \geq \delta - \mu$$

Let us proceed by contradiction and assume that (5) does not hold, i.e. $|\dot{y}(\tau)| < \varepsilon$, $\forall t \in [t, t+\delta]$, then $d(\delta) < \varepsilon$ and $\int_t^{t+\tau} |b^\top y(\tau)|^2 d\tau \geq \delta - \mu$. This contradicts (4) according to which

$$\forall b \in S^{n-1} : \int_t^{t+\delta} |\pi_{y(\tau)} b|^2 d\tau = \delta - \int_t^{t+\delta} |b^\top y(\tau)|^2 d\tau \geq \mu$$

Therefore (5) holds true. We now show that (5) implies (4): one has $|b^\top y(t)|^2 = \cos(\theta(t))^2$ with $\theta(t)$ the angle between b and $y(t)$. Using the (assumed) uniform continuity of $\dot{y}$, (5) implies the existence of an interval $[t_1, t_2] \subset [t, t+\delta]$ such that $(t_2 - t_1) = \varepsilon_1(\varepsilon) > 0$ and $|y(t_2) - y(t_1)| \geq \varepsilon_2(\varepsilon) := \frac{\varepsilon}{2n} \varepsilon_1(\varepsilon) > 0$. This in turn implies that the distance between $\theta(t_1)$ and $\theta(t_2)$ is larger than some $\varepsilon_3(\varepsilon) > 0$. Since the uniform continuity of $\dot{y}$ implies the uniform continuity of $\dot{\theta}$, and since $\cos(\theta(t_1))$ and $\cos(\theta(t_2))$ cannot be both equal to one, one deduces the existence of a positive number $\varepsilon_4(\varepsilon)$ such that $\int_{t_1}^{t_2} \cos(\theta(\tau))^2 d\tau \leq (t_2 - t_1)(1 - \varepsilon_4(\varepsilon))$. Therefore,

$$\int_t^{t+\delta} |b^\top y(\tau)|^2 d\tau = \int_t^{t_1} |b^\top y(\tau)|^2 d\tau + \int_{t_1}^{t_2} |b^\top y(\tau)|^2 d\tau + \int_{t_2}^{t+\delta} |b^\top y(\tau)|^2 d\tau$$
$$\leq (t_1 - t) + (t_2 - t_1)(1 - \varepsilon_4(\varepsilon)) + (t + \delta - t_2) \leq \delta - \mu(\varepsilon)$$

with $\mu(\varepsilon) = \varepsilon_1(\varepsilon) \varepsilon_4(\varepsilon) > 0$.

6.2 Proof of Lemma 6

To prove that M is bounded, is suffices to ensure that, for any constant vector $b \in S^{n-1}$, $|Mb|$ is bounded. Define $u = Mb$, it follows that:

$$\dot{u} = b - KQu$$

This equation is similar to the Eq. (10) of $\tilde{x}$. Therefore

$$|M(t)b| \leq |M(0)b| + \frac{1}{\gamma}, \quad \forall b \in S^{n-1}$$

This implies that $||M||$ is bounded. To show that M is an invertible matrix, define $\Delta := \det(M)$. From Jacobi's formula, one has

$$\dot{\Delta} = \Delta \mathrm{tr}(M^{-1}\dot{M}) = \Delta \mathrm{tr}(M^{-1} - KM^{-1}QM) = \Delta \mathrm{tr}(M^{-1}) - \Delta \mathrm{tr}(KQMM^{-1})$$
$$= \Delta \mathrm{tr}(M^{-1}) - \underbrace{\sum_{i=0}^{l-1} \left(\mathrm{tr}(K) - y_i^\top K y_i\right)}_{\in [l(n-1)k_m;\, l(n-1)k_M]} \Delta \tag{25}$$

Note that this equation holds even if M is not invertible. Indeed, using the fact that $\det(M) = \prod_{i=1}^n \lambda_i$ and $\mathrm{tr}(M^{-1}) = \sum_{i=1}^n \frac{1}{\lambda_i}$, with λ_i ($i = 1\ldots n$) the eigenvalues of M, one verifies that

$$\dot{\Delta} = -\sum_{i=0}^{l-1}\left(\mathrm{tr}(K) - y_i^\top K y_i\right)\Delta + \sum_{i=1}^n \prod_{j=1, j\neq i}^n \lambda_j \tag{26}$$

Since $M(0)$ is symmetric positive definite by assumption, all eigenvalues of $M(0)$ are positive and $\Delta(0) > 0$. Assume that Δ is equal to zero for the first time at the time instant $t_0 > 0$. Then, $\mathrm{tr}(M(t)) > 0$ on $[0, t_0[$ and $\mathrm{tr}(M(t)) \geq 0$. In view of (26), $\Delta(t) \geq r(t)$ with $r(t)$ the solution to the equation $\dot{r} = -l(n-1)k_M r$, with $r(0) = \Delta(0)$. Therefore, $\Delta(t) \geq y(0)\exp(-l(n-1)k_M t) > 0$, $\forall t$. This contradicts the existence of t_0 and proves that $M(t)$ is always invertible.

Let us now prove that $\Delta(t)$ is lower bounded by a positive number. Rewrite Eq. (25) as follows:

$$\dot{\Delta} = \left(\mathrm{tr}(M^{-1}) - \sum_{i=0}^{l-1}\left(\mathrm{tr}(K) - y_i^\top K y_i\right)\right)\Delta, \tag{27}$$

Using the fact that $\mathrm{tr}(M^{-1}) > \frac{n}{\Delta^{1/n}}$, this equation shows that $\dot{\Delta} \geq 0$ if $\Delta < \left(\frac{nk_M}{l(n-1)}\right)^n$. Therefore Δ is ultimately lower bounded by $\left(\frac{nk_M}{l(n-1)}\right)^n$.

Finally, since Δ is lower bounded by a positive number and M is upper bounded, it follows (by direct application of [23]) that the condition number

$$\kappa(M) = ||M||\cdot||M^{-1}|| \leq \frac{2}{\Delta}\left(\frac{||M||_F}{\sqrt{n}}\right)^n \text{ is upper bounded.}$$

6.3 Proof of Lemma 7

We have to distinguish the strong observability case from the weak observability case. When Q is uniformly positive definite (the case of strong observability case when using two or more non-collinear directions) and from properties of M pointed out in Lemma 6, it is clear that there exists $\varepsilon > 0$ such that $M(t)^\top Q(t) M(t) > \varepsilon I$ and therefore (12) is verified for $\delta' > 0$ and $\mu' = \varepsilon/\delta'$. The more involved situation is when Q is singular at time instants. Using the fact that $Q(t) - \pi_{y_i(t)} \geq 0$, for $i = \{0, \ldots, l-1\}$ and without loss of generality, it suffices to prove (12) when $Q(t) = \pi_{y_0}(t) = \pi_y(t)$. By introducing $b \in S_{n-1}$ and using, for any $(t, \tau) \in \mathbb{R}^2$, the following identity:

$$M(\tau)b = M(t)b + \int_t^\tau \dot{M}(s)b\,ds = M(t)b + (\tau - t)b - \int_t^\tau K\pi_y M b\,ds$$

one obtains (after multiplying both sides of the above equation by $\pi_{y(\tau)}$, and taking the integral of the norm of each side):

$$\int_t^{t+\delta'} |\pi_y(\tau) M(\tau) b|^2 d\tau = \int_t^{t+\delta'} \left| \pi_y(\tau) M(t) b + (\tau - t) \pi_y(\tau) b - \pi_y(\tau) \int_t^\tau K\pi_y M b\,ds \right|^2 d\tau$$

By exploiting the Schwartz inequality: $(a - b)^2 \geq 0.5a^2 - b^2$, one has:

$$\int_t^{t+\delta'} |\pi_y M b|^2 d\tau \geq \frac{1}{2} \int_t^{t+\delta'} |\pi_y M(t) b + (\tau - t)\pi_y b|^2 d\tau$$
$$- \int_t^{t+\delta'} \left| \pi_y \int_t^\tau K\pi_y M b\,ds \right|^2 d\tau \qquad (28)$$

Denoting the maximal eigenvalue of K as k_M, it follows that

$$\int_t^{t+\delta'} \left| \pi_y(\tau) \int_t^\tau K\pi_y M b\,ds \right|^2 d\tau \leq \int_t^{t+\delta'} \left| \int_t^\tau K\pi_y M b\,ds \right|^2 d\tau$$
$$\leq k_M^2 \int_t^{t+\delta'} \left(\int_t^\tau |\pi_y M b|^2 ds \right) d\tau$$

Changing the order of integration, one gets

$$\int_t^{t+\delta'} \left| \int_t^\tau K\pi_y Mb\, ds \right|^2 d\tau \le k_M^2 \int_t^{t+\delta'} |\pi_y Mb|^2 \left(\int_s^{t+\delta'} d\tau \right) ds$$

$$\le \delta' k_M^2 \int_t^{t+\delta'} |\pi_y Mb|^2 ds$$

Introducing now the above inequality in (28), one obtains

$$\int_t^{t+\delta'} |\pi_y Mb|^2 d\tau \ge \frac{\int_t^{t+\delta'} |\pi_y(\tau)M(t)b + (\tau - t)\pi_y(\tau)b|^2 d\tau}{2(1 + \delta' k_M^2)} \tag{29}$$

For any $\delta' \ge \delta$, one has

$$\int_t^{t+\delta'} |\pi_y(\tau)M(t)b + (\tau - t)\pi_y(\tau)b|^2 d\tau \ge \int_{t+\delta'-\delta}^{t+\delta'} |\pi_y(\tau)M(t)b + (\tau - t)\pi_y(\tau)b|^2 d\tau$$

$$\ge \int_{t+\delta'-\delta}^{t+\delta'} (\tau - t)^2 \left| \frac{\pi_y(\tau)M(t)b}{\tau - t} + \pi_y(\tau)b \right|^2 d\tau$$

$$\ge (\delta' - \delta)^2 \int_{t+\delta'-\delta}^{t+\delta'} \left| \frac{\pi_y(\tau)M(t)b}{\tau - t} + \pi_y(\tau)b \right|^2 d\tau$$

Now, using the fact that $\pi_y(\tau)M(t)b/(\tau - t)$ converges to zero when $\tau \in [t + \delta' - \delta, t + \delta']$ and δ' increases, it follows that

$$\lim_{\delta' \to \infty} \int_{t+\delta'-\delta}^{t+\delta'} \left| \frac{\pi_y(\tau)M(t)b}{\tau - t} + \pi_y(\tau)b \right|^2 d\tau > \lim_{\delta' \to \infty} \frac{1}{2} \int_{t+\delta'-\delta}^{t+\delta'} |\pi_y(\tau)b|^2 d\tau > \frac{\mu}{2}$$

This in turn implies that there exists $\delta' > \delta$ such that

$$\int_t^{t+\delta'} |\pi_y Mb|^2 d\tau > \frac{\mu(\delta' - \delta)^2}{8(1 + \delta' k_M^2)} \tag{30}$$

which demonstrates the desired results (12) in which $\mu' = \frac{\mu(\delta'-\delta)^2}{8(1+\delta'k_M^2)}$.

References

1. Häming, K., Peters, G.: The structure-from-motion reconstruction pipeline a survey with focus on short image sequences. Kybernetika **46**(5), 926–937 (2010)
2. Triggs, B., McLauchlan, P., Hartley, R., Fitzgibbon, A.: Bundle adjustment a modern synthesis. In: ICCV '99: Proceedings of the International Workshop on Vision Algorithms, pp. 298–372. Springer, Heidelberg (1999)

3. Strasdat, H., Montiel, J., Davison, A.: Visual slam: why filter. J. Image Vis. Comput. **30**(2), 65–77 (2012)
4. Matthies, L., Kanade, T., Szeliski, R.: Kalman filter-based algorithms for estimating depth from image sequences. Int. J. Comput. Vis. **3**(3), 209–238 (1989)
5. Soatto, S., Frezza, R., Perona, P.: Motion estimation via dynamic vision. IEEE Trans. Autom. Control **41**(3), 393–414 (1996)
6. Armesto, L., Tornero, J., Vincze, M.: Fast ego-motion estimation with multi-rate fusion of intertial and vision. Int. J. Robot. Res. **26**(6), 577–289 (2007)
7. Civera, J., Davison, A., Montiel, J.: Inverse depth parametrization for monocular slam. IEEE Trans. Robot. **24**(5), 932–945 (2008)
8. Bekris, K., Glick, M., Kavraki, L.: Evaluation of algorithms for bearing-only slam. In: Proceedings of the IEEE International Conference on Robotics and Automation, pp. 1937–1943 (2006)
9. Baldwin, G., Mahony, R., Trumpf, J.: A nonlinear observer for 6 DOF pose estimation from inertial and bearing measurements. In: Proceedings of the IEEE International Conference on Robotics and Automation (ICRA), pp. 2237–2242 (2009)
10. Rehbinder, H., Ghosh, B.: Pose estimation using line-based dynamic vision and inertial sensors. IEEE Trans. Autom. Control **48**(2), 186–199 (2003)
11. Abdursul, R., Inaba, H., Ghosh, B.K.: Nonlinear observers for perspective time-invariant linear systems. Automatica **40**(3), 481–490 (2004)
12. Abdursul, R., Inaba, H., Ghosh, B.: Nonlinear observers for perspective time-invariant linear systems. Automatica **40**, 481–490 (2004)
13. Aguiar, A., Hespanha, J.: Minimum-energy state estimation for systems with perspective outputs. IEEE Trans. Autom. Control **51**(2), 226–241 (2006)
14. Dahl, O., Wang, Y., Lynch, A.F., Heyden, A.: Observer forms for perspective systems. Automatica **46**(11), 1829–1834 (2010)
15. Dixon, E., Fang, Y., Dawson, D., Flynn, T.: Range identification for perspective vision systems. IEEE Trans. Autom. Control **48**(12), 2232–2238 (2003)
16. De Luca, A., Oriolo, G., Giordano, P.R.: On-line estimation of feature depth for image-based visual servoing schemes. In: 2007 IEEE International Conference on Robotics and Automation, pp. 2823–2828. IEEE (2007)
17. Dani, A., Fischer, N., Kan, Z., Dixon, W.: Globally exponentially stable observer for vision-based range estimation. Mechatronics **22**(4), 381–389 (2012)
18. Batista, P., Silvestre, C., Oliveira, P.: Globally exponentially stable filters for source localization and navigation aided by direction measurements. Syst. Control Lett. **62**(11), 1065–1072 (2013)
19. Le Bras, F., Hamel, T., Mahony, R., Samson, C.: Observer design for position and velocity bias estimation from a single direction output. In: 54th IEEE Conference on Decision and Control (CDC), pp. 7648–7653. IEEE (2015)
20. Sussmann, H.J.: Single-input observability of continuous-time systems. Math. Syst. Theory **12**(4), 371–393 (1979)
21. Loría, A., Panteley, E.: Uniform exponential stability of linear time-varying systems: revisited. Syst. Control Lett. **47**(1), 13–24 (2002)
22. Hamel, T., Samson, C.: Riccati observers for position and velocity bias estimation from either direction or range measurements, Technical report (2016). arXiv:1606.07735
23. Piazza, G., Politi, T.: An upper bound for the condition number of a matrix in spectral norm. J. Comput. Appl. Math. **143**(1), 141–144 (2002)

Nonlinear Camera- and GNSS-Aided INS for Fixed-Wing UAV Using the eXogenous Kalman Filter

Lorenzo Fusini, Thor I. Fossen and Tor Arne Johansen

Abstract This chapter aims at applying a recently proposed estimator, called eXogenous Kalman Filter (XKF), to the navigation of a fixed-wing unmanned aerial vehicle (UAV) using inertial sensors, GNSS, and optical flow calculated from a camera. The proposed system is a cascade interconnection between a globally exponentially stable (GES) nonlinear observer (NLO) and a time-varying Kalman filter based on a local linearization of the system equations about the output of the preceding NLO. It is very well known that the linear time-varying Kalman filter is GES and optimal in the sense of minimum variance under some conditions, but when a nonlinear approximation (e.g., the extended Kalman filter) becomes necessary, generally such positive properties cannot be guaranteed anymore. On the other hand, a NLO often comes with strong, often global, stability properties, but without attention to optimality with respect to unknown measurement and process noise. The idea behind the XKF is to combine the advantages of the two composing estimators while surpassing the drawbacks from which they individually suffer. The theory is supported by tests on both simulated and experimental data, where the XKF is compared to a state-of-the-art solution based on the extended Kalman filter (EKF).

1 Background and Motivation

The problem of estimating the states of nonlinear systems has been approached in a large number of ways. The most popular and proficient tool has been the Kalman filter (KF) with its extensions. Under certain conditions, it guarantees optimal estimates

---

L. Fusini (✉) · T.I. Fossen · T.A. Johansen
Department of Engineering Cybernetics, NTNU Norwegian University of Science and Technology, Trondheim, Norway
e-mail: lorenzo.fusini@ntnu.no

T.I. Fossen
e-mail: thor.fossen@ntnu.no

T.A. Johansen
e-mail: tor.arne.johansen@ntnu.no

© Springer International Publishing AG 2017
T.I. Fossen et al. (eds.), *Sensing and Control for Autonomous Vehicles*,
Lecture Notes in Control and Information Sciences 474,
DOI 10.1007/978-3-319-55372-6_2

in the sense of minimum variance, and its linear time-varying formulation is known to be globally exponentially stable [1, 22, 36]. When the system to observe is not linear, a linearized version is necessary. Linear approximation about a state estimate can be performed using different techniques, leading to variants of the KF such as the extended KF, unscented KF, Monte Carlo filter, and particle filter [6, 10, 13, 21]. The global stability properties of the resulting filters might not be guaranteed anymore, as they usually depend on implicit conditions that cannot be verified on the linearized system [32, 37].

The estimation of position, velocity, and attitude is a fundamental task in navigation of UAVs. Since the early 1990s [33], the field has seen the rise and development of nonlinear observers as substitutes for Kalman filters. Particular effort has been put in attitude estimation, which is typically performed by comparing a set of vectors measured in the body-fixed coordinate frame with a set of reference vectors in a reference frame [2, 3, 15, 28, 30], while others have also included the estimation of position and velocity [12, 17, 20]. A typical payload of sensors for navigation includes, but is not limited to, IMU, magnetometers, and GNSS receiver.

The use of nonlinear observers is attractive because they often come with globally exponential stability properties, which guarantee a correct behavior in the presence of uncertain initialization or unknown disturbances, and because their computational footprint is small as a consequence of the reduced number of differential equations involved. The design, however, does not take into account properties of the noise affecting both the system model and the measurements, resulting in suboptimal estimates.

The work presented here analyzes the properties and performance of a cascade interconnection between a GES NLO and a linear time-varying Kalman filter (LKF), and provides performance results based on experimental data. The objective is to build an estimator that inherits the advantages of the constituting components and discards their shortcomings. This method, called eXogenous Kalman filter (XKF), was presented and its properties formally analyzed in [19], and tested on simulated systems for both aerial and underwater vehicles [17, 39]. Here it is applied to a camera-aided inertial navigation system for a UAV: the NLO stage is a variation of the GES observer used in [12], it uses velocity and acceleration vectors in its output injection term [9], and it is then connected to a LKF to build the proposed XKF, and formally analyzed to conclude on its GES properties. Subsequently, it is tested on experimental data collected during the flight of a fixed-wing UAV and compared to the output of an EKF and of the NLO stage. As perfect reference values are not available in the experimental data, additional simulations are presented to provide another comparison between NLO, XKF, and EKF.

A fundamental component of the NLO is its body-referenced velocity, which is calculated by means of the optical flow extracted from the camera images and the information obtained by some of the other sensors available. The method was discussed in detail in [9], and a summary is reported here.

An alternative NLO is also tested, with the velocity vectors replaced by another pair of vectors that also exploits the GNSS measurements; it can be a useful, temporary solution in case faults appear in the system and invalidate the body-referenced velocity vector.

2 Notation and Measurements

Vectors and matrices are represented by lowercase and uppercase letters, respectively. X^{-1} and X^+ denote the inverse and pseudoinverse of a matrix, respectively, and X^T the transpose of a matrix or vector. The symbols $\bar{x}$ and $\hat{x}$ indicate the estimate of x as output by the NLO and XKF, respectively, and $\check{x} = x - \bar{x}$ and $\tilde{x} = x - \hat{x}$ the respective estimation errors. The operator $\|\cdot\|$ denotes the Euclidean norm for vectors and the Frobenius norm for matrices, I_n is the identity matrix of order n, and $0_{m \times n}$ is the $m \times n$ matrix of zeros. The function sat$(\cdot)$ performs a component-wise saturation of its vector or matrix argument to the interval $[-1, 1]$. The operator $S(x)$ transforms the vector x into the skew-symmetric matrix

$$S(x) = \begin{bmatrix} 0 & -x_3 & x_2 \\ x_3 & 0 & -x_1 \\ -x_2 & x_1 & 0 \end{bmatrix}$$

The inverse operation is denoted as vex$(\cdot)$, such that vex$(S(x)) = x$. For a square matrix A, its skew-symmetric part is represented by $\mathbb{P}_a(A) = \frac{1}{2}(A - A^T)$.

The reference frames considered in the paper are the body-fixed frame {B}, the North-East-Down (NED) frame {N} (Earth-fixed, considered inertial) and the camera frame {C}. The rotation from frame {B} to {N} is represented by the matrix $R_b^n \equiv R \in SO(3)$, where $SO(3)$ represents the Special Orthogonal group of dimension 3. The camera is assumed to be fixed to the body and perfectly aligned to its axes, so the camera frame and body frame represent the same coordinate system and can be identified by {B} alone.

A vector decomposed in {B} and {N} has superscript ^{b} and ^{n} respectively. The body (camera) location w.r.t. {N} is described by $c^n = [c_x^n, c_y^n, c_z^n]^T$. A point in the environment expressed w.r.t. {N} is $t^n = [x^n, y^n, z^n]^T$: note that a point located at the mean sea level corresponds to $z^n = 0$, and such it will be considered throughout the paper. The same point expressed w.r.t. {B} is $t^b = [x^b, y^b, z^b]^T$. It will also be assumed that every point representing a feature captured by the camera is fixed w.r.t. {N}. The gravity vector is defined as $g^n = [0, 0, g]$, with g the local gravitational acceleration. The greek letters ϕ, θ, and ψ represent the roll, pitch, and yaw angles respectively, defined according to the zyx convention for principal rotations [7]. A 2-D camera image has coordinates $[r, s]^T$, aligned with the y^b- and z^b-axis respectively (see Fig. 2). The derivative $[\dot{r}, \dot{s}]^T$ of the image coordinates is the optical flow (OF).

2.1 Measurements

The experimental sensor suite consists of the following units:

- *GNSS receiver*: NED position p^n and velocity v^n;
- *IMU*: biased angular velocity $\omega_m^b = \omega^b + b^b$, where b^b represent the bias, and specific force f^b, which is assumed bias-compensated by using, for example, the method in [11];
- *machine vision based on downward-facing camera*: body-fixed velocity v^b;
- *altimeter*: height over ground c_z^n;
- *inclinometers*: roll ϕ and pitch θ angles measurements.

3 Machine Vision

A machine vision system is designed to calculate the body-fixed velocity v^b from information available from camera, GNSS, altimeter, and inclinometers. A detailed description of the method is found in [9, 29], while its most relevant aspects for the present application are reported here.

3.1 Optical Flow

There exist several methods for computing the OF; for the present work, two specific methods are combined. The first one is SIFT [26], which provided the overall best performance in [29]. The total number of OF vectors in each image depends on the number of features detected and matched together. Since the transformation in Sect. 3.2 requires at least three OF vectors [9], it is necessary to make sure that this is handled. SIFT does not guarantee three distinct OF vectors since relatively homogeneous environments, such as snow or the ocean, increase the difficulty of finding distinct features. Therefore, the OF vectors created by SIFT are combined with a second method, which is based on region matching [8].

The region matching method used here is a template matching approach based on normalized cross-correlation [34]. The displacements of twelve templates, created symmetrically across the images, are used to find twelve OF vectors. Template matches below a given threshold are discarded and the corresponding OF vector is not used. Unreliable matches can occur in case of homogeneous terrain, changes in brightness or simply when the area covered by the template has disappeared from the image in the time between the capture of images. An example of OF vectors computed with SIFT and template matching is displayed in Fig. 1.

Erroneous OF vectors are detected in case of mismatches, so it is desired to locate and remove these vectors. For this reason, an outlier detector is implemented before the vectors are used to calculate the body-fixed velocity. The outlier detector utilizes

Fig. 1 Optical flow vectors. The *red ones* are generated by SIFT, the *green ones* by template matching

a histogram to find the vectors that deviate from the median with respect to direction and magnitude.

3.2 From Optical Flow to Body Velocity

For the OF computations to be useful as observer measurements, a transformation to body-fixed velocity is necessary. The pinhole camera model is used [16, 38]: despite being an ideal model, it gives results that are good enough for the purpose of this work. The camera-fixed frame is related to the body-fixed frame as illustrated in Fig. 2, where the optical axis of the downward-looking camera is aligned with the body z-axis. The focal point of the camera is assumed to coincide with the origin of $\{B\}$.

The relationship between a point on the terrain $t^b = [x^b, y^b, z^b]^T$ and its projection onto the image plane $[r, s]^T$ is, according to the pinhole camera model,

$$\begin{bmatrix} r \\ s \end{bmatrix} = \frac{f}{z^b} \begin{bmatrix} y^b \\ -x^b \end{bmatrix}, \quad z^b \neq 0 \qquad (1)$$

where f is the focal length of the camera. As t^b in itself is not available, a transformation [9, 29] is used to express t^b as a function of roll, pitch, height over ground, image features, and camera focal length

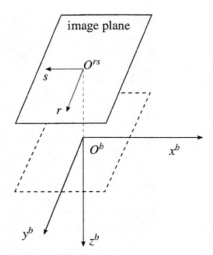

Fig. 2 Pinhole camera model. The camera is oriented downwards, while x^b is the direction of flight. O^{rs} and O^b are the origins of the image plane and body frame, respectively

$$\begin{bmatrix} x^b \\ y^b \\ z^b \end{bmatrix} = \begin{bmatrix} \frac{sc_z^n}{s\sin(\theta)+\cos(\theta)(f\cos(\phi)+r\sin(\phi))} \\ -\frac{rc_z^n}{s\sin(\theta)+\cos(\theta)(f\cos(\phi)+r\sin(\phi))} \\ -\frac{fc_z^n}{s\sin(\theta)+\cos(\theta)(f\cos(\phi)+r\sin(\phi))} \end{bmatrix} \qquad (2)$$

All features tracked by the camera are assumed to be stationary with respect to {N}, therefore the UAV's linear and angular velocities, v^b and ω^b, relative to a feature tracked by the OF algorithm will be equal for every tracked feature at a given instant in time. Furthermore, it is assumed that the terrain is flat, such that every feature is located at the same altitude: this simplifies the analysis and is a reasonable assumption for the experiment considered here, as the results show; for rough terrains, other approaches are necessary [14].

For every feature j, the relationship between OF and body-fixed linear/angular velocity is given as

$$\begin{bmatrix} \dot{r}_j \\ \dot{s}_j \end{bmatrix} = -M_j(f, r_j, s_j, \phi, \theta, c_z^n) \begin{bmatrix} v^b \\ \omega^b \end{bmatrix} \qquad (3)$$

$$M_j = \frac{f}{z_j^b} \begin{bmatrix} 0 & 1 - \frac{y_j^b}{z_j^b} & -\frac{y_j^{b^2}}{z_j^b} - z_j^b & \frac{y_j^b x_j^b}{z_j^b} & x_j^b \\ -1 & 0 & \frac{x_j^b}{z_j^b} & \frac{x_j^b y_j^b}{z_j^b} & -\frac{x_j^{b^2}}{z_j^b} - z_j^b & y_j^b \end{bmatrix} \qquad (4)$$

where $M_j \in \mathbb{R}^{2\times 6}$ is derived in [9]. The parameters in (3) on which M_j depends appear explicitly after substituting the variables in (4) with (2). If the number of features being tracked is k, then the OF vector has dimension $2k$. A matrix $M \in \mathbb{R}^{2k\times 6}$ can be created by concatenating the matrices M_j, $j = 1 \ldots k$, vertically, such that each feature j adds two rows to M, and by calculating the pseudoinverse of M, the angular and linear velocities can be computed as

$$\begin{bmatrix} v^b \\ \omega^b \end{bmatrix} = -M^+ \begin{bmatrix} \dot{r}_1 \\ \dot{s}_1 \\ \vdots \\ \dot{r}_k \\ \dot{s}_k \end{bmatrix} \tag{5}$$

All the parameters constituting M are known, since they are either measured ($r_j, s_j, \phi, \theta, c_z^n$) or known upon assembling the payload (f). M^+ exists only if $M^T M$ has full rank, such that it can be expressed as $M^+ = (M^T M)^{-1} M^T$. This can only happen if the number of flow vectors is greater than or equal to three.

4 Design and Analysis

The NLO and LKF are designed around two slightly different versions of the strapdown navigation equations, the difference being how the attitude of the UAV is parameterized. In the NLO, the attitude is parameterized as a rotation matrix with nine degrees of freedom, whereas the LKF uses Euler angles defined according to the *zyx* convention for principal rotations [7]. The NLO uses the rotation matrix in order to achieve a GES estimation error, but in the LKF this is replaced by Euler angles to take advantage of the reduced number of states, which have a direct impact on the computational footprint of the system. Additional details will be explained in the following Sections.

4.1 Nonlinear Observer

When designing the NLO, the kinematic system to observe can be divided into an attitude part $\Sigma_{A_{NLO}}$ and a translational motion part $\Sigma_{TM_{NLO}}$. Their equations are

$$\Sigma_{A_{NLO}} \begin{cases} \dot{R} = RS(\omega_m^b - b^b) \\ \dot{b}^b = 0 \end{cases} \tag{6}$$

$$\Sigma_{TM_{NLO}} \begin{cases} \dot{p}^n = v^n \\ \dot{v}^n = Rf^b + g^n \end{cases} \tag{7}$$

The following assumptions are made:

Assumption 1 A sufficient number of distinct image features are selected, such that M has full rank and its pseudoinverse can be calculated as $M^+ = (M^T M)^{-1} M^T$.

Assumption 2 The gyro bias b^b is constant, and there exists a known constant $L_b > 0$ such that $\|b^b\| \leq L_b$.

Assumption 3 There exists a constant $c_{obs} > 0$ such that, $\forall t \geq 0$, $\|f^b \times v^b\| \geq c_{obs}$.

Assumption 3 imposes the noncollinearity of the vectors v^b and f^b, i.e., the angle between them is nonzero and none of them can be identically zero. f^b is the specific force measured by the IMU, and the gravity vector is also measured by said sensor, so that a UAV flying at constant speed has $f^b = -R^T g^n$. In addition, a fixed-wing UAV always has a positive forward speed during flight and typically never accelerates just in the direction of g^n, so that Assumption 3 is never violated. If, on the other hand, a helicopter-like vehicle is used, it often finds itself hovering or moving perpendicular to the ground, situations that would violate Assumption 3. Caution must be exercised if the NLO is tested on vehicles other than fixed-wing UAVs.

Nonlinear Observer Equations

The NLO, graphically represented in Fig. 3, is based on the method proposed by [12] and is chosen as

$$\bar{\Sigma}_{A_{NLO}} \begin{cases} \dot{\bar{R}} = \bar{R} S(\omega_m^b - \bar{b}^b) + \sigma K_P J \\ \dot{\bar{b}}^b = \text{Proj}(\bar{b}^b, -k_I \text{vex}(\mathbb{P}_a(\bar{R}_s^T K_P J))) \end{cases} \quad (8)$$

$$\bar{\Sigma}_{TM_{NLO}} \begin{cases} \dot{\bar{p}}^n = \bar{v}^n + K_{pp}(p^n - \bar{p}^n) + K_{pv}(v^n - \bar{v}^n) \\ \dot{\bar{v}}^n = \bar{f}^n + g^n + K_{vp}(p^n - \bar{p}^n) + K_{vv}(v^n - \bar{v}^n) \\ \dot{\xi} = -\sigma K_P J f^b + K_{\xi p}(p^n - \bar{p}^n) + K_{\xi v}(v^n - \bar{v}^n) \\ \bar{f}^n = \bar{R} f^b + \xi \end{cases} \quad (9)$$

The subsystem $\bar{\Sigma}_{A_{NLO}}$ represents the attitude observer, in which K_P is a symmetric positive definite gain matrix, $\sigma \geq 1$ is a scaling factor that can be tuned to guarantee stability, k_I is a positive scalar gain, $\bar{R}_s = \text{sat}(\bar{R})$, and $\text{Proj}(\cdot, \cdot)$ represents a parameter

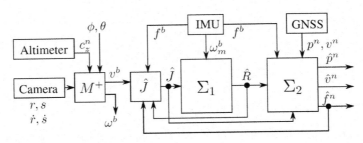

Fig. 3 Block diagram of the nonlinear observer

projection [24] that ensures that $\|\bar{b}^b\|$ does not exceed a design constant $L_{\bar{b}} > L_b$ (see Appendix). The matrix J is the output injection representing the attitude error, whose design is inspired by the TRIAD algorithm [35]. It is defined as

$$J(f^b, \bar{f}^n, v^b, v^n, \bar{R}) := \bar{A}_n A_b^T - \bar{R} A_b A_b^T \tag{10a}$$

$$A_b := \begin{bmatrix} f^b, & f^b \times v^b, & f^b \times (f^b \times v^b) \end{bmatrix} \tag{10b}$$

$$\bar{A}_n := \begin{bmatrix} \bar{f}^n, & \bar{f}^n \times v^n, & \bar{f}^n \times (\bar{f}^n \times v^n) \end{bmatrix} \tag{10c}$$

The body-fixed velocity vector v^b is calculated by means of the OF, according to (5).

The subsystem $\bar{\Sigma}_{TM_{NLO}}$ represents the translational motion observer, where K_{pp}, K_{pv}, K_{vp}, K_{vv}, $K_{\xi p}$, and $K_{\xi v}$ are observers gains yet to be defined. The presence of the term ξ lets the error dynamics of (9) be linear, which simplifies the analysis of stability.

The error dynamics of (8)–(9) can be written as

$$\check{\Sigma}_{A_{NLO}} \begin{cases} \dot{\bar{R}} = RS(\omega^b) - \bar{R}S(\omega_m^b - \bar{b}^b) - \sigma K_P J \\ \dot{\bar{b}}^b = -\text{Proj}(\bar{b}^b, \tau(J)) \end{cases} \tag{11}$$

$$\check{\Sigma}_{TM_{NLO}} \begin{cases} \dot{\check{p}}^n = \check{v}^n - K_{pp}\check{p}^n - K_{pv}\check{v}^n \\ \dot{\check{v}}^n = \check{f}^n - K_{vp}\check{p}^n - K_{vv}\check{v}^n \\ \dot{\check{f}}^n = -K_{\xi p}\check{p}^n - K_{\xi v}\check{v}^n + \check{d} \end{cases} \tag{12}$$

where $\tau(J) = -k_I \text{vex}(\mathbb{P}_a(\bar{R}_s^T K_P J))$ and $\check{d} = (RS(\omega^b) - \bar{R}S(\omega_m^b - \bar{b}^b))f^b + (R - \bar{R})\check{f}^b$. By defining the error state $\check{w} = [(\check{p}^n)^T, (\check{v}^n)^T, (\check{f}^n)^T]^T$, the error dynamics (12) can be written in a more compact form as

$$\dot{\check{w}} = (A - KC)\check{w} + B\check{d} \tag{13}$$

where

$$A = \begin{bmatrix} 0_{6\times 3} & I_6 \\ 0_{3\times 3} & 0_{3\times 6} \end{bmatrix}, \quad B = \begin{bmatrix} 0_{6\times 3} \\ I_3 \end{bmatrix},$$

$$C = \begin{bmatrix} I_6 & 0_{6\times 3} \end{bmatrix}, \quad K = \begin{bmatrix} K_{pp} & K_{pv} \\ K_{vp} & K_{vv} \\ K_{\xi p} & K_{\xi v} \end{bmatrix}.$$

The NLO just presented differs from the one in [12] in how (10) is defined, that is the magnetic field vectors are replaced by the OF velocity vectors.

Stability Analysis

Theorem 1 defines the conditions that render the equilibrium of (11)–(12) globally exponentially stable (GES).

Theorem 1 *Let σ be chosen sufficiently large and define $H_K(s) = (Is - A + KC)^{-1}B$. There exists a $\gamma > 0$ such that, if K is chosen such that $A - KC$ is Hurwitz and $\|H_K(s)\|_\infty < \gamma$, then the origin of the error dynamics (11)–(12) is GES. Moreover, K can always be chosen to satisfy these conditions.*

Proof See the proof of Theorem 3 in [12]. The magnetic field vectors are replaced by the OF velocity vectors, but since both v^b and v^n are measured quantities, the analysis remains unchanged under Assumptions 1–3. □

It is clear that the only uncertainty considered in the design and analysis of the NLO is the gyro bias, while all the high-frequency noise components affecting the system can be entirely disregarded, despite being present. This constitutes an advantage of the NLO over Kalman filter designs, as already anticipated in Sect. 1, but at the same time it offers no insight into the behavior of the variance of the estimation error. This is addressed in Sect. 4.2.

4.2 Exogenous Kalman Filter

One of the most popular tools for state estimation and filtering in navigation is the EKF, which linearizes the system to observe about the trajectory estimated by the filter itself, but whose stability properties are not clear due to its feedback structure. The XKF, conversely, linearizes the same system about the trajectory estimated by the NLO, which is guaranteed to be GES, and has a cascade structure (see Fig. 4). Since the linearization is made about an exogenous state trajectory, there is no feedback loop that can destabilize the second-stage LKF, and it follows from nonlinear stability theory that the cascade interconnection inherits the stability properties of the global NLO [19, 25, 31].

Observed System and Filter Equations

In order to reduce the number of states, the kinematic system considered for this second-stage filter is represented slightly differently from (6)–(7), as

$$\dot{\Theta} = T(\Theta)(\omega_m^b - b^b + n_\omega) \quad (14a)$$
$$\dot{b}^b = n_b \quad (14b)$$
$$\dot{p}^n = v^n \quad (14c)$$
$$\dot{v}^n = R(f^b + n_s) + g^n \quad (14d)$$

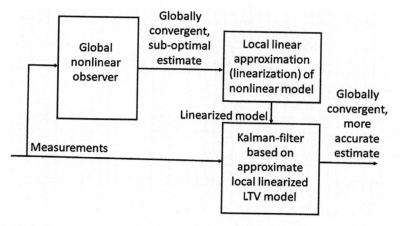

Fig. 4 Block diagram showing the cascade interconnection between an NLO and LKF, compactly called XKF, [19]

The attitude is now represented by the vector of Euler angles $\Theta = [\phi, \theta, \psi]^T$ with its dynamic equation, where $T(\Theta) \in \mathbb{R}^{3 \times 3}$ is the state-dependent transformation matrix [7]. n_b represents the process noise on the bias dynamics, and n_ω and n_s are noise on gyroscopes and accelerometers, respectively.

The output vector $y = [y_1, y_2, y_3, y_4]^T$ is necessary to implement the XKF. The measurements used to build y are position (GNSS), NED velocity (GNSS), body velocity (machine vision), and specific force (IMU), and are related to the states of the system via

$$y_1 = p^n + n_p \quad \text{GNSS} \tag{15a}$$

$$y_2 = v^n + n_{v^n} \quad \text{GNSS} \tag{15b}$$

$$y_3 = R^T v^n + n_{v^b} \quad \text{machine vision} \tag{15c}$$

$$y_4 = -R^T g^n + (\omega_m^b - \hat{b}^b + n_\omega) \times (v^b + n_{v^b}) + n_s \quad \text{IMU} \tag{15d}$$

The quantities n_p, n_{v^b}, and n_{v^n} represent noise on the sensors. All noise components in (14)–(15) are assumed to be Gaussian white noise. Note that (15d) is obtained from

$$f^b = -R^T g^n + \omega^b \times v^b + \dot{v}^b \tag{16}$$

It is common in navigation to consider only $f^b = -R^T g^n$, which might not yield accurate results in systems with high dynamics such as UAVs, for which the centripetal acceleration $\omega^b \times v^b$ is not neglectable. To verify the validity of this claim, Fig. 5 illustrates the norms of the three terms on the right-hand side of (16) as obtained during the flight test. It is clear that $\omega^b \times v^b$ contributes significantly to f^b, whereas $\dot{v}^b$ is always smaller and can be neglected. The peaks in $\omega^b \times v^b$ correspond to turns

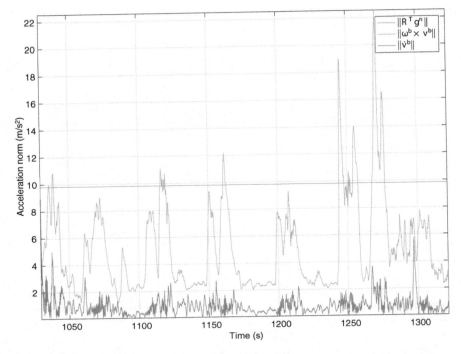

Fig. 5 Comparison of the norms of the different acceleration components

of the UAV: toward the end there are higher peaks because the UAV was preparing for landing and needed to perform some particularly sharp turns.

A more compact form for (14)–(15) is

$$\dot{x}(t) = f(x(t), t) + G(t)\, n_x(t) \quad (17a)$$
$$y(t) = h(x(t), t) + E(t)\, n_y(t) \quad (17b)$$

where f, G, and E are smooth vector- and matrix-valued functions, x represents the state vector, y the output, t the time, and n_x and n_y are vectors of process and measurement noise, respectively. Let $\bar{x}$ be an estimate for x, and assume it is bounded as given by a global NLO, with bounded error $\check{x} = x - \bar{x}$. A first-order Taylor series expansion of (17) about $\bar{x}(t)$ gives

$$\dot{x}(t) = f(\bar{x}(t), t) + F(\bar{x}(t), t)\,\check{x}(t) + G(t)\, n_x(t) + q_1(x(t), \bar{x}(t), t) \quad (18a)$$
$$y(t) = h(\bar{x}(t), t) + H(\bar{x}(t), t)\,\check{x}(t) + E(t)\, n_y(t) + q_2(x(t), \bar{x}(t), t) \quad (18b)$$

where $q_1(x(t), \bar{x}(t), t)$ and $q_2(x(t), \bar{x}(t), t)$ are higher order terms and

$$F(\bar{x}(t), t) = \frac{\partial f}{\partial x}(\bar{x}(t), t) \tag{19}$$

$$H(\bar{x}(t), t) = \frac{\partial h}{\partial x}(\bar{x}(t), t). \tag{20}$$

Building up from the theory in [19], an estimator $\hat{x}$ for x is then

$$\dot{\hat{x}}(t) = f(\bar{x}(t), t) + F(\bar{x}(t), t) x_d(t) + K(t)(y(t) - h(\bar{x}(t), t) - H(\bar{x}(t), t) x_d(t)), \tag{21}$$

where $x_d(t) = \hat{x}(t) - \bar{x}(t)$. Recall that the Kalman filter is optimal if n_x is Gaussian white noise with covariance matrix Q, n_y is Gaussian white noise with covariance matrix U, and n_x and n_y are uncorrelated. The gain matrix K can then be calculated as $K(t) = P(t)H(\bar{x}(t), t)^T U^{-1}$, with P the time-varying solution of the Riccati equation

$$\dot{P}(t) = F(\bar{x}(t), t) P(t) + P(t) F^T(\bar{x}(t), t) + G(t) Q G^T(t) - K(t) U K^T(t) \tag{22}$$

The state estimation error is $\tilde{x} := x - \hat{x} = \check{x} - \hat{x} - \bar{x}$. Combining (18) and (21) yields a linear time-varying system with a perturbation:

$$\dot{\tilde{x}}(t) = A(\bar{x}(t), t)\tilde{x}(t) + d(t), \tag{23}$$

where

$$A(\bar{x}(t), t) = F(\bar{x}(t), t) - K(t) H(\bar{x}(t), t) \tag{24}$$
$$d(t) = G(t) n_x + q_1(x(t), \bar{x}(t), t) + K(t) q_2(x(t), \bar{x}(t), t) + K(t) E(t) n_y. \tag{25}$$

The following assumptions are standard conditions that ensure boundedness and positive definiteness of the solution of the Riccati equation, and lead to nominal global convergence of the Kalman filter [1, 22].

Assumption 4 The LKF tunable parameters $P(0)$, Q, and U are positive definite and symmetric.

Assumption 5 The system $(F(\bar{x}(t), t), G(t), H(\bar{x}(t), t))$ is uniformly completely observable and controllable.

Stability of the XKF

The next theorem gives condition that ensure stability of the XKF.

Theorem 2 *Suppose Assumptions 1–5 hold. The origin $\check{x} = \tilde{x} = 0$ of the unforced error dynamics of the cascade (8)–(9) and (17), i.e., of the XKF (with $n_x = 0$ and $n_y = 0$), inherits the stability properties of the NLO.*

Proof By joining Assumptions 4 and 5 with the global exponential stability of the NLO, the conclusions of the theorem follow from Theorem 2.1 in [19]. □

Assumptions 1–3 are necessary for global exponential stability of the NLO. Assumption 4 can be satisfied by design. The requirements of Assumption 5 are hard to verify analytically a-priori. However, it is possible to calculate the observability and controllability Gramians recursively at runtime, and since they always result full rank, it can be inferred that Assumption 5 is satisfied.

If n_x and n_y are bounded inputs instead of zero, having a GES NLO allows to invoke Lemma 4.6 in [23], which implies that the origin of the error dynamics of the XKF is input-to-state stable with n_x and n_y as inputs, and that the solutions are uniformly ultimately bounded.

The stability properties of the XKF are inherited from the NLO, while the advantage over the NLO is the use of minimum-variance objectives in the design of the estimator. The linearization introduced with the LKF, however, creates an inevitable, possibly biased, random error that might lead to suboptimality and is hard to analyze in an experimental scenario. The best option in such cases is to resort to simulations to investigate the structure of the error.

5 Results

The XKF is here tested on both experimental and simulated data. Root mean square (RMS) errors are presented for both cases. In addition, a case study is presented to test the viability of the NLO with different reference vectors, should the camera system fail to provide useful OF estimates.

Fig. 6 The UAV Factory Penguin-B on the runway with the pilot, just before the experiment

5.1 Experimental Setup and Results

The UAV employed is a UAV Factory Penguin-B (Fig. 6), equipped with a custom payload that includes all the necessary sensors. The IMU is a Sensonor STIM300, a low-weight, tactical grade, high-performance sensor that includes gyroscopes, accelerometers, and inclinometers, all recorded at a frequency of 300 Hz. The chosen GPS receiver is a uBlox LEA-6T, which gives measurements at 5 Hz. The video camera is an IDS GigE uEye 5250CP provided with an 8 mm lens. The camera is configured for a hardware-triggered capture at 10 Hz: the uBlox sends a digital pulse-per-second signal whose rising edge is accurately synchronized with the time of validity of the recorded GPS position, which guarantees that the image capture is synchronized with the position measurements. The experiment has been carried out on February 6, 2015 at the Eggemoen Aviation and Technology Park, Norway, in a sunny day with good visibility, very little wind, an air temperature of about -8°C. The terrain is covered with snow and flat enough to let all features be considered as lying at zero altitude relative to altitude measurements.

In order to produce OF vectors, all the camera images are processed with a resolution of 1600×1200 (width $\times$ height) pixels and in their original state, without any filtering. The lens distortion of the camera is not accounted for, and no correction is applied to the images. SIFT is implemented with the open source computer vision library (OpenCV) [5] with default settings. Each match is tagged with a value indicating the accuracy of the match, and the smallest of these values is considered to be the best match. To increase the reliability of the OF vectors, each match is compared to the best one. Every match with an uncertainty more than double the uncertainty of the best match is not used. Also the template matching algorithm is implemented with OpenCV. The size of the templates is chosen to be 120×90 pixels and a correlation of 99% is required in order for a template match to be considered reliable and used.

The NLO is implemented using forward Euler discretization with a time-varying step depending on the interval of data acquisition of the fastest sensor, namely the STIM300, and it is typically around 0.003 seconds. The various parameters and gains are chosen as $L_{b^b} = 2°/s$, $L_{\hat{b}^b} = 2.1°/s$, $\sigma = 1$, $K_P = \text{diag}[0.08, 0.04, 0.06]$, $k_I = 0.02$, $K_{pp} = 30I_3$, $K_{pv} = 2I_3$, $K_{vp} = 0.01I_3$, $K_{vv} = 20I_3$, $K_{\xi p} = I_3$, and $K_{\xi v} = 50I_3$. All the gains are obtained by running the NLO several times and correcting the gains until a satisfactory performance was achieved. The gyro bias estimates are initialized with the standstill values, the other states with zero.

The covariance matrices Q and U are first tuned based on previous experience with the same sensors and system model, and then more finely tuned with trial and error. In both XKF and EKF, all estimates have zero initial value.

The reference provided for the position, velocity, and attitude is the output of the EKF of the autopilot mounted on the Penguin-B; the autopilot uses a different set of sensors than the one presented here. The path flown by the UAV and its NED velocity are in Figs. 7 and 8. An exact reference for the gyro bias is not available, but an approximation of the real value is calculated by averaging the gyro measurements

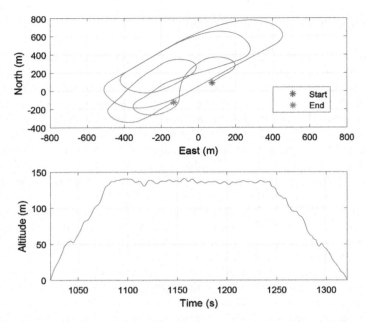

Fig. 7 Path of the UAV during the experiment. The zero is the base station from where the UAV was controlled

at standstill before and after the flight. The accelerometer bias is not estimated, but it is computed the same way as the gyro bias and subtracted from the accelerometers measurements before being used in the XKF (Figs. 7 and 8).

Ignoring for the sake of readability the time parameters, the equations implemented to run the discrete LKF in the XKF are

$$K = \bar{P}H(\bar{x})[H(\bar{x})\bar{P}H^T(\bar{x})]^{-1}$$
$$\hat{x} = x_t + K[y - \bar{y} - H(\bar{x})(\hat{x} - \bar{x})]$$
$$\hat{P} = [I - KH(\bar{x})]\bar{P}[I - KH(\bar{x})]^T + KUK^T$$
$$x_t = \hat{x} + h[f(\bar{x}) + F(\bar{x})(\hat{x} - \bar{x})]$$
$$\Phi = I + hF(\bar{x})$$
$$\Gamma = hE$$
$$\bar{P} = \Phi\hat{P}\Phi^T + \Gamma Q\Gamma^T$$

Figures 9, 10, 11 and 12 display the estimation errors of the NLO alone, of the entire XKF, and of an EKF with the same tuning parameters and initialization as the XKF. The time on the x-axes is the time elapsed since starting up the UAV; the dataset represented corresponds to the entire flight of the UAV. The RMS errors are reported in Table 1. All estimators converge, as expected, and the XKF and EKF perform better than the NLO.

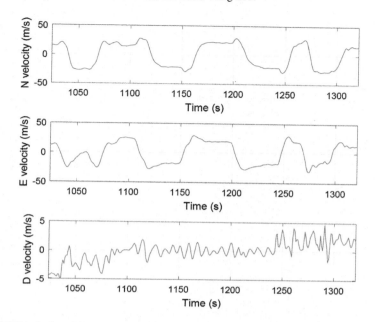

Fig. 8 NED velocity of the UAV during the experiment

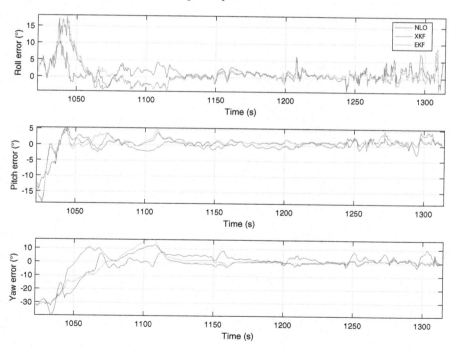

Fig. 9 Euler angles estimation error of the three methods with respect to the autopilot EKF

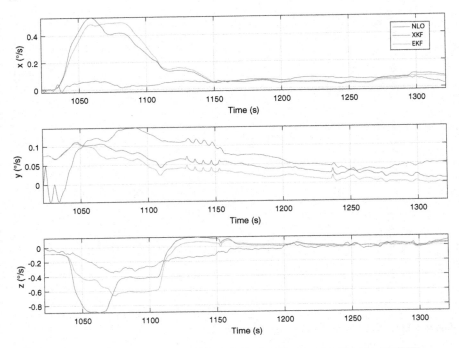

Fig. 10 Gyro bias estimation error of the three methods with respect to the autopilot EKF

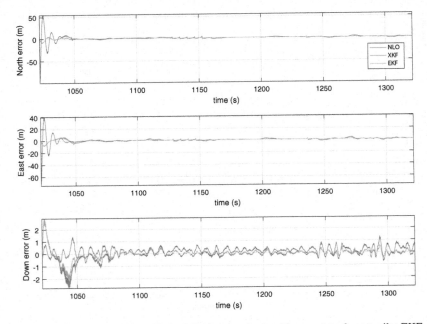

Fig. 11 NED position estimation error of the three methods with respect to the autopilot EKF

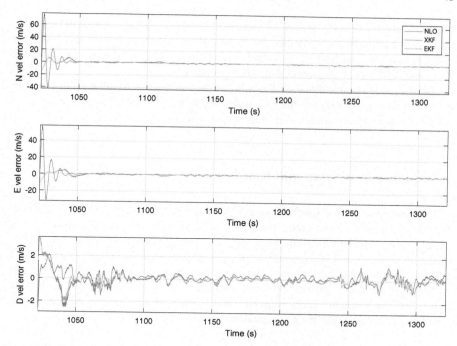

Fig. 12 NED velocity estimation error of the three methods with respect to the autopilot EKF

Table 1 RMS errors for the experimental data

State	NLO	XKF	EKF
Roll ϕ (°)	1.887	1.363	1.355
Pitch θ (°)	1.319	1.265	1.336
Yaw ψ (°)	0.055	0.018	0.021
Gyro bias x (°/s)	0.065	0.059	0.051
Gyro bias y (°/s)	0.049	0.030	0.012
Gyro bias z (°/s)	0.028	0.019	0.019
North position (m)	0.876	0.417	0.419
East position (m)	0.944	0.449	0.440
Down position (m)	0.330	0.176	0.156
North velocity (m/s)	0.291	0.286	0.257
East velocity (m/s)	0.352	0.285	0.251
Down velocity (m/s)	0.374	0.313	0.314

Another term of comparison for the estimators is their behavior with large initialization errors. If, for example, the estimators are initialized with large attitude estimation errors, the EKF yaw estimation error does not converge to zero, whereas NLO and XKF converge due to their GES property.

There is no guarantee that the EKF onboard the autopilot is the device that gives the best estimates with respect to the real, unknown values. For this reason, a simulation is run in order to have access to perfect measurements of the states.

5.2 Simulation Setup and Results

The simulated flight has several changes of direction and slight changes in altitude. Instead of simulating the entire machine vision system, a body velocity sensor is simulated by adding Gaussian white noise to the known, exact value. The GPS-measured position and velocity are modeled as Gauss–Markov processes with added Gaussian white noise, according to the specifications in [4]. Gaussian white noise is added to all the other sensors too, with standard deviations 0.135 deg/s for the rate gyros, 0.02 m/s^2 for the accelerometers, and 0.3 m/s for the body-fixed velocity. A constant bias is added to the gyro measurements.

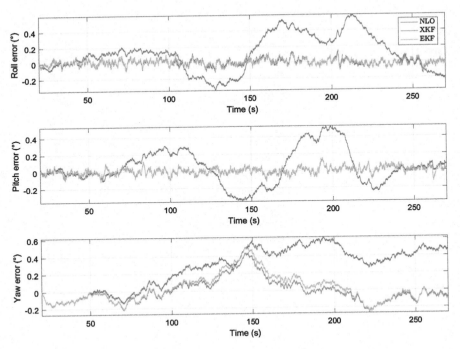

Fig. 13 Euler angles estimation error of the three methods with simulated data

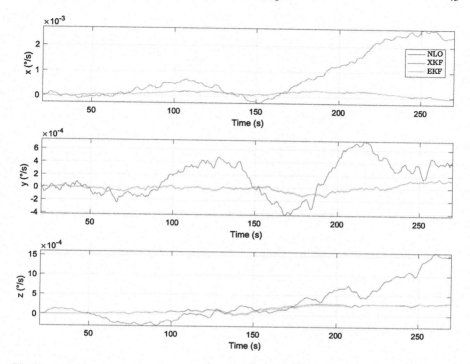

Fig. 14 Gyro bias estimation error of the three methods with simulated data

Table 2 RMS errors for the simulation

State	NLO	XKF	EKF
Roll ϕ (°)	0.288	0.044	0.040
Pitch θ (°)	0.127	0.039	0.039
Yaw ψ (°)	0.397	0.126	0.121
Gyro bias x (°/s)	0.002	1.24e–04	1.23e–03
Gyro bias y (°/s)	4.69e–04	8.22e–05	8.15e–05
Gyro bias z (°/s)	0.001	2.69e–04	2.66e–04
North position (m)	0.189	0.153	0.153
East position (m)	0.179	0.146	0.141
Down position (m)	0.310	0.274	0.276
North velocity (m/s)	0.219	0.047	0.044
East velocity (m/s)	0.226	0.051	0.053
Down velocity (m/s)	0.359	0.047	0.047

The results are in Fig. 13, 14, 15 and 16 and the RMS errors in Table 2. It is clear that the NLO performs worse than the XKF and EKF, and it is particularly evident in the position and velocity estimates that the NLO estimates have more noise.

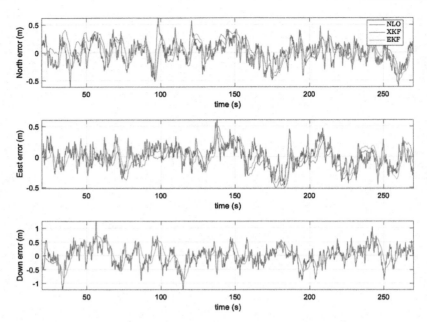

Fig. 15 NED position estimation error of the three methods with simulated data

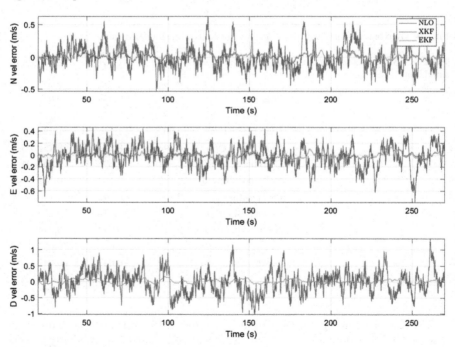

Fig. 16 NED velocity estimation error of the three methods with simulated data

Table 3 RMS errors for the NLO using acceleration and heading from GPS

State	NLO
Roll ϕ (°)	2.681
Pitch θ (°)	1.965
Yaw ψ (°)	0.391
Gyro bias x (°/s)	0.053
Gyro bias y (°/s)	0.124
Gyro bias z (°/s)	0.275
North position (m)	0.883
East position (m)	0.915
Down position (m)	0.306
North velocity (m/s)	0.852
East velocity (m/s)	0.687
Down velocity (m/s)	0.365

5.3 Case Study: Camera or OF Failure

A fundamental component of the NLO is the body-referenced velocity obtained via machine vision. If the camera optical flow algorithm fails to function, the velocity becomes unreliable and a new solution is necessary. The acceleration vectors alone can help estimate the roll and pitch angles, whereas the yaw requires an additional pair of vectors. Assuming that the course and heading of the UAV coincide, the v^b and v^n vectors in the injection term can be replaced by

$$m^n = \begin{bmatrix} 1 \\ 0 \\ 0 \end{bmatrix}, \quad m^b = \begin{bmatrix} \cos(\psi) \\ -\sin(\psi) \\ 0 \end{bmatrix} \approx \frac{1}{\|v^n\|} \begin{bmatrix} v_N^n \\ -v_E^n \\ 0 \end{bmatrix} \quad (26)$$

The vector m^b is then calculated based on the GPS velocity readings. This solution is clearly not ideal, for it would yield inaccurate results in the presence of strong crosswinds leading to a crab angle, but it would work as a degraded accuracy mode.

The method is tested on the same experimental data used in Sect. 5.1. The RMS error is in Table 3. Plots are not presented due to lack of space. The results are in general worse than those obtained using the camera, as expected, but would still be a viable solution.

6 Conclusions

This chapter presented the results and experimental verification of the exogenous Kalman filter for the estimation of a fixed-wing UAV navigation states. The theory behind the filter was presented, with theorems proving the global exponential stability properties of the method, followed by the experimental results, where the estimates from the nonlinear observer, exogenous Kalman filter, and extended Kalman filter were compared with the output (an extended Kalman filter itself) of the autopilot onboard the UAV. The two Kalman filter solutions have better noise rejection, and they were also tested on simulated data in order to be able to compare the estimates with perfect reference values. Moreover, unlike the EKF, the XKF has proven GES properties. An additional case study was presented to provide a degraded accuracy solution in case of a camera failure, and allow the UAV to land safely.

Acknowledgements This work was partly supported by the Norwegian Research Council (grant numbers 225259 and 223254) through the Center of Autonomous Marine Operations and Systems at the Norwegian University of Science and Technology.
The authors are grateful for the assistance provided by the UAV engineers at NTNU and Maritime Robotics AS, in particular Lars Semb and Carl Erik Stephansen. Significant contributions were made to the construction of the UAV payload by the rest of the navigation team at NTNU, in particular Sigurd M. Albrektsen, Jakob M. Hansen and Kasper T. Borup, and to the development of machine vision by Jesper Hosen and Håkon H. Helgesen.

Appendix

The parameter projection $\text{Proj}(\cdot, \cdot)$ is defined as

$$\text{Proj}(\hat{b}^b, \tau) = \begin{cases} \left(I - \frac{c(\hat{b}^b)}{\|\hat{b}^b\|^2} \hat{b}^b \hat{b}^{bT}\right) \tau, & \|\hat{b}^b\| \geq L_b, \ \hat{b}^{bT}\tau > 0 \\ \tau, & \text{otherwise} \end{cases}$$

where $c(\hat{b}^b) = \min\{1, (\|\hat{b}^b\|^2 - L_b^2)/(L_{\hat{b}}^2 - L_b^2)\}$. This operator is a special case of that from Appendix E of [24]. Some of its properties are reported here: (i) $\text{Proj}(\cdot, \cdot)$ is locally Lipschitz continuous, (ii) $\|\hat{b}^b\| \geq L_{\hat{b}} \Rightarrow \hat{b}^{bT}\text{Proj}(\hat{b}^b, \tau) \leq 0$, (iii) $\|\text{Proj}(\hat{b}^b, \tau)\| \leq \|\tau\|$, and (iv) $-\tilde{b}^{bT}\text{Proj}(\hat{b}^b, \tau) \leq -\tilde{b}^{bT}\tau$.

References

1. Anderson, B.D.O.: Stability properties of Kalman-Bucy filters. J. Frankl. Inst. **291**, 137–144 (1971)
2. Batista, P., Silvestre, C., Oliveira, P.: A GES attitude observer with single vector observations. Automatica **48**(2), 388–395 (2012)
3. Batista, P., Silvestre, C., Oliveira, P.: Globally exponentially stable cascade observers for attitude estimation. Control Eng. Pract. **20**, 148–155 (2012)
4. Beard, R.W., McLain, T.W.: Small Unmanned Aircraft: Theory and Practice. Princeton University Press, Princeton (2012)
5. Bradski, G.: The OpenCV library. Dr. Dobbs J. **25**(11), 120–126 (2000)

6. Brown, R.G., Hwang, P.Y.C.: Introduction to Random Signals and Applied Kalman Filtering, 3rd edn. Wiley, London (2012)
7. Fossen, T.I.: Handbook of Marine Craft Hydrodynamics and Motion Control. Wiley, London (2011)
8. Fuh, C.S., Maragos, P.: Region-based optical flow estimation. IEEE Comput. Soc. Conf. Comput. Vis. Pattern Recognit. 130–135 (1989)
9. Fusini, L., Johansen, T.A., Fossen, T.I.: Experimental validation of a uniformly semi-globally exponentially stable non-linear observer for GNSS- and camera-aided inertial navigation for fixed-wing UAVs. In: Proceedings of the International Conference on Unmanned Aircraft Systems (2015). doi:10.1109/ICUAS.2015.7152371
10. Gelb, A.: Applied Optimal Estimation. MIT Press, Cambridge (1974)
11. Grip, H.F., Fossen, T.I., Johansen, T.A., Saberi, A.: Attitude estimation using biased Gyro and vector measurements with time-varying reference vectors. IEEE Trans. Autom. Control **57**(5), 1332–1338 (2012)
12. Grip, H.F., Fossen, T.I., Johansen, T.A., Saberi, A.: Globally exponentially stable attitude and gyro bias estimation with application to GNSS/INS integration. Automatica **51**, 158–166 (2015)
13. Gustafsson, F.: Statistical Sensor Fusion, Studentlitteratur AB, Lund, Sweden (2012)
14. Hosen, J., Helgesen, H.H., Fusini, L., Fossen, T.I., Johansen, T.A.: Vision-aided nonlinear observer for fixed-wing unmanned aerial vehicle navigation. J. Guidance Control Dyn. **39**(8), 1777–1789 (2016)
15. Hua, M.D., Martin, P., Hamel, T.: Stability analysis of velocity-aided attitude observers for accelerated vehicles. Automatica **63**, 11–15 (2016)
16. Hutchinson, S., Hager, G.D., Corke, P.I.: A tutorial on visual servo control. IEEE Trans. Robot. Autom. **12**(5), 651–670 (1996)
17. Johansen, T.A., Fossen, T.I.: Nonlinear observer for tightly coupled integration of pseudo-range and inertial measurements. IEEE Trans. Control Syst. Technol. **24**, 2199–2206 (2016). doi:10.1109/TCST.2016.2524564
18. Johansen, T.A., Fossen, T.I.: Nonlinear filtering with eXogenous Kalman filter and double Kalman Filter. In: European Control Conference, Aalborg (2016)
19. Johansen, T.A., Fossen, T.I.: The eXogenous Kalman Filter (XKF). Int. J. Control (2016). doi:10.1080/00207179.2016.1172390
20. Johansen, T.A., Fossen, T.I., Goodwin, G.C.: Three-stage filter for position estimation using pseudo-range measurements. IEEE Trans. Aerosp. Electron. Syst. **52**, 1631–1643 (2016)
21. Julier, S.J., Uhlmann, J.K.: Unscented filtering and nonlinear estimation. Proc. IEEE **92**(3), 401–422 (2004)
22. Kalman, R.E., Bucy, R.S.: New results in linear filtering and prediction theory. Trans. ASME Ser. D J. Basic Eng. **83**, 95–109 (1961)
23. Khalil, H.K.: Nonlinear Systems. Prentice-Hall, Englewood Cliffs (2002)
24. Krstic, M., Kanellakopoulos, I., Kokotovic, P.V.: Nonlinear and Adaptive Control Design. Wiley, New York (1995)
25. Loria, A., Panteley, E.: Cascaded nonlinear time-varying systems: analysis and design. In: Lamnabhi-Lagarrigue, F., Loria, A., Panteley, E. (eds.) Advanced Topics in Control Systems Theory, pp. 23–64. Springer, London (2004)
26. Lowe, D.G.: Object recognition from local scale-invariant features. In: Proceedings of the International Conference on Computer Vision, pp. 1150–1157 (1999)
27. Luenberger, D.G.: Observing a state of a linear system. IEEE Trans. Mil. Electron. **8**, 74–80 (1964)
28. Mahony, R., Hamel, T., Pflimlin, J.M.: Nonlinear complementary filters on the special orthogonal group. IEEE Trans. Autom. Control **53**(5), 1203–1218 (2008)
29. Mammarella, M., Campa, G., Fravolini, M.L., Napolitano, M.R.: Comparing optical flow algorithms using 6-DOF motion of Real-World rigid objects. IEEE Trans. Syst. Man Cybern. Part C: Appl. Rev. **42**(6), 1752–1762 (2012)

30. Martin, P., Salaun, E.: Design and implementation of a low-cost observer-based attitude and heading reference system. Control Eng. Pract. **18**(7), 712–722 (2010)
31. Panteley, E., Loria, A.: On global uniform asymptotic stability of nonlinear time-varying systems in cascade. Sys. Control Lett. **33**, 131–138 (1998)
32. Reif, K., Sonnemann, F., Unbehauen, R.: An EKF-based nonlinear observer with a prescribed degree of stability. Automatica **34**, 1119–1123 (1998)
33. Salcudean, S.: A globally convergent angular velocity observer for rigid body motion. IEEE Trans. Autom. Control **36**(12), 1493–1497 (1991)
34. Sarvaiya, J.N., Patnaik, S., Bombaywala, S.: Image registration by template matching using normalized cross-correlation. In: International Conference Advances in Computing, Control, and Telecommunication Technologies (2009). doi:10.1109/ACT.2009.207
35. Shuster, M.D., Oh, S.D.: Three-axis attitude determination from vector observations. J. Guid. Control Dyn. **4**(1), 70–77 (1981)
36. Simon, D.: Optimal State Estimation: Kalman, H Infinity, and Nonlinear Approaches. Wiley, London (2006)
37. Song, Y., Grizzle, J.W.: The extended Kalman filter as a local asymptotic observer for discrete-time nonlinear systems. J. Math. Syst. Estim. Control **5**, 59–78 (1995)
38. Sonka, M., Hlavac, V., Boyle, R.: Image Processing, Analysis, and Machine Vision. Cengage Learning, Stamford (2008)
39. Stovner, B., Johansen, T.A., Fossen, T.I., Schjølberg, I.: Three-stage filter for position and velocity estimation from long baseline measurements with unknown wave speed. In: American Control Conference, Boston (2016)

Motion Control of ROVs for Mapping of Steep Underwater Walls

Stein M. Nornes, Asgeir J. Sørensen and Martin Ludvigsen

Abstract This chapter describes an equipment setup and motion control strategy for automated visual mapping of steep underwater walls using a remotely operated vehicle (ROV) equipped with a horizontally facing doppler velocity logger (DVL) to provide vehicle velocity and distance measurements relative to the underwater wall. The main scientific contribution is the development of the motion control strategy for distance keeping and adaptive orientation using measurements from a DVL mounted in an arbitrary orientation. Autonomy aspects concerning this type of mapping operation are also discussed. The still images recorded by the stereo cameras of the ROV are post-processed into a 3D photogrammetry model using a combination of commercially available software and freeware. The system was implemented on an ROV and tested on a survey of a rock wall in the Trondheimsfjord in April 2016.

1 Introduction

An ROV is a mobile underwater sensor platform connected to a host, normally a ship, using an umbilical that can provide the ROV with high bandwidth communication and (near) unlimited electrical power. Because the thick cable necessary for supplying electrical power can result in unwanted loads or drag forces on the ROV, some ROVs such as the Nereus [4] will instead rely on battery power and only use the umbilical for communication. ROVs are traditionally manually controlled by a human operator, but are increasingly automated with dynamic positioning systems like the one presented

---

S.M. Nornes (✉) · A.J. Sørensen · M. Ludvigsen
Department of Marine Technology, Centre for Autonomous
Marine Operations and Systems (NTNU AMOS), Norwegian University of Science
and Technology (NTNU), 7491 Trondheim, Norway
e-mail: stein.nornes@ntnu.no

A.J. Sørensen
e-mail: asgeir.sorensen@ntnu.no

M. Ludvigsen
e-mail: martin.ludvigsen@ntnu.no

© Springer International Publishing AG 2017
T.I. Fossen et al. (eds.), *Sensing and Control for Autonomous Vehicles*,
Lecture Notes in Control and Information Sciences 474,
DOI 10.1007/978-3-319-55372-6_3

in [8, 18]. The high bandwidth and increasing degree of automation make ROVs useful as test platforms for autonomy algorithms that can later be implemented on autonomous underwater vehicles (AUV), since the human operators are able to continuously monitor the behavior and (if needed) interact with the system.

Operating underwater vehicles close to high rock walls or other vertical environments often pose navigational challenges since DVLs are normally pointing downward, and these locations are prone to acoustical multipath for position references. At the same time, such locations are often important to investigate for multiple end-users within the marine science and oil and gas industry.

Bottom-following based on single altimeters and depth control are commonly implemented on many ROVs. [5] added a secondary altimeter, one at the front and one at the rear of the Romeo ROV, in order to estimate the incline of the sea floor in the surge direction of the vehicle. [8] continued along this route, but instead of dedicated altimeters, they relied on the four altitude measurements received from the DVL. The addition of two extra measurements allowed the incline of the sea floor to be estimated in both North and East coordinates following the surge and sway motions of the ROV.

There have been examples of underwater vehicles with unconventional DVL orientations, perhaps most prominently the Bluefin HAUV [19]. This features a pitch-tiltable DVL and is mainly used for acoustic inspection of ship hulls. The ship-relative position is calculated by integrating the DVL velocity. For the motion control strategy presented in this paper, we will instead rely on combining the DVL measurements with the absolute position updates from ultra-short base line (USBL) acoustics, with the benefits and challenges associated. These include among others the counteraction of biases in the DVL, as well as handling noise introduced by the acoustics, and georeferencing the collected data.

Visual inspection typically requires closer proximity to the object of interest than acoustic mapping, leading to smaller margins of error. Unlike a ship hull, which is a relatively structured and known environment, mapping a deep rock wall will typically present an unknown and complex task.

The use of underwater photogrammetry, combining 2D still images into a 3D model, for close-up visual mapping and visualization of results has increased greatly the past few years. This increase can be attributed to the improvement of computing power and photogrammetry software, leading to a reduced need for experts manually aligning the images. Due to limited image swath width, image acquisition for photogrammetry requires low cross-track errors to ensure proper overlap and benefits from automated lawn mower pattern maneuvering. The quality of the images also depends on the distance to the scene, both directly due to the field of focus of the camera, and more indirectly from the amount of light that is attenuated in water by the round-trip distance of the light from lamps to cameras.

In this paper we propose an automated guidance and control strategy for high-quality visual mapping of both horizontal and vertical objects and areas of interest using a properly tilted DVL on an underwater vehicle. The performance of the proposed strategy is demonstrated on a steep underwater surface using an ROV. By relying on sensors that are common to the majority of ROVs, this approach can easily be adapted to different ROVs and AUVs. This work continues the earlier work on bottom tracking using a DVL by [8], and generalizes it to cover arbitrary vehicle, DVL and surface orientations. The main scientific contribution is the development of the distance estimation framework in Sect. 2.3 "Generalizing for Tilted DVL" and the distance guidance law in Sect. 2.3 "Guidance Law", providing a motion control strategy for automated distance keeping.

Section 2 describes the equipment setup, the motion control strategy, and the processing pipeline of the collected data. The results of the test survey are presented in Sects. 3, 4 discusses the quality of the results and what lessons can be learned from them and possible future improvements paving the way for full autonomy. Finally, concluding remarks are made in Sect. 5.

2 Method

2.1 Equipment

The ROV used in the survey is the SUB-fighter 30 K ROV shown in Fig. 1 made by Sperre AS in 2004 for NTNU. It is a light work class ROV ($2.6 \times 1.5 \times 1.6$ m, 1900 kg), and is a frequently used test platform for research and science. The ROV is usually deployed from the NTNU Research Vessel (RV) Gunnerus, and it is powered

Fig. 1 ROV 30 K

Fig. 2 The stereo cameras and horizontal DVL

from and communicates with the surface vessel through a 650 m umbilical. All systems needed for operation such as power supply, navigation computers, and monitors are fitted inside a 20 foot container.

For this survey, the ROV was equipped with a combined stereo camera and DVL rig, as shown in Fig. 2. This features two Allied Vision GC1380C cameras [2] mounted horizontally 30.5 cm apart. The cameras have a resolution of 1360×1024 pixels and are capable of recording at 20 frames per second (fps). Their high light sensitivity and signal-to-noise ratio make them suitable for underwater operations. The modest resolution, combined with keeping the recording at 0.5 fps to reduce the number of redundant images, kept the amount of data at a manageable level for the post-processing stage. The DVL is a standard 1 MHz model from Nortek AS [15] with no hardware modifications.

The lighting for the operation was provided by two HMI lamps mounted on the horizontal top bar on the front of the ROV.

2.2 ROV Control System

The ROV can be controlled manually using a joystick console, or automated using a dynamic positioning (DP) system developed and continuously expanded at NTNU [6–8, 12–14]. The DP system includes a vertical-DVL-based altitude estimator and controller developed by [8]. Section 2.3 describes how this can be modified into a more general distance controller with a DVL mounted in an arbitrary orientation.

The main architecture of the control system is displayed in Fig. 3. A nonlinear output PID-controller receives the desired state from a guidance module. The guidance module is split into three parts, where the main module calculates a desired trajectory from user-generated waypoints in NED coordinates. The adaptive orientation module proposed in Sect. 2.4 can be used to modify the desired orientation of the vehicle in order to achieve the best possible sensor orientation with respect to the scene being mapped. The distance guidance module (see Sect. 2.3 "Guidance Law") then modifies the desired position in order to achieve the desired distance to the scene. The "High-level Autonomy" block will be further explained in Sect. 4.3.

For further details on the other parts of the DP system, the reader is referred to [7, 18].

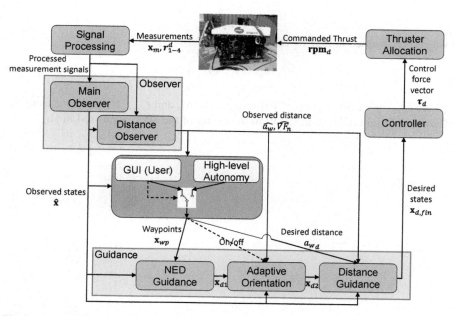

Fig. 3 An overview of the implemented control system

2.3 Distance Estimation and Guidance

The ROV position vector in the {n} (NED) frame is $\mathbf{p} = [x_p, y_p, z_p]^T$. The NED velocity vector is $\dot{\mathbf{p}}$ and can be expressed as

$$\dot{\mathbf{p}} = \mathbf{R}_b^n(\boldsymbol{\Theta}_{nb})\mathbf{v} \qquad (1)$$

where $\boldsymbol{\Theta}_{nb} = [\phi, \theta, \psi]^T$ is the attitude vector of the ROV, $\mathbf{R}_b^n(\boldsymbol{\Theta}_{nb}) \in \mathbb{R}^{3\times 3}$ is the rotation matrix from the {b} (body) frame to {n}, and $\mathbf{v} = [u, v, w]^T$ is the velocity vector in {b}.

DVL Beams

The DVL has four acoustic beams with the jth beam (shown in Fig. 4) represented in the DVL frame {d} by the vector

$$\mathbf{r}_j^d = \begin{bmatrix} x_j^d \\ y_j^d \\ a_j^d \end{bmatrix} = a_j^d \begin{bmatrix} \tan\gamma_j \cos\beta_j \\ \tan\gamma_j \sin\beta_j \\ 1 \end{bmatrix} \qquad (2)$$

which transformed to the {n} frame becomes

$$\mathbf{r}_j^n = \mathbf{R}_b^n(\boldsymbol{\Theta}_{nb})\left(\mathbf{R}_d^b(\boldsymbol{\Theta}_{bd})\mathbf{r}_j^d + \mathbf{r}_{dvl/b}^b\right) \qquad (3)$$

where $\boldsymbol{\Theta}_{bd}$ is the orientation of {d} relative to {b}, and $\mathbf{r}_{dvl/b}^b$ is the vector from the Center of Origin (CO) of the ROV to the center of the DVL expressed in {b}.

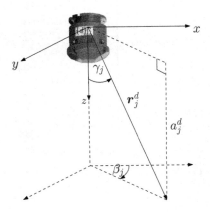

Fig. 4 The DVL beam components [8]

Seafloor Approximation for Vertical DVL Setup

As proposed in [8], it is assumed that the seafloor can be expressed as the surface given by the following equation:

$$F(x, y, z) = f(x, y) - z = 0, \quad \frac{\delta F}{\delta t} = 0 \qquad (4)$$

and that the ROV altitude rate of change can be expressed as

$$\dot{a} = \nabla F(\mathbf{p})\dot{\mathbf{p}}, \qquad (5)$$

where

$$\nabla F(\mathbf{p}) = \begin{bmatrix} \frac{\delta f}{\delta x}|_{x_p, y_p} & \frac{\delta f}{\delta y}|_{x_p, y_p} & -1 \end{bmatrix}.$$

The linear seafloor function is

$$f(x, y) = a + bx + cy \qquad (6)$$

Using the four DVL beams, we can find a least-squares approximation of the seafloor by minimizing the object function

$$\mathbf{x} = \begin{bmatrix} a \\ b \\ c \end{bmatrix} = \arg\min_{a,b,c \in \mathbb{R}} \sum_{j=1}^{4} [a_j^n - (a + bx_j^n + cy_j^n)]^2 \qquad (7)$$

which is achieved by solving

$$\mathbf{x} = (\mathbf{A}^T \mathbf{A})^{-1} \mathbf{A}^T \mathbf{b}, \qquad (8)$$

where

$$\mathbf{A} = \begin{bmatrix} 1 & x_1^n & y_1^n \\ 1 & x_2^n & y_2^n \\ 1 & x_3^n & y_3^n \\ 1 & x_4^n & y_4^n, \end{bmatrix}, \quad \mathbf{b} = \begin{bmatrix} a_1^n \\ a_2^n \\ a_3^n \\ a_4^n \end{bmatrix}$$

This yields an approximation of the ROV altitude a and the estimated seafloor gradient according to

$$\hat{\nabla} F(\mathbf{p}) = \begin{bmatrix} b & c & -1 \end{bmatrix} \qquad (9)$$

which are used as input in a Kalman filter to calculate the estimated altitude $\hat{a}$.

Generalizing for Tilted DVL

We define the wall frame {w}, such that

$$\mathbf{R}_w^n(\boldsymbol{\Theta}_{nw}) = \mathbf{R}_b^n(\boldsymbol{\Theta}_{nb_d})\mathbf{R}_d^b(\boldsymbol{\Theta}_{bd}), \tag{10}$$

where $\mathbf{R}_w^n(\boldsymbol{\Theta}_{nw}) \in \mathbb{R}^{3\times 3}$ is the rotation matrix from {w} to {n}, and $\boldsymbol{\Theta}_{nb_d} = [\phi_d, \theta_d, \psi_d]^T$ is the desired attitude vector of the ROV. Notice that the {w} and {d} frames are closely related, but while $\boldsymbol{\Theta}_{nd}$ includes all noise in $\boldsymbol{\Theta}_{nb}$, the angles $\boldsymbol{\Theta}_{bd}$ and $\boldsymbol{\Theta}_{nb_d}$ can be chosen such that $\boldsymbol{\Theta}_{nw}(t)$ is a smooth continuous function.

Choosing this frame yields the frame position vector

$$\mathbf{p}_w = \begin{bmatrix} x_w & y_w & z_w \end{bmatrix}^T = \mathbf{R}_d^{b^T}(\boldsymbol{\Theta}_{bd})\mathbf{R}_b^{n^T}(\boldsymbol{\Theta}_{nb_d})\mathbf{p} \tag{11}$$

and the frame velocity vector

$$\mathbf{v}_w = \mathbf{R}_d^{b^T}(\Theta_{bd})\mathbf{R}_b^{n^T}(\boldsymbol{\Theta}_{nb_d})\dot{\mathbf{p}} = \mathbf{R}_d^{b^T}(\boldsymbol{\Theta}_{bd})\mathbf{R}_b^{n^T}(\boldsymbol{\Theta}_{nb_d})\mathbf{R}_b^n(\boldsymbol{\Theta}_{nb})\mathbf{v}. \tag{12}$$

We assume the orientation of an underwater vehicle is either actively controlled in roll, pitch, and yaw, or is passively stabilized in roll and pitch by the buoyancy and weight distribution of the ROV and actively controlled in yaw. Like most ROVs, the ROV setup used in this chapter is assumed to be passively stabilized to 0 in roll and pitch, while the heading is controlled to the desired reference with a PID controller. Assuming $\boldsymbol{\Theta}_{nb} \to \boldsymbol{\Theta}_{nb_d}$, (12) reduces to

$$\mathbf{v}_w = \begin{bmatrix} u_w & v_w & w_w \end{bmatrix}^T = \mathbf{R}_d^{b^T}(\boldsymbol{\Theta}_{bd})\mathbf{v}. \tag{13}$$

Using (8) with

$$\mathbf{r}_j^w = \mathbf{R}_d^{b^T}(\boldsymbol{\Theta}_{bd})\mathbf{R}_b^{n^T}(\boldsymbol{\Theta}_{nb_d})\mathbf{r}_j^n \tag{14}$$

as input yields the approximate distance to the wall a_w and the gradient $\nabla \hat{F}_w$. This is translated back into the {n} frame as follows:

$$\nabla \hat{F}_n(\mathbf{p}, \boldsymbol{\Theta}_{nw}) = \nabla \hat{F}_w(\mathbf{p}_w)\mathbf{R}_d^{b^T}(\boldsymbol{\Theta}_{bd})\mathbf{R}_b^{n^T}(\boldsymbol{\Theta}_{nb_d}). \tag{15}$$

Since $\nabla F_n = \begin{bmatrix} F_{nx}, F_{ny}, F_{nz} \end{bmatrix}$ is a function of the point on the seafloor the DVL is pointed at, we need to specify both the ROV position $\mathbf{p}$ and the {w}-plane orientation $\boldsymbol{\Theta}_{nw}$. For the sake of readability, the function arguments are dropped for the remainder of the chapter.

Guidance Law

The guidance modules of the control system generate desired reference signals to the subsequent control module according to guidance laws. In [8], the altitude is controlled using the altitude guidance law according to

$$\dot{z}_d = w_d = k_{ff}\hat{\nabla}F\mathbf{R}_z(\hat{\psi})\begin{bmatrix}\hat{u} & \hat{v} & 0\end{bmatrix}^T - \dot{a}_d + k_p\tilde{a} + k_i\int_0^t \tilde{a}d\tau \qquad (16)$$

where $\tilde{a} = \hat{a} - a_d$ is the error with respect to the desired altitude, and $k_{ff}, k_p, k_i \geq 0$ are feed-forward, proportional, and integral gains, respectively. This replaces the z_d and w_d reference signals calculated by the main guidance module (see [7] for further details), which are then used as inputs for the controller module. By considering the horizontal plane in the {w}-frame instead of the {n}-frame, the new proposed distance guidance law in this chapter becomes

$$\dot{z}_{wd} = w_{wd} = k_{ff}\hat{\nabla}F_n\mathbf{R}_w^n(\boldsymbol{\Theta}_{nw})\begin{bmatrix}1 & 0 & 0\\ 0 & 1 & 0\\ 0 & 0 & 0\end{bmatrix}\left(\hat{\mathbf{v}}_w + \left(\boldsymbol{\omega}_{d/b}^d + \mathbf{R}_d^b(\boldsymbol{\Theta}_{bd})\boldsymbol{\omega}_{b/n_d}^b\right) \times \mathbf{r}_L\right)$$

$$-\dot{a}_{wd} + k_p\tilde{a}_w + k_i\int_0^t \tilde{a}_w d\tau, \qquad (17)$$

where $\tilde{a}_w = \hat{a}_w - a_{wd}$ is the error with respect to the desired distance, $\boldsymbol{\omega}_{d/b}^d$ is the rotational velocity of the DVL with respect to the vehicle frame, $\boldsymbol{\omega}_{b/n_d}^b$ is the desired rotational velocity of the vehicle, and $\mathbf{r}_L = [0, 0, \hat{a}_w]^T$ is the lever arm vector. As before, this replaces the z_{wd} and w_{wd} reference signals calculated by the main guidance module.

In contrast to [8], we have also proposed to include the orientation of the DVL and the roll and pitch of the vehicle in the calculations. The old control law in (16) is limited to only controlling the altitude along the vertical z-axis and using a vertical DVL. The new generalized guidance law in (17) can be applied to any desired vehicle and sensor orientation. As proposed in [8], the roughness of the seafloor must be considered for using a platforming or contouring control strategy.

By choosing a fixed orientation of both the DVL and vehicle, (17) reduces to

$$\dot{z}_{wd} = w_{wd} = k_{ff}\hat{\nabla}F_n\mathbf{R}_w^n(\boldsymbol{\Theta}_{nw})\begin{bmatrix}\hat{u}_w & \hat{v}_w & 0\end{bmatrix}^T - \dot{a}_{wd} + k_p\tilde{a}_w + k_i\int_0^t \tilde{a}_w d\tau, \qquad (18)$$

which is the same guidance law as proposed in [14].

2.4 Adaptive Orientation

Many sensors, both optical and acoustic, provide the best quality data when facing the scene to be recorded orthogonally. Images taken with a camera at an oblique angle with respect to the scene may have uneven lighting due to the differences in the distance the light has to travel through water. If the differences in distance are too large, parts of the image may also end up being out of focus.

The data quality from a side scan sonar (SSS) can also be reduced if the vehicle is moving along an incline while the vehicle maintains the same orientation it would have on a flat seafloor. While the acoustic beams on one side of the vehicle are fully reflected by the rising incline after only a fraction of the intended range, the beams on the other side of the vehicle may not be reflected at all.

In order to improve the quality of the collected data, it could be advantageous to have the sensor platform automatically adjust its orientation to match the orientation of the seafloor. In other words, we want to control the sensor direction vector $\mathbf{n}_{sensor}$ such that

$$\mathbf{n}_{sensor} \rightarrow \mathbf{n}_{sensor\,d} = -\frac{\nabla F_n}{|\nabla F_n|}. \tag{19}$$

This can be achieved by either adjusting the orientation of the vehicle $\boldsymbol{\Theta}_{nb}$, the sensor rig orientation $\boldsymbol{\Theta}_{bd}$, or a combination of the two. As for the distance control, the roughness of the seafloor and a compromise between a platforming and a contouring strategy must be considered for the adaptive orientation as well.

Case: ROV with Adaptive Heading

For an ROV that is passively stabilized in roll and pitch, we only want to control the heading. The heading ψ_a satisfying (19) can be found to be

$$\tan(\psi_a) = \frac{-F_{ny}}{-F_{nx}} \implies \psi_a = atan2(-F_{ny}, -F_{nx}), \tag{20}$$

where $atan2(y, x)$ is the four-quadrant inverse tangent. Assuming we already have a controller that guarantees $\psi \rightarrow \psi_d$, we can achieve the desired orientation using the guidance law

$$\dot{\psi}_d = r_d = k_{hp}(\psi_a - \psi_d) \tag{21}$$

subject to the implemented constraints $r_d \in [-\dot{\psi}_{max}, \dot{\psi}_{max}]$ and $\dot{r}_d \in [-\ddot{\psi}_{max}, \ddot{\psi}_{max}]$. $k_{hp} \geq 0$ is an adjustable gain for how quickly the heading should converge to ψ_a, and $\dot{\psi}_{max}, \ddot{\psi}_{max} > 0$ denotes the maximum turn rate and angular acceleration, respectively. These parameters need to be selected based on the dynamics of the vehicle, and should be well within what the controller is expected to be able to follow. By adding these constraints, we ensure that the desired heading $\psi_d(t)$ is a smooth continuous function, satisfying the requirement proposed in Sect. 2.3 "Generalizing for Tilted DVL".

This approach should not be employed in areas where the sea floor is close to horizontal, as ψ_a will be rapidly changing for small values of F_{ny} and F_{nx}. At the same time, there is no real benefit to have adaptive heading in areas with no discernible incline, so the problem is simply avoided by not adapting the heading unless the incline is above a certain threshold. Keep in mind that it is important to use a switching strategy that allows for smooth transitions between the fixed heading and adaptive heading, and avoids excessive switching between the modes when operating close to the selected threshold.

Case: Adaptive DVL/Camera Pitch

Assuming we have a DVL mounted on a pivot mount on the ROV in the previous case with the orientation $\Theta_{bd} = [0, \theta_{DVL}, 0]^T$, we can adjust the DVL orientation to match the incline of the seafloor. θ_{DVL} is defined as 0° when the DVL is pointing vertically downward and 90° pointing horizontally forward. The desired pitch θ_a can then be found to be

$$\tan(\theta_a) = \frac{\sqrt{F_{nx}^2 + F_{ny}^2}}{-F_{nz}} \implies \theta_a = atan2(\sqrt{F_{nx}^2 + F_{ny}^2}, -F_{nz}), \qquad (22)$$

and can be controlled using a similar approach to the one in "Case: ROV with Adaptive Heading".

Case: AUV with Adaptive Roll

For an underactuated AUV equipped with a camera or a SSS, the heading and pitch of the vehicle will need to be used to control the cross-track error and altitude, respectively. If the AUV is capable of actively controlling the roll, this can then be adapted to the seafloor orientation for better quality data.

The desired roll ϕ_a can be found to be

$$\tan(\phi_a) = \frac{-F_{nx} \sin \psi_d + F_{ny} \cos \psi_d}{-F_{nx} \sin \theta_d \cos \psi_d - F_{ny} \sin \theta_d \sin \psi_d - F_{nz} \cos \theta_d} \implies \qquad (23)$$
$$\phi_a = atan2(-F_{nx} \sin \psi_d + F_{ny} \cos \psi_d, -F_{nx} \sin \theta_d \cos \psi_d - F_{ny} \sin \theta_d \sin \psi_d - F_{nz} \cos \theta_d),$$

and can be controlled using a similar approach to the one in "Case: ROV with Adaptive Heading" and "Case: Adaptive DVL/Camera Pitch".

2.5 Photogrammetry Post-processing Procedure

When the entire area of interest has been sufficiently covered with still images, typically at least 50% overlap between the neighboring images, the photogrammetry processing can be initiated. An overview of the processing pipeline from captured 2D still images to finished 3D photogrammetry model is given in Fig. 5. The majority of the procedure is automated and can be run on an ordinary desktop computer or even a powerful laptop. The software is a combination of commercially available software and freeware.

In short, online data (including images) from the ROV is collected using Lab-VIEW. Navigation data is post-processed and used to georeference the images using MATLAB. The images are color corrected using a batch plugin [3] for GIMP 2.8 [11] (both freeware). The alignment of images and generation of the 3D model are done using Agisoft Photoscan 1.2 [1]. For further details on the processing pipeline, the reader is referred to [13].

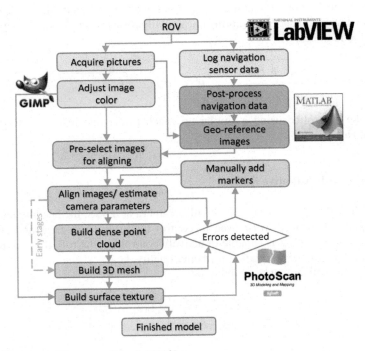

Fig. 5 Flowchart of the processing chain [13]

3 Experimental Results

The mapping survey was conducted close to Stokkbergneset (63°28'N, 9°54'E) in the Trondheimsfjord on April 22, 2016. The chosen survey area was a rock wall with an approximate incline of 70° located at 175–200m depth. The mapping strategy was conducted twice on two separate 5 m by 5 m sections of the wall, from now on referred to as Box1 and Box2.

The main guidance module of the ROV was programmed to move along a pattern of 5 parallel vertical 5 m long lines with 1 m line spacing to cover the face of the rock wall with sufficient level of overlap. The heading was set to be a constant 215°, and the speed was set to 0.1 m/s. The distance guidance module then modified these desired vertical track lines online to maintain a constant distance to the wall. The selected distance (with respect to the cameras) was 2.1 m for Box1 and 2.3 m for Box2. Due to the geometry of the ROV, the lamps were 40 cm closer to the wall, and thus the main concern for potential collision with the wall.

The resulting desired track lines (green) with the estimated ROV track (red) from Box1 are shown in Fig. 6. The blue lines denote the lawnmower pattern setup by the operator. In this case, the estimated distance to the wall deviated less than 25 cm from the desired distance at all times, with a root-mean-square error of 6.9 cm for the entire pattern (see Fig. 7 for details). For Box2 the respective figures were maximum error 31 and 8.9 cm root-mean-square error. Each pattern was completed within 6 min.

Figures 8 and 9 show the resulting 3D model of Box1 seen from the East, without and with texture, respectively. Using the procedure outlined in Sect. 2.5, the post-processing took approximately 6 h per box on an ordinary desktop computer.

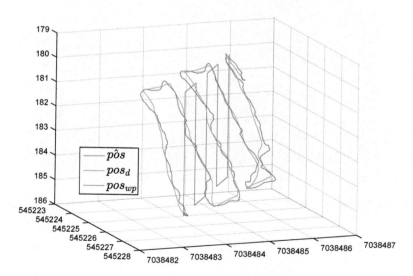

Fig. 6 Estimated (*red*), desired (*green*), and waypoint (*blue*) UTM position of the ROV

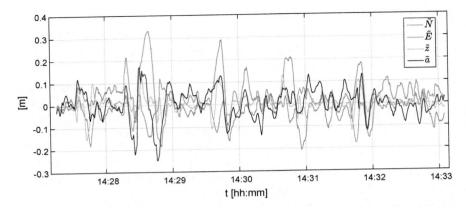

Fig. 7 Estimated errors in Northing (*blue*), Easting (*red*), depth (*yellow*), and distance (*black*)

Fig. 8 Shaded relief 3D model of Box1

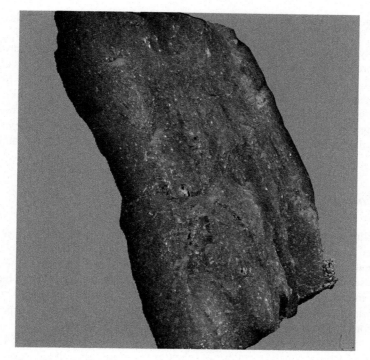

Fig. 9 Textured 3D model of Box1

4 Discussion

This section describes the experiences that can be drawn from the mapping survey, identifying issues that need to be handled and suggesting potential improvements.

4.1 Evaluating the Results

The system was able to maintain a distance error well within acceptable limits. The image quality was generally high, even when one of the lamps malfunctioned. The largest deviations from the desired distance usually coincided with a sharp change in incline (horizontal shelves in the wall). This might have been reduced somewhat by having tuned the nonlinear PID-controller more aggressively, but at the same time this plays directly into the balance between platforming and contouring the roughness of the seafloor, as mentioned in Sect. 2.3 "Guidance Law" and discussed in more detail in [8]. Having the DVL angled 20° downward to better match the incline of the wall could also have been a potential improvement. A pivot mount for the DVL would increase the flexibility of the system, allowing it to map features with varying slopes.

The system could likely have handled a higher ROV speed and thus have been even more efficient. However, the limited swath width of the DVL limits its abilities to detect obstacles and sudden changes in incline, so since this was the first test of the system it was decided to rather err on the side of caution. Determining the limitations of the full system will be an important continuation of this research, including the tolerance thresholds for error sources such as bias, noise, lever arms and misalignment of the different sensors, errors in timing, and speed of sound.

4.2 Potential Improvements

The ability to detect obstacles and sudden changes in incline in the path of travel would be an important improvement to the system. This can be achieved by adding additional sensors orientated in the direction of travel. In order to fit into the framework described in (8) and [8], the only demand is that the sensor data can be transformed into one or more vectors on the $\mathbf{r}_j^n$-form described in (2) and (3). Adding more measurement points would also allow for the possibility of using a higher order model for a seafloor function, rather than the linear model in (6).

There are several candidates for sensors to add to an improved distance-keeping system. In keeping with the mentality of selecting sensors that are already available on most ROVs, a scanning sonar and altimeter(s) located in different places on the ROV are among the top contenders. Utilizing the distance estimates that can be extracted from an image pair from the stereo cameras through for instance a disparity map is also an interesting possibility.

An additional benefit of using multiple sensors for the same purpose is the increase in redundancy. Redundancy is useful in all systems, but in particular in safety critical operations often found in the oil and gas industry.

4.3 Towards Autonomy

The motion control strategy proposed in this chapter represents a stepping stone toward full autonomy. As in [10, 12], we are adopting a bottom-up approach to autonomy, and Fig. 10 shows the control architecture in the framework proposed in [12]. The distance guidance proposed in Sect. 2.3 and the adaptive orientation scheme proposed in Sect. 2.4 are an important part of both the guidance and optimization layer and control execution layer in this framework, adapting to the terrain and optimizing the quality of the collected data.

The photogrammetry procedure from Sect. 2.5 is optimized for quality rather than speed, and is therefore poorly suited for real-time operation. However, several other mosaicking and image analysis algorithms have been demonstrated to work in (near)

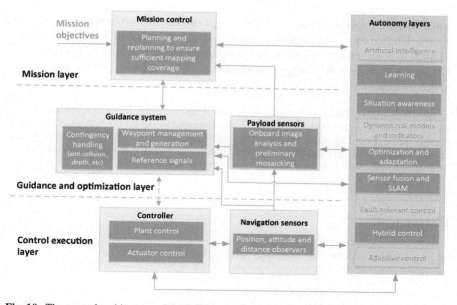

Fig. 10 The control architecture of the ROV in the framework proposed by [12]

real time, for instance in [9, 17]. By employing this, the onboard image analysis would not only provide an online quality control of the collected data, it could also be used as input for collision avoidance and planning/replanning.

When the image analysis detects lack of coverage in the preliminary mosaick, the mission layer autonomy could replan the vehicle path to cover the detected gaps. In cases of consistent under- or overcoverage, the mission layer could also learn to adjust line spacing, mapping pattern or speed in order to maximize coverage of the area to be mapped (see [16]). In extreme cases the system might also be required to sacrifice resolution for overview if the mission objectives include a limited time and/or energy budget. Alternately, the system could also prioritize mapping in areas with a high occurrence of detected features as described in [6].

The image analysis can also be used to improve the estimates of $\hat{a}_w$ and $\hat{\nabla F}_n$, either combining it with the DVL measurements through sensor fusion, or switching between the estimates from the different sensors using a hybrid control approach.

5 Conclusion

In this paper a new and more generalized method for DVL orientation and corresponding wall tracking controller were developed. The results from the test survey have demonstrated how the DVL measurements enabled the DP system to efficiently map a vertical underwater wall with still imagery, while maintaining a constant distance to the face of the wall for ideal image quality and coverage. Because this approach relied on sensors that are common to the majority of ROVs, it can easily be adapted to a wide range of ROVs, and reduce the reliance on manual operators. An efficient post-processing strategy using commercially available software was applied to the collected data, further reducing the required resources for the end user.

Acknowledgements This work has been carried out at the Centre for Autonomous Marine Operations and Systems (NTNU AMOS). This work was supported by the Research Council of Norway through the Centres of Excellence funding scheme, Project number 223254 - NTNU AMOS. The authors would also like to thank the crew of the Applied Underwater Robotics Laboratory (NTNU AUR-Lab) and RV Gunnerus for their help in carrying out the experiments.

References

1. Agisoft LLC. http://www.agisoft.com. Accessed 14 July 2016
2. Allied Vision Tehnologies. http://www.altavision.com.br/Arquivos/AVT/Manuals/GC1380_User_Manual.pdf. Accessed 15 July 2016
3. BIMP. http://www.alessandrofrancesconi.it/projects/bimp/. Accessed 15 July 2016
4. Bowen, A.D., Yoerger, D.R., Taylor, C., McCabe, R., Howland, J., Gomez-Ibanez, D., Kinsey, J.C., Heintz, M., McDonald, G., Peters, D.B., Fletcher, B., Young, C., Buescher, J., Whitcomb, L.L., Martin, S.C., Webster, S.E., Jakuba, M.V.: The *Nereus* hybrid underwater robotic vehicle. Underwater Technol. Int. J. Soc. Underwater Technol. **28**(3), 79–89 (2009)
5. Caccia, M., Bono, R., Bruzzone, G., Veruggio, G.: Bottom-following for remotely operated vehicles. Control Eng. Pract. **11**(4), 461–470 (2003). (MCMC00)
6. Candeloro, M., Mosciaro, F., Sørensen, A.J., Ippoliti, G., Ludvigsen, M.: Sensor-based autonomous path-planner for sea-bottom exploration and mosaicking. IFAC-PapersOnLine **48**(16), 31–36 (2015)
7. Dukan, F., Ludvigsen, M., Sørensen, A.J.: Dynamic positioning system for a small size ROV with experimental results. In: OCEANS 2011 IEEE - Spain (2011)
8. Dukan, F., Sørensen, A.J.: Sea floor geometry approximation and altitude control of ROVs. IFAC J. Control Eng. Pract. (CEP) **29**, 135–146 (2014)
9. Ferreira, F., Veruggio, G., Caccia, M., Bruzzone, G.: Real-time optical slam-based mosaicking for unmanned underwater vehicles. Intell. Serv. Robot. **5**(1), 55–71 (2012)
10. Fossum, T.O., Ludvigsen, M., Nornes, S.M., Rist-Christensen, I., Brusletto, L.: Autonomous robotic intervention using ROV: an experimental approach. In: OCEANS 2016 MTS/IEEE - Monterey, September 2016
11. GIMP. http://www.gimp.org/. Accessed 15 July 2016
12. Ludvigsen, M., Sørensen, A.J.: Towards integrated autonomous underwater operations for ocean mapping and monitoring. Annu. Rev. Control **42**, 145–157 (2016)
13. Nornes, S.M., Ludvigsen, M., Ødegård, Ø., Sørensen, A.J.: Underwater photogrammetric mapping of an intact standing steel wreck with ROV. In: Proceedings of the 4th IFAC Workshop on Navigation, Guidance and Control of Underwater Vehicles NGCUV, pp. 206–211, April 2015
14. Nornes, S.M., Ludvigsen, M., Sørensen, A.J.: Automatic relative motion control and photogrammetry mapping on steep underwater walls using ROV. In: OCEANS 2016 MTS/IEEE - Monterey, September 2016
15. NORTEK AS. http://www.nortek-as.com/lib/brochures/nortek-dvl-datasheet-1/at_download/file. Accessed 15 July 2016
16. Paull, L., Saeedi, S., Seto, M., Li, H.: Sensor-driven online coverage planning for autonomous underwater vehicles. IEEE/ASME Trans. Mechatron. **18**(6), 1827–1838 (2013)

17. Rossi, M., Scaradozzi, D., Drap, P., Recanatini, P., Dooly, G., Omerdić, E., Toal, D.: Real-time reconstruction of underwater environments: from 2D to 3D. In: OCEANS 2015 - MTS/IEEE Washington, pp. 1–6, October 2015
18. Sørensen, A.J., Dukan, F., Ludvigsen, M., Fernandez, D.A., Candeloro, M.: Development of dynamic positioning and tracking system for the ROV minerva. In: Further Advances in Unmanned Marine Vehicles, pp. 113–128. IET, UK (2012). Chap. 6
19. Vaganay, J., Elkins, M.L., Willcox, S., Hover, F.S., Damus, R.S., Desset, S., Morash, J.P., Polidoro, V.C.: Ship hull inspection by hull-relative navigation and control. In: Proceedings of OCEANS 2005 MTS/IEEE, vol. 1, pp. 761–766 (2005)

Part II
Localization and Mapping

Underwater 3D Laser Scanners: The Deformation of the Plane

Albert Palomer, Pere Ridao, David Ribas and Josep Forest

Abstract Development of underwater 3D perception is necessary for autonomous manipulation and mapping. Using a mirror-galvanometer system to steer a laser plane and using triangulation, it is possible to produce full 3D perception without the need of moving the sensor. If the sensor does not meet certain hardware requirements, the laser plane is distorted when it passes through the different media (air–viewport–water). However, the deformation of this plane has not been studied. In this work a ray-tracing model is presented to study the deformation of the laser plane. To validate it, two types of datasets have been used, one synthetically generated using the model presented below, and another one using real data gathered underwater with an actual laser scanner. For both datasets an elliptic cone is fitted on the data and compared to a plane fit (the surface commonly used for triangulation). In the two experiments, the elliptic cone proved to be a better fit than the plane.

1 Introduction

In the field of robotics, 3D perception and computer vision are crucial for interacting with the environment. With the development of tools and techniques based on 3D perception such as stereo vision or depth cameras, robot capabilities have increased significantly. On the robot–environment interaction, [1] presents a method for real-time textureless object recognition based on a robust template matching. Moreover, [2] demonstrates a behaviour on a humanoid attempting to catch a flying

---

A. Palomer (✉) · P. Ridao · D. Ribas · J. Forest
Computer Vision and Robotics Institute, Universitat de Girona, Girona, Spain
e-mail: apalomer@eia.udg.edu

P. Ridao
e-mail: pere@eia.udg.edu

D. Ribas
e-mail: dribas@udg.edu

J. Forest
e-mail: forest@eia.udg.edu

© Springer International Publishing AG 2017
T.I. Fossen et al. (eds.), *Sensing and Control for Autonomous Vehicles*,
Lecture Notes in Control and Information Sciences 474,
DOI 10.1007/978-3-319-55372-6_4

ball using stereo vision. In another example, [3], the present authors presented an even more refined task where they catch objects with uneven shapes using learning-by-demonstration techniques. In the field of mapping, [4, 5] present extensive reviews for different visual Simultaneous Localization And Mapping (SLAM).

An underwater robot has two main sensor modalities for perceiving the environment, acoustic sensors and optical sensors. Both types of sensor have their advantages and disadvantages. Acoustic sensors are able to penetrate deeper in the water and their performance is not influenced by water turbidity. However, the refresh rate and resolution are medium to low, in addition to them being expensive sensors. On the other hand, underwater vision needs clear water to work and the penetration range is smaller than for sonars. The underwater robotics community has been very active in the field of mapping using both technologies. Using sonar, several 2D and 3D SLAM techniques have been developed and presented during the past years [6–17]. The community has also presented 2D and 3D SLAM techniques using vision-based algorithms [18–23]. Moreover, [24, 25] presented relevant works in underwater mapping using a laser-based structured light source.

Despite all these studies, little work has been done in autonomous underwater manipulation [26–30]. If an underwater sensor could provide dense 3D point-clouds as the Kinect sensors do for mobile robotic applications, it would significantly improve underwater manipulation capabilities. Following this idea, [31] presented a structured light solution where the 3D of a scene was recovered by projecting a light pattern. Although this technique is useful for online 3D reconstruction, it has the limitation that the pattern cannot be changed online. Therefore, the resolution decreases as the distance to the scene increases. In [32, 33], the authors presented developments in underwater manipulation using laser triangulation techniques. In these publications the laser was mounted on the wrist of a robotic arm, allowing 3D data to be gathered by steering the laser with the arm, and also allowing the resolution of the sensed scene to be changed on demand. Although robotic arms usually have a very accurate positioning system, underwater manipulators are slow for this task. Therefore, detecting changes in the working environment with such a technique is not possible. To overcome this limitation, it is possible to use a laser system such as [34] where the laser is steered using a mirror and a galvanometer.

To the best of the authors' knowledge, in the underwater laser scanning literature only three published works use a triangulation method with a steered laser [34–36]. It is known that flat viewports produce distortions in the light when used to seal underwater housings [37]. Despite this, [34] used a planar interface for sealing the laser scanner assuming that the laser plane remains planar after being refracted by the viewport. In contrast, [35, 36] presented similar hardware, but instead, they attempted to minimize the deformation using spherical and cylindrical viewports respectively (see Fig. 1). To avoid such deformation, it is necessary to mount the system in a way that makes the laser plane contain the viewport's surface normal, for all the intersection points between the laser plane and the viewport (see Fig. 1a–c). For instance, in [35] (using a sphere), the laser plane will remain the same as long as the laser focal point is at the centre of the sphere and the rotation axis of the laser passes through it (see Fig. 1a, b for ideal conditions and Fig. 1d, e for non-ideal

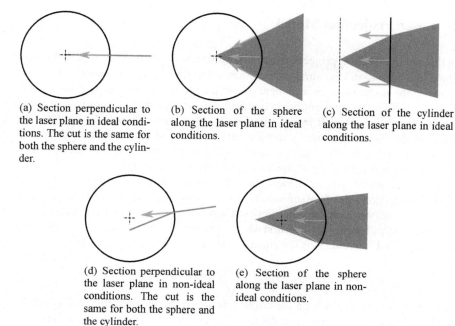

Fig. 1 Different scenarios where the laser plane is perfectly aligned or not with respect to the cylinder and sphere centre. The cylinder or sphere surface is shown in *black*, the laser plane in *red*, and the cylinder or sphere normals at the incidence points is shown in *green*. The *dashed lines* indicate the sphere and cylinder centre and revolution axis

conditions). In the case of [36], where a cylindrical viewport is used, the laser plane has to contain the rotation axis, which also has to be perfectly aligned with the cylinder revolution axis (see Fig. 1a for ideal conditions and Fig. 1d for non-ideal conditions). Nevertheless, even though this technique avoids distorting the laser plane, its aperture would still be affected (see Fig. 1c). Moreover, in both cases the sphere and the cylinder need to be manufactured ensuring concentric surfaces at both side of the viewport. Because it is very difficult to make sure that all these conditions are met, the deviations from these ideal conditions will need to be taken into account during the triangulation in order to achieve high accuracy. On the other hand, if a flat viewport is used, the manufacturing process is simplified at the cost of using a suitable surface (other different than a plane) to describe the underwater distorted laser during the triangulation. The goal of the present work is to study such a surface as well as to evaluate the improvement in accuracy with respect to the use of a more common planar beam approximation.

This report is structured as follows. Section 2 models how the laser plane is distorted when the light plane propagates through the interface between two different media with different indices of refraction. The experimental results and the surface fitting are presented in Sect. 3 followed by the conclusions and future work in Sect. 4.

2 Laser Projection Model

In this section the laser projection model is presented. This model describes the laser plane, the mirror, and the air/viewport/water interfaces in order to estimate the laser shape in the underwater environment. Laser planes are created using a laser beam and a lens. This lens can be either cylindrical, producing a non-uniformly distributed intensity line, or a Powell lens for uniformly distributed intensity lines [38]. One could model the laser plane by introducing the lens and the beam into the overall model. However, for the sake of simplicity, it is assumed that the laser plane is produced by a source of light expanding at a given aperture angle.

From a geometrical point of view, the only two optical processes that take place in the light-emission side of the sensor are: (1) reflection on the mirror and (2) refraction at both the air/viewport and the viewport/water interfaces. Light reflection is the process of light changing its direction when it reaches a given surface without crossing it (see Fig. 2). Equation (1) is the equation describing this optical process.

$$\boldsymbol{r}_l = \boldsymbol{i} - 2\,(\boldsymbol{i} \cdot \boldsymbol{n})\,\boldsymbol{n} \tag{1}$$

This process only depends on the angle α between the incident light ray $\boldsymbol{i}$ and the normal of the surface where it is reflected $\boldsymbol{n}$. Light refraction is the change in direction that a light beam suffers when crossing the surface separating two different media (see Fig. 3). This optical process follows Snell's law described in Eq. (2).

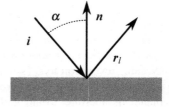

Fig. 2 Reflection

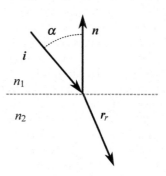

Fig. 3 Refraction

$$r_r = \frac{n_1}{n_2}(\boldsymbol{n} \times (-\boldsymbol{n} \times \boldsymbol{i})) - \boldsymbol{n}\sqrt{1 - \left(\frac{n_1}{n_2}\right)^2 (\boldsymbol{n} \times \boldsymbol{i}) \cdot (\boldsymbol{n} \times \boldsymbol{i})} \qquad (2)$$

This law involves three elements: (1) the index of refraction of the first medium n_1, (2) the index of refraction of the second medium n_2 and (3) the angle α between the incident light $\boldsymbol{i}$ and the normal of the surface separating the two media $\boldsymbol{n}$.

Both optical processes are well known and their behaviour is described using vector equations as shown in Eqs. (1) and (2). However, there is no method available

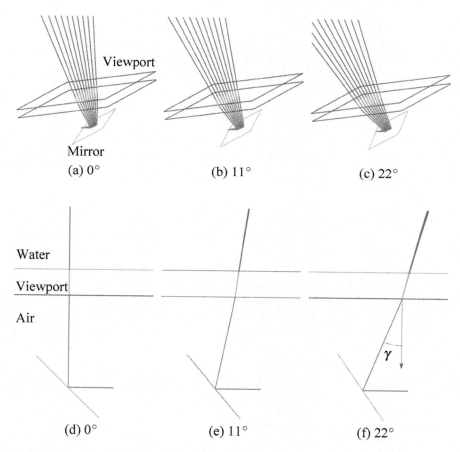

Fig. 4 3D (*top row* **a–c**) and 2D (*bottom row* **d–f**) images of 10 rays reflected and refracted using the proposed model, for three different mirror angles producing different incidence angles γ of the laser plane and the viewport: the *left column* (**a** and **d**) 0°, the *centre column* (**b** and **e**) 11°, and the *right column* (**c** and **f**) 22°. This model computed 10 rays using a laser aperture of 50° with index of refraction 1.49 for the viewport and 1.33 for the water. Each ray represents a different light path to the scene. *Red rays* represent the laser in the air while, *green* and *blue* represent it in the viewport and water, respectively. The *white* boundary represents the mirror edge and the *grey boxes* the two surfaces of the viewport (the air/viewport and the viewport/water)

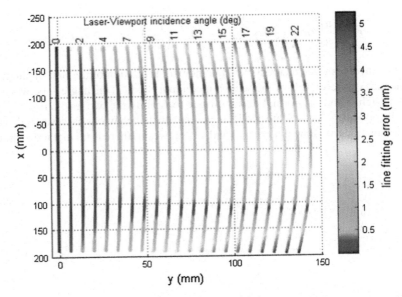

Fig. 5 Projection of the simulated laser at 1.5 m distance. Each stripe corresponds to the projection of the laser for a specific angle of incidence between the laser and the viewport (in the figure varying from 0° to 22°). Each laser projection is coloured according to the error with respect to its best fitting line

to identify each one of the rays of the laser. Instead, one can only estimate the equation of the laser plane, and Eqs. (1) and (2) cannot be used to compute how that plane is reflected and refracted. Therefore, it is not possible to work with the plane parameters of the laser to estimate the surface that it produces underwater. To produce the underwater laser surface, a number of rays are sampled evenly spread across the laser aperture. For each of them, its reflection by the mirror is computed using Eq. (1). Then it is refracted using Eq. (2) first with the air/viewport interface using $n_1 = n_{air}$ and $n_2 = n_{viewport}$, and second with the viewport/water interface using $n_1 = n_{viewport}$ and $n_2 = n_{water}$.

In Fig. 4, 10 rays computed using this model are depicted. There, it can be seen how each ray is reflected and refracted differently from the others. This is an effect produced because the laser emitter is close to both the mirror surface and the viewport. Therefore, the light rays are not parallel to each other and the angle between each one of the rays and the interaction surface normal is different. The second row of Fig. 4d–f shows an orthogonal projection of the model on the plane that contains the normal to the laser plane, the normal to the mirror and the normal to the viewport. Using this view it is possible to observe how, as the laser incidence angle γ increases, the rays of the laser in the water medium form a thicker line, meaning that these rays are not coplanar. To better observe this deformation of the plane, Fig. 5 shows the intersection of the same model, this time using 35 rays, with a plane parallel to the viewport at a distance of 1.5 m. In geometry, the intersection of two planes is a single straight line. However, in Fig. 5 it is possible to observe how each intersection appears to be more and more curved as the incidence angle increases. To get a sense of how nonlinear each one of the stripes is, a line is computed for each one. Then, the intersection is coloured according to the distance of each point on the intersection to the fitted line. According to this model, as the angle increases, using a plane to fit the laser surface would produce errors up to approximately 5 mm at 1.5 m distance.

The data obtained with this model suggests that the laser remains planar after passing the viewport only when the normal to the viewport is parallel to the laser plane. Therefore, the laser plane is distorted because of the change in the index of refraction of the different media when such condition is not met.

3 Model Validation

In this section, the validation of the model is presented. The previously described model suggests that the laser plane is distorted due to the change in the index of refraction of the different media. To validate this result, two surfaces, a plane and an elliptical cone, will be fitted to 3D points gathered using either the previously presented model or a real underwater laser scanner. Section 3.1 introduces the way the two surfaces are computed from a set of 3D points. Next, Sect. 3.2 describes how the two surfaces are fitted to 3D data produced using the previously presented model. Finally, in Sect. 3.3, the surfaces are also fitted to real 3D points gathered using an underwater laser scanner like the one presented in [34].

3.1 Surface Fitting

In the present work [34], the triangulation method used for 3D reconstruction was exploited assuming that the laser plane remains planar after going through different media using a planar sealing viewport. The method proposed in Sect. 2 suggests that such an assumption might not be accurate. Instead, we propose to fit an elliptic cone, for two reasons. First, there is a closed form expression to compute the intersection of a ray and a given elliptic cone. This is very interesting in relation to the real-time execution, because in this way, it is possible to compute the 3D triangulation of each point without using numerical methods. Second, an elliptic cone expressed by its parametric equation (see section "Elliptic Cone Fitting"), can be either a plane (by setting one of the two constants a or b in Eq. (5) to zero) or a non-flat surface with contour levels similar to the ones observed in Fig. 5.

Plane Fitting

Fitting a plane to a set of 3D points is a well-known problem which has been solved [39–43]. Using Principal Component Analysis (PCA) for plane fitting [40], it is possible to fit the plane that minimizes the error between the points and the fitted plane. However, this method is not robust to outliers, while other methods such as RANdom SAmple Consensus (RANSAC) [39] are.

Given a set of points $P = \{p_1, p_2, \ldots, p_n\}$ with $p_i = [x_i \ y_i \ z_i]^T$, the best fitting plane $\pi : n^T \cdot p = d$ (being n the normal of the plane and p a 3D point on the plane) is the one that contains the centre of gravity of P

$$c_P = [x_c \ y_c \ z_c]^T = \frac{1}{n} \sum_{i=1}^{n} p_i \qquad (3)$$

and its normal is parallel to the direction of minimum variance of the set P. This direction can be found by computing the eigenvector corresponding to the smallest eigenvalue of the covariance matrix of the points A:

$$A = \begin{pmatrix} \sum_{i=1}^{n}(x_i - x_c)^2 & \sum_{i=1}^{n}(x_i - x_c)(y_i - y_c) & \sum_{i=1}^{n}(x_i - x_c)(z_i - z_c) \\ \sum_{i=1}^{n}(y_i - y_c)(x_i - x_c) & \sum_{i=1}^{n}(y_i - y_c)^2 & \sum_{i=1}^{n}(y_i - y_c)(z_i - z_c) \\ \sum_{i=1}^{n}(z_i - z_c)(x_i - x_c) & \sum_{i=1}^{n}(z_i - z_c)(y_i - y_c) & \sum_{i=1}^{n}(z_i - z_c)^2 \end{pmatrix} \qquad (4)$$

Elliptic Cone Fitting

Here, a least squares formulation is used to fit an elliptic cone to a set of points. An elliptic cone is a surface with two parameters $c = [h \ \beta]$ in the 3D space. Its parametric equation is given by:

$$f(c) = \begin{cases} x = a \ h \ cos(\beta) \\ y = b \ h \ sin(\beta) \\ z = h \end{cases} \qquad (5)$$

a and b being the aperture of the elliptic cone in the x and y directions, respectively. This equation of the elliptic cone only represents an elliptic cone with its vertex on the origin and its main axis aligned with the z axis. Therefore, we need to add a transformation $t = [x \ y \ z \ \phi \ \theta \ \psi]^T$ to the model to represent this elliptic cone in any position in the 3D space:

$$g(c, t) = t \oplus f(c) \qquad (6)$$

with $\oplus$ being the 3D compounding operation between a 3D transformation $t = [x_t \ y_t \ z_t \ \phi_t \ \theta_t \ \psi_t]^T$ and a 3D point $p = [x_p \ y_p \ z_p]^T$:

$$t \oplus p = R_z(\psi_t) R_y(\theta_t) R_x(\phi_t) \begin{pmatrix} x_p \\ y_p \\ z_p \end{pmatrix} + \begin{pmatrix} x_t \\ y_t \\ z_t \end{pmatrix} \tag{7}$$

where R_x, R_y and R_z are the three rotation matrices along the three cartesian axis x, y and z. For further insight, please see [44].

Letting $P = \{p_1, p_2, \ldots, p_n\}$ be a set of points, and defining $d(g(c, t), p)$ as the minimum distance from a point p to the cone $g(c, t)$, the following minimization problem can be solved using least squares:

$$[c, t] = \operatorname*{argmin}_{[c, t]} \sum_{i=1}^{n} d\left(g(c, t), p_i\right) \tag{8}$$

3.2 Surface Fitting Using Synthetic Data

Using the previously explained model (see Sect. 2), a point cloud was generated to represent the laser shape (i.e. the surface that contains the laser light) in the underwater medium for a specific set of laser–viewport angles. The model has been used to compute how 35 rays are reflected and refracted into the water. For this experiment, the aperture angle for the laser has been set to 55° and the indices of refraction have been set to 1.00 for air, 1.49 for the viewport and 1.33 for water. To sample points on the surface generated by the laser, 5 points are taken for each one of the 35 rays at 100 mm intervals, starting 100 mm away from the point where the ray leaves the viewport.

Figure 6 presents the fitting error for angles ranging from 0° to 22°. The blue lines are the different results for the plane fitting error, while the red lines are those for the elliptic cone. In both cases, the mean error is represented as a solid line while the dashed line represents the two σ error range and the dash-dot line is the maximum error. It can be seen how the elliptic cone is a much better fit for the laser surface than the plane according to the synthetic model. While the mean error for the plane fit at the extreme angle is almost 0.85 mm with a standard deviation of 0.84 mm and a maximum error of 4.05 mm, for the same angle, the mean error for the elliptic cone fit is 0.027 mm with a standard deviation of 0.03 mm and a maximum error of 0.195 mm.

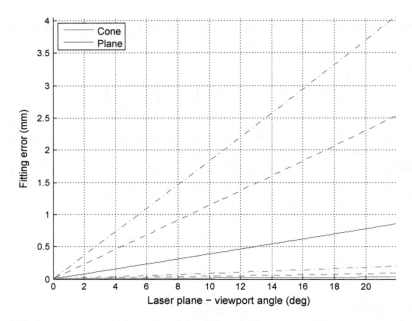

Fig. 6 Fitting error on synthetic generated data. The *solid line* represents the mean error between each point and the fitted surface (plane in *blue* and elliptic cone in *red*), while the *dashed line* represents the two standard deviations of the fitting error. The *dash-dot line* represents the maximum error for each angle

3.3 Surface Fitting Using Real Data

The real dataset has been gathered using the underwater laser scanner shown in Fig. 7. The sensor consists of three main housings: for the control, (not shown in the picture), the camera and the laser. While the camera cylinder only contains the camera, the laser housing contains the laser and a mirror that is actuated using a galvanometer. The mirror-galvanometer system is what allows for the laser movement. The control housing contains the main electronics that synchronize the laser positioning with the camera triggering. When the light passes through the viewport of the laser housing, it is refracted into the water and, if it is projected onto a surface, the projected laser light is visible in the camera.

To obtain the 3D points using the real sensor, it is not possible to sample them along each laser ray as was done using data from the model. To be able to obtain the points that are part of the laser surface, we have to intersect the rays passing through the camera pixels where the laser has been detected, with the equation of the surface where the laser has been projected. For such a computation, the camera parameters need to be known. In [37], a calibration method that takes into account the refraction of the light in an underwater camera using a flat viewport is presented. This camera calibration technique first obtains the focal length f and the image centre $c = \begin{bmatrix} c_x c_y \end{bmatrix}$ using state-of-the art techniques, in our case [45]. Then, the camera

Fig. 7 Underwater laser scanner

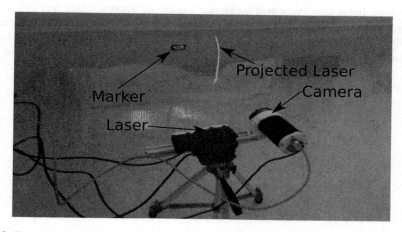

Fig. 8 Experimental setup. The marker is used to estimate the equation of the plane where the laser is projected, in this case the water tank wall. Note how the projected laser does not form a line but a curve as predicted by the model presented in Sect. 2

viewport position is computed using least squares. With the camera calibrated, using a ray-tracing technique such as the one in [46] the 3D position of the camera with respect to a known pattern can be computed. Assuming that the pattern is planar, and is on a planar surface, we can compute the equation of this planar surface with respect to the camera. Therefore, if the laser is projected onto that surface and identified in the camera image, the same ray-tracing technique used before [46], can now be used to get the 3D coordinate of the detected laser with respect to the camera.

Using the previously explained triangulation method, the laser scanner from Fig. 7 is used to gather laser points in the water tank. The whole experimental setup can be seen in Fig. 8. Using this laser scanner, it is not possible to gather points for such a wide range of laser–viewport incidence angles as were used in the synthetic dataset. Here we are restricted by the camera aperture and the distance to the projection plane. Therefore, only a subset of angles is studied in the experiment. Since the data obtained from the model showed that distortion appears when the incidence angle between the laser and the viewport is different from 0, we have chosen a configuration for the scanner to gather data in such conditions.

In this experiment, all housings, the laser and the camera, are full of air, the sealing viewports are made of Polymethyl Methacrylate (PMMA), also known as acrylic glass, and the sensor was submerged into the fresh water of a test tank. The index of refraction for each of these media is $n_{air} = 1.00$ for air [47], $n_{pmma} = 1.49$ for acrylic glass [48], and $n_{water} = 1.33$ for fresh water at 20° [49]. The laser has been projected onto the planar surface at four different distances between 0.5 and 1.5 m. Then, the two surfaces, the plane and the elliptic cone, have been fitted for each incidence angle.

The results are reported in Fig. 9. Note that the laser scanner is not calibrated. Therefore, these laser–viewport incidence angles are approximated and computed using the mechanical drawings of the design. When comparing the experimental results in Fig. 9 with the synthetic ones in Fig. 6, it is clear that the fitting error has increased for both the plane and the elliptic cone. This is due to noise in the

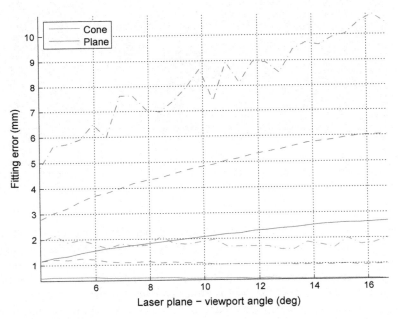

Fig. 9 Fitting error on real data. The *solid line* represents the mean error between each point and the fitted surface (plane in *blue* and elliptic cone in *red*), while the *dashed line* represents the two standard deviations of the fitting error. The *dash-dot line* represents the maximum error for each angle

measurement of each 3D point. In this case, the main source of such noise comes from the detections. First, the noise in the detection of the pattern will lead to inaccurate estimation of the equation of the projection plane. Second, the noise in the detection of the laser in the image produces an error in the rays used for the ray tracing and triangulation with the projection plane, adding more error to the measured 3D points. Moreover, it is possible to observe how the plane fitting error is not linear in contrast to the synthetic dataset. This might be because of the assumptions made in the process of obtaining this real data. These assumptions are: (1) the wall of the water tank where the laser is projected is flat, and (2) the two viewport faces are parallel. However, these experimental results still confirm that the proposed surface better describes the underwater laser. For all the studied angles, the mean error between the points and the elliptic cone is around 0.5 mm, while this error for the plane is always grater than 1 mm and increasing to more than 2.5 mm. Moreover, the two sigma error interval for the elliptic cone fit is more or less constant and around 1 mm for the studied angles. In contrast to this, the two sigma error interval of the plane error increases from 3 to 6 mm. Finally it is possible to see that the maximum error of the plane increases with the angle, reaching errors of more than 1 cm while the one for the elliptic cone remains more or less stable below 2 mm.

4 Conclusions and Future Works

This chapter presented a model based on ray tracing which has been used to study how a laser plane is distorted when passing through a flat viewport separating media with different indices of refraction (in this case air and water). The results obtained with this model show how the surface of the underwater laser becomes less planar when the incidence angle between the laser and the viewport increases. The only angle where the laser plane is not distorted is the one in which the planar laser surface contains the viewport's normal: in other words, the laser plane is perpendicular to the viewport. The model has been used to produce 3D points within the laser underwater surface after travelling through the air and the viewport. Using this synthetically generated data the fit of two surfaces, a plane and an elliptic cone, has been studied. Then, the same has been done using a real laser scanner in an underwater environment. It has been shown that the demonstrated elliptic cone is a better surface to describe the laser in the underwater medium than the plane.

Future works will include repeating the experiments under improved conditions. This includes using a mechanized calibration plane instead of a man-made wall to project the laser scanner. This will allow us to determine if the differences between the model and the real data are due to the assumption that the wall was planar, or whether the model needs some refinement. Moreover, future work will also include developing a method to find the different elements of the presented model so that, we can compute the cones for which no data has been gathered. This helps to develop a calibration method that finds all the parameters of the model.

References

1. Hinterstoisser, S., Cagniart, C., Ilic, S., Sturm, P., Navab, N., Fua, P., Lepetit, V.: Gradient response maps for real-time detection of textureless objects. IEEE Trans. Pattern Anal. Mach. Intell. **34**(5), 876–888 (2012)
2. Riley, M., Atkeson, C.G.: Robot catching: Towards engaging human-humanoid interaction. Auton. Robots **12**(1), 119–128 (2002)
3. Kim, S., Shukla, A., Billard, A.: Catching objects in flight. IEEE Trans. Robot. **30**(5), 1049–1065 (2014)
4. Fuentes-Pacheco, J., Ruiz-Ascencio, J., Rendón-Mancha, J.M.: Visual simultaneous localization and mapping: a survey. Artif. Intell. Rev. **43**(1), 55–81 (2012)
5. Hitomi, E.E., Silva, J.V.L., Ruppert, G.C.S.: 3D scanning using RGBD imaging devices: a survey. Developments in Medical Image Processing and Computational Vision, vol. 19, pp. 379–395. Springer International Publishing, Berlin (2015)
6. Carpenter, R.N.: Concurrent mapping and localization with FLS. In: Proceedings of the 1998 workshop on Autonomous Underwater Vehicles, 1998. AUV'98, vol. 02841, pp. 133–148 (1998)
7. Leonard, J.J., Carpenter, R.N., Feder, H.J.S.: Stochastic mapping using forward look sonar. Robotica **19**(5), 467–480 (2001)
8. Newman, P., Leonard, J.: Pure range-only sub-sea SLAM. In: Proceedings of the IEEE International Conference on Robotics and Automation (ICRA), pp. 1921–1926 (2003)
9. Eliazar, A., Parr, R.: DP-SLAM: Fast, Robust simultaneous localization and mapping without predetermined landmarks. Proc. IEEE Int. Conf. Robot. Autom. (ICRA) **18**, 1135–1142 (2003)
10. Fairfield, N., Jonak, D., Kantor, G.A., Wettergreen, D.: Field results of the control, navigation, and mapping systems of a hovering AUV. In: Proceedings of the 15th International Symposium on Unmanned Untethered Submersible Technology (2007)
11. Roman, C., Singh, H.: A self-consistent bathymetric mapping algorithm. J. Field Robot. **24**(1–2), 23–50 (2007)
12. Ribas, D., Ridao, P., Domingo, J.D., Neira, J.: Underwater SLAM in man-made structured environments. J. Field Robot. **25**(11–12), 898–921 (2008)
13. Barkby, S., Williams, S.B., Pizarro, O., Jakuba, M.: A featureless approach to efficient bathymetric SLAM using distributed particle mapping. J. Field Robot. **28**(1), 19–39 (2011)
14. Aykin, M.D., Negahdaripour, S.: On feature matching and image registration for two-dimensional forward-scan sonar imaging. J. Field Robot. **30**(4), 602–623 (2013)
15. Hurtós, N., Ribas, D., Cufí, X., Petillot, Y., Salvi, J.: Fourier-based registration for robust forward-looking sonar mosaicing in low-visibility underwater environments. J. Field Robot. **32**(1), 123–151 (2015)
16. Mallios, A., Ridao, P., Ribas, D., Carreras, M., Camilli, R.: Toward autonomous exploration in confined underwater environments. J. Field Robot. (2015)
17. Palomer, A., Ridao, P., Ribas, D.: Multibeam 3D underwater SLAM with probabilistic registration. Sensors **16**(560), 1–23 (2016)
18. Eustice, R., Pizarro, O., Singh, H.: Visually augmented navigation in an unstructured environment using a delayed state history. Proc. IEEE Int. Conf. Robot. Autom. (ICRA) **1**(April), 25–32 (2004)
19. Williams, S., Mahon, I.: Simultaneous localisation and mapping on the great barrier reef. Proc. IEEE Int. Conf. Robot. Autom. (ICRA) **2**, 1771–1776 (2004)
20. Eustice, R., Singh, H., Leonard, J., Walter, M., Ballard, R.: Visually navigating the RMS Titanic with SLAM information filters. Proc. Robot. Sci. Syst. (2005)
21. Johnson-Roberson, M., Pizarro, O., Williams, S.B., Mahon, I.: Generation and visualization of large-scale three-dimensional reconstructions from underwater robotic surveys. J. Field Robot. **27**(1), 21–51 (2010)
22. Gracias, N., Ridao, P., Garcia, R., Escartin, J., Cibecchini, F., Campos, R., Carreras, M., Ribas, D., Magi, L., Palomer, A., Nicosevici, T., Prados, R., Neumann, L., Filippo, F.D., Mallios,

A.: Mapping the moon: using a lightweight AUV to survey the site of the 17th Century ship La Lune. In: Proceedings of the MTS/IEEE OCEANS conference (2013)
23. Campos, R., Gracias, N., Palomer, A., Ridao, P.: Global alignment of a multiple-robot photo-mosaic using opto-acoustic constraints. In: NGCUV2015 Girona, vol. 48, pp. 20–25. Elsevier Ltd. (2015)
24. Roman, C., Inglis, G., Rutter, J.: Application of structured light imaging for high resolution mapping of underwater archaeological sites. In: Oceans'10 Ieee Sydney, pp. 1–9. IEEE (2010)
25. Inglis, G., Smart, C. Vaughn, I., Roman, C.: A pipeline for structured light bathymetric mapping. In: IEEE/RSJ International Conference on Intelligent Robots and Systems, no. Figure 1, pp. 4425–4432 (2012)
26. Ridao, P., Carreras, M., Ribas, D., Sanz, P.J., Oliver, G., Vision, C.: Intervention AUVs: The Next Challenge. In: World IFAC Congress (2014)
27. Marani, G., Choi, S.K., Yuh, J.: Underwater autonomous manipulation for intervention missions AUVs. Ocean Eng. **36**(1), 15–23 (2009)
28. Sanz, P.J., Ridao, P., Oliver, G., Casalino, G., Petillot, Y., Silvestre, C., Melchiorri, C., Turetta, A.: TRIDENT: An European Project Targeted to Increase the Autonomy Levels for Underwater Intervention. Oceans 2013 San Diego (2013)
29. Carrera, A., Palomeras, N., Hurtós, N., Carreras, M.: Free-floating panel intervention by means of learning by demonstration. In: IFAC Proceedings Volumes (IFAC-PapersOnline), vol. 48, no. 2, pp. 38–43 (2015)
30. Ridao, P., Palomer, A., Youakim, D., Petillot, Y.: Control of an Underwater Redundant Robot Arm in Presence of Unknown Static Obstacles. World IFAC Congress (2017). Submitted
31. Massot-Campos, M., Oliver-Codina, G.: Underwater laser-based structured light system for one-shot 3D reconstruction. In: IEEE SENSORS 2014 Proceedings, pp. 1138–1141 (2014)
32. Prats, M., Fernandez, J.J., Sanz, P.J.: Combining template tracking and laser peak detection for 3D reconstruction and grasping in underwater environments. In: IEEE International Conference on Intelligent Robots and Systems, pp. 106–112 (2012)
33. Sanz, P.J., Peñalver, A., Sales, J., Fornas, D., Fernández, J.J., Pérez, J., Bernabé, J.: GRASPER: A multisensory based manipulation system for underwater operations. In: Proceedings - 2013 IEEE International Conference on Systems, Man, and Cybernetics, SMC 2013, pp. 4036–4041 (2013)
34. Chantler, M.J., Clark, J., Umasuthan, M.: Calibration and operation of an underwater laser triangulation sensor: the varying baseline problem. Opt. Eng. **36**(9), 2604 (1997)
35. Kocak, D., Caimi, F., Das, P., Karson, J.: A 3-D laser line scanner for outcrop scale studies of seafloor features. Proc. Oceans Conf. **3**, 1105–1114 (1999)
36. Hildebrandt, M., Kerdels, J., Albiez, J., Kirchner, F.: A practical underwater 3D-Laserscanner. Oceans (2008)
37. Treibitz, T., Schechner, Y.Y., Kunz, C., Singh, H.: Flat refractive geometry. IEEE Trans. Pattern Anal. Mach. Intell. **34**(1), 51–65 (2012)
38. Powell, I.: Design of a laser beam line expander 1. Appl. Opt. **26**(17), 3705–3709 (1987)
39. Fischler, M.a., Bolles, R.C., Boller, R.C.: Random sample consensus: a paradigm for model fitting with applications to image analysis and automated cartography. Commun. ACM **24**(6), 381–395 (1981)
40. Weingarten, J., Gruener, G., Siegwart, R.: Probabilistic plane fitting in 3D and an application to robotic mapping. In: IEEE International Conference on Robotics and Automation, 2004. Proceedings. ICRA'04. 2004, vol. 1, pp. 927–932 (2004)
41. Triebel, R., Burgard, W., Dellaert, F.: Using hierarchical EM to extract planes from 3D range scans. In: IEEE International Conference on Robotics and Automation (ICRA), pp. 4437–4442 (2005)
42. Pathak, K., Birk, A., Vaskevicius, N., Pfingsthorn, M., Schwertfeger, S., Poppinga, J.: Online three-dimensional SLAM by registration of large planar surface segments and closed-form pose-graph relaxation. J. Field Robot. **27**(1), 52–84 (2010)
43. Xiao, J., Zhang, J., Zhang, J., Zhang, H., Hildre, H.P.: Fast plane detection for SLAM from noisy range images in both structured and unstructured environments. In: 2011 IEEE International Conference on Mechatronics and Automation, pp. 1768–1773 (2011)

44. Smith, R., Self, M., Cheeseman, P.: Estimating uncertain spatial relationships in robotics. Auton. Robot Veh. **1**, 167–193 (1990)
45. Zhang, Z.: A flexible new technique for camera calibration (Technical Report). IEEE Trans. Pattern Anal. Mach. Intell. **22**(11), 1330–1334 (2002)
46. Bosch, J., Gracias, N., Ridao, P., Ribas, D.: Omnidirectional underwater camera design and calibration. Sensors **15**(3), 6033–6065 (2015)
47. Ciddor, P.E.: Refractive index of air: new equations for the visible and near infrared. Appl. Opt. **35**(9), 1566–1573 (1996)
48. Sultanova, N., Kasarova, S., Nikolov, I.: Dispersion properties of optical polymers. Acta Phys. Pol., A **116**(4), 585–587 (2009)
49. Daimon, M., Masumura, A.: Measurement of the refractive index of distilled water from the near-infrared region to the ultraviolet region. Appl. Opt. **46**(18), 3811–3820 (2007)

Advances in Platforms and Algorithms for High Resolution Mapping in the Marine Environment

R. Thomas Sayre-McCord, Chris Murphy, Jeffrey Kaeli, Clayton Kunz, Peter Kimball and Hanumant Singh

Abstract A confluence of technologies is changing the manner in which we approach the use of Autonomous Underwater Vehicles (AUVs) in the marine environment. In this paper we review the role of several of these technologies and the way interactions between them will now enable the use of adaptive methodologies for mapping and exploring the underwater environment. We focus primarily on imaging sensors but these methodologies are widely applicable for other types of sensing modalities as well. We look at the role of acoustic telemetry, multi-hop underwater data transmission, in-situ machine learning techniques, and mapping in highly dynamic environments such as under sea ice. In addition, we discuss the role of "hobby" robotics for surface and aerial vehicles in the marine environment.

R.T. Sayre-McCord (✉)
Massachusetts Institute of Technology, Woods Hole Oceanographic Institution, Cambridge, MA, USA
e-mail: rtsm@mit.edu

C. Murphy
General Dynamics Mission Systems - Bluefin Robotics, Quincy, MA, USA
e-mail: chris.murphy@gd-ms.com

J. Kaeli · C. Kunz · P. Kimball
Woods Hole Oceanographic Institution, Woods Hole, MA, USA
e-mail: jkaeli@whoi.edu

C. Kunz
e-mail: clayton.kunz@gmail.com

P. Kimball
e-mail: pkimball@whoi.edu

H. Singh
Northeastern University, Boston, MA, USA
e-mail: ha.singh@northeastern.edu

© Springer International Publishing AG 2017
T.I. Fossen et al. (eds.), *Sensing and Control for Autonomous Vehicles*,
Lecture Notes in Control and Information Sciences 474,
DOI 10.1007/978-3-319-55372-6_5

1 Introduction

Over the course of the last three decades AUV technology has moved from a developing technology into a reliable set of systems where basic vehicle performance can be taken for granted. The vehicle primarily used for the purpose of this paper, SeaBED [67], has ten vehicles of its class in use around the globe. With the task of vehicle operation mostly established, our research has focused on working in more difficult areas [77], on the fly replanning of missions based on observed data [54], and more recently, adaptive mission planning with minimal human intervention [32]. Here we focus primarily on improved AUV–human interaction over the course of a mission. As autonomous underwater robotics moves toward the study of dynamic processes with multiple vehicles, there is an increasing need for AUVs to compress and convey large volumes of data to human operators, and to other vehicles. Acoustic telemetry presents the only general method of communication with AUVs, however, it is inherently severely limited in throughput (see Sect. 3).

Below, we review our recent work that aims to solve these problems with a multi-pronged approach. Specifically, we review our methods for on board data selection, compression, transmission, and rapid understanding, as well as methods of working in dynamic environments. Finally, we examine the role of hobby robotics and how they have lead to a new generation of cheap, easy to use, marine surface and aerial vehicles that can be brought to bear on a number of problems of interest to oceanographers.

While our algorithms work across the board for all kinds of sensors, we choose to focus on imaging data as it is often the most challenging of standard oceanographic data. We note (and document at the start of each section) that individual components of this work have been published in far greater detail in the literature. The value of this publication lies in bringing the different pieces together into a coherent system.

2 Underwater Imaging[1]

Imaging underwater involves significant limitations; the need for external lighting in deeper waters, short range due to high attenuation, scattering from suspended particles, and color distortion due to wavelength dependent attenuation [25]. One of the basic constraints that roboticists face in the use of underwater imaging data is to compensate for illumination and color distortions to allow for better human interpretation and for the automatic processing of large data sets. Automatic processing algorithms in particular typically rely on accurate and consistent contrast and color information for texture and feature extraction.

Various approaches have been taken for single image correction: Markov Random Fields with statistical priors learnt from training images [74], frame averaging [57], adaptive histogram equalization over image sub-regions [68], homomorphic methods

---

[1] For a more detailed discussion of the methods in this section see [38, 40].

assuming slowly varying illumination and highly varying reflectance [69], and white balancing [38].

Each of these techniques can achieve aesthetically pleasing results for a single image, but the color and texture consistency of these corrected images across datasets is often unreliable [38]. These techniques also focus on the images themselves without taking advantage of the unique lighting constraints and additional sensors often present on underwater imaging platforms. One such sensor, the Doppler Velocity Log (DVL), is a ubiquitous oceanographic sensor used extensively for vehicle navigation. It emits four beams of sound, and measures the phase shift in the returns from the seafloor to determine both relative velocity and range to the bottom. In the method below, DVL ranges are used in conjunction with measured sensor offsets to estimate real physical parameters, specifically the attenuation coefficients and the beam pattern of the strobe, from the images themselves. Subsequently, color and illumination artifacts in the images are corrected for based on a physical model of the imaging environment.

By taking advantage of the available data from the DVL, combined with using the overlap between adjacent images, a consistent and physically revealing correction can be performed to generate corrected images.

2.1 Imaging Model

To estimate and correct for the physical effects occurring in the underwater imaging environment a simple model of the imaging physics is used. Ignoring the effects of backscatter (i.e., assuming that the water is relatively clear and the imaging distance is within a few meters), and that the only light source is artificial (i.e., assuming that the imaging occurs at night or below several hundred meters depth), then the imaging intensity is modeled as a combination of four multiplicative components that are dependent on the pixel location p within the image and the color channel (RGB) λ. The four components are:

(1) The intensity of the light source on the image plane $E(p, \lambda)$, given by Eq. 1, where $S(\lambda)$ is the color spectrum of the light source (assumed equal), $B(p)$ is the beam pattern of the light source, l_s is the distance from the light source to the image plane, $\alpha(\lambda)$ is the attenuation coefficient of the medium (following the Beer–Lambert Law), and γ is the angle between the source and the surface normal (assumes a Lambertian surface).

$$E(p, \lambda) = S(\lambda) B(p) \frac{e^{-\alpha(\lambda) l_s}}{l_s^2} \cos(\gamma) \qquad (1)$$

(2) The attenuation of light over a distance l_α from image plane to camera, given by Eq. 2:

$$R(p, \lambda) = \frac{e^{-\alpha(\lambda)l_\alpha}}{l_\alpha^2} \qquad (2)$$

(3) The effects of the camera, $GL(p)$, where G is the gain of the camera and $L(p)$ is the lens effect of the camera, and

(4) the true reflectance of the imaging plane $r(p, \lambda)$ which is the value we seek to recover.

Combining the four components together gives the full model of the image seen by the camera $I(p, \lambda)$, Eq. 3, and the solution for the reflectance of the image in log space, Eq. 4.

$$I(p, \lambda) = S(\lambda)B(p)GL(p)r(p, \lambda)\frac{e^{-\alpha(\lambda)(l_\alpha+l_s)}}{l_s^2 l_\alpha^2} \cos(\gamma) \qquad (3)$$

$$\log r(p, \lambda) = \log I(p, \lambda) - [\log B(p) + \log L(p) - \alpha(\lambda)(l_\alpha + l_s) - 2\log l_s l_\alpha + C] \qquad (4)$$

The right hand side of Eq. 4 has terms that describe, in order; the image seen by the camera, the beam pattern of the light source, the lens distortion of the camera, the wavelength dependent attenuation of light, the spherical spreading of light, and constant terms under the assumption of a white light source and small angular offset deviation between light source and image plane. Based on the results from [40], the effect of spherical spreading of the light is dominated by the attenuation of light in water and may be ignored. In addition, the constant terms in the equation, while important for recovering the exact properties of the imaged area, are unimportant for creating a consistent image set for automatic processing or for a visually pleasing image set for human analysis. There remain three unknown terms; lens distortion, color attenuation, and beam pattern that must be accounted for to generate the true reflectance of the seafloor $r(p, \lambda)$ from the camera image $I(p, \lambda)$.

2.2 Correcting Underwater Imagery

To back out the true reflectance of the seafloor a multi-step process is performed. First, we correct for the lens distortion using a four step calibration method [34] to get $L(p)$. Next, we correct for the beam pattern. Due to the varying height of the AUV and varying seafloor level the beam pattern is irregular in the image space, making simple frame averaging ineffective. By estimating the distance and slope of the seafloor with the DVL, the image can be converted to the angular space of the light source, where a standard average provides a good approximation of the beam pattern $B(p)$. Finally, by substituting $L(p)$ and $B(p)$ into Eq. 4 the only nonconstant unknowns are $r(p, \lambda)$ (the true image) and $\alpha(\lambda)$ (attenuation which must be corrected for). While this single equation can not be solved, using standard feature matching techniques a single feature may be found in two images, providing a pair of equations with a pair of unknowns. Using many feature matches the overall $\alpha(\lambda)$ can be approximated, allowing for Eq. 4 to be solved exactly for the underlying reflectance of the seafloor.

Using known information about the physical environment and AUV imaging geometry, rather than only the image data, the color correction process is both visually appealing and consistent across image sets.

3 Image Compression

To allow for human understanding and intervention during an AUV mission the color corrected imagery must be transmitted back to the surface. Except in cases where the vehicle is operating in a vertical acoustic channel, as in [37], the ocean imposes severe limitations on acoustically communicating data to the surface. Along with the inherently low throughput of acoustic communications, AUV and surface ship noise combine with environmental conditions such as multipath to cause frequent packet loss, making low available bandwidth and long propagation delays the norm [2]. Effective data rates for acoustic modems used to communicate underwater are routinely as low as tens of bits per second [28]. Connections may be unpredictably intermittent, with long periods of no communication. Time-multiplexing of the channel for Long Baseline (LBL) navigation, or round-robin communication of multiple vehicles, lowers effective bit-rates for each vehicle even further (Fig. 1).

To accommodate the peculiarities of the medium, channel coding methods with high rates of error-correction are typically employed. While underwater acoustic communications has achieved rates up to hundreds of kilobits per second [2], reliable acoustic communications over long distances currently requires the use of low-rate communications with high error tolerance, such as frequency-hopping frequency shift keying (FH-FSK) or highly error-corrected phase shift keying (PSK). Since the ocean is a shared broadcast medium, time-multiplexing of the channel for navigation or communication with other vehicles may be required, which lowers effective bit-rates even further. The WHOI Micro-Modem [28], used by SeaBED-class AUVs, uses low-frequency bandwidth to allow for multiple senders and receivers. It is capable of sending one 256-bit FH-FSK packet in slightly over 3 s, or one 1536-bit error-tolerant PSK packet in slightly over 6 s, delivering an effective bit-rate between 80 and 256 bits per second. Commercially available, general-purpose underwater acoustic modems typically advertise up to 10,000 bits per second although environmental factors may prevent them from achieving those speeds.

Over the course of a dive, a single AUV can easily collect one million samples of scalar environmental data, such as temperature, salinity, methane concentration, or vehicle depth. The same vehicle may easily capture tens of thousands of visual and sonar images. This sensor data is typically inaccessible until after the vehicle has been recovered, since low transmission rates have largely limited vehicle telemetry to vehicle state and health information.

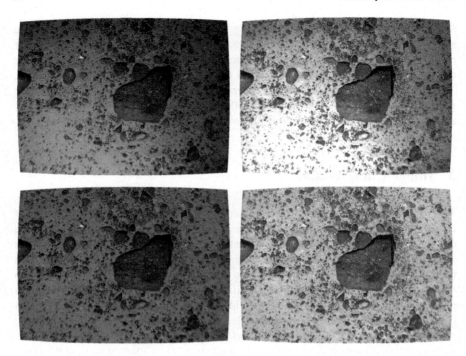

Fig. 1 Original image (*top left*) corrected for attenuation alone (*top right*), beam pattern alone (*bottom left*), and both (*bottom right*) (originally published in [38])

One widely used standard for underwater vehicle communications is the Compact Control Language (CCL) [72], which defines a number of simple methods for encoding individual samples of depth, latitude, bathymetry, altitude, salinity, and other data. CCL relies only upon quantization to provide compression and makes no use of the inherent correlation between successive samples from most instruments.

In 1996, Eastwood et al. proposed predictive coding methods that could be used in concert with these methods to improve performance [26] and, Schneider and Schmidt have incorporated predictive coding into their work [64], sending up a mean value followed by smaller, quantized delta values. While this provides some compression, transform codes allow higher efficiency in exchange for more computational effort.

Transform compression methods typically follow a standard two-step pattern. First, a source coder such as the Discrete Cosine Transform (DCT) or Discrete Wavelet Transform (DWT) exploits the inherent correlation within most data, and concentrates the energy of the signal into a sparse set of coefficients. Second, these coefficients are quantized and entropy encoded for transmission [63]. Wavelet compression is described by Donoho et al. as being especially appropriate for functions that are "piecewise-smooth away from discontinuities" [23]. While not all sensors emit signals of this form, this is an apt description for many oceanographic sensors.

3.1 SPIHT[2]

One method that has shown great success for use on underwater datasets is a compression technique based upon the Set Partitioning in Hierarchical Trees (SPIHT) embedded coding method. SPIHT is particularly well suited for the underwater environment due to its fully embedded coding method; truncating the encoded bitstream at any point produces the optimal encoding for that data length. This allows fine-resolution imagery to build on previously transmitted low-resolution thumbnails.

SPIHT was originally designed for photo compression, and can be used on higher dimensional datasets as well. Two-dimensional data like imagery is transformed with the 2D form of the DWT into approximation coefficients (low-frequency information) and sets of detail coefficients (high frequency information). SPIHT is then used to transform the coefficients into sorting and refinement bits, which are ordered to progressively describe the full image.

Examples of different compression levels for an image using the SPIHT compression method are shown in Fig. 2. Further examples of the compression pipeline can be found in [53].

3.2 Tile-Based Compression[3]

Traditional compression techniques such as the SPIHT-based method described in Sect. 3.1 are designed to transmit a compressed version of data that is as faithful as possible (in a mathematical sense) to the underlying data. In the case of AUV operations the original data will be downloaded once the AUV returns to the surface, so often the goal of real-time transmission of data is to give the operator the best understanding (in a heuristic sense) of the mission environment. To achieve that goal, a different compression method, designed to encode "high-resolution thumbnails" of seafloor imagery is described below. This method provides extremely high compression of seafloor imagery which, unlike traditional image compression methods, exploits the high levels of inter-dive redundancy. The method is inspired by the vector quantization literature, but differentiates itself from past work by using large quantization vectors, or "tiles," much like an image mosaic. Compression artifacts are reduced during decoding, resulting in a caricature of the original image.

The approach decomposes each image into a grid of tiles, and encodes each tile as the index of a visually or semantically similar, but previously captured, image tile. This method is inspired by the widely used JPEG standard, which encodes images as a grid of 8×8 patches. Similar to JPEG, the technique below is also a patch-based approach, however, the patches used are larger and instead of transmitting quantized wavelet coefficients, the descriptive coefficients are used to look up the

---

[2]For a more detailed discussion of the methods in this section see [53].
[3]For a more detailed discussion of the methods in this section see [55].

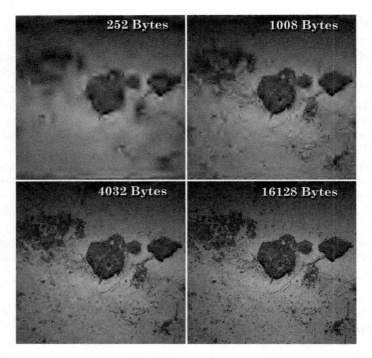

Fig. 2 The same image (in 1024 × 1024 Y'UV) encoded at four different sizes using SPIHT. 81% of the transmitted data is used to reconstruct the luminance, the rest describes the image color channels (originally published in [53])

most similar entry in a database of tiles taken from other related images—a form of vector quantization.

The tile library is shared between the AUV and the receiver prior to the dive. Tile similarity can be calculated using a variety of methods; L_2 distance was found to be effective, though computationally challenging for an embedded system. As a computationally cheaper alternative, Principle Components Analysis (PCA) is used to accelerate tile comparisons, and to identify the best match for each source tile. Tile indices are transmitted to the surface, where indexes are transformed back into tiles, and a caricature, or "high-resolution thumbnail," of the original image is generated. To reduce compression artifacts in the resulting image a gradient-preserving blur Poisson editing technique is used.

Tile Database

The first step of the technique is to construct a database of tiles which describe the space of images that will be encoded. Computing a tile database given an image set amounts to sampling sub-images along a uniform grid. The image set is ideally of the same type of scene (e.g., coral reefs) as the images to be compressed (perhaps gathered from the first dive in the area), and in practice may contain upwards of 100,000 tiles.

In addition to a large database of tiles, a notion of similarity or distance between tiles is required that is fast to compute. This metric is used to determine the nearest neighbors to a query tile. Ideally, the L_2 distance metric across all of the pixels in an image would be used so that the nearest neighbor selected minimizes the root mean square (RMS) reconstruction error. However, the L_2 distance is too expensive to use due to the high dimensionality of the tiles, and the large library size. Instead, the L_2 distance is approximated using PCA to optimally (in a L_2 sense) project each tile into a low-dimensional space, allowing encoding on commodity hardware onboard an AUV.

First, each RGB color image tile x with dimension $n \times m$ is flattened into $3\,nm \times 1$ vectors. These vectors are packed side-by-side into a matrix X of size $3\,nm \times N$ where N is the size of the training set. Next, this matrix is mean centered by subtracting the mean value from each column, generating matrix X_m.

The Nonlinear Iterative Partial Least Squares (NIPALS) algorithm is used to compute a basis $P = [p_1, p_2, \ldots, p_k]$ such that,

$$X_m = PT + E \qquad (5)$$

where $T \in R^{k \times N}$ is a matrix containing the low-dimensional feature vector representation of the training set and E is the residual.

Each new tile query y is projected onto the principal components P to generate a compressed representation $t(y) = PTy$ comparable with the low-dimensional representation of the training set. The result is that the L_2 distance metric can be efficiently evaluated according to Eq. 6 by comparing only a few components of low-dimensional vectors instead of taking the pixel-wise difference of two image tiles:

$$d(x_1, x_2) = \|x_1 - x_2\|_2 \approx \|Pt(x_1) - Pt(x_2)\|_2 = \|t(x_1) - t(x_2)\|_2 \qquad (6)$$

As the number of components approaches the dimensionality of the data set, the PCA distance metric is guaranteed to converge to the L_2 distance metric. For transmission, the nearest image tile indices are packed bitwise, and transmitted over the acoustic link offering significant levels of compression over standard compression techniques.

Block Artifact Reduction

Due to the tile-based nature of the compression algorithm and the constraints of extremely low bit rates, the images have noticeable "block artifacts" along the grid structure. Post-processing is used to ameliorate this issue. These post-processing techniques are performed by the receiver, typically a surface ship where computation power is less restricted, after an image has been received.

One approach to reducing the block artifacts is to blur the image along the grid. However, this produces an equally disturbing artifact where the pixels along the grid are noticeably too smooth. The goal is to smooth the less noticeable low-frequency content while preserving sharp boundaries. To achieve this goal, the Poisson image editing technique as described by Perez and colleagues [56] is used. Poisson image editing interpolates a region of an image $f \in \Omega$ constrained at the boundary $\partial \Omega$ with Direchlet boundary conditions f^*. The interpolated region is made (by enforcing Eq. 7) to follow a vector guidance field $\mathbf{v}$, which in this case is composed of the image gradients of each tile.

$$\min_f \iint_\Omega \|\nabla f - \mathbf{v}\|^2 \text{ with } f|_{\partial \Omega} = f^*|_{\partial \Omega} \qquad (7)$$

For images, the above equation is discretized along the pixel grid. Let f_p be the value at pixel p and $\langle p, q \rangle$ be the set of neighboring pixel pairs.

$$\min_{f|\Omega} \sum_{\langle p,q \rangle \cup \Omega \neq \emptyset} (f_p - f_q - v_{pq})^2 \text{ with } f_p = f_p^*, \text{ for all } p \in \partial \Omega \qquad (8)$$

For all $\langle p, q \rangle$, define v_{pq}:

$$v_{pq} = \begin{cases} f_p - f_q & \text{if tile}(p) = \text{tile}(q) \\ 0 & \text{otherwise.} \end{cases} \qquad (9)$$

That is, respect all image gradients as best as possible except those between two pixels from different tiles. The minimization above is given as the solution to a system of sparse linear equations which can be solved using direct LU factorization.

Testing of the full compression pipeline was performed in [55] on a SeaBED AUV during dives near Channel Islands National Park, off the California coast. A comparison of the compressed image of a skate from these trials is shown in Fig. 3 and a selection of images compressed at various tile sizes is shown in Fig. 4.

4 Data Transmission and Multi-vehicle Networking[4]

While the compression algorithms described in the previous section provide an effective manner of packing data for transmission in the limited acoustic band, they are only one part of the picture. For human-in-the-loop AUV operations there must be a well managed, and ideally generalizable, architecture for when and what data is sent in increasingly complex scenarios.

---

[4]For a more detailed discussion of the methods in this section see [54].

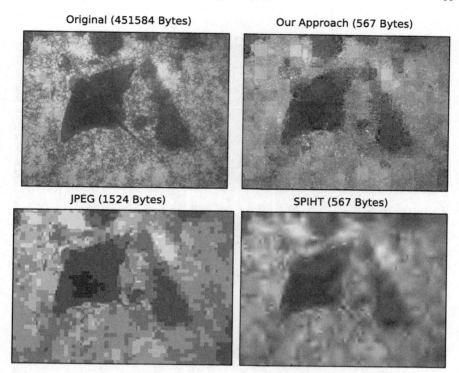

Fig. 3 A comparison of tile-based (*top right*), standard JPEG (*bottom left*), and the SPIHT wavelet coder (*bottom right*) on a 432 × 336 image of a ray (originally published in [55])

AUV missions may now involve multiple vehicles working toward a loosely defined set of goals, often in dangerous and unconstrained environments such as under ice. The mission goals can be defined during the mission and may be based on complex analysis of science data, including imagery. Architectural advances in autonomy, evident in architectures such as MOOS-IvP [9], have not been met with similar advances in AUV telemetry handling.

Photographic and SONAR imagery are only transmitted by a few special-purpose communication systems [52]. These systems rely on specific vehicle geometries, and do not scale to support multiple vehicles. Missions involving multiple vehicles may extend beyond the effective range of a single acoustic link, yet existing systems do not support relaying large data across multiple AUV "hops."

CAPTURE—**C**ommunications **A**rchitecture that delivers arbitrary science data using **P**rogressive **T**ransmission, via multiple **U**nderwater **R**elays and **E**avesdropping—aims to help fill that void. In concert with an abstracted physical layer, CAPTURE provides an end-to-end communication architecture for interactively obtaining data across an acoustic network of marine vehicles. CAPTURE employs progressively encoded compression of telemeter imagery, SONAR, and time-series

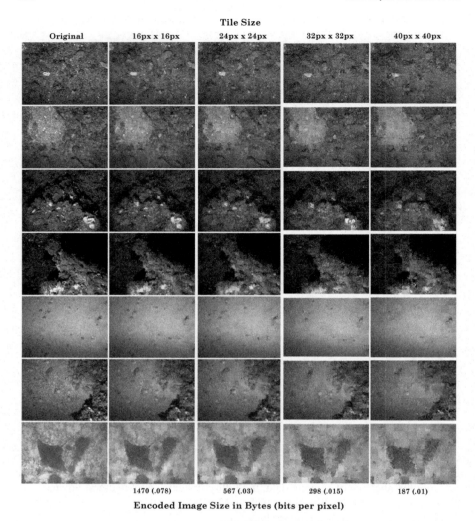

Fig. 4 Results from encoding a series of images using tiles at a range of compression levels (originally published in [55])

sensor data from underwater vehicles. These resources are automatically selected by the vehicle, and transmitted as a sequence of gradually improving data "previews." High-quality versions of these previews, up to an error-free reconstruction, can be requested by operators immediately, or at any later time over the course of a mission. CAPTURE has been designed to operate on multiple vehicle architectures and supports multihop relay communication across several vehicles. CAPTURE has been deployed in multiple field operations on diverse vehicle platforms.

While there are some robust physical communication layers available commercially, there does not exist a common cross-manufacturer interface such as the Hayes/AT Command Set that dictated the course of terrestrial telephone modems. The Goby Autonomy Project [65] has made advances in developing an open source generic abstraction for acoustic modems and implementing drivers for physical modem hardware, as have various proprietary solutions [51]. These drivers allow software to operate independent of the modem's underlying proprietary languages.

Numerous medium access control (MAC) protocols such as multiple access with collision avoidance for wireless (MACAW) [11] have been developed to mediate between multiple communicating nodes at the data link layer. Research at higher networking layers exists as well [15], yet few field experiments have involved multiple autonomous vehicles communicating high-bandwidth data across multiple hops. Perhaps the best known such experiment, SeaWeb [60], was performed by Benthos in concert with the U.S. Navy, utilizing fixed nodes. Still, there currently exist no transport or application layer protocols in widespread use for underwater vehicles. Common terrestrial protocols are largely unfit for underwater use as they use headers that approach the size of acoustic packets, and rely on the rapid forwarding of messages by network nodes, using bandwidth to replace history. Relative to the bandwidth available with modern acoustic modems, underwater vehicles can be considered to have nearly infinite storage. CAPTURE exploits that storage by having every node in the network permanently store each piece of data that is transmitted.

CAPTURE consists of four distinct components, shown in Fig. 5a. First, a set of data is acquired by the AUV and registered as a transmittable resource with the

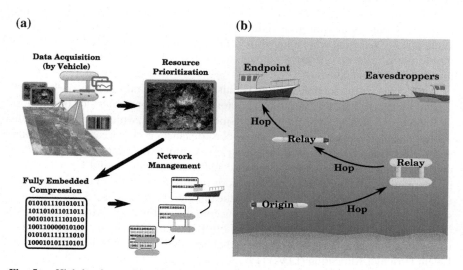

Fig. 5 a High-level overview of data flow through the four main components of CAPTURE. b A depiction of a possible network configuration for CAPTURE. Multiple vehicles are underwater relaying messages to the surface where the operator and eavesdroppers reside (originally published in [54])

telemetry system, via a platform-specific driver. Examples of possible resources include a single image, or a time series of measurements from a single sensor. The platform-specific drivers isolate the telemetry system from the specific capabilities or limitations of each host vehicle. Second, new resources are automatically selected for compression and transmission to the surface, or existing resources are selected for further transmission based on requests from the surface. Automatic selection provides an avenue for high-level algorithms, such as mine identification or interest operators, to guide the selection of interesting telemetry. Third, selected resources are compressed using progressive coding methods such as SPIHT. Progressive coding methods, specifically those that are fully embedded, ensure that an approximation of the data can be reconstructed with each newly received bit of data. Finally, the transmission of the resource to the surface is managed to ensure end-to-end delivery. When multiple underwater vehicles are available, intermediate vehicles can relay data to the surface as hops in the route, or help through "eavesdropping."

5 Image Selection and Rapid Image Understanding for Adaptive Missions[5]

The combination of image compression and architectures such as CAPTURE still leaves an operator only able to view a small fraction of the imagery that an AUV produces during a standard mission. This leads to a high *latency of understanding* between the capture of image data and the time at which operators are able to gain a visual understanding of the survey environment. To aid the rapid understanding of an AUV survey mission two additional elements are desired: (a) a method of automatically selecting which images to transmit during a mission, and (b) a method of informing a surface operator of the mission as a whole based on those images that the operator has received.

At the core of the methods below is clustering images into groups of "similar" images which are meant to describe similar seafloor environments. One of the most well-known clustering algorithms is the K-means algorithm, which seeks to find a set of cluster centers that minimize the within-class distances between each cluster center and the members of its representative class [33]. A similar algorithm, k-medoids, only considers data points as potential cluster centers, and is more useful for generating representative images. Both of these methods require the number of cluster to be set a priori. Other methods seek to determine the number of clusters based on the natural structure of the data. Affinity propagation accomplishes this by picking "exemplars" that are suggested by nearby data points [29]. Hierarchical methods have also been used to learn objects [71] and underwater habitats [58] based on topic models using Latent Dirichlet Allocation (LDA) [13]. A drawback of all methods mentioned thus far is that they operate upon a static dataset. This "offline" approach is ill-suited to

---

[5]For a more detailed discussion of the methods in this section see [38, 39].

real-time robotic imaging because it offers no way to characterize the dataset until after all the data has been collected.

Clustering data in an "online" fashion provides two important benefits. Firstly, it allows data to be processed continuously throughout the mission, reducing the overall computational load. Secondly, at any point in time it provides a summary of the imagery captured thus far by the vehicle. A drawback to online methods is that they offer less guarantees of stability and are ultimately dependent upon the order in which images are presented to the algorithm.

As with other recent work in the field of navigation summaries [31, 32], the algorithms below use the concept of "surprise" to determine both the subset of images that best describe the mission as a whole, and those images that describe the most unusual elements of the survey environment.

5.1 Surprise-Based Summaries

Here, an event is defined as surprising if it is not expected based on previous data. The idea of what is expected can be modeled as a probability distribution over a set of variables called the prior knowledge about the world. When a novel event occurs, it augments this body of knowledge and creates a slightly different posterior knowledge of the world. If the amount of knowledge added by any single event is large enough, that event can be said to be unexpected and thus is "surprising."

This concept has been formalized in a Bayesian framework as the difference between the posterior and prior models of the world [36]. For measuring this difference, the Kullback–Leibler (KL) divergence, or relative entropy, was shown to correlate with an attraction of human attention [36].

Rather than modeling the prior knowledge Π^- as a single distribution $P(F)$ over a set of features F, we follow [31] and model it over each member of summary set $\mathbf{S}$ containing M members.

$$\Pi^- = \{P(F|S_1), \ldots, P(F|S_M)\} \qquad (10)$$

The posterior knowledge Π^+ is simply the union of prior knowledge with the new observation Z

$$\Pi^+ = \{P(F|S_1), \ldots, P(F|S_M), P(F|Z)\} \qquad (11)$$

The set theoretic surprise ξ is defined as the Hausdorff distance between the posterior and prior distribution using the KL divergence as a distance metric [31]. The Hausdorff metric is a measure of the distance between two sets based on the greatest possible difference between one point in the first set to the nearest point on the other sets. Since the prior and posterior sets differ only by Z, the surprise can be simply expressed as the KL distance between observation Z and the nearest summary image in $\mathbf{S}$.

$$\xi(Z|\mathbf{S}) = \inf_{\pi^- \in \Pi^-} d_{KL}(P(F|Z) \| \pi^-) \qquad (12)$$

When a new observation's surprise exceeds a threshold, it is added to the summary set. The threshold is generally set as the lowest value of surprise in the current summary. That member of the old summary set with the lowest surprise is then removed and replaced by the new observation, and the surprise threshold is set to the next least-surprising member of the summary set. In this manner, a temporally global summary of the images is maintained at all times [31].

5.2 Image Clustering

With the previous section describing a metric of "surprise," there is a method of characterizing which elements are new and unexpected in an image set. For actual computation of the "surprise" for an image, each image is characterized by 1000 keypoints quantized into 14 binary QuAHOG patterns [38] which gives a 14 bin histogram for each image. To create a true (symmetric) distance metric between two images the average of the one-directional KL divergences is used, as done in [32].

Since the primary goal of the algorithm is to provide an in mission summary of the image data so far collected, the set of summary images should be changed as new images are taken such that the summary images best describe all currently available data. To do so, the summary set is initialized with the first N images, where N is the size of the summary set and their corresponding surprise values are set to the smallest surprise measured relative to the set of images before it. Progress continues throughout the rest of the data until the surprise threshold is exceeded by a novel image. When this happens, the novel surprising image is incorporated into the summary set, the least surprising image is removed, and the surprise threshold is augmented to the new lowest surprise value within the set.

To create clusters out of the data, each non-summary image is assigned to the summary image that it is closest to according to the symmetric KL divergence. Combined with navigation data this clustering allows for surveyed areas to be assigned to a certain class of seafloor, with that class described by the associated summary image, see Fig. 6.

While the AUV maintains its summary set at the seafloor, it must also make decisions about which elements of the summary set to send to the surface. Early in the mission the surprise threshold grows rapidly as the algorithm determines which images best represent the data. Thus, the first summary image is not sent until the surprise threshold does not change for a specified number of images, implying that the vehicle is imaging consistent terrain that could be represented well by a single image.

For all subsequent summary images sent, there must be a balance between choosing a summary image that is different enough from the previously transmitted

Fig. 6 Photomosaic (*left*) and bathymetry (*middle left*) of the entire mission. The final semantic map (*middle right*) using 9 images which have been heuristically merged into 5 distinct classes (*right*) and color coded (originally published in [38])

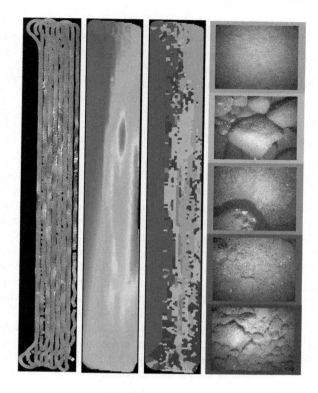

summary images to be informative, while at the same time representing enough non-summary images to make it a worthwhile choice for a map basis. Hence, the summary subset is selected that both minimizes the Haussdorff distance when the summary set is partitioned into subsets, as well as represents enough non-summary images to exceed a given threshold.

Selecting good summary images to transmit is important because these images will be used to represent the entire dataset for the duration of the mission. Furthermore, this means that as new summary images are added to the summary set, previously transmitted summary images should not be removed from the summary set given the high cost of transmitting an image. Therefore, after a summary image is transmitted, it becomes "static," as opposed to the other "dynamic" summary images.

Ideally, online summary methods do not require distances to be recomputed for all existing data points when new data appears which is one quality that makes them attractive for power-limited underwater robots. Thus, when a new summary image is added to the set, it is best to not lose the information already computed by simply removing the least-surprising summary image and the non-summary images that it represents. Instead, the low-surprise summary image is merged with the nearest summary image by merging all the images in the two clusters and representing them by the summary image that originally had more non-summary images.

The navigation summary pipeline has been implemented on a 3000+ image dataset collected by the towed camera system SeaSLED in the Marguerite Bay area of the west Antarctic Peninsula in December 2010 and [38] on a 2800 image dataset collected by the SeaBED AUV in 2003 in the Stellwagen Marine Sanctuary. Further discussion of the results of these test can be found in [38, 39].

6 Mapping Dynamic Environments[6]

While in the past mapping AUVs have typically been used to build wide area bathymetric maps, most of the areas under consideration have been relatively static. A few objects may change, but the map itself remains consistent. With the increased use of AUVs under sea ice there has been considerable interest in extending AUV mapping capabilities to environments that are quasi dynamic, that is, slowly changing over the course of a survey. Sea ice for instance, may be rotating and translating at rates that are comparable to the speed of the AUV. Regardless of the mapping modality, the desire is to build a self-consistent map, geo-referenced if possible, and the primary problem is that the navigation accuracy of the vehicle is much worse than the resolution of the mapping sensors being used.

To address the localization error that occurs in standard AUV navigation the problem of mapping and navigation can be combined using techniques from the simultaneous localization and mapping (SLAM) literature [73]. Specifically, the formulation of a pose graph allows for the combination of general AUV sensors with a single camera and a multibeam sonar into a single flexible sparse graphical framework. By optimizing over this graph, both a higher accuracy AUV trajectory estimate and a centimeter-scale 3D bathymetric map are created. For a general discussion of pose graphs used for localization see Dellaert and Kaess [20], and for a brief overview with respect to AUV navigation and this methodology see [44]. In addition to the the AUV trajectory and bathymetric map, this methodology allows for the calibration of the navigation sensors through the use of specialized nodes in the pose graph.

The techniques presented here are related to earlier work in the fields of multibeam bathymetric mapping, visual SLAM, and bundle adjustment, synthesizing these disparate fields into a single consistent framework for use on AUVs. The post-processing SLAM algorithm in [42] uses iceberg-relative AUV velocity measurements and a simple model of iceberg motion to create self-consistent multibeam maps of free drifting icebergs. In the field of bathymetric range SLAM, the algorithm in [61] uses the idea of fixed submaps that are registered using the iterative closest point (ICP) algorithm [80] to impose relative pose constraints in a delayed-state extended Kalman filter. The approach below reformulates the idea into the pose graph framework, but it relies on the same assumption that local improvements in a map will imply a global improvement. A simplified model for submap matching is also used that does not rely on the full ICP algorithm. More distantly related to the approach described here

---

[6]For a more detailed discussion of the methods in this section see [44, 77].

are the efforts in [7, 27], both of which rely on particle filters and use an occupancy grid representation to directly optimize over the consistency of the generated terrain map, and the method described in [76], which uses a sparse extended information filter. Using the multibeam as a source of relative pose constraints in the optimization, rather than explicitly modeling the multibeam measurements in the pose graph, allows for the use of the pose graph framework for all constraints—visual, acoustic, and proprioceptive—in a straightforward manner. Moreover, the unified approach allows the resolution of the generated maps to approach the advertised resolution (2 cm at 10 m range) of the 245 kHz multibeam sensor (Imagenex DeltaT, 2012), exceeding the current common practice of gridding on the order of 50 cm [61, 79] by an order of magnitude.

The fusion of IMU, camera, and velocity sensors is a well studied problem in SLAM, and discussion of using those sensors in a factor graph framework can be found in [44]. Here, the focus is on the inclusion of multibeam data within the factor graph framework for dynamic environments.

6.1 Incrementally Building a Multibeam Map

Building a map with multibeam data reduces to a question of geometry—if the pose S_t of the multibeam sonar head is known at the time t of each ping, then the range from each beam can be placed into a globally referenced point cloud, which can eventually be gridded into a map. Determining S_t requires knowing the AUV's position $\mathbf{X}_t$ at time t and the 6-DOF pose offset $\mathbf{P}$ between the AUV's navigation frame and the multibeam head. If R_j is the rotation describing the beam angle for beam j and $b_{t,j}$ is the range for beam j at time t, then the corresponding point $p_{t,j}$ is

$$P_{t,j} = \mathbf{X}_t \oplus \mathbf{P} \oplus \mathbf{R}_j \oplus b_{t,j} \quad (13)$$

and the mapping problem requires estimating X_t and P. The first of these parameters is computed in the pose graph optimization, assuming good knowledge of the synchronization between the navigation sensors and the multibeam. The second requires extrinsic calibration of the multibeam head, which is performed automatically by minimizing the variance of gridded maps built with hypothesized offsets, outside of the pose graph framework. See [43, 70] for details.

Multibeam sonars are generally used to build maps incrementally, one ping at a time, and successive pings are accumulated into a range map. Because each ping has relatively low resolution (relative to laser scanners, for example), distinctive landmarks are difficult to identify and differentiate from one another in single scans. For this reason, multibeam mapping efforts using SLAM, such as that of [7, 61], do not rely on landmarks, but instead either make use of relative pose constraints (in the former case) or directly use an occupancy grid representation that dispenses with the need for the explicit connection of mapping regions (in the latter case). The pose graph framework is much better suited for handling relative pose constraints than full occupancy grids, and therefore is used here. Small submaps are built in which the AUV's trajectory crosses over itself, and the transformation that best aligns them is found which determines a relative pose constraint induced by the match. In the pose graph, these constraints are very similar to the 6-DOF odometry constraints linking poses at adjacent time steps, but they link two poses that are arbitrarily far from each other in time. Because the constraint only links two poses in the graph, the smoothing step does not force the individual submaps to remain static—rather the whole trajectory (and hence the map) is warped to minimize the overall error described by all of the constraints.

The submaps themselves are produced using the starting trajectory estimate from navigation aided by visual bundle adjustment described in [44], taking into account the footprint of the sensor on the terrain, as well as the shape of the planned mission. Crossover points are planned for, and tracklines are designed to be close enough to each other to allow significant sensor footprint overlap from leg to leg. Both "intersections" in the trajectory and parallel tracklines with overlapping sensor footprints can be used to produce relative pose constraints. Once two gridded submaps are built, they are aligned by scanning over a horizontal displacement between the two of them, and finding the (x, y) displacement that minimizes the average difference in depths between the two maps M_1 and M_2, namely

$$\frac{1}{N_{overlap}} \sum_{(x,y) \in M_1 \wedge (x+\Delta x, y+\Delta y) \in M_2} (\bar{z}_{(1,x,y)} - \bar{z}_{(2,x+\Delta x, y+\Delta y)})^2 \quad (14)$$

where the sum is over all cells that contain data in both maps, and $\bar{z}_{k,x,y}$ is the mean depth for the cell at location (x, y) in map k. The additional step of matching submaps using full ICP is not used (in comparison to [61], for example) for several reasons. By searching only for a horizontal displacement that aligns the maps, the assumption is made that the attitude and depth of the AUV are well known at this point in the mapping process, which should be the case given the refinement of the trajectory using visual features. Furthermore, ICP is minimizing a different local error function, typically the sum of distances between points and locally computed planes, rather than displacement in meshes in the z direction. Experimentally the ICP-optimized transformation matching two submaps was often found to result in higher overall depth variance once the combined point clouds are binned [44].

A confidence estimate of the depth image correlation is provided by fitting a 2D quadratic locally to the error surface being minimized. Equation 14 provides an error term for each (x, y) offset estimate, to which a quadratic is then fitted using linear least squares. The shape of the quadratic around the minimum is governed by the quadratic's Hessian matrix, which is in turn used as the information matrix to weight the relative pose constraint's influence on the pose graph. This is because the Hessian describes how quickly the error function changes as the 2D displacement is changed, which indicates the confidence in the match. If the determinant of the Hessian is too close to zero (or is negative), then the match is unreliable and the link is not added to the pose graph. The use of the Hessian here means that matching submaps of flat terrain will only weakly influence the pose graph, because the correlation error score at the best displacement will have a small second derivative. More importantly, linear features such as troughs and ridges will have strong minimum errors along one dimension only, which is reflected by the magnitude of the eigenvalues of the Hessian: the resulting information matrix will constrain the AUV poses only in the directions in which the match is well-conditioned.

Once the local submap match has been found and the information matrix determined, the displacement and information matrix are transformed into a relative pose constraint in the graph by orienting the displacement and Hessian relative to the first of the two AUV poses being compared. For each pair of matched submaps, a single relative pose constraint of the type $C_{i,j}$ is added to the graph in this way, using the multibeam pings nearest the center of each submap to "anchor" the constraint—if the multibeam ping times do not correspond exactly to the times of the pose nodes in the graph (e.g., when using the DVL as the base sensor instead of the multibeam), then the constraint is adjusted slightly so that it correctly constrains the relationship between the two closest pose nodes to these times. The number of factors added to the graph depends on the trajectory of the dive, but as long as the number of these cross-constraints does not increase more than linearly with the size of the graph, the sparsity of the overall system remains intact.

6.2 Quasi Dynamic Mapping—Mapping Sea Ice

For an AUV, navigating with respect to drifting terrain appears much the same as navigating in a current: the vehicle might "crab" to compensate for moving in a direction different from the vehicles heading. Once the floe starts to rotate, however, the AUV's onboard navigation will deteriorate, because heading relative to the terrain (typically the Earth) is provided by the north-seeking gyrocompass, so the estimated location of goal points will be incorrect from trackline to trackline. While floe rotation rates are often small for the size of surveys carried out on under-ice mapping missions, rotation rates as low as as 2.5 degrees per hour are sufficient to confound the AUVs navigation estimate during the course of a dive, and much higher rates (e.g., tens of degrees per hour) are commonly seen with smaller ice floes in turbulent environments [77]. There has been little work using AUVs to map drifting and rotating ice. Our approach calls for the orientation of the ice to be explicitly estimated as part of the SLAM solution, but we do so using the already-described framework of pose graph optimization, so little of the algorithm has to change in order to account for the rotating world frame.

There are a few possible ways to change the pose graph to move from a global reference frame to a terrain-relative frame. One possibility is to add a seventh degree of freedom to each AUV pose node, and then to change the attached factors to account for the transformation from world heading to terrain-relative heading. This is a rather intrusive approach, and unnecessarily reduces the sparsity of the system, as the measurements of AUV attitude are independent of the external measurements of terrain orientation. Instead, we attach pose nodes with a single degree of freedom representing terrain orientation to each AUV attitude measurement factor. Since all the factors connecting AUV pose nodes in the graph only measure relative pose (odometry), only the attitude measurement factors constrain the global orientation of the whole trajectory, so it makes sense to inject a terrain-relative orientation there in the graph. Each attitude measurement factor is modified so that it attaches to a pose node estimating the terrain orientation as well as an AUV pose node estimating Earth-relative attitude. In other words, the attitude measurement factor is trying to enforce a terrain-relative attitude in the AUV pose node, rather than a world-relative attitude. The augmented pose graph contains additional factors connecting the terrain orientation pose nodes which limit the allowed rotation rate of the terrain. This prevents solutions which might use the additional degree of freedom for each AUV pose to overfit the multibeam submap matches. Finally, measurement factors are added when an external sensor is available to measure the terrain orientation relative to the global reference frame. This might be provided by instrumenting an ice floe with a compass, or as in the case here, by mooring a GPS-equipped ship to the floe throughout the course of the dive. Because the connection between the ship and the ice is not entirely rigid and GPS is not an ideal heading measurement system, small variations in the measured heading are to be expected, which are allowed by using a variance value of one degree in the terrain orientation measurement factors. Our results are still reasonably accurate and have made a marked difference in mapping and understanding the properties of sea ice [77].

7 Hobby Components in Marine Robotics

One of the major changes in robotics over the last half decade has been the emergence of hobby and commercial technologies in robotics. For example, the release of the Kinect significantly changed terrestrial robotic sensing both in quality and affordability [81]. Similarly, the recent availability of cheap yet well designed hobby technologies has significantly lowered both the cost and complexity required for robotic vehicles in marine environments. Where marine robotics has long meant complex (typically underwater) vehicles that require dedicated engineering teams, hobby technologies have opened the door for scientific groups to own and operate their own research vehicles. By shifting from underwater vehicles to surface or air vehicles, they become easier to design (no water pressure), implement (access to signals such as GPS), and use (constant high-bandwidth communication). Below, two examples of hobby technologies being used for scientific mapping are presented, the first (the WHOI Jetyak [41]) is a custom built system with most of the robotics computing performed by an off-the-shelf model airplane controller, and the second (the

DJI Phantom [21]) is a commercially available quadcopter with a built in stabilized camera.

8 The WHOI Jetyak

A good demonstration of the power of off-the-shelf hobby parts for marine robotics is the WHOI Jetyak [41], a multifunctional autonomous surface vehicle (ASV) for marine science. The WHOI Jetyak is a Mokai [50] gasoline powered kayak equipped with servos for thrust and steering control, a long range radio for communications, an ArduPilot controller [3] for control and autonomy, and a reconfigurable science payload (see Fig. 7a). While there are many other scientific ASVs available on the market, (e.g., [5, 16]), the Jetyak is perhaps the first to take advantage of widely available open-source hobbyist technology. While typical robotic vehicles require a long development time to design, build, and implement, the use of the ArduPilot gave ready made solutions for control (ArduPlane comes with thrust and rudder controllers), sensing (cheap integrated GPS and heading sensors), communication (through model airplane remotes and a standardized communication protocol), and visualization and mission planning (multiple well developed mission planners, e.g., [4, 62]). Because the ArduPilot software is entirely open source, the user does not sacrifice any access to data or control, allowing for implementation of more advanced algorithms, see Sect. 8.2. By combining several off-the-shelf products—a Mokai kayak, a ArduPilot controller, and servos, a fully fledged autonomous science platform can be made with a small development cost, low overall price, and easy operation.

8.1 Example Applications

There are currently six Jetyaks in use for a variety of scientific purposes. The Jetyak excels as an oceanographic platform for several reasons: (a) its air-breathing gasoline

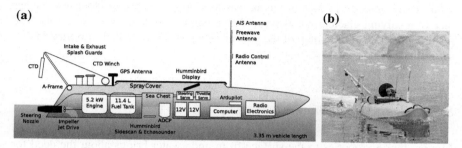

Fig. 7 a Schematic of a WHOI Jetyak. b A WHOI Jetyak returns safely from imaging the face of a calving glacier in Greenland (originally published in [41])

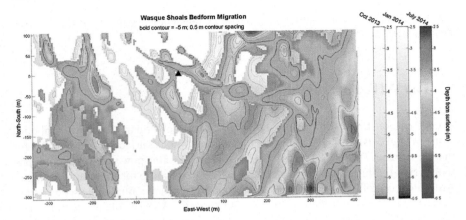

Fig. 8 Data from three Jetyak surveys over a nine month period show dune migration in Wasque Shoals off of Martha's Vineyard, MA. The most recent survey is color contoured, and the older surveys are gray shaded (originally published in [41])

engine allows for a top speed of 10 m/s making it well suited for energetic environments, (b) its draft of less than 20 cm allows it access to very shallow tidal regions, (c) its access to high accuracy trajectory information from Real Time Kinematic (RTK) GPS allows for high accuracy mapping beyond what can be achieved underwater, even with the use of advanced algorithms such as those described in Sect. 6, and (d) high-bandwidth radio communications allow for direct control of the vehicle through a remote control or mission planning software giving a highly user friendly interface. In comparison to human operated vehicles, the Jetyak can perform precision mapping maneuvers difficult for human pilots to maintain over long periods (shown to have trackline following capability within a few meters in currents [41]), has a shallower draft than standard human operated boats, and can be driven into dangerous environments (e.g., near calving glaciers, see Fig. 7b). The power of the Jetyak to produce high resolution maps using multibeam sonar is shown in Fig. 8, which shows the same area of seafloor surveyed over a nine month period giving detailed maps of the seafloor over time, providing information on the movement of sand on the seafloor. In [48], a Jetyak was used to survey the seafloor and glacier face of a calving glacier—an area too dangerous for human operated vessels. The resulting high resolution map of the glacier face provided detailed information of a subglacial plume.

8.2 Advanced Autonomy

Moving marine robotics to the surface from underwater brings about the need for enhanced autonomy algorithms in the new environment. While for the most part

AUV operations can be considered to take place in a static environment, ASVs must account for a dynamic set of obstacles, i.e., other boats with human operators. Currently ASVs are operated with human supervision or in an unoccupied ocean. The ability to operate ASVs without human supervision would drastically open up their potential; scientific studies such as the one found in Fig. 8 could be performed on a daily scale, rather than once every three months. ASV autonomy can be broken down into two steps—mapping the world and making decisions based upon that map; for the most part ASV localization and vehicle control have been assumed due to the reliable presence of GPS on the ocean and the scale on which decisions are currently made.

The first element of any ASV mapping system can be (and should be) the use of marine charts, which provide accurate descriptions of land, and in the case of deeper keeled vessels, a map of impassible shallows. Because marine charts provide a good description of the large scale static environment, ASV mapping systems are primarily concerned with detecting other boats in the environment. These detection systems may be based upon computer vision [45] or laser scanners [12] (mimicking approaches commonly used in terrestrial robotics) or radar [66] (mimicking the sensors commonly used on human controlled surface vessels). While computer vision and laser scanners are well studied for robotic obstacle detection, they suffer in the foggy conditions which often come up unexpectedly on the ocean. Standard radar systems are high powered and have a large dead-zone making them ill-suited for small ASV operation, however, broadband radar systems, such as the Lowrance 3G Radar [47], are low power (18 W) and have a dead-zone of only a few meters around the vehicle. Thanks to the work of [18] the raw data may be accessed, and a Lowrance broadband radar has been placed on the Jetyak for obstacle avoidance, as well as used in [5, 66]. Radars provide range and bearing measurements of other obstacles, and are well studied in a marine environment, working reliably in weather and in darkness.

Once detected, individual measurements must be combined over time using data association to identify and track specific vehicles. Most radar manufacturers come with this capability through automatic radar plotting aids (ARPA), however, these are typically not exposed to the user for work in robotics. A significant amount of work has been done on radar tracking [6], forming the underpinnings of ARPA, which may be reimplemented by robotic vehicles. Schuster et al. [66] show an effective pipeline using the Joint Probabilistic Data Association (JPDA) technique to associate measurements with specific tracks, and an interacting multiple model (IMM) Kalman filter to track and predict the motion of each vessel.

With a projection of where other vehicles in the environment are and a prediction of where they are going, the ASV must make decisions on what actions to take to avoid other boats. While this problem is similar to the common 2D+Time obstacle avoidance problem often seen in terrestrial robotics [46], it has the added element that all boats must act under the constraints of COLREGs [35]. Similar to the rules of driving, COLREGs defines which boat has the right-of-way, and what actions the non-right-of-way boat must perform. Unlike driving on the road, boats operate in a significantly more unstructured environment and the rules of COLREGs are specifically written to include human judgement. See [17, 78] for further discussion of the implications of COLREGs for autonomous vehicles.

Particularly successful methods to date have used velocity obstacles with rule-based constraints [45] and used interval programming to optimize over a multi-objective function with both goal point and COLREGs compliant objectives [8, 78]. Both of these methods take advantage of COLREGs natural enforcement in velocity space, working in a reactive manner to adjust a globally computed path with local deviations to avoid other vessels according to COLREGs. While there has been numerous other efforts to incorporate COLREGs within a path planner, see [14, 17], a planner that can account for COLREGs in a multi-vehicle environment with static obstacles has yet to be effectively demonstrated.

9 UAVs

Another recent advance in marine robotics has been the availability of easy to use off-the-shelf unmanned aerial vehicles (UAVs), most notably quadcopters due to their stability and ease of flight. While UAVs have been used for scientific purposes previously (e.g., [10]), recent products such as the DJI Phantom quadcopter [21] have taken UAVs from specialist items requiring engineers to easy to use items affordable within almost any science budget. Products such as the Phantom provide stable easy to fly platforms that can be used to generate high-quality geo-referenced imagery of an area. Unlike custom built AUVs and the ArduPilot driven Jetyak, most of the top quadcopters on the market come with closed source software making them effective tools for marine science but poor platforms for robotics development.

9.1 Large Scale Imaging

One of the most effective and most common uses of small UAVs like the Phantom is to create high resolution large scale maps. By flying the UAV in a lawn-mower pattern above an area, taking images at regular intervals, a set of images is generated with high overlap making it possible to create a single high resolution map, or depending on the area being imaged, a 3D model of the area (e.g., Fig. 9c). Similar to the UAVs themselves, recent commerial software has made both the imaging of areas (e.g., [24]) and the creation of image mosaics and 3D models (e.g., [1]) easy for non-experts. These methods were tested to map penguin super-colonies in the Danger Islands off Antarctica in December 2015, using a DJI Phantom 3 as a platform, Maps Made Easy [24] for automated flight, and Photoscan [1] for image processing. An example image is shown in Fig. 9a and a section of a 3D point cloud is shown in Fig. 9c. The pace at which commerical products in the UAV field are advancing should also be noted – the two primary technical problems experienced during the Danger Islands

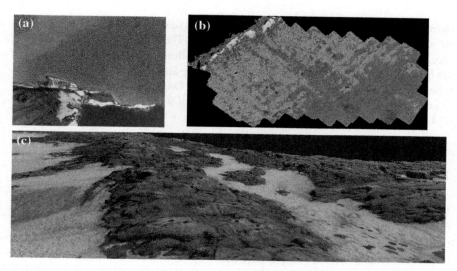

Fig. 9 a Raw image of a penguin colony on Heroina Island. **b** Image mosaic of a penguin colony with penguins detected using morphological processing shown in green. **c** Side view of a 3D point cloud of Brash Island, generated using Photoscan [1]. All data collected in the Danger Islands in Antarctica in December, 2015 using a Phantom 3 Professional quadcopter [21]

expedition; the inability to automatically maintain a height above ground level and the lack of built in collision avoidance, have both been solved in less than a year by more advanced software [62] and more advanced hardware [22].

9.2 Automatic Processing

Along with the proliferation of imaging platforms comes a need for algorithms to automatically process the imagery to get desired information out of it. For example, the Danger Island expedition sought to count, as well as get specific locations for, all nesting penguins on the islands. This requires either a large quantity of manual labor or an automatic "penguin detector." There is a huge amount of work in computer vision on the problem of detecting specific objects in images (e.g., [19, 75]), which focus on taking advantage of patterns of detail in imagery which typically does not exist in aerial imagery [59]. While various techniques have been proposed for obstacle detection in aerial imagery, they have been almost exclusively focused on the problem of human or vehicle detection (e.g., [30, 59]). The objects to be counted during large scale mapping are often of a very different shape and distribution from humans or vehicles, and therefore require different algorithms for their detection.

For the problem of detecting penguins (see Fig. 9a for an example image), we have found that a fairly simple morphological detection scheme is effective. Penguins are detected by thresholding images in gray scale, and selecting the resulting blobs that lie within a defined range for size and aspect ratio. The parameters for thresholding, size, and aspect ratio are tuned via a genetic algorithm [49] on a training data set of six hand labeled 1000 × 1000 images. Results of the algorithm can be seen in Fig. 9b, where each green square marks a detected penguin. On this mosaic, of 1475 penguins manually counted there was a true detection rate of 93.7% and a false positive rate of 4.5%.

10 Conclusion

In this publication we have examined a host of technologies that should allow us to work towards the goal of adaptive missions underwater. We have also looked at the rapidly evolving nature of hobby electronics and how that is affecting the next generation of platforms on the surface and in the air in marine environments. We note that though the platforms are being revolutionized we still need to continue to work on the data processing techniques for the large amounts of data that can now be collected very easily.

References

1. Agisoft. PhotoScan. www.agisoft.com (2015)
2. Akyildiz, I.F., Pompili, D., Melodia, T.: Underwater acoustic sensor networks: research challenges. Ad hoc Netw. **3**(3), 257–279 (2005)
3. ArduPilot. ArduPilot. http://ardupilot.org (2016)
4. ArduPilot. Mission Planner. www.ardupilot.org/planner/index.html (2016)
5. ASV Global. ASV Global. www.asvglobal.com (2016)
6. Bar-Shalom, Y., Daum, F., Huang, J.I.M.: The probabilistic data association filter. IEEE Control Syst. Mag. **29**(6), 82–100 (2009)
7. Barkby, S., Williams, S., Pizarro, O., Jakuba, M.: An efficient approach to bathymetric SLAM. In: 2009 IEEE/RSJ International Conference on Intelligent Robots and Systems, pp. 219–224. IEEE (2009)
8. Benjamin, M.R., Leonard, J.J., Curcio, J.A., Newman, P.M.: A method for protocol-based collision avoidance between autonomous marine surface craft. J. Field Robot. **23**(5), 333–346 (2006)
9. Benjamin, M.R., Schmidt, H., Newman, P.M., Leonard, J.J.: Nested autonomy for unmanned marine vehicles with MOOS-IvP. J. Field Robot. **27**(6), 834–875 (2010)
10. Berni, J.A.J., Zarco-Tejada, P.J., Suárez, L., Fereres, E.: Thermal and narrowband multispectral remote sensing for vegetation monitoring from an unmanned aerial vehicle. IEEE Trans. Geosci. Remote Sens. **47**(3), 722–738 (2009)
11. Bharghavan, V., Demers, A., Shenker, S., Zhang, L.: MACAW: a media access protocol for wireless LAN's. ACM SIGCOMM Comput. Commun. Rev. **24**(4), 212–225 (1994)
12. Blaich, M., Köhler, S., Schuster, M., Schuchhardt, T., Reuter,J., Tietz, T.: Mission integrated collision avoidance for USVs using laser range finder. In: OCEANS 2015-Genova, pp. 1–6. IEEE (2015)
13. Blei, D.M., Ng, A.Y., Jordan, M.I.: Latent dirichlet allocation. J. Mach. Learn. Res. **3**, 993–1022 (2003)

14. Campbell, S., Naeem, W., Irwin, G.W.: A review on improving the autonomy of unmanned surface vehicles through intelligent collision avoidance manoeuvres. Annu. Rev. Control **36**(2), 267–283 (2012)
15. Chitre, M., Shahabudeen, S., Freitag, L., Stojanovic, M.: Recent advances in underwater acoustic communications & networking. In: OCEANS 2008, vol. 2008, pp. 1–10. IEEE (2008)
16. Curcio, J., Leonard, J., Patrikalakis, A.: SCOUT-a low cost autonomous surface platform for research in cooperative autonomy. In: Proceedings of OCEANS 2005 MTS/IEEE, pp. 725–729. IEEE (2005)
17. Curcio, J.A.: Rules of the road for unmanned marine vehicles. Springer Handbook of Ocean Engineering, pp. 517–526. Springer, Berlin (2016)
18. Dabrowski, A., Busch, S., Stelzer, R.: A digital interface for imagery and control of a Navico/Lowrance broadband radar. In: Robotic Sailing, pp. 169–181. Springer (2011)
19. Dalal, N., Triggs, B.: Histograms of oriented gradients for human detection. In: 2005 IEEE Computer Society Conference on Computer Vision and Pattern Recognition (CVPR'05), vol. 1, pp. 886–893. IEEE (2005)
20. Dellaert, F., Kaess, M.: Square Root SAM: simultaneous localization and mapping via square root information smoothing. Int. J. Robot. Res. **25**(12), 1181–1203 (2006)
21. DJI. Phantom 3 professional. www.dji.com/phatom-3-pro (2015)
22. DJI. Phantom 4. www.dji.com/phatom-4 (2016)
23. Donoho, D.L., Vetterli, M., DeVore, R.A.: Data compression and harmonic analysis. IEEE Trans. Inf. Theory **44**(6), 2435–2476 (1998)
24. Drones Made Easy. Map Pilot for DJI. http://www.dronesmadeeasy.com/Articles.asp?ID=254 (2015)
25. Duntley, S.Q.: Light in the sea. JOSA **53**(2), 214–233 (1963)
26. Eastwood, R.L., Freitag, L.E., Catipovic, J.A.: Compression techniques for improving underwater acoustic transmission of images and data. In: OCEANS'96. MTS/IEEE. Prospects for the 21st Century. Conference Proceedings. IEEE (1996)
27. Fairfield, N., Kantor, G., Wettergreen, D.: Real-time SLAM with octree evidence grids for exploration in underwater tunnels. J. Field Robot. **24**(1–2), 03–21 (2007)
28. Freitag, L., Grund, M., Singh, S., Partan, J., Koski, P., Ball, K.: The WHOI micro-modem: an acoustic communications and navigation system for multiple platforms. In: Proceedings of OCEANS 2005 MTS/IEEE, pp. 1086–1092. IEEE (2005)
29. Frey, B.J., Dueck, D.: Clustering by passing messages between data points. Science **315**(5814), 972–976 (2007)
30. Gaszczak, V., Breckon, T.P., Han, J.: Real-time people and vehicle detection from UAV imagery. In: IS&T/SPIE Electronic Imaging, pp. 78780B–78780B. International Society for Optics and Photonics (2011)
31. Girdhar, Y., Dudek, G.: ONSUM: a system for generating online navigation summaries. In: 2010 IEEE/RSJ International Conference on Intelligent Robots and Systems (IROS), pp. 746–751. IEEE (2010)
32. Girdhar, Y., Dudek, G.: Efficient on-line data summarization using extremum summaries. In: 2012 IEEE International Conference on Robotics and Automation (ICRA), pp. 3490–3496. IEEE (2012)
33. Hart, P.E., Stork, D.G., Duda, R.O.: Pattern Classification. Willey, New York (2001)
34. Heikkila, J., Silvén, O.: A four-step camera calibration procedure with implicit image correction. In: 1997 IEEE Computer Society Conference on Computer Vision and Pattern Recognition, 1997. Proceedings, pp. 1106–1112. IEEE (1997)
35. International Maritime Organization: COLREGS - International Regulations for Preventing Collisions at Sea. In: Convention on the International Regulations for Preventing Collisions at Sea, vol. 1972, pp. 1–74 (1972)
36. Itti, L., Baldi, P.F.: Bayesian surprise attracts human attention. In: Advances in Neural Information Processing Systems, pp. 547–554 (2005)
37. Pelekanakis, C., Stojanovic, M., Freitag, L.: High rate acoustic link for underwater video transmission. In: OCEANS 2003. Proceedings, vol. 2, pp. 1091–1097. IEEE (2003)

38. Kaeli, J.W.: Computational strategies for understanding underwater optical image datasets. Ph.D. thesis, Massachusetts Institute of Technology (2013)
39. Kaeli, J.W., Singh, H.: Online data summaries for semantic mapping and anomaly detection with autonomous underwater vehicles. In: OCEANS 2015-Genova, pp. 1–7. IEEE (2015)
40. Kaeli, J.W., Singh, H., Murphy, C., Kunz, C.: Improving color correction for underwater image surveys. In: Proceedings IEEE/MTS Oceans 11, Kona, Hawaii, 19–22 September 2011, pp. 805–810 (2011)
41. Kimball, P., Bailey, J., Das, S., Geyer, R., Harrison, T., Kunz, C., Manganini, K., Mankoff, K., Samuelson, K., Sayre-McCord, T., et al.: The WHOI Jetyak: an autonomous surface vehicle for oceanographic research in shallow or dangerous waters. In: 2014 IEEE/OES Autonomous Underwater Vehicles (AUV), pages 1–7. IEEE (2014)
42. Kimball, P.W., Rock, S.M.: Mapping of translating, rotating icebergs with an autonomous underwater vehicle. IEEE J. Ocean. Eng. **40**(1), 196–208 (2015)
43. Kunz, C.: AUV navigation and mapping in dynamic, unstructured environments. Ph.D. thesis, Massachusetts Institute of Technology (2011)
44. Kunz, C., Singh, H.: Map building fusing acoustic and visual information using autonomous underwater vehicles. J. Field Robot. **30**(5), 763–783 (2013)
45. Kuwata, Y., Wolf, M.T., Zarzhitsky, D., Huntsberger, T.L.: Maritime autonomous navigation with COLREGS, using velocity obstacles. IEEE J. Ocean. Eng. **39**(1), 110–119 (2014)
46. LaValle, S.M.: Planning Algorithms. Cambridge University Press, Cambridge (2006)
47. Lowrance. Lowrance 3G Broadband Radar. http://www.lowrance.com/en-US/Products/Radar/Broadband-3G-Radar-en-us.aspx (2016)
48. Mankoff, K.D., Straneo, F., Cenedese, C., Das, S.B., Richards, C.G., t Singh, H.: Structure and dynamics of a subglacial discharge plume in a Greenlandic Fjord. J. Geophys. Res.: Oceans (2016)
49. Mitchell, M.: An Introduction to Genetic Algorithms. MIT press, Cambridge (1998)
50. Mokai Manufacturing Inc. Mokai. www.mokai.com (2015)
51. Murphy, C.: TOPICS: a modular software architecture for high-latency communication channels. In: OCEANS 2013-San Diego, pp. 1–6. IEEE (2013)
52. Murphy, C.: Data quality monitoring with witness. In: 2014 IEEE/OES Autonomous Underwater Vehicles (AUV), pp. 1–4. IEEE (2014)
53. Murphy, C., Singh, H.: Wavelet compression with set partitioning for low bandwidth telemetry from AUVs. In: Proceedings of the Fifth ACM International Workshop on UnderWater Networks, p. 1. ACM (2010)
54. Murphy, C., Walls, J.M., Schneider, T., Eustice, R.M., Stojanovic, M., Singh, H.: CAPTURE: a communications architecture for progressive transmission via underwater relays with eavesdropping. IEEE J. Ocean. Eng. **39**(1), 120–130 (2014)
55. Murphy, C.. Wang, R.Y., Singh, H.: Seafloor image compression with large tilesize vector quantization. In: 2010 IEEE/OES Autonomous Underwater Vehicles, pp. 1–8. IEEE (2010)
56. Pérez, P., Gangnet, M., Blake, A.: Poisson image editing. In: ACM Transactions on Graphics (TOG), vol. 22, pp. 313–318. ACM (2003)
57. Pizarro, O., Singh, H.: Toward large-area mosaicing for underwater scientific applications. IEEE J. Ocean. Eng. **28**(4), 651–672 (2003)
58. Pizarro, O., Williams, S.B., Colquhoun, J.: Topic-based habitat classification using visual data. In: OCEANS 2009-EUROPE, pp. 1–8. IEEE (2009)
59. Reilly, V., Solmaz, B., Shah, M.: Shadow casting out of plane (SCOOP) candidates for human and vehicle detection in aerial imagery. Int. J. Comput. Vis. **101**(2), 350–366 (2013)
60. Rice, J.: SeaWeb acoustic communication and navigation networks (2005)
61. Roman, C., Singh, H.: A self-consistent bathymetric mapping algorithm. J. Field Robot. **24**(1–2), 23–50 (2007)
62. SPH Engineering. UgCS. www.ugcs.com (2016)
63. Saha, S.: Image compression-from DCT to wavelets: a review. Crossroads **6**(3), 12–21 (2000)
64. Schneider, T., Schmidt, H.: The dynamic compact control language: a compact marshalling scheme for acoustic communications. In: Oceans 2010 IEEE-Sydney, pp. 1–10. IEEE (2010)

65. Schneider, T., Schmidt, H.: Unified command and control for heterogeneous marine sensing networks. J. Field Robot. **27**(6), 876–889 (2010)
66. Schuster, M., Blaich, M., Reuter, J.: Collision avoidance for vessels using a low-cost radar sensor. Int. Fed. Autom. Control **2009**, 9673–9678 (2014)
67. Singh, H., Armstrong, R., Gilbes, F., Eustice, R., Roman, C., Pizarro, O., Torres, J.: Imaging coral I: imaging coral habitats with the SeaBED AUV. Subsurf. Sens. Technol. Appl. **5**(1), 25–42 (2004)
68. Singh, H., Howland, J., Pizarro, O.: Advances in large-area photomosaicking underwater. IEEE J. Ocean. Eng. **29**(3), 872–886 (2004)
69. Singh, H., Roman, C., Pizarro, O., Eustice, R., Can, A.: Towards high-resolution imaging from underwater vehicles. Int. J. Robot. Res. **26**(1), 55–74 (2007)
70. Singh, H., Whitcomb, L., Yoerger, D., Pizarro, O.: Microbathymetric mapping from underwater vehicles in the deep ocean. Comput. Vis. Image Underst. **79**(1), 143–161 (2000)
71. Sivic, J., Russell, B.C., Zisserman, A., Freeman, W.T., Efros, A.A.: Unsupervised discovery of visual object class hierarchies. In: IEEE Conference on Computer Vision and Pattern Recognition, 2008. CVPR 2008, pp. 1–8. IEEE (2008)
72. Stokey, R.P., Freitag, L.E., Grund, M.D.: A compact control language for AUV acoustic communication. In: Europe Oceans 2005, vol. 2, pp. 1133–1137. IEEE (2005)
73. Thrun, S., Burgard, W., Fox, D.: Probabilistic robotics. In: Intelligent Robotics and Autonomous Agents, vol. 2. The MIT Press, Cambridge (2005)
74. Torres-Méndez, L.A., Dudek, G.: Color correction of underwater images for aquatic robot inspection. In: International Workshop on Energy Minimization Methods in Computer Vision and Pattern Recognition, pp. 60–73. Springer, Berlin (2005)
75. Viola, P., Jones, M.J.: Robust real-time face detection. Int. J. Comput. Vis. **57**(2), 137–154 (2004)
76. Walter, M., Hover, F., Leonard, J.: SLAM for ship hull inspection using exactly sparse extended information filters. In: IEEE International Conference on Robotics and Automation (ICRA), 2008, pp. 1463–1470. IEEE (2008)
77. Williams, G., Maksym, T., Wilkinson, J., Kunz, C., Murphy, C., Kimball, P., Singh, H.: Thick and deformed Antarctic sea ice mapped with autonomous underwater vehicles. Nat. Geosci. **8**(1), 61–67 (2015)
78. Woerner, K.: Multi-contact protocol-constrained collision: avoidance for autonomous marine vehicles. Ph.D. thesis, Massachusetts Institute of Technology (2016)
79. Yoerger, D.R., Bradley, A.M., Walden, B.B., Cormier, M.H., Ryan, W.B.F.: High resolution mapping of a fast spreading mid ocean ridge with the Autonomous Benthic Explorer. In: International Symposium on Unmanned Untethered Submersible Technology, pp. 21–31 (1999)
80. Zhang, Z.: Iterative point matching for registration of free-form curves and surfaces. Int. J. Comput. Vis. **13**(2), 119–152 (1994)
81. Zhang, Z.: Microsoft Kinect sensor and its effect. IEEE MultiMed. **19**(2), 4–10 (2012)

New Design Techniques for Globally Convergent Simultaneous Localization and Mapping: Analysis and Implementation

Pedro Lourenço, Bruno Guerreiro, Pedro Batista, Paulo Oliveira and Carlos Silvestre

Abstract This chapter presents an overview of algorithms deeply rooted in a sensor-based approach to the SLAM problem that provide global convergence guarantees and allow for the use of partially observable landmarks. The presented algorithms address the more usual range-and-bearing SLAM problem, either in 2-D using a LiDAR or in 3-D using an RGB-D camera, as well as the range-only and bearing-only SLAM problems. For each of these formulations a nonlinear system is designed, for which state and output transformations are considered together with augmented dynamics, in such a way that the underlying system structure can be regarded as linear time-varying for observability analysis and filter design purposes. This naturally allows for the design of Kalman filters with, at least, globally asymptotically stable error dynamics, for which several experimental and simulated trials are presented to highlight the performance and consistency of the obtained filters.

C. Silvestre on leave from the Instituto Superior Técnico, Universidade de Lisboa, Lisbon, Portugal

---

P. Lourenço · B. Guerreiro · P. Batista
Laboratory for Robotics and Engineering Systems, Institute for Systems and Robotics, Lisbon, Portugal
e-mail: plourenco@isr.tecnico.ulisboa.pt

B. Guerreiro
e-mail: bguerreiro@isr.tecnico.ulisboa.pt

P. Batista (✉)
Department of Electrical and Computer Engineering, Instituto Superior Técnico, Universidade de Lisboa, Lisbon, Portugal
e-mail: pbatista@isr.tecnico.ulisboa.pt

P. Oliveira
Department of Mechanical Engineering, Instituto Superior Técnico, Universidade de Lisboa, Lisbon, Portugal
e-mail: pjcro@isr.tecnico.ulisboa.pt

C. Silvestre
Department of Electrical and Computer Engineering of the Faculty of Science and Technology, University of Macau, Macao, China
e-mail: csilvestre@umac.mo

1 Introduction

When navigating in an unknown environment, the mapping of that environment and the localization within that map have been shown to be dependent on each other, and a more intricate solution than the traditional navigation strategies has to be considered: the simultaneous localization and mapping (SLAM). The research community has devoted significant effort to the study of this problem, for which the seminal works that established the statistical foundations for describing the relationships between landmarks and their correlations include [16, 34, 35]. Further research showed that a full and consistent solution to this problem would require all the vehicle pose and map variables to be considered together, which renders the problem intrinsically nonlinear. Among the many technical solutions emerging from this challenge are the extended Kalman filter (EKF) [11], the use of Rao-Blackwellized particle filters as in FastSLAM [31], or the use of information filters [37]. A detailed survey on most of the used techniques, sensors, applications, and challenges can be found in [9], wherein several other more specialized surveys are referenced.

An important component of most SLAM algorithms is the association between the landmarks measurements and the state landmarks, or when it is necessary to close a loop, being one of the major sources of inconsistency in SLAM algorithms. Several strategies are widely used, such as the simplistic nearest neighbor (NN) validation gating, the joint compatibility branch and bound (JCBB) [33], and the combined constrained data association (CCDA) [3], while other strategies such as those in [2] use sensors that provide unique characteristics of each measured landmark.

The SLAM problem can also be characterized by the fundamental type of measurements available for filtering, usually referred to as landmarks. When the landmark measurements have a lower dimension than the considered mapping space (a single noise-free observation provides only a line or surface as an estimate for the relative position of the landmark), the resulting subproblems are usually divided into range-only SLAM (RO-SLAM) and bearing-only SLAM (BO-SLAM), while the more usual SLAM problem is sometimes referred to as range-and-bearing SLAM (RB-SLAM) to underline the case where all the relative coordinates of measured landmarks are readily available.

A fundamental aspect of the RO-SLAM problem is the absence of association errors, as the information carried by the ranging signals allows the unambiguous association of measurements and the corresponding states at all times, which also enables error-free loop closing. Conversely, the initialization of a RO-SLAM strategy may represent a challenge, and most RO-SLAM solutions rely on some form of initializing procedure in order to create a new landmark in the state, such as the trilateration (in 2-D) with ranges from different instants [1]. As the RO-SLAM problem bears resemblance to the sensor networks problem, in the sense that an agent receives signals from a network of sensors, the two ideas have been used in conjunction in works such as [14, 26].

The BO-SLAM problem is more challenging than the RO-SLAM, because an observation of the former corresponds to an unbounded set. Besides triangulation, more advanced probabilistic approaches can be used to address this issue, such as a sum of Gaussians [24], deterring the initialization until the obtained estimate is well approximated by a Gaussian [4], or using multiple hypothesis [36]. Nowadays, BO-SLAM is most associated with monocular vision [21], as there has been an intense research effort in this particular application. One of the most relevant developments is presented in [12], being the first real-time SLAM algorithm using only a camera as data source. Other interesting approaches include [15] which introduces a closed-form pose-chain optimization algorithm that uses sparse graphs as well as appearance-based loop detection, and ORB-SLAM [32], that uses ORB features, which are rotation invariant and have faster extraction than SURF features [7].

In general, the SLAM problem has a nonlinear nature which can be tackled using EKF-based solutions, as well as other similar filters that usually imply the linearization of the dynamics or measurement models, resulting in lack of consistency and convergence guarantees [20, 22]. To address these issues, several authors have analyzed the unobservable space of the error system obtained after linearization, showing that it has smaller dimensionality than that of the underlying nonlinear error system [19]. This yields erroneous gains of information, and a possible solution proposed by the same authors is to select the linearization points of the EKF that ensure the same dimensions between the mentioned unobservable spaces. While focusing on automatic calibration, [23] addresses the observability properties of nonlinear SLAM based on differential geometry, as it is a necessary (but not sufficient) condition for the convergence of any filtering algorithm. Another approach to this issue is to use the so-called robocentric or sensor-based approach, as firstly proposed in the robocentric map joining algorithm [10]. This algorithm can improve the consistency of the regular EKF, yet, as it still considers the estimation of the (unobservable) incremental pose in the filter state, it cannot provide guarantees of convergence. Other algorithms that provide some guarantees of convergence include methods that usually assume that the linearized system matrices are evaluated at the ideal values of the state variables [13, 20], or other that resort to stochastic stability concepts assuming access to both linear and angular velocity measurements [8]. Nevertheless, a formal theoretical result on global convergence for EKF-based SLAM algorithms is still absent from the literature.

This chapter presents an overview of algorithms that are rooted in a sensor-based approach to the SLAM problem that can be used in aerial robots. The usual SLAM approach requires a single filter to maintain estimates of the map and vehicle pose along with the respective covariances and cross-covariances. However, it is possible to use an alternate formulation that uses a Kalman filter (KF) that explores the linear time-varying (LTV) nature of the sensor-based SLAM system and analyzing its observability. This formulation is deeply rooted in the theory of sensor-based control, exploring the fact that vehicle-fixed sensors provide measurements that in this approach do not need to be transformed to an Earth-fixed frame [38]. The first of these purely sensor-based SLAM filters was proposed for the two-dimensional RB-SLAM problem [18], suppressing pose representation in the state and therefore avoiding singularities and nonlinearities, subsequently extended for 3-D in [29] using an RGB-D camera. These works have provided necessary and sufficient conditions for observability of the nonlinear error system, for which a KF can be designed that yields global asymptotic stability (GAS).

Regarding the less straightforward formulations of the SLAM problem, the work presented in [27] introduces a novel RO-SLAM algorithm that eliminates the landmark initialization problem through the establishment of global convergence results. As in the previous algorithms, the proposed 3-D sensor-based formulation avoids the representation of the pose of the vehicle in the state, allowing the direct use of odometry-like information that is usually expressed in body-fixed coordinates. Finally, [25] proposes a 3-D BO-SLAM algorithm with exponentially fast global convergence and allows for undelayed initialization at any depth. Building on the previous approaches, this algorithm uses a state augmentation and an output transformation that lead to the design of an LTV system whose observability conditions are given in a constructive analysis with clear physical insight. These two solutions are influenced by the source localization algorithms proposed in [5, 6], as a similar state augmentation is used to achieve the global convergence results.

Building on the results mentioned above, the main contributions of this chapter include: (i) the consolidation and definition of a class of sensor-based SLAM problems such as 2-D and 3-D RB-SLAM, RO-SLAM, and BO-SLAM; (ii) a collection of physically intuitive and constructive observability results; (iii) the filter implementation details that ensure a global asymptotic stability of the respective error dynamics; (iv) an alternative method to obtain the Earth-fixed quantities from the results of the sensor-based filters; (v) a collection of experimental and simulation results that validate and illustrate the main properties and performance of the proposed filters.

The remaining of the chapter is organized as follows. Section 2 introduces the sensor-based SLAM problems while their observability analysis is detailed in Sect. 3. The sensor-based filter implementation details are provided in Sect. 4 along with the Earth-fixed trajectory and map algorithm. Finally, the main results stemming from the proposed algorithms are depicted in Sect. 5 and some concluding remarks are given in Sect. 6.

2 Sensor-Based SLAM

Following the previous discussion, this section details the design of dynamical systems as part of the sensor-based simultaneous localization and mapping filters using only one source of external environment perception, capable of either measuring relative positions, ranges, or bearings, apart from vehicle motion information. Let $\{B\}$ denote the body-fixed frame and $\{E\}$ denote the inertial/Earth-fixed frame, whereas $\left(\mathbf{R}(t), {}^E\mathbf{p}(t)\right)$ represents the transformation from $\{B\}$ to $\{E\}$ and, therefore, the pose of the vehicle. The attitude is given by the rotation matrix $\mathbf{R}(t) \in \mathrm{SO}(n)$ and the position is given by ${}^E\mathbf{p}(t) \in \mathbb{R}^n$, with $n = 2, 3$. The former satisfies $\dot{\mathbf{R}}(t) = \mathbf{R}(t)\mathbf{S}(\boldsymbol{\omega}(t))$, where $\boldsymbol{\omega}(t) \in \mathbb{R}^{n_\omega}, n_\omega = 1, 3$, is the angular velocity of the vehicle expressed in body-fixed coordinates. The environment is characterized by point landmarks that may be

Fig. 1 For RB-SLAM, sensors measure the position of a landmark relative to the vehicle, for RO-SLAM, the distance to a landmark, and for BO-SLAM, the relative direction to a landmark

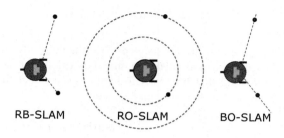

naturally extracted or artificially placed. These N landmarks constitute the set $\mathcal{M}$ and are denoted by ${}^E\mathbf{p}_i(t) \in \mathbb{R}^n$ or $\mathbf{p}_i(t) \in \mathbb{R}^n$, respectively describing the landmark location in frame $\{E\}$ or in frame $\{B\}$. As in the Earth-fixed frame landmarks are assumed static, considering the motion of the landmark in $\{B\}$, it is possible to write

$$\dot{\mathbf{p}}_i(t) = -\mathbf{S}(\boldsymbol{\omega}(t))\mathbf{p}_i(t) - \mathbf{v}(t) \tag{1}$$

where $\mathbf{S}(.)$ is a skew-symmetric matrix that encodes the cross product for $n_\omega = 3$, $\mathbf{S}(\omega) = \begin{bmatrix} 0 & -\omega \\ \omega & 0 \end{bmatrix}$ for $n_\omega = 1$, and $\mathbf{v}(t) \in \mathbb{R}^n$ is the linear velocity of the vehicle expressed in $\{B\}$. Typically, $\boldsymbol{\omega}(t)$ and $\mathbf{v}(t)$ are both known inputs and need not be estimated. However, this situation may change, depending on the information provided by the system outputs.

Consider now that the N landmarks are divided in two different sets depending on their visibility status: $\mathcal{M}_O := \{1, \ldots, N_O\}$ containing the N_O observed or visible landmarks and $\mathcal{M}_U := \{N_O + 1, \ldots, N\}$ containing the N_U unobserved, or non-visible, ones. Landmarks belonging to $\mathcal{M}_O$ will have some kind of system output associated, which leads to the definition of

$$\mathbf{y}_i(t) = \mathbf{f}(\mathbf{p}_i(t)), \quad i \in \mathcal{M}_O \tag{2}$$

where $\mathbf{y}_i(t)$ can be equal to $\mathbf{p}_i(t)$, $\|\mathbf{p}_i(t)\|$, or $\frac{\mathbf{p}_i(t)}{\|\mathbf{p}_i(t)\|}$ according to the version of sensor-based SLAM to be designed (see Fig. 1). Combining this information it is now possible to write the generic nonlinear system

$$\begin{cases} \dot{\mathbf{p}}_i(t) = -\mathbf{S}(\boldsymbol{\omega}(t))\mathbf{p}_i(t) - \mathbf{v}(t) & i \in \mathcal{M} \\ \mathbf{y}_j(t) = \mathbf{f}(\mathbf{p}_j(t)) & j \in \mathcal{M}_O \end{cases} \tag{3}$$

If $\boldsymbol{\omega}(t)$ is an input and $\mathbf{v}(t)$ is either added as a state with constant dynamics or kept as an input, then the first equation in (3) can be considered linear for observability analysis purposes. The main problem rests with the output equation that may be nonlinear. In that case, further action is necessary to obtain a linear-like system: state augmentation and/or output transformation. This subject will be addressed in the sequel.

2.1 Range-and-Bearing SLAM

Range-and-bearing sensors provide the most information possible in terms of point-based maps, and, therefore, may be exploited in order to estimate more quantities. Following that line of reasoning, consider that the measured angular velocity is corrupted with a constant bias, i.e.,

$$\boldsymbol{\omega}_m(t) = \boldsymbol{\omega}(t) + \mathbf{b}_\omega(t). \tag{4}$$

Further consider that the linear velocity is not directly measured, and, as such, needs to be estimated. Then, the generic nonlinear system (3) becomes

$$\begin{cases} \dot{\mathbf{p}}_i(t) = -\mathbf{S}(\boldsymbol{\omega}_\mathbf{m}(t))\,\mathbf{p}_i(t) - \mathbf{g}(\mathbf{p}_i(t), \mathbf{b}_\omega(t)) - \mathbf{v}(t) & i \in \mathcal{M} \\ \dot{\mathbf{v}}(t) = \mathbf{0} \\ \dot{\mathbf{b}}_\omega(t) = \mathbf{0} \\ \mathbf{y}_j(t) = \mathbf{p}_j(t) & j \in \mathcal{M}_O \end{cases}, \tag{5}$$

where

$$\mathbf{g}(\mathbf{p}_i(t), \mathbf{b}_\omega(t)) := \begin{cases} -\mathbf{S}(1)\,\mathbf{p}_i(t) b_\omega(t), & n = 2 \\ \mathbf{S}(\mathbf{p}_i(t))\,\mathbf{b}_\omega(t), & n = 3 \end{cases}. \tag{6}$$

For the visible landmarks, the nonlinear term $\mathbf{g}(\mathbf{p}_i(t), \mathbf{b}_\omega(t))$ can be written as $\mathbf{g}(\mathbf{y}_i(t), \mathbf{b}_\omega(t))$, which, even though it is still nonlinear it may be considered as linear time-varying for observability purposes as $\mathbf{p}_i(t)$ is available. For the remaining landmarks, this term is still nonlinear.

It should be mentioned that each 2-D landmark could be accompanied by additional features in the form of directional landmarks. That specific case is addressed in [18], where more information on both the system and its observability analysis can be found.

2.2 Range-only SLAM

In the case of range-only external perception, it is necessary to have some measure of the linear movement, and therefore the linear velocity is here also used as an output (it must be measured). Considering the measurement model, $y_i(t) = \|\mathbf{p}_i(t)\|$ for all $i \in \mathcal{M}_O$, which in this case is nonlinear, a state augmentation is proposed, yielding the new state

$$\begin{cases} \mathbf{x}_{L_i}(t) := \mathbf{p}_i(t) \\ \mathbf{x}_V(t) := \mathbf{v}(t) \\ x_{R_i}(t) := \|\mathbf{x}_{L_i}(t)\| \end{cases}. \tag{7}$$

It is a simple matter of computation to then write the full system dynamics for this extended state, resulting in the new system

$$\begin{cases} \dot{\mathbf{x}}_{L_i}(t) = -\mathbf{S}(\boldsymbol{\omega}(t))\,\mathbf{x}_{L_i}(t) - \mathbf{x}_V(t) & i \in \mathcal{M} \\ \dot{\mathbf{x}}_V(t) = \mathbf{0} & \\ \dot{x}_{R_i}(t) = -\dfrac{\mathbf{y}_V^T(t)}{x_{R_i}(t)}\mathbf{x}_{L_i}(t) & i \in \mathcal{M}. \\ \mathbf{y}_v(t) = \mathbf{x}_V(t) & \\ \mathbf{y}_j(t) = \mathbf{p}_j(t) & j \in \mathcal{M}_O \end{cases} \qquad (8)$$

As in range-and-bearing SLAM, there are still nonlinear terms in the dynamics that need to be taken care of. In this case, the term $\frac{\mathbf{y}_V^T(t)}{x_{R_i}}\mathbf{x}_{L_i}(t)$ can be rewritten for the visible landmarks, as the denominator is one of the system outputs, i.e., $\frac{\mathbf{y}_V^T(t)}{x_{R_i}(t)}\mathbf{x}_{L_i}(t) = \frac{\mathbf{y}_V^T(t)}{y_i(t)}\mathbf{x}_{L_i}(t)$ for all $i \in \mathcal{M}_O$. With this change, the system pertaining to this subset of the landmarks is now linear time-varying for observability analysis purposes, as the dynamics only depend on known system inputs and outputs.

2.3 Bearing-only SLAM

As in the previous situation, bearing-only measurements require the linear velocity to be measured. In this case, it will be introduced as an input, accompanying the angular velocity as the inputs of the dynamical system. As for the measurement model, the output of the nonlinear system is now $\mathbf{y}_i(t) = \frac{\mathbf{p}_i(t)}{\|\mathbf{p}_i(t)\|} := \mathbf{b}_i(t)$. As this output is nonlinear, the simple output transformation $\mathbf{p}_i(t) - \mathbf{b}_i(t)\|\mathbf{p}_i(t)\| = \mathbf{0}$ is considered together with the state augmentation,

$$\begin{cases} \mathbf{x}_{L_i}(t) := \mathbf{p}_i(t) \\ x_{R_i}(t) := \|\mathbf{x}_{L_i}(t)\| \end{cases}, \qquad (9)$$

making it possible to avoid the nonlinearity in the output. This yields the new system

$$\begin{cases} \dot{\mathbf{x}}_{L_i}(t) = -\mathbf{S}(\boldsymbol{\omega}(t))\,\mathbf{x}_{L_i}(t) - \mathbf{v}(t) & i \in \mathcal{M} \\ \dot{x}_{R_i}(t) = -\dfrac{\mathbf{x}_{L_i}^T(t)}{x_{R_i}(t)}\mathbf{v}(t) & i \in \mathcal{M} \\ \mathbf{0} = \mathbf{x}_{L_j} - \mathbf{b}_j(t) x_{R_j}(t) & j \in \mathcal{M}_O \end{cases} \qquad (10)$$

which is still nonlinear in the dynamics. Notice that here $\boldsymbol{\omega}(t)$ and $\mathbf{v}(t)$ are both inputs, and that $\mathbf{b}_i(t)$ is a measurement. With this in mind, it is possible to replace the nonlinear term in the dynamics of the state $x_{R_i}(t)$ with information that is measured, thus yielding a linear time-varying structure to the system part that relates to the visible landmarks. This is performed by noting that $\frac{\mathbf{x}_{L_i}^T(t)}{\mathbf{x}_{R_i}(t)}\mathbf{v}(t) = \mathbf{b}_i^T(t)\mathbf{v}(t)$ for all $i \in \mathcal{M}_O$.

With these manipulations, it is possible to obtain dynamical systems for sensor-based SLAM that, when looking only at the visible landmarks, resemble linear time-varying systems that mimic the original nonlinear systems. The following section deals with their observability. It must be noted, however, that the state augmentations are not enforced in any way, i.e., the relation $x_{R_i}(t) = \|\mathbf{x}_L(t)\|$ is not explicitly used in the dynamics.

3 Observability

The systems designed in the previous section resemble LTV systems. However, due to the presence of non-visible landmarks in all of them, there are still nonlinear terms that prevent the use of linear tools for analysis. The quantities associated with the non-visible landmarks are, by definition, not observable. Therefore, it is reasonable to discard them from the state when analyzing observability (see [18, 27, 29] for previous successful applications of this approach by the authors). This yields a reduced system of the form

$$\begin{cases} \dot{\mathbf{x}}(t) = \mathbf{A}(t, \mathbf{y}(t), \mathbf{u}(t))\mathbf{x}(t) + \mathbf{B}(t, \mathbf{y}(t), \mathbf{u}(t))\mathbf{u}(t) \\ \mathbf{y}(t) = \mathbf{C}(t, \mathbf{y}(t), \mathbf{u}(t))\mathbf{x}(t) \end{cases}, \quad (11)$$

whose dynamics, even though still nonlinear, do not depend on the state itself but on inputs and outputs. As both the inputs and outputs are known functions of time, then, for observability analysis and observer design purposes, the system (11) can in fact be considered as linear time-varying.

Having reached this stage in the system design and analysis, it is possible to use [5, Lemma 1] to ascertain whether and in what conditions the system is observable by studying its observability Gramian. Figure 2 summarizes the whole process presented in the previous section and further explained here. After analyzing the observability of the LTV system, the next step is to study the observability of the original nonlinear system, starting with a comparison between the state of the LTV system and that of the nonlinear system. This includes investigating the conditions in which the state augmentation relations become naturally imposed by the dynamics. It is in that case that the states of the two systems become equivalent, which validates the augmentation approach in the sense that a state observer with uniformly globally asymptotically/exponentially stable error dynamics for the LTV system is also a state observer for the underlying nonlinear system, and the estimation error converges asymptotically/exponentially fast for all initial conditions.

The final step before proceeding to observer design is to study the uniform complete observability (UCO) of the LTV system. This is a stronger form of observability that is necessary to guarantee the global asymptotical/exponential stability of the

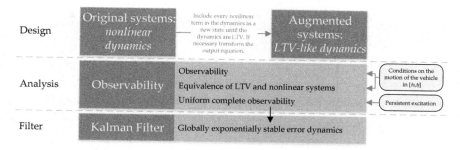

Fig. 2 Schematic description of the process of designing a globally convergent sensor-based SLAM filter

Kalman filter. For this purpose, uniform bounds on the observability Gramian calculated on a moving time interval are investigated. Due to this uniformity, the resulting conditions are more demanding, which can be regarded as persistent excitation-like conditions.

In range-and-bearing SLAM, the conditions for observability depend on the number of landmarks observed, due to the fact that the nonmeasured quantities (linear velocity and rate-gyro bias) affect all the landmarks. On the other hand, the conditions for the observability of both bearing-only and range-only SLAM do not depend on the number of landmarks, as they are independent of each other. There is another important distinction between the two classes of problems pertaining the quantity of information made available with each measurement. In range-and-bearing SLAM one single measurement provides all the necessary information to estimate the position of a landmark, even though several landmarks are needed for immediate full state recovery, whereas in range-only and bearing-only SLAM measurements from several viewpoints have to be acquired to allow for landmark estimation. This is also the case for range-and-bearing SLAM when the number of available landmarks is not enough to guarantee observability without motion.

The remainder of this section summarizes the theoretical results corresponding to each version of SLAM designed previously and the associated necessary and sufficient conditions.

2-D range-and-bearing SLAM [18, Theorems 2–4 and 6]

Observability and Equivalence	Two landmarks are visible or one landmark is visible and there is an instant when its derivative is nonzero.
UCO	Two landmarks are visible or one landmark is visible and its derivative is sufficiently away from zero, uniformly in time.

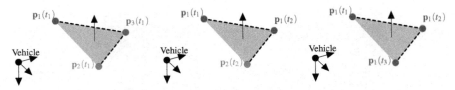

Fig. 3 Geometrical interpretation of the observability conditions for range-and-bearing SLAM in 3-D

3-D range-and-bearing SLAM [29, Theorems 1-4]

Observability and Equivalence — Three landmarks form a plane in one observation, two observations of two landmarks form a plane or three observations of one landmark form a plane (see Fig. 3).

UCO — The vectors defined by the three landmarks that form a plane (regardless of the observation moment) are sufficiently away from collinearity, uniformly in time.

3-D Range-only SLAM [27, Theorems 1-4]

Observability and Equivalence — The linear velocity in three observation moments spans $\mathbb{R}^3$ (see left side of Fig. 4 for an example of the bidimensional case).

UCO — The vectors defined by the three velocity measurements are sufficiently away from co-planarity so that the spanned space does not degenerate in time.

3-D Bearing-only SLAM [25, Theorems 1-3]

Observability and Equivalence — Two different absolute bearings to one landmark are measured (see right side of Fig. 4).

UCO — The variation in the bearing measurement is sufficiently away from zero to not degenerate in time.

These results allow for the design and implementation of a Kalman filter with globally asymptotically/exponentially stable error dynamics.

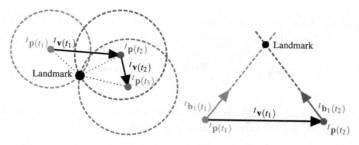

Fig. 4 Examples of trilateration (*left*) and triangulation (*right*) for positioning a landmark in 2-D. These demonstrate the geometric interpretation of the derived observability conditions, i.e., the importance of proper vehicle motion for univocal determination of the coordinates of a landmark

4 Filter Design and Implementation

Considering a discrete time implementation of the filter, let T_s denote the sampling period of the synchronized array of sensors used in each solution, noting that a multirate implementation can be devised. The system is discretized using the forward Euler discretization, with special care when there is a rotation of a landmark from one instant to the following, where it is considered that the angular velocity is constant over each sampling interval, i.e., $\mathbf{R}_{k+1}^T \mathbf{R}_k = \exp(-\mathbf{S}(\boldsymbol{\omega}_k) T_s)$. Considering additive disturbances, the generic discretized system is given by

$$\begin{cases} \mathbf{x}_{F_{k+1}} = \mathbf{F}_{F_k}\mathbf{x}_{F_k} + \mathbf{G}_{F_k}\mathbf{v}_k + \boldsymbol{\xi}_k \\ \mathbf{y}_{k+1} = \mathbf{H}_{F_{k+1}}\mathbf{x}_{F_{k+1}} + \boldsymbol{\theta}_{k+1} \end{cases}, \qquad (12)$$

where the dynamics matrices have the structure

$$\mathbf{F}_{F_k} = \begin{bmatrix} \mathbf{F}_k & \mathbf{0}_{n_O \times n_U} \\ \mathbf{F}_{UO_k} & \mathbf{F}_{U_k} \end{bmatrix}, \ \mathbf{G}_{F_k} = \begin{bmatrix} T_s \mathbf{B}_k \\ T_s \mathbf{B}_{U_k} \end{bmatrix}, \ \mathbf{H}_{F_k} = \begin{bmatrix} \mathbf{C}_k & \mathbf{0}_{n_O \times n_U} \end{bmatrix}, \qquad (13)$$

and where $\mathbf{F}_k$ is the discretized version of matrix $\mathbf{A}$ in (11), which accounts for the observable part of the system dynamics, whereas matrices with subscripts $(.)_U$ and $(.)_{UO}$ denote the unobservable states and cross terms, respectively. Also, the vectors $\boldsymbol{\xi}_k$ and $\boldsymbol{\theta}_k$ represent the model disturbance and measurement noise, respectively, assumed to be zero-mean discrete white Gaussian noise with covariances $\boldsymbol{\Xi}_k$ and $\boldsymbol{\Theta}_k$. These will depend on the actual formulation of SLAM in question. However, in all the formulations described in this chapter, the dynamics depends on the actual inputs and outputs, which means that the noise characterization is not exact. As a general rule of thumb, these noise parameters can be calibrated a priori with a Monte Carlo analysis and with actual measurements to better cope with this issue.

In particular, recalling the three presented SLAM problems, both the RB-SLAM and BO-SLAM have similar structure for the system matrix, defined as

$$\hat{\mathbf{F}}_{F_k} = \begin{bmatrix} \mathbf{F}_{L_k} & \mathbf{0}_{n_L \times (n_V + n_R)} \\ \mathbf{0}_{(n_V + n_R) \times n_L} & \mathbf{I}_{(n_V + n_R)} \end{bmatrix}, \quad (14)$$

although the state vectors have diverse variables and dimensions. On the other hand, the RO-SLAM has a more intricate structure, defined as

$$\mathbf{F}_k = \begin{bmatrix} \mathbf{F}_{L_k} & T_s \mathbf{A}_{LV} & \mathbf{0}_{n_L \times n_R} \\ \mathbf{0}_{n_V \times n_L} & \mathbf{I}_3 & \mathbf{0}_{n_L \times n_R} \\ T_s \mathbf{A}_{RL_k} & \mathbf{0}_{n_R \times n_V} & \mathbf{I}_{n_R} \end{bmatrix}, \quad (15)$$

with $\mathbf{F}_{L_k} = \text{diag}\left(\mathbf{R}_{k+1}^T \mathbf{R}_k, \ldots, \mathbf{R}_{k+1}^T \mathbf{R}_k\right)$, see [25, 27] for the remaining matrices.

From these discrete LTV systems, the filter prediction and update steps are computed using the standard equations of the Kalman filter for LTV systems [17], with the detail that the non-visible landmarks must be propagated in open loop. Nevertheless, particularly in RB-SLAM and BO-SLAM, prior to the update step it might be necessary to associate the landmark measurements with the state landmarks, either for a simple update step or for a more intricate loop closing procedure.

To complement the sensor-based filter, the authors proposed a strategy to obtain estimates of the pose of the vehicle and the Earth-fixed landmark map, with uncertainty characterization, denoted as Earth-fixed trajectory and map (ETM) estimation algorithm. Considering the relation between landmarks expressed in the two working frames and noting that $^E\mathbf{p}_{i_k}$ is constant, it is possible to write the error function

$$^E\mathbf{e}_{i_k} = {}^E\hat{\mathbf{p}}_{i_{k-1}} - \hat{\mathbf{R}}_k \hat{\mathbf{p}}_{i_k} - {}^E\hat{\mathbf{p}}_k, \quad (16)$$

which can be minimized as in the optimization problem presented in [30, Sect. 3]. This yields the optimal rotation and translation given a map expressed in the Earth-fixed frame and in the body-fixed frame and the combined uncertainty of (16). With this information, the new Earth-fixed map can be computed using

$$^E\hat{\mathbf{p}}_{i_k} = \hat{\mathbf{R}}_k \hat{\mathbf{p}}_{i_k} + {}^E\hat{\mathbf{p}}_k. \quad (17)$$

An important step in this algorithm is the initialization of the Earth-fixed map, which can be computed directly from the sensor-based map in the first instant by assuming that the transformation between Earth-fixed and sensor frames is known at time k_0, a traditional assumption in most SLAM strategies.

The pose estimates provided by this strategy are accompanied with uncertainty characterizations, following a perturbation theory approach as described in [30]. The same reasoning can be employed to obtain estimates with uncertainty description for the Earth-fixed map (see [28]).

5 Practical Examples

This section aims to provide several examples of practical implementations of the sensor-based algorithms detailed along this chapter. In particular, the results presented in this section are obtained using four sensor-based SLAM algorithms: (i) 2-D Range-and-Bearing SLAM, (ii) 3-D Range-and-Bearing SLAM, (iii) Range-Only SLAM, and (iv) Bearing-Only SLAM. It should also be stressed that each of these experiments was originally designed as a proof of concept, and as such, alternative sensors or processes for obtaining measurements can be employed. Table 1 summarizes the typical sensors used for each one of the SLAM variants discussed here, pinpointing those employed in the first three examples shown below (further details can be found in [18, 25, 27, 29] and references therein). The BO-SLAM example consists of simulation results.

The Earth-fixed estimates of the maps and vehicle trajectories are also provided for the range-and-bearing algorithms, which are obviously dependent on the performance of the ETM algorithm, affected by the nonlinearity intrinsic to the problem of translating and rotating a map arbitrarily between coordinate frames, also found in EKF-based SLAM algorithms. As the sensor-based SLAM filter does not depend on the ETM algorithm, it is argued that using this separate approach it may be possible to obtain a less uncertain Earth-fixed trajectory and landmark map. As such, all the landmark association, loop closing, control, and decision procedures can be made in the sensor-based frame, minimizing the effects of nonlinearities in the consistency of the filter estimates.

5.1 Range-and-Bearing SLAM

This subsection presents experimental results from two different implementations of the sensor-based range-and-bearing SLAM filter coupled with the ETM algorithm, one in two dimensions using a LiDAR [18] and other in three dimensions [29] using an RGB-D camera.

Table 1 Measurements and their respective sensors

Quantities	Sensors
Landmark position	LiDAR (i) / RGB-D camera (ii) / Stereo or trinocular camera
Landmark range	Radio/acoustic transceivers (iii)
Landmark bearing	Radio/acoustic transceivers / Single camera
Linear velocity	Odometry / Optical flow (iii)
Angular velocity	IMU (i-iii)

When analyzing the convergence properties of any navigation filter, one of the main goals is to observe a decreasing uncertainty in all variables. This can be seen in Fig. 9 for the 3-D RB-SLAM filter, where the uncertainty of all the vehicle related variables decreases over time, whereas the uncertainty of each landmark decreases whenever visible and increases otherwise, as shown in Fig. 5 for the 2-D RB-SLAM filter. In addition, both the sensor-based map and the result of the ETM algorithm are presented in Fig. 6, featuring the final results of the 2-D RB-SLAM experimental trials. It can be seen that both the sensor-based and ETM maps are consistent, noting also the small sensor-based uncertainty for the landmarks that were recently visible, and the large uncertainty for the landmarks that are not visible for a long time, as the vehicle progresses along environment.

As can be inferred from Fig. 5, there are several loop closure procedures during the trials, adding relevant information about the consistency of the proposed algorithms.

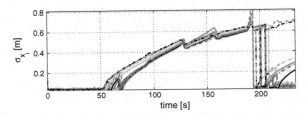

Fig. 5 2-D RB-SLAM: Uncertainty convergence in the sensor-based filter. STD of first 15 landmark positions. ©2013 IEEE. Reprinted, with permission, from [18]

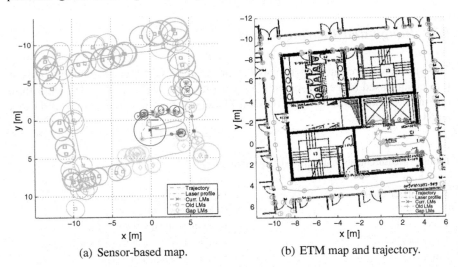

(a) Sensor-based map. (b) ETM map and trajectory.

Fig. 6 2-D RB-SLAM: Map of environment in the sensor and Earth frame. This figure shows the current laser profile (in *gray*), the current/old/older landmarks in *magenta/yellow/light blue*, along with their 95% confidence bounds. ©2013 IEEE. Reprinted, with permission, from [18]

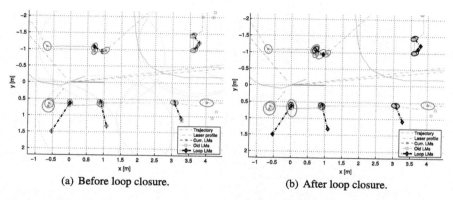

(a) Before loop closure. (b) After loop closure.

Fig. 7 2-D RB-SLAM: illustration of a loop closure. ©2013 IEEE. Reprinted, with permission, from [18]

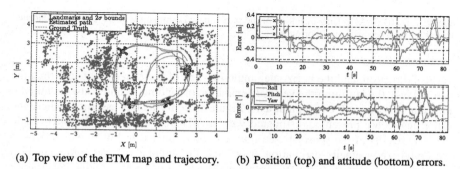

(a) Top view of the ETM map and trajectory. (b) Position (top) and attitude (bottom) errors.

Fig. 8 3-D RB-SLAM: Earth-fixed estimates with ground truth. Reprinted from [29] with permission with permission of Springer

In one of the occasions, the moments just before and right after the loop closure are captured in Fig. 7, where a detailed version of the map of Fig. 6a at the relevant time instant is presented. The landmark associations between the current and older landmarks are shown in solid black and the fused landmarks positions and uncertainty bounds obtained after the loop closure are also depicted in solid black.

In Fig. 8a, a top view of the Earth-fixed map is shown along with the estimated trajectory (solid line) and the ground truth trajectory (dashed line) obtained from a *VICON* motion tracking system. The colored squares, that coincide by construction, and triangles indicate the start and end of the run, respectively. The ellipses are the 2-D projection of the 2σ uncertainty ellipsoids. The small quadrotors represent the pose of the vehicle in several instants, both with ground truth (dashed red) and SLAM estimates (solid green).

The evolution of the position estimation error is depicted in the top of Fig. 8b. It is noticeable that, even though the horizontal estimates are quite accurate, the vertical ones are worse. These results are not unexpected since there was no motion in the

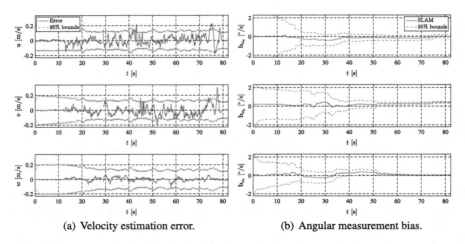

Fig. 9 3-D RB-SLAM: Time evolution of the sensor-based estimates with 2σ bounds. Reprinted from [29] with permission of Springer

vertical direction, and the horizontal angle-of-view of the *Kinect* is greater than the vertical one, thus limiting the vertical separation of landmarks and consequently the information extractable from that axis. Figure 8b confirms this assertion, where the Euler angles are presented, estimated with small error, except for the pitch angle (θ). The results of the sensor-based filter can be evaluated in Fig. 9, which depicts the body-fixed velocity estimation errors and the angular rate measurement bias estimates. The velocity estimation error is depicted alongside the 95% uncertainty bounds, and, even though the velocity is modeled to be constant, it follows the velocity accurately (standard deviation of 0.02 m/s in the vertical axis and 0.05 m/s in the horizontal ones). Furthermore, its uncertainty converges while generally maintaining the consistency throughout the run. The measurement bias on the right is obviously presented without ground truth, but its uncertainty can be seen to converge, confirming the results of Sect. 3.

5.2 Range-only SLAM

In opposition to the range-and-bearing and bearing-only examples presented here, the landmarks used in the range-only experiments are artificially placed beacons in the environment. Therefore, and taking into account the observability requirements, the trajectory of the vehicle was intended to maximize the exposure to each of the beacons, as well as to provide sufficient excitation to the filter.

As explained in Sect. 1, the initialization of the landmarks is one of the more challenging issues in RO-SLAM procedures. In this work, however, the global convergence results imply that this issue is solved as whichever the initial guess the

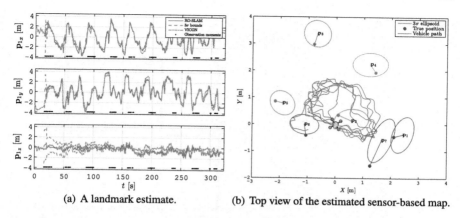

Fig. 10 RO-SLAM: One estimated landmark through time against the ground truth and the observation moments (*left*). The *top* view of the sensor-based map at the end of the run. Both include 3σ uncertainty bounds. Reprinted from [27] with permission from Elsevier

filter will converge. Figure 10a depicts exactly this, showing the estimated position (solid blue) with 3σ uncertainty bounds of one landmark against the ground truth provided by *VICON* (dashed red), and it can be seen that the convergence is very fast in the horizontal plane. Moreover, after converging, the estimates are very close to ground truth. However, in the vertical axis, the estimation is much worse, and the convergence is also slower, which is due to the less rich trajectory in that axis. The optical flow procedure employed is somewhat noisy, and as its measurements of the linear velocity are directly used in the dynamics matrix as if they were the true value, the noise can make that direction appear observable, even if the information is sparse.

Finally, an example of the estimated map in the body-fixed frame is presented in Fig. 10b. The top view of the sensor-based map is shown along with the true landmark positions and the vehicle path rotated and translated to the body-fixed frame. The colored ellipses represent orthogonal cross sections of 3σ uncertainty ellipsoids, i.e., the estimation uncertainty, and the small circles mark the true landmark positions. These experiments show the good performance of the proposed algorithm in realistic conditions, especially in the horizontal variables. The filter has some problems in the vertical coordinates due to the less rich velocity profile and noisy optical flow measurements, although with a proper trajectory the algorithm was shown to behave well [27]. Therefore, these experiments underpin the need for appropriate trajectories.

5.3 Bearing-only SLAM

This example of a sensor-based bearing-only application is performed on a simulated scenario with known association between measurements and landmarks. Figure 11

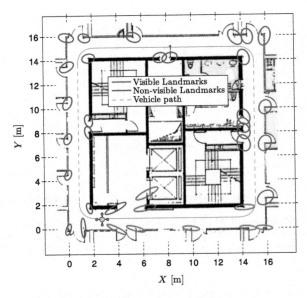

Fig. 11 BO-SLAM: The estimated map at the end of the experiment, along with the true path. ©2015 IEEE. Reprinted, with permission, from [25]

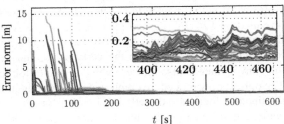

Fig. 12 BO-SLAM: The evolution of the norm of the estimation error for all the 36 landmarks. ©2015 IEEE. Reprinted, with permission, from [25]

depicts a top view of the estimated sensor-based map at the end of the run with the 95% uncertainty ellipses in green and blue depending whether they are observed in that instant or not, including the real trajectory of the vehicle in dashed red, and the pose of the vehicle at that moment, that is represented by the yellow quadrotor. Note that the ellipses surround the true values, as they should in a consistent filter. Furthermore, it can be seen that the recently or currently (re-)observed landmarks have much tighter uncertainty ellipses than older ones, demonstrating the uncertainty convergence when the observability conditions are satisfied. Finally, in Fig. 12 the estimation error for all the 36 landmarks is shown. It is noticeable that even though the initial estimate may be far off, the error will converge until after 2 laps, it is under 40 centimeters depending on how long each landmark is observed.

In an experimental application using natural landmarks extracted from the environment, it is not straightforward to guarantee observability while moving as was attempted for the range-only counterpart and many features will not have enough time to converge. However, preliminary results have shown that the landmark state still converges without being initialized with any special care, while also showing that loop closures occur naturally throughout the runs. Hence, the algorithm recognizes previously visited places, even without a specially tailored procedure, providing a good measure of the consistency and validity of the algorithm.

6 Concluding Remarks

This chapter aimed at a broad presentation of the fundamentals behind a class of sensor-based simultaneous localization and mapping filters with global convergence guarantees, providing the necessary and sufficient conditions for observability, and thus convergence, in a constructive and physical intuitive manner. Several experimental examples of practical implementations were provided, illustrating the performance and consistency of the proposed sensor-based SLAM strategies, together with the ETM algorithm that also provides the vehicle trajectory and map in Earth-fixed coordinates.

Acknowledgements This work was supported by the Fundação para a Ciência e a Tecnologia (FCT) through ISR under LARSyS UID/EEA/50009/2013, and through IDMEC, under LAETA UID/EMS/50022/2013 contracts, by the University of Macau Project MYRG2015-00126-FST, and by the Macao Science and Technology Development Fund under Grant FDCT/048/2014/A1. The work of P. Lourenço and B. Guerreiro were supported respectively by the Ph.D. Student Grant SFRH/BD/89337/2012 and by the Post-doc Grant SFRH/BPD/110416/2015 from FCT.

References

1. Ahmad, A., Huang, S., Wang, J.J., Dissanayake, G.: A new state vector for range-only SLAM. In: Proceedings of the 2011 Chinese Control and Decision Conference (CCDC), pp. 3404–3409 (2011)
2. Bacca, B., Salvi, J., Cufí, X.: Long-term mapping and localization using feature stability histograms. Robot. Auton. Syst. **61**(12), 1539–1558 (2013)
3. Bailey, T.: Mobile Robot Localisation and Mapping in Extensive Outdoor Environments. Ph.D. thesis, University of Sydney, Australian Center of Field Robotics (2002)
4. Bailey, T.: Constrained initialisation for bearing-only SLAM. In: Proceedings of the 2003 IEEE International Conference on Robotics and Automation, vol. 2, pp. 1966–1971. IEEE (2003)
5. Batista, P., Silvestre, C., Oliveira, P.: Single range aided navigation and source localization: observability and filter design. Syst. Control Lett. **60**(8), 665–673 (2011)
6. Batista, P., Silvestre, C., Oliveira, P.: Globally exponentially stable filters for source localization and navigation aided by direction measurements. Syst. Control Lett. **62**(11), 1065–1072 (2013)
7. Bay, H., Ess, A., Tuytelaars, T., Van Gool, L.: Speeded-up robust features (SURF). Comput. Vis. Image Underst. **110**(3), 346–359 (2008)
8. Bishop, A.N., Jensfelt, P.: A stochastically stable solution to the problem of robocentric mapping. In: IEEE International Conference on Robotics and Automation, 2009. ICRA '09, pp. 1615–1622 (2009)
9. Cadena, C., Carlone, L., Carrillo, H., Latif, Y., Scaramuzza, D., Neira, J., Reid, I., Leonard, J.J.: Past, present, and future of simultaneous localization and mapping: toward the robust-perception age. IEEE Trans. Robot. **32**(6), 1309–1332 (2016)

10. Castellanos, J.A., Martinez-Cantin, R., Tardós, J.D., Neira, J.: Robocentric map joining: improving the consistency of EKF-SLAM. Robot. Auton. Syst. **55**(1), 21–29 (2007)
11. Csorba, M., Durrant-Whyte, H.F.: A new approach to simultaneous localisation and map building. In: SPIE Aerosense (1996)
12. Davison, A.J., Reid, I.D., Molton, N.D., Stasse, O.: MonoSLAM: real-time single camera SLAM. IEEE Trans. Pattern Anal. Mach. Intell. **29**(6), 1052–1067 (2007)
13. Dissanayake, G., Newman, P., Durrant-Whyte, H.F., Clark, S., Csobra, M.: A solution to the simultaneous localisation and mapping (SLAM) problem. IEEE Trans. Robot. Autom. **17**(3), 229–241 (2001)
14. Djugash, J., Singh, S.: Motion-aided network SLAM with range. Int. J. Robot. Res. **31**(5), 604–625 (2012)
15. Dubbelman, G., Browning, B.: COP-SLAM: closed-form online pose-chain optimization for visual SLAM. IEEE Trans. Robot. **31**(5), 1194–1213 (2015)
16. Durrant-Whyte, H.F.: Uncertain geometry in robotics. IEEE J. Robot. Autom. **4**(1), 23–31 (1988)
17. Gelb, A.: Applied Optimal Estimation. MIT Press, Cambridge (1974)
18. Guerreiro, B.J., Batista, P., Silvestre, C., Oliveira, P.: Globally asymptotically stable sensor-based simultaneous localization and mapping. IEEE Trans. Robot. **29**(6), 1380–1395 (2013)
19. Huang, G., Mourikis, A.I., Roumeliotis, S.I.: Observability-based rules for designing consistent EKF SLAM estimators. Int. J. Robot. Res. **29**(5), 502–528 (2010)
20. Huang, S., Dissanayake, G.: Convergence and consistency analysis for extended Kalman filter based SLAM. IEEE Trans. Robot. **23**(5), 1036–1049 (2007)
21. Jensfelt, P., Kragic, D., Folkesson, J., Bjorkman, M.: A framework for vision based bearing only 3D SLAM. In: Proceedings of the 2006 IEEE International Conference on Robotics and Automation, pp. 1944–1950 (2006)
22. Julier, S.J., Uhlmann, J.K.: A counter example to the theory of simultaneous localization and map building. In: Proceedings of the 2001 IEEE International Conference on Robotics and Automation (ICRA), Seul, South Korea, vol. 4, pp. 4238–4243 (2001)
23. Kelly, J., Sukhatme, G.S.: Visual-inertial sensor fusion: localization, mapping and sensor-to-sensor self-calibration. Int. J. Robot. Res. **30**(1), 56–79 (2011)
24. Lemaire, T., Lacroix, S., Solà, J.: A practical 3D bearing-only SLAM algorithm. In: Proceedings of the 2005 IEEE/RSJ International Conference on Intelligent Robots and Systems, pp. 2449–2454 (2005)
25. Lourenço, P., Batista, P., Oliveira, P., Silvestre, C.: A globally exponentially stable filter for bearing-only simultaneous localization and mapping in 3-D. In: Proceedings of the 2015 European Control Conference, Linz, Austria, pp. 2817–2822 (2015)
26. Lourenço, P., Batista, P., Oliveira, P., Silvestre, C.: Simultaneous localization and mapping in sensor networks: a GES sensor-based filter with moving object tracking. In: Proceedings of the 2015 European Control Conference, Linz, Austria, pp. 2359–2364 (2015)
27. Lourenço, P., Batista, P., Oliveira, P., Silvestre, C., Philip Chen, C.L.: Sensor-based globally exponentially stable range-only simultaneous localization and mapping. Robot. Auton. Syst. **68**, 72–85 (2015)
28. Lourenço, P., Guerreiro, B.J., Batista, P., Oliveira, P., Silvestre, C.: 3-D inertial trajectory and map online estimation: building on a GAS sensor-based SLAM filter. In: Proceedings of the 2013 European Control Conference, Zurich, Switzerland, pp. 4214–4219 (2013)
29. Lourenço, P., Guerreiro, B.J., Batista, P., Oliveira, P., Silvestre, C.: Simultaneous localization and mapping for aerial vehicles: a 3-D sensor-based GAS filter. Auton. Robot. **40**, 881–902 (2016)
30. Lourenço, P., Guerreiro, B.J., Batista, P., Oliveira, P., Silvestre, C.: Uncertainty characterization of the orthogonal procrustes problem with arbitrary covariance matrices. Pattern Recognit. **61**, 210–220 (2017)
31. Montemerlo, M., Thrun, S., Koller, D., Wegbreit, B.: FastSLAM: a factored solution to the simultaneous localization and mapping problem. In: Eighteenth National Conference on Artificial Intelligence, pp. 593–598. American Association for Artificial Intelligence (2002)

32. Mur-Artal, R., Montiel, J.M.M., Tardós, J.D.: ORB-SLAM: a versatile and accurate monocular SLAM system. IEEE Trans. Robot. **31**(5), 1147–1163 (2015)
33. Neira, J., Tardós, J.D.: Data association in stochastic mapping using the joint compatibility test. IEEE Trans. Robot. Autom. **17**(6), 890–897 (2001)
34. Smith, R.C., Cheeseman, P.: On the representation and estimation of spatial uncertainty. Int. J. Robot. Res. **5**(4), 56–68 (1986)
35. Smith, R.C., Self, M., Cheeseman, P.: Estimating uncertain spatial relationships in robotics. pp. 167–193. Springer, New York (1990)
36. Solà, J., Monin, A., Devy, M., Lemaire, T.: Undelayed initialization in bearing only SLAM. In: Proceedings of the 2005 IEEE/RSJ International Conference on Intelligent Robots and Systems, pp. 2499–2504 (2005)
37. Thrun, S., Liu, Y., Koller, D., Ng, A., Durrant-Whyte, H.: Simultaneous localization and mapping with sparse extended information filters. Int. J. Robot. Res. **23**(7–8), 693–716 (2004)
38. Weiss, L., Sanderson, A.C., Neuman, C.P.: Dynamic sensor-based control of robots with visual feedback. IEEE J. Robot. Autom. **3**(5), 404–417 (1987)

Pose-Graph SLAM for Underwater Navigation

Stephen M. Chaves, Enric Galceran, Paul Ozog,
Jeffrey M. Walls and Ryan M. Eustice

Abstract This chapter reviews the concept of pose-graph simultaneous localization and mapping (SLAM) for underwater navigation. We show that pose-graph SLAM is a generalized framework that can be applied to many diverse underwater navigation problems in marine robotics. We highlight three specific examples as applied in the areas of autonomous ship hull inspection and multi-vehicle cooperative navigation.

1 Introduction

Simultaneous localization and mapping (SLAM) is a fundamental problem in mobile robotics whereby a robot uses its noisy sensors to collect observations of its surroundings in order to estimate a map of the environment while simultaneously localizing itself within that same map. This is a coupled chicken-and-egg state estimation problem, and remarkable progress has been made over the last two decades in the formulation and solution to SLAM [5].

One of the resulting key innovations in the modeling of the SLAM problem has been the use of pose graphs [10, 24], which provide a useful probabilistic representation of the problem that allows for efficient solutions via nonlinear optimization methods. This chapter provides an introduction to pose-graph SLAM as a unifying

---

S.M. Chaves (✉) · P. Ozog · J.M. Walls · R.M. Eustice
University of Michigan, 2600 Draper Dr, Ann Arbor, MI 48105, USA
e-mail: schaves@umich.edu

P. Ozog
e-mail: paulozog@umich.edu

J.M. Walls
e-mail: jmwalls@umich.edu

R.M. Eustice
e-mail: eustice@umich.edu

E. Galceran
ETH Zurich, Leonhardstrasse 21, LEEJ304, 8092 Zürich, Switzerland
e-mail: enricg@ethz.ch

© Springer International Publishing AG 2017
T.I. Fossen et al. (eds.), *Sensing and Control for Autonomous Vehicles*,
Lecture Notes in Control and Information Sciences 474,
DOI 10.1007/978-3-319-55372-6_7

framework for underwater navigation. We first present an introduction to the general SLAM problem. Then, we show how challenging SLAM problems stemming from representative marine robotics applications can be modeled and solved using these tools. In particular, we present three SLAM systems for underwater navigation: a visual SLAM system using underwater cameras, a system that exploits planarity in ship hull inspection using sparse Doppler velocity log (DVL) measurements, and a cooperative multi-vehicle localization system. All of these examples showcase the pose graph as a foundational tool for enabling autonomous underwater robotics.

2 Simultaneous Localization and Mapping (SLAM)

Over the last several decades, robotics researchers have developed probabilistic tools for fusing uncertain sensor data in order to localize within an *a priori* unknown map—the SLAM problem. These tools have reached a level of maturity where they are now widely available [3, 10, 12].

Early approaches to the SLAM problem tracked the most recent robot pose and landmarks throughout the environment in an extended Kalman filter (EKF) [31, 40]. Here, the SLAM estimate is represented as a multivariate Gaussian with mean vector and fully dense covariance matrix. Complexity of the Kalman filter, however, grows with the size of the map, as the measurement updates and memory requirements are quadratic in the state dimension. Thrun et al. [41] observed that the information matrix (inverse of the covariance matrix) of the estimate is approximately sparse, leading to more efficient solutions using an *information* filtering approach that forced sparsity. The information filtering approach features constant-time measurement updates and linear memory requirements. Extending the seminal work of Lu and Milios [33], Eustice et al. [15] showed that by considering a delayed-state information filter, the information matrix of the SLAM problem is *exactly* sparse, leveraging the benefits of the information parameterization without sparse approximation errors. Most SLAM systems today formulate the problem in the exactly sparse sense by optimizing over the entire robot trajectory.

2.1 SLAM Formulation

The *full* SLAM formulation considers optimizing over the entire history of robot poses and landmarks. This problem can be solved using the maximum *a posteriori* (MAP) estimate, given the prior observations of the robot motion and landmarks in the environment:

$$X^*, \mathcal{L}^* = \arg\max_{X, \mathcal{L}} p(X, \mathcal{L}|\mathcal{U}, \mathcal{Z}), \qquad (1)$$

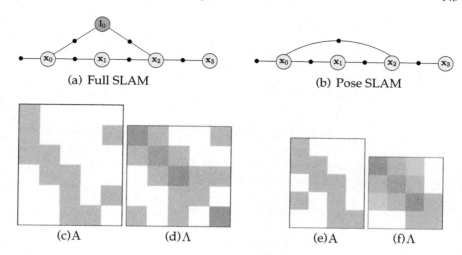

Fig. 1 Factor graph representations of the full SLAM (**a**) and pose SLAM (**b**) formulations. The corresponding measurement Jacobian (A) and information matrix ($\Lambda = A^\top A$) for each system are shown below the factor graphs, with matrix block entries corresponding to poses in yellow and landmarks in purple. In the full SLAM system (**a**)(**c**)(**d**), loop closures include measurements to landmarks. The columns of the measurement Jacobian A correspond to the following ordering: $\{\mathbf{x}_0, \mathbf{x}_1, \mathbf{x}_2, \mathbf{x}_3, \mathbf{l}_0\}$. In the pose SLAM system (**b**)(**e**)(**f**), the columns of A correspond to an ordering of $\{\mathbf{x}_0, \mathbf{x}_1, \mathbf{x}_2, \mathbf{x}_3\}$

where $\mathbf{x}_i \in X$ are the robot poses, $\mathbf{l}_k \in \mathcal{L}$ are the landmark poses, $\mathbf{u}_i \in \mathcal{U}$ are the control inputs (or motion observations), and $\mathbf{z}_j \in \mathcal{Z}$ are the perceptual observations of map features. The full SLAM formulation is shown in Fig. 1a in the form of a factor graph.

Often, when building a SLAM system, the SLAM problem is divided into two sub-problems: (*i*) a "front-end" system that parses sensor measurements to build an optimization problem, and (*ii*) a "back-end" solver that optimizes over robot poses and map features.

The formulation considered in this work is *pose* SLAM (Fig. 1b), where there is no explicit representation of landmarks, but rather features observed in the environment are used to construct a relative measurement between robot poses [33]. In this case, the MAP estimate becomes

$$X^* = \operatorname*{argmax}_{X} p(X|\mathcal{U}, \mathcal{Z}), \tag{2}$$

and the model of the environment is derived from the robot trajectory itself. This formulation is especially beneficial when the main perceptual sensors are cameras or laser scanners and the environment features are difficult to repeatedly detect or are too numerous to track.

Assuming measurement models with additive Gaussian noise, the optimization of (1) or (2) leads to the following nonlinear least-squares problem:

$$X^*, \mathcal{L}^* = \underset{X,\mathcal{L}}{\mathrm{argmax}}\ p(X, \mathcal{L}|\mathcal{U}, \mathcal{Z})$$

$$= \underset{X,\mathcal{L}}{\mathrm{argmin}}\ -\log p(X, \mathcal{L}|\mathcal{U}, \mathcal{Z})$$

$$= \underset{X,\mathcal{L}}{\mathrm{argmin}} \Big[\sum_i \|\mathbf{x}_i - f_i(\mathbf{x}_{i-1}, \mathbf{u}_{i-1})\|^2_{\Sigma^i_w} + \sum_j \|\mathbf{z}_j - h_j(\mathbf{x}_{i_j}, \mathbf{l}_{k_j})\|^2_{\Sigma^j_v} \Big], \quad (3)$$

where f_i and h_j are the measurement models with zero-mean additive Gaussian noise with covariances Σ^i_w and Σ^j_v, and we define $\|\mathbf{e}\|^2_\Sigma = \mathbf{e}^\top \Sigma^{-1} \mathbf{e}$. Linearizing about the current estimate, the problem (3) collapses into a linear least-squares form for the state update vector, solved with the normal equations:

$$\underset{\Delta\Theta}{\arg\min}\ \|A\Delta\Theta - \mathbf{b}\|^2,$$
$$\Delta\Theta = (A^\top A)^{-1} A^\top \mathbf{b}, \quad (4)$$

where the vector Θ includes the poses and landmarks, A is the stacked whitened measurement Jacobian, and **b** is the corresponding residual vector. Under the assumption of independent measurements, this formulation leads to an information matrix ($\Lambda = A^\top A$) that is inherently sparse, as each observation model depends on only a small subset of poses and landmarks. Thus, modern back-end solvers leverage sparsity patterns to efficiently find solutions.

We solve the nonlinear problem by re-linearizing about the new solution and solving again, repeating until convergence (with Gauss-Newton, for instance). Each linear problem is most commonly solved by direct methods such as Cholesky decomposition of the information matrix or QR factorization of the measurement Jacobian [10, 24]. Aside from direct methods, iterative methods, e.g., relaxation-based techniques [11] and conjugate gradients [8, 42], have also been applied to solve large linear systems in a more memory-efficient and parallelizable way.

2.2 Graphical Representations of SLAM

The SLAM problem introduced above can also be viewed as a probabilistic graphical model known as a *factor graph* (or *pose graph* in the case of pose SLAM, that is, with no explicit representation of landmarks). A factor graph is a bipartite graph with two types of components: nodes that represent variables to be estimated (poses along the robot's trajectory) and factors that represent constraints over the variables (noisy sensor measurements), as shown in Fig. 1. If each measurement is encoded in a factor, $\Psi_i(\mathbf{x}_i, \mathbf{l}_i)$, where $\mathbf{x}_i$ and $\mathbf{l}_i$ are the robot and landmark poses corresponding to measurement i (and we assume all measurement noise terms are independent), the nonlinear least-squares problem can be written as

$$X^*, \mathcal{L}^* = \arg\min_{X,\mathcal{L}} \sum_i \Psi_i(\mathbf{x}_i, \mathbf{l}_i), \tag{5}$$

such that the optimization minimizes the sum of squared errors of all the factor potentials. This graphical model view of SLAM is equivalent to the optimization view presented above.

2.3 Advantages of Graph-Based SLAM Methods

Indeed, recent research in SLAM has turned to graph-based solutions in order to avoid drawbacks associated with filtering-based methods [10, 24]. Notably, EKF-SLAM has quadratic complexity per update, but graph-based methods that parameterize the entire robot trajectory in the information form feature constant-time updates and linear memory requirements. Hence, they are faster on large-scale problems. In addition, unlike filtering-based methods, these optimization-based solutions avoid the commitment to a static linearization point and take advantage of re-linearization to better handle nonlinearities in the SLAM problem.

Despite their advantages, nonlinear least-squares SLAM methods present some important challenges. First, since they operate on the information matrix, it is expensive to recover the joint covariances of the estimated variables. Nonetheless, some methods have been developed to improve the speed of joint covariance recovery [23]. Second, since these methods smooth the entire trajectory of the robot, the complexity of the problem grows unbounded over time and performance degrades as the robot explores. However, the examples presented in this chapter are made possible by online *incremental* graph-based solvers like incremental smoothing and mapping (iSAM) [24] and iSAM2 [25] that leverage smart variable ordering and selective re-linearization, and only update the solutions to the parts of the pose graph that have changed. As we will see in section "Multi-Session SLAM", the generic linear constraints (GLC) method [6] can additionally be used to compress the representation of the problem and enable tractable operation in large environments over long durations.

Several open source factor graph libraries are available to the community including Ceres solver [1], iSAM [24], GTSAM [9], and g2o [30].

3 Underwater Pose Graph Applications

In this section, we outline several representative applications for underwater navigation where the use of pose graphs has extended the state of the art. Underwater SLAM can take on many forms depending on the sensors available, the operating environment, and the autonomous task to be executed. As we will show, the pose graph formulation is applicable to many of these forms.

For the remainder of the chapter, let $\mathbf{x}_{ij} = [x_{ij}, y_{ij}, z_{ij}, \phi_{ij}, \theta_{ij}, \psi_{ij}]^\top$ be the 6-degree-of-freedom (DOF) relative pose of frame j as expressed in frame i, where x, y, z are the Cartesian translation components, and ϕ_{ij}, θ_{ij}, and ψ_{ij} denote the roll (x-axis), pitch (y-axis), and yaw (z-axis) Euler angles, respectively. A pose with a single subscript (e.g., $\mathbf{x}_i$) is expressed with respect to a common local frame.

One foundational sensor that enables underwater SLAM is the Doppler velocity log (DVL), central to all applications presented below. As the robot explores the environment, pose nodes are added to the graph and the dead-reckoned navigation estimate from the DVL is constructed into odometry constraints between consecutive robot poses. In this way, the DVL provides an odometric backbone for various pose SLAM formulations. When available and applicable, absolute prior measurements from, for example, pressure depth, inertial measurement unit (IMU), gyroscope, compass, or GPS can be added to the pose graph as unary factors.

In the sections that follow, we highlight other factor types derived for specific underwater applications, as well as describe methods centered around pose SLAM that are state of the art in marine autonomy.

3.1 Visual SLAM with Underwater Cameras

Cameras are prevalent perceptual sensors in robotics research because of their low cost but also highly accurate and rich data. Their popularity has led to research in visual SLAM, where measurements derived from the camera are included in the inference.[1] Within the visual pose SLAM formulation, the robot poses in the pose graph represent discrete image capture events during the underwater mission, and feature-based registrations between overlapping images [21] produce pairwise constraints between the poses. These camera-derived constraints often occur between sequential poses in the graph; however, they can also serve as loop-closure constraints between non-sequential poses when the robot revisits a portion of the environment that it has previously seen, enabling large reductions in its navigation uncertainty.

The visual registration process searches for overlapping images within the pose graph, proposes a camera registration hypothesis given two image candidates, and adds the camera-derived constraint to the graph upon a successful registration. A typical pairwise registration pipeline is shown in Fig. 2 and is described as follows:

1. Given two overlapping images collected by the robot, first undistort each image and enhance with contrast-limited adaptive histogram search (CLAHS) [14].
2. Extract features such as scale-invariant feature transform (SIFT) [32] or speeded up robust features (SURF) [4] from each image.
3. Match features between the images using a nearest-neighbors search in the high-dimensional feature space assisted by pose-constrained correspondence search (PCCS) [16].

---

[1]More background on visual SLAM can be found in [16, 27].

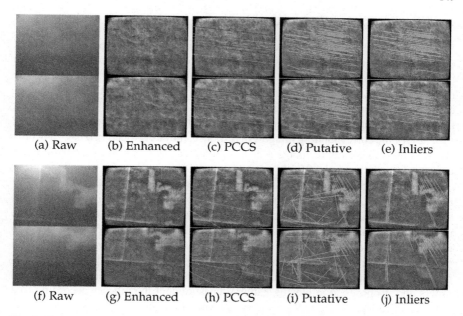

Fig. 2 Figures courtesy of Kim and Eustice [27]. Underwater visual SLAM: The pairwise image registration pipeline is shown for two registration hypotheses. The top row shows a feature-poor image set that registers successfully because of strong relative constraints between poses that guide feature-matching via PCCS. The *bottom* row is also a successful registration, but largely due to the strong features in the images. Steps in the registration pipeline are shown from *left* to *right*: **a, f** Raw overlapping images. **b, g** Undistorted and enhanced, before extracting features. **c, h** Feature matching is guided by PCCS. **d, i** Putative correspondences. **e, j** Geometric consensus is used to identify inliers. Finally, a two-view bundle adjustment solves for the 5-DOF relative pose constraint

4. Fit a projective model among feature-matching inliers using a geometric consensus algorithm such as random sample consensus (RANSAC) [19].
5. Perform a two-view bundle adjustment problem to solve for the 5-DOF bearing-only transformation between camera poses and its first-order covariance estimate [20].

The camera measurement produces a low-rank (modulo scale) relative pose constraint between two robot poses i and j in the SLAM graph. This measurement, $h^{5\text{dof}}$, therefore has five DOFs: three rotations, and a vector representing the direction of translation that is parameterized by azimuth and elevation angles. We denote the camera measurement as

$$h^{5\text{dof}}\left(\mathbf{x}_i, \mathbf{x}_j\right) = [\alpha_{ij}, \beta_{ij}, \phi_{ij}, \theta_{ij}, \psi_{ij}]^\top, \tag{6}$$

consisting of the baseline direction of motion azimuth α_{ij}, elevation β_{ij}, and the relative Euler angles $\phi_{ij}, \theta_{ij}, \psi_{ij}$.

Saliency-Informed Visual SLAM

The underwater environment is particularly challenging for visual SLAM because it does not always contain visually useful features for camera-derived measurements. In the case of featureless images, the registration pipeline spends much time attempting registrations that are very likely to fail, despite overlap between the image candidates. For autonomous ship hull inspection, Kim and Eustice [27] discovered that the success of camera registrations was correlated to the texture richness, or *visual saliency*, of the corresponding images. In response, they developed two bag-of-words (BoW)-based measures of image registrability, *local saliency* and *global saliency*, to better inform the visual SLAM process.

In a framework known as *saliency-informed visual SLAM*, Kim and Eustice [27] augmented the SLAM system with the knowledge of visual saliency in order to design a more efficient and robust loop-closure registration process. This system first limits the number of poses added to the graph by only adding poses corresponding to images that pass a local saliency threshold. This thresholding ensures that the graph predominantly contains poses expected to be useful for camera-derived measurements and eliminates poses with a low likelihood of registration. Second, the system orders and proposes loop-closing camera measurement hypotheses according to a measure of saliency-weighted geometric information gain:

$$\mathcal{I}_L^{ij} = \begin{cases} \frac{S_{L_j}}{2} \ln \frac{|R + H^{5dof} \Sigma_{ii,jj} H^{5dof\,T}|}{|R|}, & \text{if } S_{L_j} > S_L^{\min} \\ 0, & \text{otherwise} \end{cases}, \quad (7)$$

where R is the five-DOF camera measurement covariance, H^{5dof} is the measurement Jacobian of (6), $\Sigma_{ii,jj}$ is the joint marginal covariance of current pose i and target pose j from the current SLAM estimate, S_{L_j} is the local saliency of image j, and $S_L^{\min}$ is the minimum local saliency threshold.

Proposing loop closures in this way leads to registration hypotheses that both induce significant information in the pose graph and are likely to be successfully registered, thereby focusing computational resources during SLAM to the most worthwhile loop-closure candidates. The saliency-informed visual SLAM process is shown in Fig. 3 for autonomous ship hull inspection with the HAUV. This result features a total mission time of 3.40 h and 8,728 poses. The image registration component totaled 0.79 h and the (cumulative) optimization within iSAM [24] totaled 0.52 h. Thus, the cumulative processing time for this system was 1.31 h, which is 2.6 times faster than real time.

Active Visual SLAM

The benefit of saliency-informed visual SLAM can be extended to the *active* SLAM paradigm, where the robot makes decisions about which actions to execute in order to improve the performance of SLAM. Recent works [7, 28] performed belief-space planning for active SLAM with the saliency-informed pose SLAM formulation outlined above.

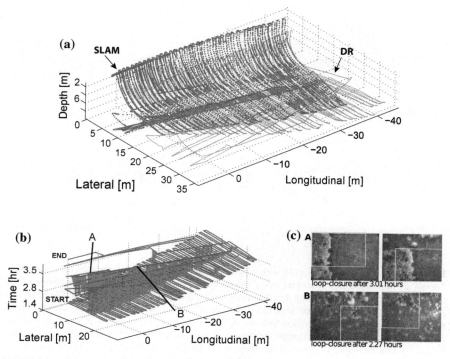

Fig. 3 Figures courtesy of Kim and Eustice [27]. Real-time saliency-informed visual SLAM with the HAUV on the *SS Curtiss*. The pose graph resulting from SLAM is shown in (**a**) in *blue*, with *red* links representing camera-derived constraints between poses. For comparison, the trajectory estimate from dead-reckoned navigation based on the DVL alone is displayed in *gray*. The SLAM pose graph is shown again in (**b**) with the z-axis scaled by time. In this view, *larger red* links correspond to larger loop closures in time. Two loop-closure image pairs are shown in (**c**)

We can view the active SLAM framework through the lens of the pose graph by treating each candidate trajectory (or action) as a set of predicted virtual poses and factors that are added to the existing pose graph built by the robot up to planning time. The robot can then evaluate the solution of this simulated SLAM system within an objective function that quantifies some information-theoretic measure, like navigation uncertainty as described by the covariance matrix. Results from active SLAM methods for autonomous ship hull inspection are given in Fig. 4.

3.2 Planar SLAM from Sparse DVL Points

One of the main benefits of factor graphs is the ease of including additional sources of information. In the case of visual SLAM, the vehicle can constrain the estimate of its trajectory with non-visual perceptual data, such as sonar data or acoustic ranges.

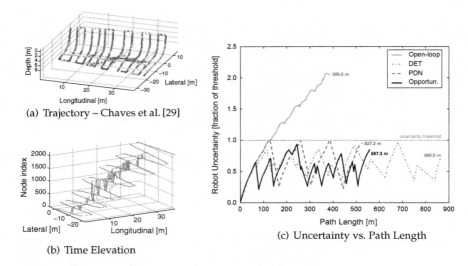

Fig. 4 Active visual SLAM results with a hybrid simulation of autonomous ship hull inspection. Shown are four inspection strategies: open-loop coverage survey, a deterministic pre-planned strategy (DET), the PDN method [28], and the opportunistic active SLAM algorithm [7]. The trajectory resulting from the opportunistic approach [7] is shown in (**a**), with poses color coded by visual saliency (red = salient, blue = non-salient). A time-elevation plot is shown in (**b**) with camera-derived constraints in the pose graph displayed by *red* links. The uncertainty versus path length plot of the four strategies is shown in (**c**). Built on the saliency-informed visual SLAM framework, both the PDN and opportunistic active SLAM methods perform favorably in bounding navigation uncertainty while maintaining efficient area coverage for the inspection task

One interesting source of information is the extraction of coarse perceptual cues using the DVL. In this section, we describe how the DVL can model locally planar environments using a factor graph SLAM back-end.[2]

In addition to measuring the velocity of an underwater vehicle, the raw DVL sensor data contains the range of each of the four beams. These three-dimensional (3D) points provide sparse perceptual information that a few researchers have leveraged in prior work, with a particular focus on terrain-aided localization and bathymetric SLAM. Underwater terrain-aided techniques are typically performed with a multi-beam sonar, which is much denser than a DVL. Despite this trend, Eustice et al. [13] and Meduna et al. [34] proposed methods for a vehicle equipped with a DVL to localize with respect to a prior bathymetric map derived from a large surface vessel equipped with a multibeam sonar with high spatial resolution.

More recently, Ozog et al. [36] leveraged this information in the context of automated underwater ship hull inspection with the HAUV, which establishes hull-relative navigation using a DVL pointed nadir to the ship hull surface [22] (Fig. 5). In particular, they used the sparse DVL range returns to model a large ship hull as a collection of locally planar features, greatly improving the robustness of long-term underwater

---

[2] A more detailed description can be found in [35, 36].

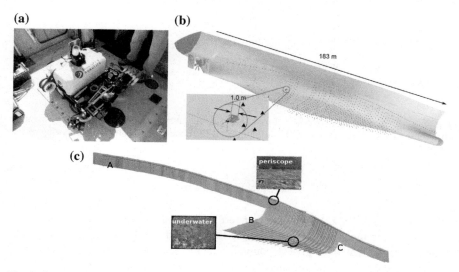

Fig. 5 Size comparison of the HAUV and a typically sized surveyed vessel ((**a**) and (**b**)). Using factor graph SLAM, the surface of the ship hull can be estimated as a collection of locally planar patches, shown as *gray* patches in (**c**)

visual SLAM across multiple sessions. In this section, we briefly summarize this approach and describe factors necessary for inclusion into a factor graph SLAM back-end.

Pose-to-Plane Factors

As the HAUV inspects a ship hull, it fits a least-squares 3D plane to a collection of DVL points in the vehicle frame. Suppose this plane, indexed by k and denoted π_k, is observed with respect to the vehicle at time i. The corresponding observation model for this measurement is

$$\mathbf{z}_{\pi_{ik}} = \mathbf{x}_i \boxminus \pi_k + \mathbf{w}_{ik}, \tag{8}$$

where $\mathbf{x}_i$ is a pose indexed by i, $\mathbf{w}_k \sim \mathcal{N}(\mathbf{0}, \Sigma_{\mathbf{w}_{ik}})$, and $\boxminus$ is a nonlinear function that expresses plane π_k with respect to pose frame i. For this section, a plane $\pi^\top = [n_x, n_y, n_z]^\top$ is a 3D column vector consisting of the surface normal of the plane in Euclidean coordinates scaled by the distance of the plane to the local origin.

Piecewise-Planar Factors

Ship hulls inspected by the HAUV typically exhibit curvature, both in the bow-to-stern and side-to-centerline directions. Therefore, Ozog et al. noted that the HAUV will observe neighboring planar patches that are themselves not coplanar. To address this, they adapted a ternary factor that can constrain two neighboring planes that do not necessarily overlap. The corresponding observation model for a piecewise-planar ("pwp") measurement, $\mathbf{z}^{\text{pwp}}_{\pi_{ik}}$, of two neighboring planes, k and l are as follows:

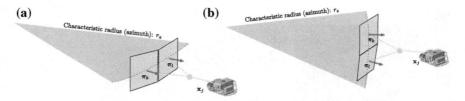

Fig. 6 Characteristic radii overview for side-to-side curvature (**a**) and top-to-bottom curvature (**b**). These radii account for allowable variations in the surface normals of two neighboring planar patches

$$\mathbf{z}_{\pi_{kl}}^{\text{pwp}} = (\mathbf{x}_i \boxminus \boldsymbol{\pi}_k) - (\mathbf{x}_i \boxminus \boldsymbol{\pi}_l) + \mathbf{w}_{kl}^{\text{pwp}}, \qquad (9)$$

where $\mathbf{w}_{kl}^{\text{pwp}} \sim \mathcal{N}(\mathbf{0}, \Sigma_{\mathbf{w}_{kl}^{\text{pwp}}})$ is an error term that accounts for the curvature of the ship hull being inspected. By introducing this term, the differences between planes k and l are weighted to give account for them being non-coplanar. In addition, the measurement is conditionally Gaussian by construction and so can be easily incorporated into the factor graph. The curvature model is based on two characteristic radii that are briefly described in Fig. 6.

Multi-Session SLAM

The planar factors described in this section are particularly useful in the context of multi-session SLAM. Ozog et al. showed that these observation models can be incorporated to a visual localization pipeline using a combination of particle filtering and visual SLAM techniques described in Sect. 3.1. Once localized, the HAUV further adds factors into the SLAM graph using the method inspired from [26]. With this process, multiple sessions can be automatically aligned into a common reference frame in real time. This pipeline is illustrated in Fig. 7, along with the keyframes used for the visual re-acquisition of the hull.

The HAUV can maintain real-time performance of multi-session SLAM by marginalizing redundant nodes in the pose graph. Once a pose node is marginalized, however, it induces dense connectivity to other nodes. The GLC framework alleviates this by replacing the target information, Λ_t with a n-ary factor:

$$\mathbf{z}_{glc} = \mathbf{G}\mathbf{x}_c + \mathbf{w}',$$

where $\mathbf{w}' \sim \mathcal{N}(\mathbf{0}, \mathbf{I}_{q \times q})$, $\mathbf{G} = \mathbf{D}^{1/2}\mathbf{U}^\top$, $\Lambda_t = \mathbf{U}\mathbf{D}\mathbf{U}^\top$, q is the rank of Λ_t, and $\mathbf{x}_c$ is the current linearization of nodes contained in the elimination clique. $\mathbf{U}\mathbf{D}\mathbf{U}^\top$ is the eigendecomposition of Λ_t, where U is a $p \times q$ matrix of eigenvectors, and D is a $q \times q$ diagonal matrix of eigenvalues. To preserve sparsity in the graph, the target information Λ_t is approximated using a Chow–Liu Tree (CLT) structure, where the CLT's unary and binary potentials are represented as GLC factors.[3] Thus, GLC serves as an approximate marginalization method for reducing computational complexity

---

[3] A more detailed description of GLC can be found in [6].

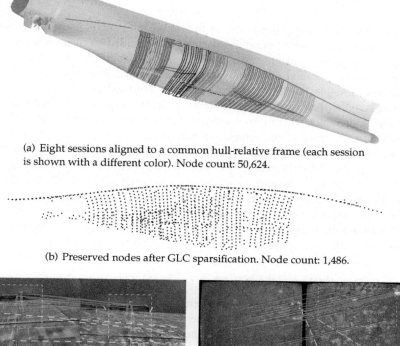

(a) Eight sessions aligned to a common hull-relative frame (each session is shown with a different color). Node count: 50,624.

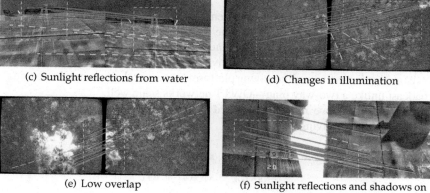

(b) Preserved nodes after GLC sparsification. Node count: 1,486.

(c) Sunlight reflections from water

(d) Changes in illumination

(e) Low overlap

(f) Sunlight reflections and shadows on hull, 2011 (left) to 2014 (right)

Fig. 7 Planar-based factor potentials and GLC graph sparsification play a key part in the HAUV localization system. This method works in conjunction with the visual SLAM techniques from Sect. 3.1 to allow for long-term automated ship hull surveillance. Successful instances of localization in particularly challenging hull regions are shown in (**c**) through (**f**), with visual feature correspondences shown in *red*

of the pose graph. In the example of Fig. 7, the multi-session pose graph is reduced from 50,624 to 1,486 nodes.

3.3 Cooperative Localization

Underwater localization with AUVs in the mid-depth zone is notoriously difficult [29]. For example, both terrain-aided and visually-aided navigation assume that vehicles are within sensing range of the seafloor. Underwater vehicles typically employ acoustic beacon networks, such as narrowband long-baseline (LBL) and ultra-short-baseline (USBL), to obtain accurate bounded-error navigation in this regime. Acoustic beacon methods, however, generally require additional infrastructure and limit vehicle operations to the acoustic footprint of the beacons.

Acoustic modems enable vehicles to both share data and observe their relative range; however, the underwater acoustic channel is unreliable, exhibits low bandwidth, and suffers from high latency (sound is orders of magnitude slower than light) [37]. Despite these challenges, cooperative localization has been effectively implemented among teams of underwater vehicles (Fig. 8). Each vehicle is treated as a mobile acoustic navigation beacon, which requires no additional external infrastructure and is not limited in the range of operations by static beacons. In this section, we show that an effective cooperative localization framework can be built by exploiting the structure of the underlying factor graph.[4]

Acoustic Range Observation Model

The use of synchronous-clock hardware enables a team of vehicles to observe their relative range via the one-way-travel-time (OWTT) of narrowband acoustic broadcasts [17]. The OWTT relative range is measured between the transmitting vehicle at the time-of-launch (TOL) and the receiving vehicle at the time-of-arrival (TOA). Since ranging is passive—all receiving platforms observe relative range from a single broadcast unlike a two-way ping—OWTT networks scale well.

The OWTT relative range is modeled as the Euclidean distance between the transmitting and receiving vehicles

$$z_r = h_r(\mathbf{x}_{\mathsf{TOL}}, \mathbf{x}_{\mathsf{TOA}}) + w_r$$
$$= \|\mathbf{x}_{\mathsf{TOL}} - \mathbf{x}_{\mathsf{TOA}}\|_2 + w_r,$$

where $w_r \sim \mathcal{N}(0, \sigma_r^2)$ is an additive noise perturbation. Since attitude and depth are typically instrumented with small bounded error, we often project the 3D range measurement into the horizontal plane.

Multiple Vehicle Factor Graph

Representing correlation that develops between individual vehicle estimates as a result of relative observations has been a challenge for cooperative localization algorithms [2, 39]. Factor graphs explicitly represent this correlation by maintaining a distribution over the trajectories of *all* vehicles.

---

[4] A more detailed description of cooperative localization with factor graph-based algorithms appears in [43, 44].

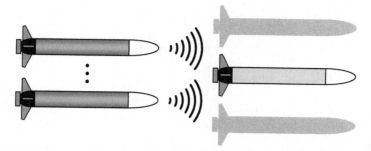

Fig. 8 Cooperative multiple vehicle network. *Yellow* vehicles (*right*) benefit from the information shared from *blue* vehicles (*left*). Note that communication may be bidirectional

Earlier, we showed the pose SLAM formulation citing a single vehicle (2). We can expand this formulation to represent the posterior distribution of a network of vehicles given relative constraints. Consider, for example, an M vehicle network. The posterior can be factored

$$p(X_1, \ldots, X_M | Z_1, \ldots, Z_M, Z_r) \propto \prod_{i=1}^{M} \underbrace{p(X_i | Z_i)}_{\mathcal{C}_{\text{local}_i}} \prod_k \underbrace{p(\mathbf{z}_k | \mathbf{x}_{i_k}, \mathbf{x}_{j_k})}_{\text{range factors}}, \qquad (10)$$

where X_i is the set of ith vehicle poses, Z_i is the set of ith vehicle observations, and Z_r is the set of relative vehicle OWTT range constraints and each $\mathbf{z}_k$ represents a constraint between poses on vehicles i_k and j_k. We see that the posterior factors as a product of each vehicle's local information ($\mathcal{C}_i$) and the set of all relative range observations. Therefore, in order to construct (and perform inference on) the full factor graph, the ith vehicle must have access to the set of local factors from all other vehicles, $\{\mathcal{C}_{\text{local}_j}\}_{j \neq i}$, and the set of all relative vehicle factors. Distributed estimationhms can leverage the sparse factor graph structure in order to broadcast information across the vehicle network. The factor graph for a three-vehicle network is illustrated in Fig. 9.

Recently, several authors have proposed real-time implementations that exploit this property [18, 38, 43]. Figure 10 illustrates an example of a three-vehicle network consisting of two AUVs and a topside ship implementing the algorithm proposed in [43]. AUV-A had intermittent access to GPS during brief surface intervals (highlighted in green). AUV-B remained subsea during the duration of the trial. Figure 10b shows the resulting position uncertainty for AUV-B is bounded and nearly identical to that of the post-process centralized estimator.

Cooperative localization provides a means to improve navigation accuracy by exploiting relative range constraints within networks of vehicles. An architecture based around factor graphs addresses challenges endemic to cooperative localizationes a sparse representation ideal for low-bandwidth communication networks, and is extensible to new factor types.

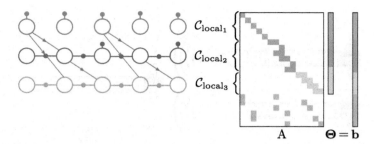

Fig. 9 Example cooperative localization factor graph. Empty circles represent variable pose nodes, *solid* dots are odometry and prior factors, and arrows illustrate range-only factors and the direction of communication. In this example, red represents a topside ship with access only to GPS, while *blue* and *orange* represent AUVs

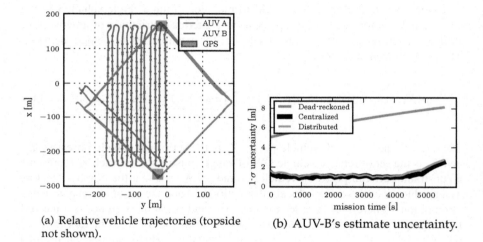

(a) Relative vehicle trajectories (topside not shown).

(b) AUV-B's estimate uncertainty.

Fig. 10 Summary of field trial and performance comparison. **a** An XY view of the vehicle trajectories. *Blue* dots indicate where AUV-B received range observations. **b** The smoothed uncertainty in each AUV-B pose as the fourth root of the determinant of the pose marginal covariance

4 Conclusion

This chapter has shown how state-of-the-art knowledge about the SLAM problem can enable key marine robotics applications like ship hull inspection or cooperative navigation. We reviewed the general concept of pose SLAM and its associated mathematical framework based on factor graphs and nonlinear least-squares optimization. We then presented three diverse underwater SLAM systems applied to autonomous inspection and cooperative navigation tasks.

References

1. Agarwal, S., Mierle, K., Others: Ceres solver. http://ceres-solver.org
2. Bahr, A., Walter, M.R., Leonard, J.J.: Consistent cooperative localization. In: Proceedings of the IEEE International Conference on Robotics and Automation, Kobe, Japan, pp. 3415–3422 (2009)
3. Bailey, T., Durrant-Whyte, H.: Simultaneous localization and mapping (SLAM): part II. IEEE Robot. Autom. Mag. **13**(3), 108–117 (2006)
4. Bay, H., Ess, A., Tuytelaars, T., Van Gool, L.: Speeded-up robust features (SURF). Comput. Vis. Image Underst. **110**(3), 346–359 (2008)
5. Cadena, C., Carlone, L., Carrillo, H., Latif, Y., Scaramuzza, D., Neira, J., Reid, I.D., Leonard, J.J.: Past, present, and future of simultaneous localization and mapping: toward the robust-perception age. IEEE Trans. Robot. **32**(6), 1309–1332 (2016)
6. Carlevaris-Bianco, N., Kaess, M., Eustice, R.M.: Generic node removal for factor-graph SLAM. IEEE Trans. Robot. **30**(6), 1371–1385 (2014)
7. Chaves, S.M., Kim, A., Galceran, E., Eustice, R.M.: Opportunistic sampling-based active visual SLAM for underwater inspection. Auton. Robot. **40**(7), 1245–1265 (2016)
8. Dellaert, F., Carlson, J., Ila, V., Ni, K., Thorpe, C.E.: Subgraph-preconditioned conjugate gradients for large scale SLAM. In: Proceedings of the IEEE/RSJ International Conference on Intelligent Robots and Systems, Taipei, Taiwan, pp. 2566–2571 (2010)
9. Dellaert, F., Others: Gtsam. https://borg.cc.gatech.edu/download
10. Dellaert, F., Kaess, M.: Square root SAM: simultaneous localization and mapping via square root information smoothing. Int. J. Robot. Res. **25**(12), 1181–1203 (2006)
11. Duckett, T., Marsland, S., Shapiro, J.: Fast, on-line learning of globally consistent maps. Auton. Robot. **12**(3), 287–300 (2002)
12. Durrant-Whyte, H., Bailey, T.: Simultaneous localization and mapping: part I. IEEE Robot. Autom. Mag. **13**(2), 99–110 (2006)
13. Eustice, R., Camilli, R., Singh, H.: Towards bathymetry-optimized Doppler re-navigation for AUVs. In: Proceedings of the IEEE/MTS OCEANS Conference and Exhibition, Washington, DC, USA, pp. 1430–1436 (2005)
14. Eustice, R., Pizarro, O., Singh, H., Howland, J.: UWIT: underwater image toolbox for optical image processing and mosaicking in Matlab. In: Proceedings of the International Symposium on Underwater Technology, Tokyo, Japan, pp. 141–145 (2002)
15. Eustice, R.M., Singh, H., Leonard, J.J.: Exactly sparse delayed-state filters for view-based SLAM. IEEE Trans. Robot. **22**(6), 1100–1114 (2006)
16. Eustice, R.M., Pizarro, O., Singh, H.: Visually augmented navigation for autonomous underwater vehicles. IEEE J. Ocean. Eng. **33**(2), 103–122 (2008)
17. Eustice, R.M., Singh, H., Whitcomb, L.L.: Synchronous-clock one-way-travel-time acoustic navigation for underwater vehicles. J. F. Robot. **28**(1), 121–136 (2011)
18. Fallon, M.F., Papadopoulis, G., Leonard, J.J.: A measurement distribution framework for cooperative navigation using multiple AUVs. In: Proceedings of the IEEE International Conference on Robotics and Automation, Anchorage, AK, pp. 4256–4263 (2010)
19. Fischler, M.A., Bolles, R.C.: Random sample consensus: a paradigm for model fitting with application to image analysis and automated cartography. Commun. ACM **24**(6), 381–395 (1981)
20. Haralick, R.M.: Propagating covariance in computer vision. In: Proceedings of the International Conference Pattern Recognition, vol. 1, Jerusalem, Israel, pp. 493–498 (1994)
21. Hartley, R., Zisserman, A.: Multiple View Geometry in Computer Vision, 2nd edn. Cambridge University Press, Cambridge (2004)
22. Hover, F.S., Vaganay, J., Elkins, M., Willcox, S., Polidoro, V., Morash, J., Damus, R., Desset, S.: A vehicle system for autonomous relative survey of in-water ships. Mar. Technol. Soc. J. **41**(2), 44–55 (2007)

23. Ila, V., Polok, L., Solony, M., Smrz, P., Zemcik, P.: Fast covariance recovery in incremental nonlinear least square solvers. In: Proceedings of the IEEE International Conference on Robotics and Automation, Seattle, WA, USA, pp. 4636–4643 (2015)
24. Kaess, M., Ranganathan, A., Dellaert, F.: iSAM: incremental smoothing and mapping. IEEE Trans. Robot. **24**(6), 1365–1378 (2008)
25. Kaess, M., Johannsson, H., Roberts, R., Ila, V., Leonard, J.J., Dellaert, F.: iSAM2: incremental smoothing and mapping using the Bayes tree. Int. J. Robot. Res. **31**(2), 216–235 (2012)
26. Kim, B., Kaess, M., Fletcher, L., Leonard, J., Bachrach, A., Roy, N., Teller, S.: Multiple relative pose graphs for robust cooperative mapping. In: Proceedings of the IEEE International Conference on Robotics and Automation, Anchorage, Alaska, pp. 3185–3192 (2010)
27. Kim, A., Eustice, R.M.: Real-time visual SLAM for autonomous underwater hull inspection using visual saliency. IEEE Trans. Robot. **29**(3), 719–733 (2013)
28. Kim, A., Eustice, R.M.: Active visual SLAM for robotic area coverage: theory and experiment. Int. J. Robot. Res. **34**(4–5), 457–475 (2015)
29. Kinsey, J.C., Eustice, R.M., Whitcomb, L.L.: Underwater vehicle navigation: recent advances and new challenges. In: IFAC Conference on Manoeuvring and Control of Marine Craft, Lisbon, Portugal (2006)
30. Kummerle, R., Grisetti, G., Strasdat, H., Konolige, K., Burgard, W.: g2o: a general framework for graph optimization. In: Proceedings of the IEEE International Conference on Robotics and Automation, Shanghai, China, pp. 3607–3613 (2011)
31. Leonard, J.J., Durrant-Whyte, H.: Simultaneous map building and localization for an autonomous mobile robot. In: Proceedings of the IEEE/RSJ International Conference on Intelligent Robots and Systems, Osaka, Japan, pp. 1442–1447 (1991)
32. Lowe, D.: Distinctive image features from scale-invariant keypoints. Int. J. Comput. Vis. **60**(2), 91–110 (2004)
33. Lu, F., Milios, E.: Globally consistent range scan alignment for environment mapping. Auton. Robot. **4**(4), 333–349 (1997)
34. Meduna, D., Rock, S., McEwen, R.: Low-cost terrain relative navigation for long-range AUVs. In: Proceedings of the IEEE/MTS OCEANS Conference and Exhibition, Woods Hole, MA, USA, pp. 1–7 (2008)
35. Ozog, P., Eustice, R.M.: Real-time SLAM with piecewise-planar surface models and sparse 3D point clouds. In: Proceedings of the IEEE/RSJ International Conference on Intelligent Robots and Systems, Tokyo, Japan, pp. 1042–1049 (2013)
36. Ozog, P., Carlevaris-Bianco, N., Kim, A., Eustice, R.M.: Long-term mapping techniques for ship hull inspection and surveillance using an autonomous underwater vehicle. J. F. Robot. **33**(3), 265–289 (2016)
37. Partan, J., Kurose, J., Levine, B.N.: A survey of practical issues in underwater networks. ACM SIGMOBILE Mob. Comput. Commun. Rev. **11**(4), 23–33 (2007)
38. Paull, L., Seto, M., Leonard, J.J.: Decentralized cooperative trajectory estimation for autonomous underwater vehicles. In: Proceedings of the IEEE/RSJ International Conference on Intelligent Robots and Systems, Chicago, IL, pp. 184–191 (2014)
39. Roumeliotis, S.I., Bekey, G.A.: Distributed multirobot localization. IEEE Trans. Robot. Autom. **18**(5), 781–795 (2002)
40. Smith, R., Self, M., Cheeseman, P.: Estimating uncertain spatial relationships in robotics. In: Cox, I., Wilfong, G. (eds.) Autonomous Robot Vehicles, pp. 167–193. Springer, Berlin (1990)
41. Thrun, S., Liu, Y., Koller, D., Ng, A.Y., Ghahramani, Z., Durrant-Whyte, H.: Simultaneous localization and mapping with sparse extended information filters. Int. J. Robot. Res. **23**(7–8), 693–716 (2004)
42. Thrun, S., Montemerlo, M.: The graph SLAM algorithm with application to large-scale mapping of urban structures. Int. J. Robot. Res. **25**(5–6), 403–429 (2006)
43. Walls, J.M., Cunningham, A.G., Eustice, R.M.: Cooperative localization by factor composition over faulty low-bandwidth communication channels. In: Proceedings of the IEEE International Conference on Robotics and Automation, Seattle, WA (2015)
44. Walls, J.M., Eustice, R.M.: An origin state method for communication constrained cooperative localization with robustness to packet loss. Int. J. Robot. Res. **33**(9), 1191–1208 (2014)

Exploring New Localization Applications Using a Smartphone

Fredrik Gustafsson and Gustaf Hendeby

Abstract Localization is an enabling technology in many applications and services today and in the future. Satellite navigation often works fine for navigation, infotainment, and location-based services, and it is the dominating solution in commercial products today. A nice exception is the localization in Google Maps, where radio signal strength from WiFi and cellular networks are used as complementary information to increase accuracy and integrity. With the ongoing trend with more autonomous functions being introduced in our vehicles and with all our connected devices, most of them operated in indoor environments where satellite signals are not available; there is an acute need for new solutions. At the same time, our smartphones are getting more sophisticated in their sensor configuration. Therefore, in this chapter we present a freely available Sensor Fusion app developed in house, how it works, how it has been used, and how it can be used based on a variety of applications in our research and student projects.

1 The Smartphone as a Sensor Platform

The modern smartphone has during the last few years developed into a highly capable sensor platform. The start of this development was with the introduction of the iPhone by Apple in 2007. The iPhone identified a new need for connected people in the modern world. Apple was followed by other companies, predominantly using the Android operating system backed by Google. Most smartphone manufacturers release new devices on a regular basis. In their strive to gain market shares, they push their developers to provide for new interactive services and increasingly immersive user experiences. This has led to better screens, faster CPUs, and, not least, more and better sensors being integrated in smartphones.

F. Gustafsson (✉) · G. Hendeby
Linköping University, 581 83 Linköping, Sweden
e-mail: fredrik.gustafsson@liu.se

G. Hendeby
e-mail: hendeby@isy.liu.se

© Springer International Publishing AG 2017
T.I. Fossen et al. (eds.), *Sensing and Control for Autonomous Vehicles*,
Lecture Notes in Control and Information Sciences 474,
DOI 10.1007/978-3-319-55372-6_8

We will from now on focus on Android platforms, since it allows third parties to access radio (WiFi) measurements, which are very important in many applications.

1.1 Available Sensors

The first Android smartphone released to the public, the HTC Dream 2009, included only a minimal set of sensors: an accelerometer, a magnetometer, and a GPS receiver. Since then, the number and variety of sensors included in smartphones have increased dramatically; and today the flagship models include not only accelerometer, gyroscope, magnetometer, and GPS, but also several other sensors, such as barometer, thermometer, humidity sensor, heart rate, oxygen saturation, etc. (the exact setup depends on the brand and the model). These sensors are intended to provide information about the user as well of his/her surroundings. See Table 1 for examples of different phones and the sensors they include. The android platform allows for extracting measurements from these sensors. The intention is to allow app developers to provide the user with an immersive experience where he/she interacts with his/her immediate environment.

Except for the type of sensors mentioned above, a smarthphone also hosts several microphones and cameras which are needed for the phone's primary usage, but from which it is also possible to extract interesting measurements. A smartphone furthermore has several radio stacks to enable usage of the cellular network, WiFi, Bluetooth, *near field communication* (NFC), etc. Direct access to the radio measurements is not provided, but metadata from the channels can be extracted, i.e., information about available base stations, access points, the *received signal strength* (RSS), which in itself are interesting measurements.

Finally, concept phones which profile themselves by providing additional sensors exist on the market. Examples of this include:

- The SpoonPhone by BeSpoon,[1] which comes equipped with an *ultra-wide band* (UWB) unit for positioning of special UWB tags.
- The Phab 2 Pro from Lenovo,[2] which provides real-time depth images from the built-in stereo camera using Google's Tango.
- The CAT S60, which is one example of a smartphone with integrated thermal camera.

---

[1] URL: http://spoonphone.com/en/.
[2] URL: http://shop.lenovo.com/se/sv/tango/.

Table 1 Different generations of Android smartphones and which sensors they include

Device	HTC Dream	Nexus One	Nexus S	Galaxy Nexus	Nexus 4	Nexus 5	Samsung Galaxy S4	Nexus 6	Nexus 5X	Nexus 6P	Samsung Galaxy S6	HTC One M9	Motorola Moto G^4
Release year	2009	2010	2010	2011	2012	2013	2013	2014	2015	2015	2015	2015	2016
Android version	1.6	2.1	2.3	4.0	4.2	4.4	4.2	5.0	6.0	6.0	5.0	5.0	6.0
Accelerometer	✓	✓	✓	✓	✓	✓	✓	✓	✓	✓	✓	✓	✓
Ambient light		✓	✓	✓	✓	✓	✓	✓	✓	✓	✓	✓	✓
Barometer				✓	✓	✓	✓	✓	✓	✓	✓		
Fingerprint scanner									✓	✓	✓		✓
GPS	✓	✓	✓	✓	✓	✓	✓	✓	✓	✓	✓	✓	✓
Gyroscope				✓	✓	✓	✓	✓	✓	✓	✓	✓	✓
Heart rate											✓		
Humidity							✓						
Magnetometer	✓	✓	✓	✓	✓	✓	✓	✓	✓	✓	✓	✓	✓
NFC				✓	✓	✓	✓	✓	✓	✓	✓	✓	
O^2 saturation											✓		
Proximity		✓	✓	✓	✓	✓	✓	✓	✓	✓	✓	✓	✓
Thermometer							✓						

Including these new sensors in smartphones, or as add-ons to them, provides a source of highly interesting sensor data. Price, size, and other practical considerations are what so far limit these sensors from gaining wide adaptation from smartphone manufacturers.

As illustrated above and indicated in evaluation studies such as [10], the sensory data available from smartphones is impressive, and diverse enough to be considered an interesting source of information for researchers, teachers, and *do-it-yourselfers* (DIYers). The reasons to consider the smartphone are manyfold. A smartphone comes as a portable and wireless consumer product where all the sensors have been properly integrated. The package also includes computing power to perform proper preprocessing (quad core processors with sufficient RAM is standard today) and is prepared to distribute the data in several different ways. Another benefit is the price. Even the most high-end devices typically cost a few hundred dollars, which is considerably cheaper than buying the components and then putting together the equivalent laboratory equipment. However, the smartphone is designed to provide the user with experiences and not as a sensor, hence software is needed to make use of the available hardware. The Sensor Fusion app is one example of a tool making this easy.

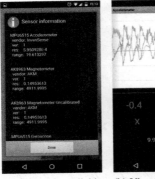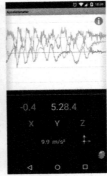

(a) List of available sensors. (b) Visualizing the accelereations measured by the accelerometer. (c) Visualizing the position measured by the GPS. (d) Information view for the accelerometer.

Fig. 1 Illustration of the Sensor Fusion app

1.2 Sensor Fusion App

The Sensor Fusion app[3] is an app for Android devices that gives easy access to available sensor data. The Sensor Fusion app has the following three main functionalities:

- Visualizing measurements directly on the device.
- Logging measurements to file for offline analysis.
- Streaming measurements from the device to a computer via a network connection in real time.

The app does also provide information about which sensors (brand, resolution and range, and sometimes chip type) are available in the device it is installed on, as illustrated in Fig. 1a.

In October 2016, the most current version of the Sensor Fusion app publicly available from Google Play store is version 2.0b8. It supports the sensors listed in Table 2, but is under constant development to include more sensors. For more details about the app see Table 2.

Visualization of sensor data is illustrated in Fig. 1b, c. The purpose of the visualization views is to give the user intuition and a fast overview of the properties of the quantity she is studying. For educational purposes, each view comes with an information view explaining the physical significance and suggesting simple experiments to understand it better. See Fig. 1d for an illustration of this for the accelerometer.

The primary purpose of logging and streaming measurements from the Sensor Fusion app is to allow for online and offline analysis of the measurements. Data can either be saved directly to the device or streamed via any available network connection to a minimal server. The intuitive interface for this is illustrated in Fig. 2a. The app

---

[3]The Sensor Fusion app is available from the Google Play store: https://goo.gl/0qNyU.

Table 2 Facts about the Sensor Fusion app as of October 2016. Please see the supporting homepage and Google Play Store for up-to-date information

Name:	Sensor Fusion app
Current Version:	2.0b8
Google Play Store:	https://goo.gl/0qNyU
Supporting Page:	http://sensorfusion.se/sfapp/
Android Support:	10+
Rating:	4.3/5
Cost:	Free
Installs:	∼ 7000
Supported sensors:	• acceleration, [m/s^2]
	• angular rates from the gyroscope (raw and calibrated) [rad/s]
	• magnetic field (raw and calibrated) [μT]
	• pressure [hPa]
	• proximity [cm] (some devices only provide a binary near/far)
	• light [lx]
	• ambient temperature [°C]
	• position from the GPS (latitude, longitude, altitude) [°, °, m]
	• orientation [unit quaternion] (this is a soft sensor)
	• received signal strength (RSS) and other relevant information about WiFi networks in the environment
	• RSS and other relevant information from cellular providers in the area

makes use of the Android networking capability, which allows the user to use WiFi or cellular networks to transfer the measurements, as well as allows for utilizing existing VPN connections. This makes the app very flexible even in corporate settings with network restrictions.

A Java library has been made available for free[4] that provides the server functionality needed to receive the measurements. Providing the server as a Java library makes it easy to develop applications around the streamed data, and allows for straightforward integration in common scientific tools such as MATLAB and Mathematica.

The log file format and the communication protocol are readable formatted text. Each individual measurement is represented by one row, consisting of: a time stamp (in milliseconds), a data source identifier, followed by the measurement itself, as exemplified in Fig. 2b. This design is not optimized for bandwidth utilization, but makes it easy to implement servers to receive the streamed data in other languages if necessary. It is also beneficial when trying to get a quick initial understanding for the data being collected.

---

[4] Available from the supporting homepage: http://sensorfusion.se/sfapp/.

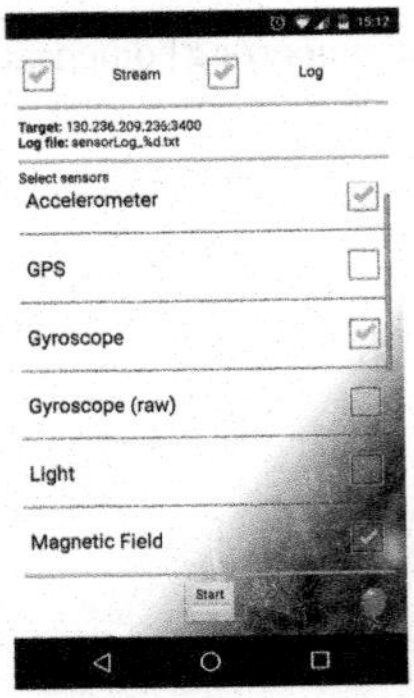

(a) Interface for logging and streaming measurements.

```
78270241  ACC -1.5330963  8.94664 3.073944
78270241  GYR 0.00340271  0.012924194 -0.014724731
78270250  ACC -1.5283356  8.94426 3.009674
78270250  GYR 0.0012664795  0.007583618 -0.020065308
78270247  MAG 2.8381348 -41.152954  -23.033142
78270261  ACC -1.5664215  8.968063  3.0525208
78270261  GYR 0.00340271  0.0033111572  -0.024337769
78270264  MAG 1.6082764 -39.01825 -22.434998
78270270  ACC -1.6164093  8.982346  2.9882507
78270270  GYR 1.9836426E-4  -0.0052337646 -0.026473999
78270280  ACC -1.640213 9.003769  3.0239563
78270280  GYR -0.005142212  -0.0052337646 -0.025405884
78270290  ACC -1.6497345  8.94664 3.0501404
78270290  GYR -0.012619019  0.0033111572  -0.022201538
78270280  MAG 1.4709473 -38.760376  -22.317505
78270300  ACC -1.6449738  8.9561615 3.040619
78270300  GYR -0.011550903  0.0054473877  -0.021133423
```

(b) Example of the data format used. The first column is a time stamp in millisecond since an unspecified point in time, followed by a tag indicating the producing sensor, followed by the measurements.

Fig. 2 Illustration of the logging and streaming from the Sensor Fusion app

1.3 Example: Implementing an Orientation Estimation Lab

In this section, a laboratory exercise is used to exemplify how the Sensor Fusion app can be used. The laboratory exercise is used in a course in Sensor Fusion given at advanced level at Linköping University [7]. The Sensor Fusion app is instrumental for the way the laboratory exercise is designed. By using the app, it is possible to reach a level of student involvement and interaction that would have been impossible to achieve with traditional educational sensor platforms. A more detailed description about the lab and how the Sensor Fusion app is used can be found in [6, 8].

The task of the laboratory exercise is to estimate the orientation of a smartphone in real time, using measurements from the inertial sensors and the magnetometer available in the phone. The result is then compared to the built-in solution provided by the phone. Robustness to disturbances is a key feature, and different methods to reject outliers are demonstrated. See Fig. 3. In this way, the students get hands-on experience of standard filtering techniques and typical problems associated with actual measurements. Those students who wish to do so (a majority of the students) can perform the exercise using their own phone.

Within the first few minutes of the laboratory session, the students install the Sensor Fusion app on their phones, and connect it to the university WiFi. They are first asked to play around with the app to get a feeling for the measurements they should use. They then implement a filter in MATLAB for orientation estimation based on the available measurements. The filter is fed with measurements in real time from their phone, which makes it possible to easily demonstrate the significance of the different measurements and different design choices. The accelerometer measurements provide information about the down direction, the magnetometer can be used to provide a forward direction, whereas the gyroscope helps handle quick changes but is

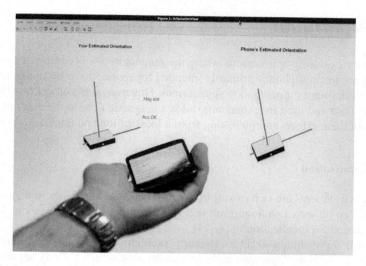

Fig. 3 Student engaged in comparing solutions to the laboratory exercise

prone to drift over time. It is also possible to add external disturbances, and based on this discuss about different methods to detect and mitigate outliers. The flexibility of the platform, and its user friendliness, have been key factors to make the lab successful in achieving its learning objectives, and well appreciated judging from the student feedback and course evaluations. Another important factor is that the students can use their own phones, which seems to motive them more and make the subject more tangible.

2 Examples of Demonstrators and Rapid Prototypes

The remaining sections are devoted to illustrating how the app can be used in rapid prototyping or demonstrators of research. There are of course numerous examples in the literature that fit into this scope. However, we focus on selected examples from our own research. All examples are quite recent, but some of them were performed before the Sensor Fusion app was released the first time. These applications have been the motivation for the design of the app, with the main idea that the app could have been used if it had only existed at that time, and it would have saved a lot of implementation time. Some of the examples are performed with the app more recently, and these show how a smartphone with the app actually works as a rapid prototyping platform. Many of the examples are from master's thesis projects, which in many cases is a convenient way to test new concepts.

2.1 Magnetometers

We start by exploring applications where the magnetometer is central. The magnetometers are in smartphones primarily intended for an electronic compass, which is of course important for navigation applications. However, soft iron deflects the earth magnetic field and hard iron (magnets) adds a magnetic field to the earth magnetic field. Both these effects are interesting from a localization and tracking perspective.

Train Localization

Indoor environments are rich of soft iron which provides variations in the magnetic field that can be seen as a fingerprint that is a function of the 3D position. This has been explored for localization in [4, 14].

We will here explain a 1D localization problem, namely the positioning of a train along a railway. This also includes determining which railway segment it is, and which of a number of parallel railways that the train is following. The latter is important for autonomous trains, since GPS does not provide the resolution or integrity required to distinguish between parallel rails.

The main idea is that the iron in the railway is magnetized, and the variations in its magnetic filed yields a fingerprint, which seen over an interval becomes unique. Figure 4a shows the spectrogram of the longitudinal magnetometer m_x as a function of time. A fingerprinting-based localization algorithm can be constructed with the following steps:

1. Collect inertial and GPS data of the same railway segment many times.
2. To each measurement set, apply a navigation algorithm (a standard Kalman smoother application) to get an accurate position estimate as a function of time.
3. Resample the magnetometer measurements from time to space.
4. Average the magnetometer signal as a function of position, and compute the standard deviation for each position.
5. The fingerprinting algorithm consists of a particle filter with position and speed as states and the magnetometer measurement (or the spectrum over a space interval) as measurement. This particle filter application falls into the framework presented in [1], using the fingerprint fit as a likelihood function in the measurement update.

The details of this approach can be found in [5].

Tracking Vehicles

Vehicles have a lot of iron, both soft and hard, and its influence on the magnetic field can be observed in a magnetometer at up to tens of meters distance. This makes it possible to detect and even track single vehicles, for the purpose of traffic monitoring on an autonomous highway for instance. Figure 5 shows the result from one test. Usually a particle filter is used when there are multiple hypotheses, but since there

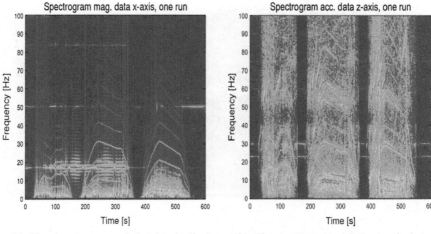

(a) The spectrogram of the longitudinal magnetometer m_x.

(b) The spectrogram of the vertical accelerometer a_z.

Fig. 4 Spectrogram captured from magnetometer and accelerometer from a train. Both plots show the same ten minutes journey, including three full stops, which is traversed many times to build the fingerprint. Some of the frequencies depend on the fundamental frequency of the wheels and its harmonics, which is only speed dependent and not position dependent. These frequencies have to be canceled out when building the fingerprint and computing the match with an existing fingerprint. Otherwise, it is the magnitude of the 3D magnetometer and amplitude of vertical acceleration, respectively, which is used as a fingerprint

is a limited number of ways a car can enter the scene in this case, a Kalman filter bank was used. The details are found in [16], and an application to traffic monitoring in [18].

Tracking Magnets

A network of magnetometers can also be used to track magnets in small volumes corresponding to a desk, as demonstrated with many use cases in [17]. A single permanent magnet has a magnetic field that is rotational invariant, so tracking must be limited to five degrees of freedom (3D position and 2D orientation).

A particular case is shown in Fig. 6, where the puck in a table hockey game is tracked. The symmetry of the puck (5 degrees of freedom), its high speed, where its position is often occluded from the viewer, makes this an excellent application of magnetic tracking. The magnetometers used in this demonstrator are of the same kind that can be found in smartphones, so it would be possible to use four smartphones under the board instead.

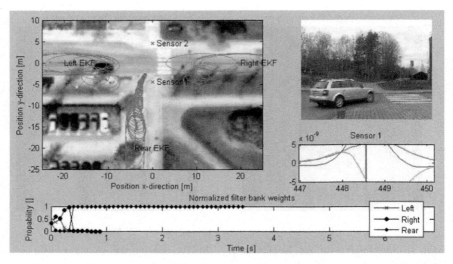

Fig. 5 Snapshot of a scenario where a vehicle is tracked with two three-axis magnetometers. The upper left figure shows the estimate with confidence bounds from three separate Kalman filters, each one conditioned on a different prior (car from *left*, *right*, *below*, respectively). The lower figure shows the relative probability for each Kalman filter, where the correct hypothesis dominates after 0.5 s. The upper right figure shows a snapshot from a video camera, and the figure below the magnetometer signal at this time instance. *Youtube* https://youtu.be/Q-UEO5aq8fs

Fig. 6 Tracking the puck in a table hockey game. A small magnet is mounted inside the puck, and four three-axis magnetometers are placed in a rectangular shape below the surface. From the 12 measurements, a 5D pose (magnetic field is rotation invariant) of the puck can be computed at a high sampling rate. *Youtube* https://youtu.be/oDqFIC6MIFE

2.2 Accelerometer

The accelerometer is primarily used as a gravity sensor in smartphone applications, where the tilt angle relative to the horizontal plane is computed. Together with a gyroscope and a magnetometer, it provides an orientation estimate as discussed in Sect. 1.3. We will here describe a couple of unconventional applications.

Localization of Trains

This section outlines an alternative to the method described in section "Train Localization" to localize a train along a railway. The physical property used here is that the railway is not perfectly straight and even, but contains small deformations that give rise to measurable variations in acceleration. In [5], it is demonstrated that this signature is as unique as the magnetic field variations, and gives the same accuracy. The localization algorithm is quite similar to the one described in more detail in section "Train Localization". First, the acceleration needs to be resampled as a function of distance rather than time, which requires a separate smoothing filter, and then the acceleration profile from multiple runs is averaged to form a fingerprint. The localization algorithms need at least the position and speed as states, to be able to match the measured 3D acceleration with the correct fingerprint. Figure 4b shows one example of measured acceleration as a function of time, which can be used to build up the fingerprint, or be matched with an existing fingerprint database.

Roller Coaster Anomaly Detection

As a variation of the same theme as above, the acceleration profile can also be used to detect deviations from what is normal according to the fingerprint. Figure 7 shows how the sensors are mounted on the train, and examples of measurements from the accelerometers and gyroscopes, respectively. Again, the measurements need to be resampled to position, rather than time. GPS does not provide accurate position in this environment, probably due to high speeds and acceleration. Therefore, a smoother based on inertial information has to be used. Here, the standstill before and after the ride can be used to synchronize the filter. The ride time depends on the weight of the passengers and the humidity, primarily. The data collection is performed during normal testings in the morning, when the train is empty, so the weight of the train set is the same each run. Contrary to the application above with train localization, there is not a particular data set with training data for building the fingerprint. Here, rather all data sets can be used to build the fingerprint, taking some forgetting into account. A new run that has a temporal peak or new frequency that is outside the confidence bounds of the fingerprint, indicates an anomaly. For instance, a temporal peak can depend on a broken weld in the rail, and a new frequency can depend on an imbalance in the wheel sets of the train.

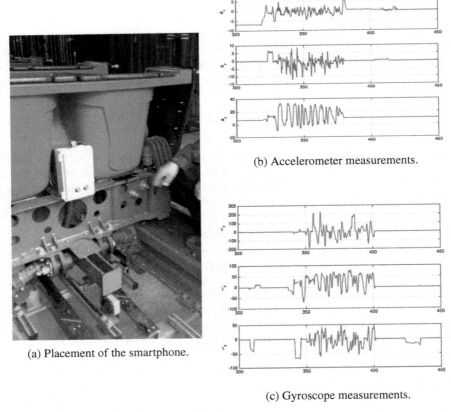

Fig. 7 A smartphone with the Sensor Fusion app is placed in the rear end of the roller coaster train (**a**). During the morning testing routines, the accelerations (**b**) and angular rates (**c**) are measured during three runs with an empty train. *Youtube* https://youtu.be/6FLPGeCHkrk

Pothole Detection

Potholes in the surface of our roads are a nuisance to drivers. When hitting a pothole, the vehicle is subject to an impulse force that gives a peak in acceleration. This can be used to detect and grade a pothole, and the GPS (or any other localization algorithm) is then used to geotag this information.

There are several publicly available apps for pothole detection. We here outline the work [12], where a pothole detector in the vehicle is combined with a cloud-based clustering algorithm. The idea is to collect a database of accurate locations of potholes, and the grade of each pothole, to be used by other drivers or by road administration authorities to fix the potholes in order of severity (how big they are, and how often drivers hit them). Figure 8 shows an example of a partial map of reported pot holes from a fleet of vehicles.

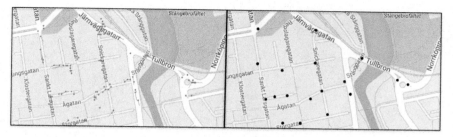

Fig. 8 A fleet of vehicles has been equipped with pothole detection algorithms and each detection is reported to the cloud together with GPS position. Since the GPS is not perfect, the reports are scattered around the true position. Therefore, a clustering algorithm is implemented in the cloud that adaptively updates a map where each detection is associated with a position

2.3 GPS

GPS suffers from multipath in urban areas due to a lack of clear sky and line of sight to the satellites. This is commonly referred to as the Manhattan problem, see Fig. 9a. Also vegetation gives scattering in the foliage that deteriorates the accuracy. In many applications, there is a need to correct the GPS position by means of sensor fusion.

Consider for example the detections in Fig. 8. Since these are taken from vehicles, the position is restricted to the streets. An ad hoc solution used in some simple navigation systems is to just project the position to the closest street segment. A thorough exposition of this problem is offered in [2].

Sprint Orienteering

Here we point out a challenging application of estimating the trajectory of sprint orienteers. Figure 9b illustrates how inaccurate GPS positions can be corrected with a filter that also incorporates map information and the rules of the game. The algorithm uses raw GPS measurements and the knowledge that the sprinter follows the rules, that is, only runs on the white areas on the map and passes all controls.

Rhino Tracking

Tools for wildlife monitoring are today a well-established market segment for animal conservation. The most common solution is to put a GPS collar around the neck of the animal. The collar contains besides GPS a battery and some radio communication ability. For cattle monitoring, one alternative could be to put smartphones in the collars, because here regular charging is plausible.

(a) Illustration of the Manhattan problem, where the buildings disturb the GPS measurements. A few large outliers and the zigzag pattern within blocks contribute to the overestimated distance.

(b) Snapshot of a video showing the trajectory of a sprinter comparing his own view in blue (can be seen as ground truth), what is shown live and after the completion on the organisers web site (red) and the sensor fusion algorithm (green). This example shows the trajectory of the winner of the world cup competition in Venice 2014.
Youtube https://youtu.be/tu2HfpJne5M

Fig. 9 Illustration of problems with GPS positioning in urban environments in standard apps, and a solution to counteract multipath effects (the Manhattan problem)

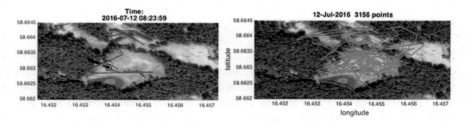

Fig. 10 The right figure shows the raw GPS measurements sampled every 10 s during one day. There are frequent outliers, and a high noise level (compared to what is obtained in a smartphone). The *left figure* shows a snapshot from a Kalman filter with outlier detection. The filter also naturally estimates heading. *Youtube* https://youtu.be/oO5dpC_NmNQ

In project Ngulia [3], a GPS foot band has been evaluated on a rhino in the Kolmården Wildlife Park, see Fig. 10. The advantage of processing the GPS positions by a Kalman filter is to reject outliers and also estimate the velocity vector, including the heading direction.

2.4 Camera

The smartphone captures images with the ability to also tag them with position (GPS) and pan/tilt (orientation filter), in contrast to standard cameras. This opens up new possibilities for tracking objects in ground coordinated, and synchronizing camera networks.

Using a terrain model, or assuming flat ground, the camera pose can be used to compute an accurate image footprint on the ground. This facilitates processing of images from camera networks, where a point on the ground is covered by several cameras. As one example, six smartphones were placed on the savannah part of Kolmården Wildlife Park during one day, and tens of thousands of images were collected. A large number of the images were annotated and used to build a classifier that can distinguish rhinos from humans and other animals, see Fig. 11.

2.5 Received Signal Strength

The RSS from different radio sources in the environment can be a valuable source of information when trying to perform localization. One way to use the measurements is to SLAM using RSS fingerprints, as in [11]. Here another usage of RSS is presented.

Search and Rescue

One interesting use case of the Sensor Fusion app is to evaluate how useful RSS is in a search and rescue situation. The scenario is that a person is lost, or victim of an avalanche or earthquake, but has a telephone and working battery. The rescue personnel can then use RSS measurements to locate the victim. Figure 12 shows an experiment where one smartphone was hidden on the ground with WiFi hot spot mode on (WiFi gives a higher sampling rate than the cellular network). Another smartphone was mounted on a drone, and the Sensor Fusion app collects position (GPS) and RSS, and transmits these measurements together with its GPS position, to the operator. On the operator side, a localization algorithm estimates the position of the hidden phone recursively, and shows this in the operator interface. This feedback enables the operator to efficiently steer the drone to maximize information and minimize time to convergence. The final estimation error is only a couple of meters. The details can be found in [15].

2.6 Microphone

A smartphone is often equipped with several microphones, one for sound and at least another one for noise cancelation. The microphones can also be used for acoustic localization. One example is presented in [13], where the speaker in the smartphone

Fig. 11 Illustration of using machine learning to classify rhinos. First, different steps in the learning chain are illustrated, then a snapshot of online classification and labeling is shown. *Youtube* https://youtu.be/xzyCbWGitOc

sends out a wideband acoustic signature, and the microphone records its echo. The time delay corresponds to the round trip time to the closest object, which can be used for localization purposes.

Figure 13 illustrates another application from [9].

Exploring New Localization Applications Using a Smartphone

Fig. 12 A view (*left*) from a drone carrying a smartphone with the Sensor Fusion app. The app streams WiFi RSS from a hidden phone together with the drone's GPS position (*yellow circles*) to a computer. The position of the hidden phone (*green symbol*) is estimated adaptively, and shown in the operator interface (*red line*) in the user interface

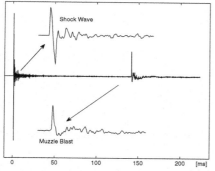

(a) The microphone can detect first the impulsive shockwave from a supersonic bullet, and then the explosive muzzle blast.

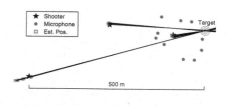

(b) A network of microphones provide sufficient information for lozalizing the sniper position and the aiming direction.

Fig. 13 A network of microphones, each one estimating one time difference, is used to localize a sniper

3 Summary

The Sensor Fusion app has proved to be an important tool for us both in research and when teaching, and it is under constant development. The next step is probably to incorporate Bluetooth RSS. Another feature in the pipeline is the ability to start and stop recordings of video and sound. In this way, the time interval and the file names can be synchronized with the other sensor data, and eventually streamed over the network over different protocols.

Another vision is to create a repository with desktop demonstrators based on streamed sensor data. One can imagine all of the examples in this chapter to be implemented as standalone applications in, e.g., MATLAB.

The need for a flexible and versatile platform based on a smartphone with the Sensor Fusion app will become even more important in the *internet of things* (IoT) era. Localization is a kind of third enabling technology for IoT (after power and communication), and the sensor suit in a smartphone is perfect for localization in almost all kind of environments, as richly illustrated in this chapter. IoT requires application specific sensors and actuators, and several of these can be found on the smartphone itself, or as Bluetooth connected add-ons. This makes the proposed platform ideal for rapid prototyping sand cheap demonstrators.

Acknowledgements The first version of the Sensor Fusion app; it was developed in collaboration with HiQ (URL: http://www.hiq.se), funded by a SAAB award in the name of former CEO Åke Svensson received by Prof. Fredrik Gustafsson. These contributions are gratefully acknowledged. The authors also want to thank their colleagues and the students who have contributed to the research described in this chapter, as well as the Swedish Research Council (VR), Vinnova, and the Swedish Foundation for Strategic Research (SSF) for funding the described projects.

References

1. Gustafsson, F., Gunnarsson, F., Bergman, N., Forssell, U., Jansson, J., Karlsson, R., Nordlund, P.-J.: Particle filters for positioning, navigation, and tracking. IEEE Trans. Signal Process. **50**(2) (2002)
2. Gustafsson, F., Orguner, U., Schön, T.B., Skoglar, P., Karlsson, R.: Navigation and tracking of road-bound vehicles. Handbook of Intelligent Vehicles, pp. 397–434. Springer, Berlin (2012)
3. Gustafsson, F., Bergenäs, J., Stenmarck, M.: Project Ngulia: tracking rangers, rhinos, and poachers. ISIF Perspect. Inf. Fusion **1**(1), 3–15 (2016)
4. Haverinen, J., Kemppainen, A.: Global indoor self-localization based on the ambient magnetic field. Robot. Auton. Syst. **57**(10), 1028–1035 (2009). Appeared at 5th International Conference on Computational Intelligence, Robotics and Autonomous Systems, Incheon, Korea
5. Hedberg, E., Hammar, M.: Train localization and speed estimation using on-board inertial and magnetic sensors. Master's thesis no LiTH-ISY-EX–15/4893–SE, Department of Electrical Engineering, Linköpings universitet, Sweden (2015)
6. Hendeby, G.: Orientation estimation using smartphone sensors. http://www.control.isy.liu.se/student/tsrt14/file/orientation.pdf (2014). Accessed 28 Oct 2016
7. Hendeby, G.: TSRT14 Sensor Fusion. http://www.control.isy.liu.se/en/student/tsrt14/ (2016). Accessed 28 Oct 2016
8. Hendeby, G., Gustafsson, F., Wahlström, N.: Teaching sensor fusion and Kalman filtering using a smartphone. In: Proceedings of 19th IFAC World Congress, Cape Town, South Africa (2014)
9. Lindgren, D., Wilsson, O., Gustafsson, F., Habberstad, H.: Shooter localization in wireless microphone networks. EURASIP J. Adv. Signal Process. **1**, 2010 (2010)
10. Ma, Z., Qiao, Y., Lee, B., Fallon, E.: Experimental evaluation of mobile phone sensors. In: 24th IET Irish Signals and Systems Conference, Letterkenny, Ireland (2013)
11. Nilsson, M., Rantakokko, J., Skoglund, M., Hendeby, G.: Indoor positioning using multi-frequency RSS measurements with foot-mounted IMU. In: Proceedings of Fifth International Conference on Indoor Positioning and Indoor Navigation, Busan, Korea, Oct. 27–30 (2014)

12. Norén, O.: Monitoring of road surface conditions. Master's thesis no LiTH-ISY-EX–14/4773–SE, Department of Electrical Engineering, Linköpings universitet, Sweden (2014)
13. Nyqvist, H.E., Hendeby, G., Gustafsson, F.: On joint range and velocity estimation in detection and ranging sensors. In: 19th International Conference on Information Fusion, pp. 1674–1681, Heidelberg, Germany (2016)
14. Robertson, P., Frassl, M., Angermann, M., Doniec, M., Julian, B.J., Puyol, M.G., Khider, M., Lichtenstern, M., Bruno, L.: Simultaneous localization and mapping for pedestrians using distortions of the local magnetic field intensity in large indoor environments. In: International Conference on Indoor Positioning and Indoor Navigation, Montbeliard-Belfort, France (2013)
15. Sundqvist, J., Ekskog, J.: Victim localization using RF-signals and multiple agents in search & rescue. Master's thesis no LiTH-ISY-EX–15/4871–SE, Department of Electrical Engineering, Linköpings universitet, Sweden (2014)
16. Wahlström, N.: Target tracking using Maxwell's equations. Master's thesis no LiTH-ISY-EX–10/4373–SE, Department of Electrical Engineering, Linköpings universitet, Sweden (2010)
17. Wahlström, N.: Modeling of Magnetic Fields and Extended Objects for Localization Applications. Dissertations no 1723, Linköping Studies in Science and Technology, SE-581 83 Linköping, Sweden (2015)
18. Wahlström, N., Hostettler, R., Gustafsson, F., Birk, W.: Rapid classification of vehicle heading direction with two-axis magnetometer. In: The 37th International Conference on Acoustics, Speech and Signal Processing (ICASSP), Kyoto, Japan (2012)

Part III
Path Planning

Model-Based Path Planning

Artur Wolek and Craig A. Woolsey

Abstract Model-based path planning for autonomous vehicles may incorporate knowledge of the dynamics, the environment, the planning objective, and available resources. In this chapter, we first review the most commonly used dynamic models for autonomous ground, surface, underwater, and air vehicles. We then discuss five common approaches to path planning—optimal control, level set methods, coarse planning with path smoothing, motion primitives, and random sampling—along with a qualitative comparison. The chapter includes a brief interlude on optimal path planning for kinematic car models. The aim of this chapter is to provide a high-level introduction to the field and to suggest relevant topics for further reading.

1 Introduction

Path planning supports vehicle guidance, the second of three canonical tasks required to maneuver a vehicle: navigation, guidance, and control. Briefly, *navigation* is the process of estimating a vehicle's state of motion, *guidance* is the process of determining the desired state of motion (the *guidance command*), and *control* is the process of realizing the guidance command. A desired path can provide the reference condition needed to determine a guidance command.

For human-operated vehicles, guidance is managed by the operator. For an automobile driver maintaining her lane or for a ship's pilot maneuvering his vessel toward its mooring, the guidance task may be second nature. For automated vehicles, however, effective guidance requires an algorithm which solves a clearly defined problem. To enable an autonomous underwater vehicle to dock with a recharging station or an unmanned aircraft to land on a moving ship, one must automate the guidance process.

A. Wolek (✉)
American Society for Engineering Education, Washington D.C., USA
e-mail: artur.wolek.ctr@nrl.navy.mil

C.A. Woolsey
Virginia Tech, Blacksburg, VA, USA
e-mail: cwoolsey@vt.edu

This automated guidance algorithm must accommodate the resources available for sensing, computation, communication, and control and respect the limitations and uncertainty imposed by the vehicle dynamics and the environment.

Two categories of vehicle guidance—homing guidance and midcourse guidance—are especially common and a variety of algorithms are available for these tasks [1]. In homing guidance, the focus is on the terminal state: The vehicle should converge to a (possibly moving) target and guidance command updates are based on this objective. In midcourse guidance, the vehicle monitors and corrects its motion relative to a prescribed path. This pre-planned path is typically designed to minimize some cost such as time, energy, or risk, to maximize some value such as information gained, or both. Here, we focus on path planning to support midcourse guidance. Although we briefly survey the spectrum of path planning methods, we focus especially on those that incorporate a vehicle motion model. Model-based methods incorporate vehicle performance capabilities and limitations, ensuring the vehicle is able to follow the resulting path.

In discussing path planning, it is important to distinguish between a *path* and a *trajectory*. A *trajectory* is a time-parameterized state history which identically satisfies the equations of motion. For a smooth motion model, trajectories are smooth. A *path*, on the other hand, is simply a continuous, parametric curve through configuration space. (The notions of state and configuration are discussed in more detail in the following section.) While a path is parametric, the curve may not be parameterized by time and it may not be smooth.

Paths are a more general class of curves than trajectories, and they provide additional freedom that can be used to an advantage in vehicle motion control. One obvious advantage is that a vehicle's speed may be regulated independently of its motion along a path in order to ensure stability, respect performance limitations, or optimize a given criterion [2]. In fact, path following may even eliminate performance limitations that can arise in trajectory tracking, such as the bandwidth limit imposed by unstable zero dynamics [3].

While a given path may not correspond to a dynamically feasible trajectory, a trajectory does define a dynamically feasible path. Indeed, trajectory generation is a common approach to path planning when the path must be dynamically feasible. If perfect tracking of the parametric path is not required, however, there are alternative path planning approaches that address environmental constraints, computational efficiency, or other concerns. Having obtained a path which addresses these concerns, but which is dynamically infeasible, one can take further steps to ensure the actual vehicle path remains suitably close to the desired one.

In this chapter, we consider *model-based path planning*, recognizing that the guidance problem of interest will suggest a particular motion model. In Sect. 2, we review the vehicle dynamic models that are most commonly used in model-based guidance.

2 Vehicle Motion Models

2.1 General Principles

The *state* of a system is defined by a minimal set of time-dependent variables whose evolution according to some physical process characterizes the system's behavior for a given purpose. In vehicle navigation, guidance, and control, the appropriate definition of the vehicle's state depends on the problem at hand. To guide an autonomous underwater glider across an ocean, for example, it may be sufficient to define the vehicle state by its three-dimensional position. To dock an unmanned surface vessel with a moving host vessel requires more information: relative position and orientation, at least, and perhaps their rates of change. In discussing vehicle motion, we will occasionally distinguish between the state and the *configuration*, which is defined solely by position and orientation. The configuration may or may not define the complete state, depending on whether one adopts a kinematic or a dynamic motion model, as discussed below.

In modeling vehicle motion, the configuration is typically defined as an element in a smooth *configuration manifold* Q [4]. For translation in two-or three-dimensional space, the configuration manifold is simply $Q = \mathbb{R}^n$ where $n \in \{2, 3\}$. For rigid motion (i.e., translational and rotational motion), the configuration manifold is $Q = SE(n)$, the special Euclidean group in the plane ($n = 2$) or in three dimensions ($n = 3$). For multi-body systems, such as biomimetic vehicles, $Q = G \times S$ where G represents the configuration of the base body and S describes the relative configuration of the appendages, that is, the shape. In practice, the configuration of a given system may be constrained to some subset of the configuration manifold, which in turn constrains the feasible paths.

For kinematic models, where the velocity can be directly prescribed, the configuration manifold is the state space. For dynamic models, where forces and moments are incorporated explicitly, velocity is also part of the system state. In these cases, the state space is the *tangent bundle* TQ of the configuration manifold [4], that is, the space of all configurations and velocities.

For vehicles subject to motion constraints, the state space can sometimes be reduced. In the case of holonomic constraints, for example, one may simply eliminate unneeded configuration variables. For a nonholonomic constraint, such as the "no-slip" condition for a wheeled ground vehicle, the attainable velocities are restricted in a way that does not allow such a simple reduction of the state dimension. Nonholonomic control systems are discussed in [4, 5].

Table 1 summarizes several common state-space descriptions for vehicle motion. Figure 1 illustrates the models. In general, planning over a large space and time scale can be accomplished using a low-order model, but greater fidelity may be needed at smaller scales.

Having identified the relevant state space for model-based path planning, it remains to develop the motion model. While kinematic equations are straightforward, dynamic modeling can be challenging. Representing external forces and moments,

Table 1 Summary of state spaces for common vehicle motion models

Motion model	State space	Description
Kinematic particle	$Q = \mathbb{R}^n$	Massless particle in planar ($n = 2$) or three-dimensional ($n = 3$) motion. Input: Velocity **V**
Dynamic particle	$TQ = T\mathbb{R}^n$	Mass particle in planar ($n = 2$) or three-dimensional ($n = 3$) motion. Input: Force **F**
Kinematic body	$Q = SE(n)$	Inertia-less rigid body in planar ($n = 2$) or three-dimensional ($n = 3$) motion. Input: Velocity **V** and angular rate $\mathit{\Omega}$
Dynamic body	$TQ = TSE(n)$	Rigid body in planar ($n = 2$) or three-dimensional ($n = 3$) motion. Input: Force **F** and moment **M**

which are influenced by the ambient environment and the vehicle's motion through it, requires disciplinary expertise. To accurately model the maneuvering of a surface vessel in an irregular seaway, for example, requires knowledge of ocean wave mechanics, ship resistance and propulsion, vessel hydrostatics and hydrodynamics, and ship stability and control.

Without diminishing the modeling challenge, we observe that the guidance and control system developer will typically work with a system of ordinary differential equations

$$\dot{\mathbf{x}} = \mathbf{f}(\mathbf{x}, \mathbf{u}, t), \tag{1}$$

where $\mathbf{x}(t) \in \mathbb{R}^n$ is the state vector expressed in local coordinates and $\mathbf{u}(t) \in \mathbb{R}^m$ is a vector of inputs. The vector field **f** that defines the motion model may also be influenced by random disturbances. One may incorporate knowledge of these disturbances (e.g., the average ambient flow velocity) in planning a path, with the expectation that midcourse guidance corrections and a well-designed feedback control law will compensate for random variations.

2.2 Vehicle-Specific Concerns

Autonomous Ground Vehicles. Considerable work on path planning has focused on wheeled mobile robots navigating over terrain. In early efforts, this work focused on basic challenges like reaching a goal while avoiding obstacles. As methods evolved, the guidance and control challenges became more difficult—prior maps might be uncertain or incomplete (e.g., [6]); obstacles might move (e.g., [7]); multiple vehicles

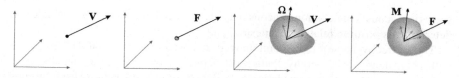

Fig. 1 Depictions of common state-space models

might need to coordinate their motion (e.g., [8]); loose, three-dimensional terrain might require more sophisticated motion models (e.g., [9]).

Most recently, advances in path planning for mobile robots have supported the emergence of self-driving cars, which promise to revolutionize modern ground transportation. Self-driving cars pose unique challenges, though, because of the wide range of environments in which they must operate. Ground transportation infrastructure ranges from highly structured road networks with active traffic management systems to "semi-structured" environments such as a sparsely occupied parking lot [10] to unstructured environments such as a construction site or an open-pit mine. These disparate environments place distinctly different constraints on the path planning problem, suggesting that the planning algorithm should account for the structure of the environment.

Autonomous Surface Vessels. Path planning for surface vessels is similar to path planning for ground vehicles. Indeed, for maneuvering in calm water, surface vessels are sometimes modeled as ground vehicles. Like ground vehicles in loose terrain, vessels slip laterally while turning. Sideslip is especially pronounced at low speeds [11], which may arise in safety critical maneuvers such as docking. For motions where the vehicle slips, a more sophisticated model is required.

Unlike the situation for ground vehicles, however, the "terrain" supporting a surface vessel may itself be moving. Currents affect a vessel's inertial motion and must be considered in path planning. Moreover, simply appending the flow velocity in the translational kinematic equations is not always sufficient; nonuniform currents can exert additional forces and moments [12]. Wind and waves may also perturb a vessel's motion in ways that must be accounted for [13].

The operating environment for surface vessels is much less structured than a road network. There are regulations for preventing collisions at sea ("Colregs"), but ocean vessels have far more freedom to plan paths around hazards. Moreover, these hazards are sparsely distributed, except when the vessel is operating in or near a port.

Autonomous Underwater Vehicles. AUVs are commonly used for ocean sampling and bathymetric mapping. They have also been used to inspect undersea pipelines and to dispense fiber optic cable. In these and similar applications, AUVs move slowly and operate well below the ocean surface. Simple motion models generally suffice for path planning, except in special scenarios where environmental constraints or mission objectives require more accurate dynamic models. For example, operating an AUV within a port or docking an AUV with a communication and power node requires a well-characterized rigid body dynamic model. In these scenarios, acoustic navigation beacons may also be needed to ensure the vehicle can determine its position relative to a desired path.

The ineffectiveness of radio communication in water requires that AUVs operate with a high degree of autonomy. Paths must be planned over long times and distances and may require explicit consideration of the environment: bathymetry, currents, tides, shipping lanes, etc.

In scenarios where it is not feasible to provide an independent undersea navigation system, such as an array of acoustic transponders, navigation may involve dead reckoning from attitude and heading reference system (AHRS) data. Water- or ground-relative odometry may be available if the vehicle is equipped with an acoustic Doppler velocity log but, in any case, navigation errors will accrue with time. In typical applications, the AUV must resurface periodically to obtain a position fix (e.g., using GPS) to correct these navigation errors. The vehicle might also upload data, while on the surface, to a supervisory agent which can revise the AUV's tasking based on this and other information. The challenge of uncertain navigation motivates risk-aware path planning, where the "risk" may be associated with hazards to navigation or even hazards to mission data; a bathymetric survey may be of little value if, for example, the pixel coordinates cannot be correlated to absolute positions.

Unmanned Aircraft Systems. UAS come in a variety of configurations but can be broadly categorized as fixed-wing, rotary-wing, or lighter-than-air. Fixed-wing aircraft travel efficiently at higher speeds, but generally require more infrastructure to support launch and recovery (a runway, a catapult, etc.). Rotary-wing aircraft, including helicopters and multirotors, can takeoff and land vertically but have restricted payload, range, and endurance relative to fixed-wing aircraft. Lighter-than-air craft (such as balloons or blimps) are large buoyancy-supported vehicles with a long endurance and ability to carry heavy payloads.

In nominal flight, a fixed-wing aircraft will establish and maintain wings-level flight at constant speed and altitude, making small deviations from this flight condition to attain waypoints, avoid conflicts, etc. For this reason, fixed-wing aircraft are often modeled as ground vehicles. Like surface vessels, aircraft can slip in turning flight, but human and automated pilots typically "coordinate" control surface deflections to eliminate sideslip. Modeling a fixed-wing aircraft as a ground vehicle that can climb and descend is often reasonable for short-term path planning. Over longer times and distances, a dynamic particle model (an "aircraft performance model") may suffice. For tasks requiring agility and precision, such as 3D terrain-following or recovery to a moving host, a rigid body "stability and control model" may be needed.

Because rotary-wing aircraft can hover, admissible paths are less constrained than are those for fixed-wing aircraft. On the other hand, these aircraft typically operate at lower altitudes where collisions with terrain are more likely. The expected proliferation of small UAS in the coming years will pose the additional challenge of increased traffic density, which may require coordinated strategies for path planning and control. Operating at low altitude while avoiding fixed and moving obstacles will presumably increase the need for three-dimensional planning strategies, a departure from the conventional approach of segregating aircraft by altitude. Integrating UAS into commercial airspace will require interaction with air traffic controllers.

Lighter-than-air craft are susceptible to wind because of their buoyancy. Depending on the design, the vehicle may not be able to navigate under its own power or may have limited control authority that makes path following challenging. These vehicles may be used in applications where accurate path following is not a high priority (e.g., advertising, cargo transport, surveillance).

3 Path Planning

3.1 Parameterizations

Given the system in Eq. 1, with some prescribed choice of input **u**, a trajectory $\tilde{\mathbf{x}}(t; \mathbf{x}_0, \mathbf{x}_1)$ is a time-parametrized state history that joins an initial state $\mathbf{x}(t_0) = \mathbf{x}_0$ with a terminal state $\mathbf{x}(t_1) = \mathbf{x}_1$. Trajectory planning is often used when the objectives have a clear temporal component (e.g., avoiding a nearby moving obstacle). Because trajectory tracking errors are time-dependent, trajectories can be difficult to track in the presence of disturbances, especially when the vehicle has limited control authority.

Letting **q** denote the vehicle's configuration, a path $\gamma(s; \mathbf{q}_0, \mathbf{q}_1)$, is a curve through configuration space that is parameterized by arclength $s \in \mathbb{R}$, or some other monotonically increasing parameter, an initial configuration $\mathbf{q}(s_0) = \mathbf{q}_0$, and a terminal configuration $\mathbf{q}(s_1) = \mathbf{q}_1$.

For particle models, where the configuration is just the vehicle's position, a path is a curve in two- or three-dimensional space. In developing guidance laws to track *smooth* paths, it is sometimes helpful to describe vehicle motion relative to a reference frame that evolves along the desired path. A common choice is the Frenet–Serret frame, which is defined at a given point along the path by the unit vectors that are tangent **T**, normal **N** and binormal **B** to the curve; see Fig. 2. Formally,

$$\mathbf{T} = \frac{d\gamma}{ds} \qquad \mathbf{N} = \frac{1}{\|\frac{d\mathbf{T}}{ds}\|} \frac{d\mathbf{T}}{ds} \qquad \mathbf{B} = \mathbf{T} \times \mathbf{N}$$

where the **T**, **N**, and **B** are interpreted as column vectors.

The geometry of a planar curve is defined entirely by its curvature $\kappa(s)$. For a three-dimensional curve, the torsion $\tau(s)$ that measures the out-of-plane twisting of the curve is also required. These quantities can be computed, for a given curve, and the evolution of the Frenet–Serret frame along the path satisfies the following equations:

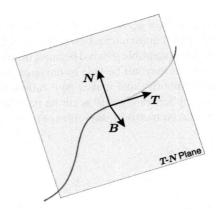

Fig. 2 The Frenet–Serret frame

$$\frac{d\mathbf{T}}{ds} = \kappa \mathbf{N} \qquad \frac{d\mathbf{N}}{ds} = -\kappa \mathbf{T} + \tau \mathbf{B} \qquad \frac{d\mathbf{B}}{ds} = -\tau \mathbf{N}.$$

The Frenet–Serret frame is undefined when the second derivative of γ vanishes. To avoid this issue, Kaminer et al. [14] adopt the "parallel transport" frame $\{\mathbf{T}, \mathbf{N}_1, \mathbf{N}_2\}$ whose orthonormal vectors satisfy the following equations:

$$\frac{d\mathbf{T}}{ds} = k_1 \mathbf{N}_1 + k_2 \mathbf{N}_2 \qquad \frac{d\mathbf{N}_1}{ds} = -k_1 \mathbf{T} \qquad \frac{d\mathbf{N}_2}{ds} = -k_2 \mathbf{T}$$

where $k_1(s)$ and $k_2(s)$ are related to the curvature and torsion as follows:

$$\kappa = \sqrt{k_1^2 + k_2^2} \qquad \tau = -\frac{d}{ds} \arctan(k_2/k_1).$$

Paths are easier to follow than trajectories because the path following error is geometric; the vehicle need not attain a given point along the desired path at a specific time.

Smoothly parameterized, curvilinear paths may be unnecessary, in practice. Reference paths are often generated using a simple sequence of waypoints. In this case, the reference path is piecewise linear; the vehicle tracks a given segment until it reaches a small region around the next waypoint and then begins tracking the next linear segment. For that matter, one may dispense with paths altogether by treating each consecutive waypoint as a target to be attained. This approach transforms the midcourse guidance problem of tracking a linear path into a simpler homing guidance problem. Homing is simple and effective if one is unconcerned about crosstrack deviations between waypoints.

3.2 Approaches

Path planning for mobile robots has been considered within multiple disciplines, resulting in many approaches. In early efforts, computer scientists and roboticists considered planning for structured environments that may be partially observable, sometimes in the context of a higher level decision making process. Because of the broad scope of this problem, many such approaches simplified the vehicle dynamics by either relying on discrete state and/or control representations or by ignoring dynamics altogether. Instead, these efforts focused on addressing environmental challenges using graph search, combinatorial optimization, decision theory, and related methods.

In parallel, control theorists worked on path planning and trajectory generation methods grounded in systems theory and the calculus of variations. These efforts focused on obtaining guidance laws with provable stability properties, analytically characterizing the conditions for optimality, and developing numerical methods to solve such problems. These methods can handle complex dynamics, as found in aeronautic and astronautic applications, but their ability to handle obstacles and other constraints is limited.

Recent progress has blurred the distinction among the various techniques, with both communities now addressing a similar class of model-based path planning problems. Before surveying the various approaches, we enumerate some of the key elements of path planning approaches used in practice:

1. **Dynamics**. The vehicle dynamic model should be rich enough to capture the salient features of the desired motion without over-complicating the planning problem. The selection of an appropriate model is related to the time and space scale of interest. Generally, higher dimensional models are required when precise vehicle response over a short distance or time horizon is critical (e.g., in docking or landing).
2. **Environment**. The environment may have static or moving obstacles or regions with varying levels of risk, such as difficult terrain or hostile areas. Some environments have an inherent structure that must be accounted for, such as road networks and traffic regulations for urban ground vehicles. In highly cluttered environments, obstacles must be explicitly accounted for by the path planner. In more sparsely cluttered environments, however, it may suffice to ignore obstacles in planning and rely on reactive behaviors. Strong disturbances in the environment can affect the vehicle path in ways that must be considered in path planning. Weaker disturbances that can be easily countered using feedback control can be ignored.
3. **Objective**. The objective of path planning is to generate paths that drive the vehicle to a desired terminal state (or terminal set). One possible objective is to simply determine whether the path planning problem can be solved and return a *feasible path* (if one exists). More commonly, one seeks a unique path that is optimal in some sense (e.g., it minimizes the associated travel time, energy expenditure, risk). Further, depending on the planning approach, the solution may be expressed

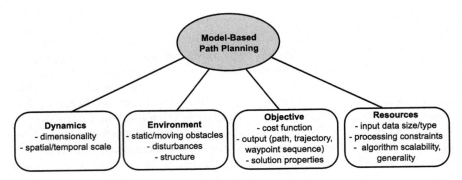

Fig. 3 Aspects of model-based path planning approaches

as either a path, trajectory, or a sequence of waypoints. In some cases it may be possible to plan paths in a lower dimensional space by making some simplifying assumptions (e.g., planning a planar path for an aircraft, assuming that it maintains a constant altitude). Finally, it is necessary to consider the desired properties of the planning algorithm in terms of convergence, sensitivity to inputs, complexity, generality, completeness and optimality.

4. **Resources**. Many robotic systems have limited power and are required to meet real-time processing requirements. However, the increasingly data-driven nature of autonomous vehicles demands that path planners use larger and more comprehensive databases to store information about the world. Thus the trade-off between complexity and performance must be balanced when evaluating the utility of a path planning approach. Another important factor to consider is the scalability and generalizability of a particular planning framework (i.e., whether the planning algorithm can be extended to different vehicle dynamics, environments, and objectives).

These various aspects of path planning are sketched in Fig. 3. In the following we will briefly review some of the major path planning approaches. Our aim is to introduce the methods at a high-level, comparing their capabilities and limitations, and to point out relevant work for further reading. For a full and comprehensive treatment of the subject refer to the classic texts [15–17] and recent surveys [18, 19].

Optimal Control

Trajectory optimization is one model-based method for generating a path. The traditional approach to trajectory optimization is to formulate an optimal control problem of the following form:

$$\text{minimize} \quad J(\mathbf{u}(\cdot)) = \phi(\mathbf{x}_1, t_1) + \int_{t_0}^{t_1} L(\mathbf{x}(t), \mathbf{u}(t), t)\, dt \tag{2}$$

$$\text{subject to} \quad \dot{\mathbf{x}}(t) = \mathbf{f}(\mathbf{x}(t), \mathbf{u}(t), t), \tag{3}$$

$$\mathbf{x}(t_0) = \mathbf{x}_0, \quad \mathbf{x}(t_1) = \mathbf{x}_1, \quad t_0 \leq t \leq t_1 \tag{4}$$

$$\mathbf{g}(\mathbf{x}(t), \mathbf{u}(t), t) \leq 0 \tag{5}$$

$$\mathbf{u}(t) \in \boldsymbol{\Omega} \subset \mathbb{R}^p \tag{6}$$

$J(\mathbf{u}(\cdot))$ in Eq. 2 is known as a Bolza type cost *functional* that maps admissible control functions $\mathbf{u}(\cdot)$ (and their resulting state history) to a scalar cost that is to be minimized. Alternate (and equivalent) formulations may include only a terminal cost $\phi(\mathbf{x}_1, t_1)$ (Mayer type) or only an integral cost (Lagrange type, with integrand $L(\mathbf{x}(t), \mathbf{u}(t), t)$). Equation 3 is an ordinary differential equation that defines the vehicle's dynamics, and Eq. 4 gives the boundary conditions the path or trajectory must satisfy. The constraint function $\mathbf{g}(\mathbf{x}(t), \mathbf{u}(t), t)$ is used in Eq. 5 to enforce general state and control variable constraints. The p-dimensional control input $u(t)$ takes values in the admissible control set $\boldsymbol{\Omega}$.

In principle, obstacles can be represented by Eq. 5, but this introduces significant complexity and therefore environments with many obstacles have traditionally not been considered in trajectory optimization. Optimal path planners are sometimes called "local planners" or "steering methods" for this reason. Pseudospectral optimal control methods have been used to address this limitation; see [20, 21]. In such an approach, obstacle constraints are only enforced at discrete nodes, and therefore the continuous path (in between nodes) is not necessarily collision free. To remedy this problem either the fineness of the discretization can be increased (at the cost of computational complexity) or obstacle boundaries can be artificially enlarged (at the cost of optimality). Further, complex environments require a suitable initial guess for convergence (as generated, for example, using a higher level roadmap planner).

A central result in optimal control theory is Pontryagin's Minimum Principle which gives necessary conditions for a control history to be optimal. Controls that satisfy this necessary condition are called *extremals*. This *indirect approach* of analysis leads to a two-point boundary value problem. Analytical solutions can be obtained for some simple problems, but more generally solutions must be obtained numerically. In the cases where analytical solutions are available they are often very efficient to compute and are widely used (e.g., Dubins [22] or Reeds-Shepp [23] paths). Alternately, *direct methods* can be used to transcribe the problem into a nonlinear program by discretizing state and/or control inputs. Direct methods include pseudospectral, collocation, and shooting methods. Many other powerful analytical techniques exist to study optimal control problems. For a detailed discussion of optimal control theory, refer to [24–26].

Side Note: Path Planning for a Planar Kinematic Body

Suppose that an autonomous vehicle is modeled as a kinematic body moving in an inertial, horizontal plane. Define coordinates (x, y, ψ) on the configuration manifold $Q = SE(2)$. That is, the pair (x, y) represents position and $\psi \in [0, 2\pi)$ represents orientation. Assume that

- The controls are speed v and turn rate r.
- The turn rate is symmetrically bounded $r \in [-r_{\max}, r_{\max}]$.
- In the absence of a flow field, the vehicle moves without slipping (i.e., it moves in the direction ψ in which it is pointed).
- In the presence of a flow field, winds or currents of magnitude w act in the direction ψ_w as additive velocities that disturb the vehicle.

The equations of motion are then

$$\dot{x} = v \cos \psi + w \cos \psi_w \qquad (7)$$
$$\dot{y} = v \cos \psi + w \sin \psi_w \qquad (8)$$
$$\dot{\psi} = r. \qquad (9)$$

The objective is to connect an initial state (x_0, y_0, ψ_0) to a nearby terminal state (x_f, y_f, ψ_f). In the following we discuss several types of paths that exist for the above problem:

- If the speed is fixed (e.g., $v = v_{\max}$ where $v_{\max}$ is the vehicle's maximum speed), and there is no flow field ($w = 0$), then the minimum-time path is a *Dubins path* [22] consisting of straight segments and/or circular arcs of radius $R_0 = v_{\max}/r_{\max}$.
- If only an upper bound on the ambient flow velocity is known (i.e., $w \leq w_{\max}$ where $w_{\max} < v_{\max}$), but the flow is unsteady and nonuniform, then a *feasible Dubins path* may be constructed [27] by artificially increasing the turn radius used for (Dubins) path planning to $R'_0 = (1 + w_{\max}/v_{\max})^2 R_0$. The feasible path is not optimal but can be perfectly tracked by a disturbance-canceling controller.
- If the flow-field magnitude w and direction ψ_w are known then the minimum-time path is a *convected Dubins path* [28]. In the flow-relative frame this path resembles a Dubins path but in the inertial frame it appears as a sequence of trochoidal and straight segments.
- If the vehicle's flow-relative speed ranges from a minimum speed (e.g., an idle or stall speed) to a maximum speed $v \in [v_{\min}, v_{\max}]$, then, in the absence of a flow field ($w = 0$) the minimum-time path is a *variable-speed Dubins path* [29]. In this case, time-optimal paths are not constructed solely from maximum speed segments; they are not simply Dubins paths. They may also include turns at minimum speed to enable a sharp turn radius $R_{\text{corner}} = v_{\min}/r_{\max} < R_0$.

Side Note: Path Planning for a Planar Kinematic Body (continued)
As an example, consider a hypothetical unmanned aircraft modeled as a planar, kinematic body with the following performance parameters:

$$v_{max} = 20 \text{ m/s} \quad v_{min} = 10 \text{ m/s} \quad r_{max} = 15 \text{ deg/s}.$$

Consider the problem of planning a path from $(x_0, y_0, \psi_0) = (0, 0, 0)$ to the terminal state $(x_f, y_f, \psi_f) = (150 \text{ m}, -150 \text{ m}, \pi/3)$. The four types of paths discussed earlier are illustrated in Fig. 4 for this problem.

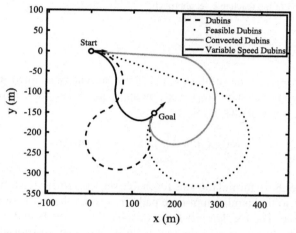

Fig. 4 Comparison of planning approaches

The Dubins and variable-speed Dubins paths are planned assuming that there is no flow field. They both begin with a right turn toward the goal. The feasible Dubins path is planned assuming that only an upper bound on the flow-field magnitude is known ($w \leq w_{max} = 5$ m/s). The planner accounts for the possibility of random disturbances, resulting in a qualitatively distinct path. The convected Dubins path is planned assuming that the flow field is completely known (with $w = 5$ m/s and $\psi_w = 3\pi/2$). The appropriate choice of path will depend on the availability (and accuracy) of flow-field information and the mission objectives. Since the kinematic body model is only an approximation of the true vehicle dynamics, the actual closed-loop path following performance of the vehicle should also be considered.

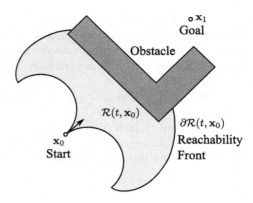

Fig. 5 Illustration of the reachable set approach

Level Sets

The *reachable set* is defined as the set of all states that can be attained at a time $t > t_0$, starting from an initial state $\mathbf{x}(t_0) = \mathbf{x}_0$, by applying admissible control inputs over the time interval $\sigma \in [t_0, t]$:

$$\mathcal{R}(t; \mathbf{x}_0) := \bigcup_{\mathbf{u}(\sigma) \in \Omega} \mathbf{x}(t; \mathbf{x}_0, t_0, \mathbf{u}(\cdot)).$$

Note that in the above formulation the reachable set is parametrized by time assuming the problem is to find a minimum-time path; however, other cost parameterizations are also possible. The *reachability front*, denoted $\partial \mathcal{R}$, is the boundary of this set as shown in Fig. 5. As t increases the reachable set will expand from the initial state until, at some instant $t^* > t_0$, it first reaches the goal $\mathbf{x}_1$ (assuming the goal is reachable from $\mathbf{x}_0$). It is intuitive that a path which first passes through $\mathbf{x}_1 \in \partial \mathcal{R}(\mathbf{x}_0, t^*)$ corresponds to an optimal (minimum-time) path. Once $\mathcal{R}(\mathbf{x}_0, t^*)$ is determined, this optimal path can be determined by following the gradient of the reachable sets in reverse time, from the goal backward to the initial state. In some cases reachability sets can be characterized analytically [30] or through computational geometry, aided by knowledge of the extremal controls [31, 32]. When constructed numerically, propagation of the reachability front can be constrained to respect obstacle boundaries.

A related approach involves calculating a *value function* that assigns to each point in the configuration space the minimum cost to reach the goal. (The value at the goal itself is then zero.) Computing the value function requires solving a Hamilton–Jacobi (HJ) partial differential equation to obtain a viscosity solution. This problem is formulated over a mesh representing the environment and obstacle boundaries can be respected by appropriately modifying the value function at those points. For the case of unconstrained (model-free) planning the value function $V(\mathbf{x})$

is obtained by solving the Eikonal equation

$$||\nabla V(\mathbf{x})|| = c(\mathbf{x})$$

with boundary condition $V(\mathbf{x}_1) = 0$, and where $c(\mathbf{x})$ is the cost rate at state $\mathbf{x}$ (e.g., a vehicle's spatially varying speed) [33, 34]. Once the value function is determined, gradient descent is used to arrive at the optimal path. When path constraints (e.g., curvature) are considered, a more general HJ equation must be solved to obtain the value function [35]. One advantage of this approach is that the solution procedure allows for the optimal path to the goal from *any* starting configuration to be easily found. Further, solutions are globally optimal. Algorithms to numerically compute the value function include fast marching and level set methods. For examples and further discussion, see [6, 36–39].

Coarse Planning with Path Smoothing

Path smoothing refers to the process of refining a nominal path. If the nominal path is already feasible, the purpose of path smoothing is to further improve its quality. For example, to remove erratic turns that are suboptimal [40]. Alternately, path smoothing can be used to construct a feasible path from a coarse, infeasible path. Here, we consider the later case. Many methods that generate a coarse path rely on representing the obstacle-free configuration space $\mathcal{C}_{\text{free}}$ with a graph (sometimes called a *roadmap* or *mesh*).

- *Cell decomposition*: A series of disjoint sets of the configuration space, called *cells*, are used to represent $\mathcal{C}_{\text{free}}$. One approach is to define cell boundaries at critical points where the connectivity of a curve moving through the free space changes (e.g., using Morse decomposition). Another approach is to connect the start, goal, and nearby obstacle vertices with obstacle-free lines (i.e., using a visibility graph); or cells can be generated from generalized Voronoi regions that identify a unique nearest obstacle with each point in the configuration space. Approximate cell decomposition uses simpler shapes, such as rectangles, to represent $\mathcal{C}_{\text{free}}$ but cannot accurately represent the boundary of arbitrarily shaped obstacles. For example, a uniform grid can be imposed on the entire space or a more efficient quadtree spatial decomposition can be used. Once the cell decomposition is generated, points within the cell (e.g., in the middle) can be selected as vertices. Then nearby vertices are joined to form the edges of the graph.
- *Random Sampling*: The probabilistic roadmap (PRM) is a method of generating an obstacle-free graph by randomly sampling the configuration space [41]. Samples that fall inside a region occupied by an obstacle are discarded and the remaining samples are stored as vertices. Rapidly exploring random trees (RRTs) [42] can also be used to explore the free space with coarse paths (e.g., straight line segments).

Fig. 6 A coarse path (*dotted line*) that respects obstacle boundaries (*gray-shaded regions*) is smoothed to produce a path (*black line*) that satisfies a more accurate vehicle dynamic model

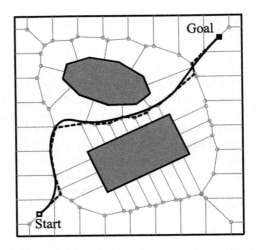

Once the coarse graph is constructed the minimum cost path between the start and the goal can be found using numerous search algorithms such as Dijkstra's algorithm or the A* algorithm.

The *potential field* method is another popular approach to produce a coarse path. The vehicle is viewed as a positively charged particle subject to the sum of an attractive potential at the goal and repulsive potentials at obstacles. By following the gradient of this potential function the vehicle is steered away from obstacles and toward the goal. Although the potential field method is easy to implement, and efficient to compute, the potential functions often produce local minima other than the goal. Local minima can be escaped using various techniques such as random walk methods. Alternatively, the attractive and repulsive potentials can be chosen such that the resulting potential is a *navigation potential* (with only one minimum at the goal).

Once a coarse path is found, it is refined using a new path with a known structure that satisfies a more accurate vehicle dynamic model (as illustrated in Fig. 6). For example, the refined path can be a connected series of locally optimal paths or parameterized curves. Often, control nodes along the coarse path are chosen (not necessarily vertices) and are used to refine the path through approximation, interpolation, optimization or the use of local planners. Previous work has used clothoids [43], Fermat's spirals [44], cubic splines [45], and motion primitives based on the vehicle model [46] among other types of curves for path smoothing. A coarse path can also be used to generate an initial guess for a parametric optimal control problem that is then solved using nonlinear programming techniques (for example, using sequential quadratic programming) [47].

Motion Primitives

Planning in the vehicle's continuous state and control space can be simplified by constraining solutions to be composed of a finite set of *motion primitives*. A motion primitive is a feasible trajectory (i.e., a state and control history) that is used as a fundamental building block to construct more complex paths. A collection of motion primitives is called a *library* and it is used to approximate the diverse set of possible motions the vehicle is capable of executing. The motion primitives can correspond to steady or transient motions; however, it is necessary to ensure that the vehicle can easily transition from one primitive to the next.

One approach to planning with motion primitives is to construct a *state lattice* [48]. The state lattice is a graph with vertices that are deterministically sampled from the configuration space. Once the vertices are chosen, a motion primitive library is generated to connect nearby vertices. (For a directed graph $G = (V, E)$, edges in E are ordered, two-element subsets (v_i, v_j) of the vertices in V. In the context of graph-based path planning, each edge corresponds to a motion primitive that has an associated cost c_{ij} to travel from vertex v_i to vertex v_j.) The motion primitives within the library can be applied at each vertex in the state lattice to establish a graph structure over the entire obstacle-free space. The vertices may be chosen with prior knowledge of the corresponding motion primitives. This is not required, however; one can instead generate a motion primitive connecting an arbitrary pair of vertices using a local planning method. The state lattice can be formulated to be time-dependent and to accommodate the environment, e.g., to conform to the geometry of a road [49].

Alternatively, motion primitives can be constructed by a judicious choice of control inputs or determined experimentally by capturing the control inputs of a human operator and the corresponding vehicle state history. Having constructed a library of motion primitives, this library is used to expand a tree structure, sometimes called a *reachability tree*, that begins from the initial state. (A tree is a type of directed acyclic graph in which every pair of vertices is joined by a unique path.) For examples of motion-primitive tree approaches see [50–52].

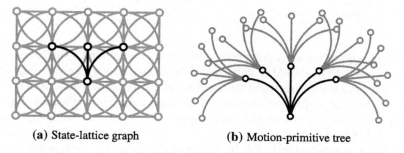

(a) State-lattice graph   (b) Motion-primitive tree

Fig. 7 Approaches to generating a motion-primitive graph. *Circles* denote vertices and *dark lines* indicate motion primitives emanating from an initial vertex. *Gray lines* indicate edges emanating from other vertices

Both these approaches are sketched in Fig. 7. In both cases the resulting structure can be efficiently searched using graph search algorithms. Many common graph search algorithms use a heuristic (i.e., a method for estimating the cost of an unexplored path) to guide the search. The use of the motion primitives reduces the complexity of the planning problem; however, the feasibility and optimality of a solution depend on the resolution at which the motion primitives are constructed, e.g., the density of vertices and the number of motion primitives contained in the library. A planner is said to be *resolution complete* if it is guaranteed to find a feasible path given sufficiently fine resolution. The A* algorithm is a commonly used graph search algorithm that is resolution complete and optimal. Dynamic graph search algorithms, such as D* lite [53], can accommodate environmental changes.

Random Sampling

Many sampling-based motion planning algorithms that generate dynamically feasible paths use tree-based planners, such as the rapidly exploring random tree (RRT) algorithm [42]. To discuss this class of algorithms it is useful to consider the basic RRT pseudocode shown in Algorithm 1.

Algorithm 1 Model-Based RRT

1: **procedure** RRT(q_0, q_{goal})
2:   G.init(q_0)   ▷ Initialize a tree with the initial state q_0
3:   **while** $q_{goal} \notin V$ **do**   ▷ Sample until the goal state q_{goal} is reached
4:     $q_{rand} \leftarrow$ SAMPLE_FREE(C_{free})   ▷ Generate a random configuration
5:     $q_{near} \leftarrow$ NEAREST_CONFIG(G, q_{rand}) ▷ Find the node in G nearest to q_{rand}
6:     $q_{new} \leftarrow$ STEER(q_{near}, q_{rand})   ▷ Attempt to connect q_{near} to q_{rand} with a path
7:     **if** OBSTACLE_FREE(q_{near}, q_{new}) **then**   ▷ Check resulting path for collisions
8:       G.add_vertex(q_{new})   ▷ Add new vertex (configuration) to the graph
9:       G.add_edge(q_{near}, q_{new})   ▷ Add new edge (path) to the graph
10:    **end if**
11: **end while**
12: **return** G ▷ Return a tree containing a feasible path to the goal
13: **end procedure**

The algorithm begins by initializing a tree with no edges and a single vertex q_0. (Alternatively, the tree could be initialized with the goal state q_{goal} and the algorithm run in reverse. Or a bi-directional tree can be constructed that expands from both the start and the goal simultaneously.) In each iteration a random configuration q_{rand} is generated by sampling the obstacle-free space C_{free}. The choice of sampling strategy (i.e., the distribution used to sample the space) will determine to what extent the environment is explored and the performance of that particular implementation [54]. A common sampling strategy is to slightly bias samples toward the goal. Goal biasing can be efficient in some cases but can also lead to pitfalls (e.g., in pathological obstacle environments). Once a sample is drawn, the nearest node in the tree is

Fig. 8 Sketch of RRT expansion procedure

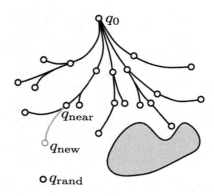

found, according to some heuristic. (Although it is less common, an admissible nearest node may also be a point along the edges connecting two vertices.) The motion of the vehicle attempting to steer from $\mathbf{q}_{near}$ to $\mathbf{q}_{rand}$ is simulated for some time Δt (generally much larger than the integration time step). The resulting state $\mathbf{q}_{new}$ and the corresponding collision-free path are added to the tree as a vertex and an edge, respectively. The procedure is sketched in Fig. 8. The algorithm iterates until the goal (or a goal region) is reached. The RRT algorithm is *probabilistically complete*—the probability that a solution is returned, if one exists, approaches unity as the number of samples tends to infinity.

In some critical applications that require hard computation-time guarantees, probabilistic completeness is inadequate. Another shortcoming of RRT algorithm is that its performance is sensitive to the heuristic used to determine nearest nodes (Line 5 in Algorithm 1) [55]. The ideal heuristic is the optimal cost to go, but this is difficult to compute. Instead, a simpler heuristic that can be efficiently computed is used, such as the Euclidean distance. However, a poor choice of heuristic can cause regions of the state space to remain unexplored—this can be a problem in dynamic or constrained environments.

Nonetheless, sampling-based motion planning has been successfully used to plan paths for numerous mobile robots. A major advantage of sampling-based approaches is the generality with which the vehicle dynamics are accounted for—one needs only to numerically integrate the equations of motion and a local guidance law (e.g., the STEER function on Line 6 of Algorithm 1). This enables various motion models to be used in a single planning framework and makes it straightforward to incorporate environmental data (e.g., from onboard sensors or forecast models). Sampling-based motion planning does not require obstacles to be represented explicitly. (Instead, a collision checking routine can be invoked incrementally during planning to check for feasibility of candidate paths [56].) However, collision checking is still a bottleneck in sampling-based algorithms and recent research is aimed at addressing this issue [57]. The RRT algorithm described above may return a feasible path that is erratic because of the random nature of the solution. However, this can be remedied using path smoothing or other heuristic techniques if additional resources are available

Table 2 Qualitative comparison of path planning approaches in four categories according to the following scale: ◐ = fair, ◕ = good, ● = excellent

Approach	Dynamics	Environment	Optimality	Generality
Optimal control	●	◐	●	◐
Level sets	◕	◕	●	◐
Path smoothing	◐	◕	◐	◕
Motion primitives	◕	●	◕	◕
Random sampling	●	●	◕*	●

*Assuming an optimal algorithm is used

during the planning cycle. Alternatively, by modifying the expansion of the tree the RRT* algorithm [56] can be employed to obtain asymptotically optimal paths. (That is, as the number of samples tends toward infinity the cost of the path returned by the algorithm approaches the optimal value.) RRT$^\#$ is another optimal planning algorithm that achieves a faster initial convergence rate than RRT* [58]. A discrete search through a coarse representation of the environment can be used to guide continuous, tree-based expansion to improve efficiency [59]. Sampling-based planning has been incorporated in the higher level problem of task and motion planning [60]. An efficient algorithm for re-planning in the presence of dynamic (moving or unexpected) obstacles, RRT^X, is presented in [61]. Other extensions include sampling the control space [7]. An interesting example of the confluence of ideas from systems theory and robotic motion planning is the LQR-Tree algorithm [62]. The algorithm grows a tree of trajectories that are each stabilized by a linear quadratic regulator (LQR). Directly-computed Lyapunov functions are used to estimate conservative regions of attraction for each trajectory. This approach results in verified feedback policies for high-dimensional, nonlinear systems (with controllable linearizations) subject to constraints. Refer to [63] for a recent survey of sampling-based planning methods.

Comparison

We summarize the above discussion by comparing the relative capabilities of the planning frameworks in Table 2. Each of the four planning approaches was assigned a score: fair (◐), good (◕), or excellent (●) in four categories:

a. **Dynamics**. Ability to account for complex vehicle dynamics.
b. **Environment**. Ability to handle obstacles, environmental disturbances, and partially observable environments.
c. **Optimality**. Ability to obtain local or globally optimal solutions (in continuous space).
d. **Generality**. Ability to accommodate different vehicle models and environmental data (e.g., changing numerical forecast models or obstacle maps).

The scoring of these methods is based on the earlier discussion, but is certainly subjective. In any case, Table 2 suggests that each of the planning approaches has its strengths and weaknesses but that random sampling performs particularly well in all of the categories. Regardless of this qualitative comparison, each approach has its own unique attributes that make it suitable for certain problems. For example, the level set method is a *multiple query* approach since it provides the cost to goal from every point in the configuration space and facilitates planning many paths to the goal. The remaining approaches only provide a single query plan.

4 Concluding Remarks

In this chapter we have reviewed the basic principles of model-based path planning: from vehicle dynamic models to planning approaches. Despite dramatic advances in the past few decades, path planning remains a central challenge in robotics and continues to receive attention from academia and industry. As autonomous systems become more capable, the path planning problem will become more challenging. Modern applications require vehicles to operate in complex, cluttered environments that require rapid re-planning algorithms. Partially observable environments increase reliance on environmental maps and/or advanced sensors. Advances in autonomy and artificial intelligence will accentuate rather than eliminate the need for corresponding advances in path planning methods. For example, machine learning can be used to identify specific obstacles, providing more information for the path planning algorithm which may improve performance. Many questions remain unanswered, and many new ones are being revealed, as autonomous systems technology continues to advance.

Acknowledgements We thank Thomas Battista, Mazen Farhood, David Grymin, James McMahon, and Michael Otte for reviewing a draft of this chapter and providing helpful comments. The first author would also like to acknowledge support from the American Society for Engineering Education's NRL Postdoctoral Fellowship program. The second author gratefully acknowledges the support of the ONR under Grant Nos. N00014-14-1-0651 and N00014-16-1-2749.

References

1. Kabamba, P.T., Girard, A.R.: Fundamentals of Aerospace Navigation and Guidance. Cambridge University Press, New York (2014)
2. Yakimenko, O.A.: Direct method for rapid prototyping of near-optimal aircraft trajectories. J. Guid. Control Dyn. **23**(5), 865–875 (2000)
3. Aguiar, A.P., Hespanha, J.P., Kokotović, P.V.: Path-following for nonminimum phase systems removes performance limitations. IEEE Trans. Autom. Control **50**(2), 234–239 (2005)
4. Bullo, F., Lewis, A.D.: Geometric Control of Mechanical Systems. Springer, New York (2005)
5. Bloch, A.M.: Nonholonomic Mechanics and Control. Springer, New York (2005)

6. Xu, B., Stilwell, D.J., Kurdila, A.J.: Fast path re-planning based on fast marching and level sets. J. Intell. Rob. Syst. **71**(3), 303–317 (2013)
7. Hsu, D., Kindel, R., Latombe, J.-C., Rock, S.: Randomized kinodynamic motion planning with moving obstacles. Int. J. Robot. Res. **21**(3), 233–255 (2002)
8. Purwin, O., D'Andrea, R., Lee, J.-W.: Theory and implementation of path planning by negotiation for decentralized agents. Robot. Auton. Syst. **56**(5), 422–436 (2008)
9. Helmick, D., Angelova, A., Matthies, L.: Terrain adaptive navigation for planetary rovers. J. Field Robot. **26**(4), 391–410 (2009)
10. Dolgov, D., Thrun, S., Montemerlo, M., Diebel, J.: Path planning for autonomous vehicles in unknown semi-structured environments. Int. J. Robot. Res. **29**(5), 485–501 (2010)
11. Sonnenburg, C.R., Woolsey, C.A.: Modeling, identification, and control of an unmanned surface vehicle. J. Field Robot. **30**(3), 371–398 (2013)
12. Thomasson, P., Woolsey, C.A.: Vehicle motion in currents. IEEE J. Ocean. Eng. **38**(2), 226–242 (2013)
13. Perez, T.: Ship Motion Control: Course Keeping and Roll Stabilisation Using Rudder and Fins. Springer, Berlin (2005)
14. Kaminer, I., Pascoal, A., Xargay, E., Hovakimyan, N., Cao, C., Dobrokhodov, V.: Path following for unmanned aerial vehicles using $\mathcal{L}_1$ adaptive augmentation of commercial autopilots. J. Guid. Control Dyn. **33**(2), 550–564 (2010)
15. Latombe, J.-C.: Robot Motion Planning. Springer Science & Business Media, New York (2012)
16. LaValle, S.M.: Planning Algorithms. Cambridge University Press, New York (2006)
17. Choset, H., Lynch, K.M., Hutchinson, S., Kantor, G., Burgard, W., Kavraki, L.E., Thrun, S.: Principles of Robot Motion: Theory, Algorithms, and Implementation. MIT Press, Cambridge (2005)
18. Paden, B., Cap, M., Yong, S.Z., Yershov, D., Frazzoli, E.: A survey of motion planning and control techniques for self-driving urban vehicles. IEEE Trans. Intell. Veh. **1**(1), 33–55 (2016)
19. Goerzen, C., Kong, Z., Mettler, B.: A survey of motion planning algorithms from the perspective of autonomous UAV guidance. J. Intell. Rob. Syst. **57**(1–4), 65–100 (2010)
20. Gong, Q., Lewis, L.R., Ross, I.M.: Pseudospectral motion planning for autonomous vehicles. J. Guid. Control Dyn. **32**(3), 1039–1045 (2009)
21. Lewis, L.R.: Rapid motion planning and autonomous obstacle avoidance for unmanned vehicles. Master's thesis, Naval Postgraduate School (2006)
22. Dubins, L.E.: On curves of minimal length with a constraint on average curvature, and with prescribed initial and terminal positions and tangents. Am. J. Math. **79**(3), 497–516 (1957)
23. Reeds, J., Shepp, L.: Optimal paths for a car that goes both forwards and backwards. Pac. J. Math. **145**(2), 367–393 (1990)
24. Burns, J.: Introduction to the Calculus of Variations and Control with Modern Applications. CRC Press, Boca Raton (2013)
25. Leitmann, G.: The Calculus of Variations and Optimal Control. Plenum Press, New York (1981)
26. Bryson, A.E., Ho, Y.-C.: Applied Optimal Control. CRC Press, Boca Raton (1975)
27. Wolek, A., Woolsey, C.: Feasible Dubins paths in presence of unknown, unsteady velocity disturbances. J. Guid. Control Dyn. **38**(4), 782–787 (2015)
28. Techy, L., Woolsey, C.A.: Minimum-time path planning for unmanned aerial vehicles in steady uniform winds. J. Guid. Control Dyn. **32**(6), 1736–1746 (2009)
29. Wolek, A., Cliff, E.M., Woolsey, C.A.: Time-optimal path planning for a kinematic car with variable speed controls. J. Guid. Control Dyn. **39**(10), 2374–2390 (2016)
30. Cockayne, E.J., Hall, G.W.C.: Plane motion of a particle subject to curvature constraints. SIAM J. Control **13**(1), 197–220 (1975)
31. Fedotov, A., Patsko, V., Turova, V.: Recent Advances in Mobile Robotics: Reachable Sets for Simple Models of Car Motion. InTech, Rijeka, Croatia (2011)
32. Bakolas, E., Tsiotras, P.: Optimal synthesis of the asymmetric sinistral/dextral Markov-Dubins problem. J. Optim. Theory Appl. **150**(2), 233–250 (2011)
33. Mitchell, I.M., Sastry, S.: Continuous path planning with multiple constraints. In: IEEE International Conference on Decision and Control (2003)

34. Kimmel, R., Sethian, J.A.: Optimal algorithm for shape from shading and path planning. J. Math. Imaging Vis. **14**(3), 237–244 (2001)
35. Takei, R., Tsai, R.: Optimal trajectories of curvature constrained motion in the Hamilton-Jacobi formulation. J. Sci. Comput. **54**(2–3), 622–644 (2013)
36. Gomez, J.V., Vale, A., Garrido, S., Moreno, L.: Performance analysis of fast marching-based motion planning for autonomous mobile robots in ITER scenarios. Robot. Auton. Syst. **63**(1), 36–49 (2015)
37. Lolla, T., Haley, P.J., Lermusiaux, P.F.J.: Time-optimal path planning in dynamic flows using level set equations: realistic applications. Ocean Dyn. **64**(10), 1399–1417 (2014)
38. Rhoads, B., Mezic, I., Poje, A.C.: Minimum time heading control of underpowered vehicles in time-varying ocean currents. Ocean Eng. **66**, 12–31 (2013)
39. Petres, C., Pailhas, Y., Patron, P., Petillot, Y., Evans, J., Lane, D.: Path planning for autonomous underwater vehicles. IEEE Trans. Robot. **23**(2), 331–341 (2007)
40. Choi, J.-W., Huhtala, K.: Constrained global path optimization for articulated steering vehicles. IEEE Trans. Veh. Technol. **65**(4), 1868–1879 (2016)
41. Kavraki, L.E., Svestka, P., Latombe, J.-C., Overmars, M.H.: Probabilistic roadmaps for path planning in high-dimensional configuration spaces. IEEE Trans. Robot. Autom. **12**(4), 566–580 (1996)
42. LaValle, S.M., Kuffner, J.J.: Randomized kinodynamic planning. Int. J. Robot. Res. **20**(5), 378–400 (2001)
43. Fleury, S., Soueres, P., Laumond, J.-P., Chatila, R.: Primitives for smoothing mobile robot trajectories. IEEE Trans. Robot. Autom. **11**(3), 441–448 (1995)
44. Lekkas, A.M., Dahl, A.R., Breivik, M., Fossen, T.I.: Continuous-curvature path generation using Fermat's spiral. Model. Identif. Control **34**(4), 183–198 (2013)
45. Judd, K.B., McLain, T.W.: Spline Based Path Planning for Unmanned Aerial Vehicles. In: AIAA Guidance, Navigation and Control Conference (2001)
46. Bottasso, C.L., Leonello, D., Savini, B.: Path planning for autonomous vehicles by trajectory smoothing using motion primitives. IEEE Trans. Control Syst. Technol. **16**(6), 1152–1168 (2008)
47. Hull, D.G.: Conversion of optimal control problems into parameter optimization problems. J. Guid. Control Dyn. **20**(1), 57–60 (1997)
48. Pivtoraiko, M., Knepper, R.A., Kelly, A.: Differentially constrained mobile robot motion planning in state lattices. J. Field Robot. **26**(3), 308–333 (2009)
49. Ziegler, J., Stiller, C.: Spatiotemporal state lattices for fast trajectory planning in dynamic on-road driving scenarios. In: IEEE/RSJ International Conference on Intelligent Robots and Systems (2009)
50. Grymin, D.J., Neas, C.B., Farhood, M.: A hierarchical approach for primitive-based motion planning and control of autonomous vehicles. Robot. Auton. Syst. **62**(2), 214–228 (2014)
51. Frazzoli, E., Dahleh, M.A., Feron, E.: Maneuver-based motion planning for nonlinear systems with symmetries. IEEE Trans. Robot. **21**(6), 1077–1091 (2005)
52. Barraquand, J., Latombe, J.-C.: Nonholonomic multibody mobile robots: controllability and motion planning in the presence of obstacles. Algorithmica **10**(2–4), 121–155 (1993)
53. Koenig, S., Likhachev, M.: D* Lite. In: American Association for Artificial Intelligence Conference (2002)
54. Urmson, C., Simmons, R.: Approaches for heuristically biasing RRT growth. In: IEEE/RSJ International Conference on Intelligent Robots and Systems (2003)
55. Cheng, P., LaValle, S.M.: Reducing metric sensitivity in randomized trajectory design. In: IEEE/RSJ International Conference on Intelligent Robots and Systems (2001)
56. Karaman, S., Frazzoli, E.: Sampling-based algorithms for optimal motion planning. Int. J. Robot. Res. **30**(7), 864–894 (2011)
57. Bialkowski, J., Karaman, S., Otte, M., Frazzoli, E.: Efficient collision checking in sampling-based motion planning. Int. J. Robot. Res. **35**(7), 767–796 (2016)
58. Arslan, O., Tsiotras, P.: Use of relaxation methods in sampling-based algorithms for optimal motion planning. In: IEEE International Conference on Robotics and Automation (2013)

59. Plaku, E., Kavraki, L.E., Vardi, M.Y.: Discrete search leading continuous exploration for kinodynamic motion planning. In: Robotics: Science and Systems (2007)
60. McMahon, J., Plaku, E.: Mission and motion planning for autonomous underwater vehicles operating in spatially and temporally complex environments. IEEE J. Ocean. Eng. **3**, 1–20 (2016)
61. Otte, M., Frazzoli, E.: RRT^X: asymptotically optimal single-query sampling-based motion planning with quick replanning. Int. J. Robot. Res. **35**(7), 797–822 (2015)
62. Tedrake, R., Manchester, I.R., Tobenkin, M., Roberts, J.W.: LQR-trees: feedback motion planning via sums-of-squares verification. Int. J. Robot. Res. **29**(8), 1038–1052 (2010)
63. Elbanhawi, M., Simic, M.: Sampling-based robot motion planning: a review. IEEE Access **2**, 56–77 (2014)

Constrained Optimal Motion Planning for Autonomous Vehicles Using PRONTO

A. Pedro Aguiar, Florian A. Bayer, John Hauser, Andreas J. Häusler, Giuseppe Notarstefano, Antonio M. Pascoal, Alessandro Rucco and Alessandro Saccon

Abstract This chapter provides an overview of the authors' efforts in vehicle trajectory exploration and motion planning based on PRONTO, a numerical method for solving optimal control problems developed over the last two decades. The chapter reviews the basics of PRONTO, providing the appropriate references to get further details on the method. The applications of the method to the constrained optimal motion planning of single and multiple vehicles is presented. Interesting applications that have been tackled with this method include, e.g., computing minimum-time trajectories for a race car, exploiting the energy from the surrounding environment

---

A.P. Aguiar · A. Rucco
University of Porto, Porto, Portugal
e-mail: pedro.aguiar@fe.up.pt

A. Rucco
e-mail: alessandrorucco@fe.up.pt

F.A. Bayer
University of Stuttgart, Stuttgart, Germany
e-mail: florian.bayer@ist.uni-stuttgart.de

J. Hauser
University of Colorado, Boulder, CO, USA
e-mail: john.hauser@colorado.edu

A.J. Häusler
Salt and Pepper Technology GmbH & Co. KG, Bremen, Germany
e-mail: a.haeusler@ieee.org

G. Notarstefano
University of Lecce, Lecce, Italy
e-mail: giuseppe.notarstefano@unisalento.it

A.M. Pascoal
Instituto Superior Técnico, Univ. Lisbon, Lisbon, Portugal
e-mail: antonio@isr.tecnico.ulisboa.pt

A. Saccon (✉)
Eindhoven University of Technology, Eindhoven, The Netherlands
e-mail: a.saccon@tue.nl

© Springer International Publishing AG 2017
T.I. Fossen et al. (eds.), *Sensing and Control for Autonomous Vehicles*,
Lecture Notes in Control and Information Sciences 474,
DOI 10.1007/978-3-319-55372-6_10

for long endurance missions of unmanned aerial vehicles (UAVs), and cooperative motion planning of autonomous underwater vehicles (AUVs) for environmental surveying.

1 Introduction

The development of efficient tools to compute optimal trajectories for autonomous vehicle has been receiving growing attention for robotic applications. These tools allow the designer to study and analyze – that is, *explore* – the dynamic performance of autonomous vehicles and, at the same time, give useful insight on how to improve vehicle performance. They are also important in developing motion planning algorithms for single and multiple vehicles acting in cooperation, in particular when both detailed vehicle models and complex geometric constraints have to be accounted for. Motion planning is a core robotics technology that, together with other control, perception, and actuation technologies enables robot autonomy.

This chapter provides an overview of the authors' efforts in vehicle trajectory exploration and motion planning based on PRONTO, a numerical method for solving optimal control problems developed over the last two decades. The name PRONTO stands for PRojection Operator based Netwon's method for Trajectory Optimization. The method is also known, for short, as the projection operator approach. Interesting applications that have been tackled with this method include, e.g., computing minimum-time trajectories for a race car, exploiting the energy from the surrounding environment for long endurance missions of unmanned aerial vehicles (UAVs), and cooperative motion planning of autonomous underwater vehicles (AUVs) for environmental surveying.

Due to the extended literature in the field of motion planning, we make no attempt in this chapter to provide an overview or comparison with other path planning and trajectory generation approaches other than PRONTO. We refer the interested reader to [8, Chap. 1] and references therein for a recent literature review.

In its basic formulation, a motion planner is a method to generate a trajectory to connect a given initial state of a robotic system to a desired final state. In the context of this chapter, robotic system means a single or a group of multiple vehicles. Robot state means the position of the robotic system and, possibly, the orientation (of each vehicle) relative to a reference coordinate system. Depending on the level of complexity of the mathematical model employed to describe the robotic system and the specific application and problem at hand, other quantities such as linear and angular velocities can be part of the system state.

Other factors that play a role and increase complexity are the presence of obstacles in the environment and the typical need to avoid contact with these obstacles as well as inter-vehicle collisions in a multi-vehicle situation.

Every constrained optimal motion problem starts from a careful selection of the vehicle model(s), constraints (e.g., collision avoidance and actuator saturation), and cost criteria (e.g., energy consumption and maneuvering time) to arrive at a well formulated optimal control problem. The problem formulation requires some experience in order to avoid defining a problem with no solutions or with too many solutions so as to allow numerical solution using the chosen optimization method. As mentioned above, in this chapter we focus our attention on PRONTO, a specific numerical method to solve constrained optimal control problems.

PRONTO has the unusual feature of working directly in *continuous-time*, constructing a sequence of trajectories with descending cost. More commonly, numerical optimal control (see, e.g., [6]) is to discretize the system dynamics, constraints, and cost functional giving a constrained nonlinear optimization problem that is subsequently solved by employing off-the-shelf constrained nonlinear solvers, specialized to handle the sparsity of the constraints resulting in the discretization of the continuous-time dynamics. PRONTO skips this transcription phase, employing instead an infinite-dimensional Newton method achieving second-order convergence.

This chapter is organized as follow. Section 2 reviews the basic of the projection operator approach, providing the appropriate references to get further details on the method. The applications of the method to the constrained optimal motion planning of single and multiple vehicles is reviewed in Sect. 3, mainly taking a historical perspective. Conclusions are drawn in Sect. 4, together with a discussion on current efforts to further extend the capability and applicability of the method.

2 A Review of PRONTO: Basics and Constraints Handling

This section provides a concise introduction to PRONTO, following the presentation given in the papers [11, 14]. For further details, the interested reader is referred to the recent and comprehensive explanation provided in [8, Chap. 4].

2.1 Basics and Geometric Interpretation

In its simplest form, PRONTO is an iterative numerical algorithm for the minimization, by a Newton descent method, of the cost functional

$$h(x(\cdot), u(\cdot)) := \int_0^T l(x(\tau), u(\tau), \tau) \, d\tau + m(x(T)) \tag{1}$$

over the set of trajectories of the nonlinear control system $\dot{x} = f(x, u)$, $x(0) = x_0$. As usual, the state and input vectors live in $\mathbb{R}^n$ and $\mathbb{R}^m$, respectively. The incremental and terminal costs l and m, defining the cost functional h in (1), as well as the system vector field f are assumed to be sufficiently smooth (e.g., C^2 in (x, u) and C^0 in t). By *trajectory*, we mean a bounded state-control curve $\eta = (x(\cdot), u(\cdot))$ satisfying $\dot{x}(t) = f(x(t), u(t))$ for all $t \geq 0$. We use the *placeholder* notation for curves so that, e.g., $u(\cdot)$ denotes the (bounded) curve $\{u(t) \in \mathbb{R}^m : t \in [0, T]\}$.

The set $\mathscr{T}$ of trajectories of the nonlinear control system $\dot{x} = f(x, u)$, $f \in C^r$, $r \geq 1$, has the structure of a (infinite dimensional) Banach manifold [13]. To work on the trajectory manifold $\mathscr{T}$, one *projects* state-control curves in the ambient Banach space onto $\mathscr{T}$ by using a linear time-varying trajectory tracking controller. To this end, note that the nonlinear feedback system

$$\begin{aligned} \dot{x} &= f(x, u), \quad x(0) = x_0, \\ u &= \mu(t) + K(t)[\alpha(t) - x], \end{aligned} \qquad (2)$$

defines a nonlinear operator (C^r when $f \in C^r$)

$$\mathscr{P} : \xi = (\alpha(\cdot), \mu(\cdot)) \mapsto \eta = (x(\cdot), u(\cdot))$$

mapping bounded curves (in its domain) to trajectories. From (2) it is easy to see that ξ is a trajectory, $\xi \in \mathscr{T}$, if and only if ξ is a fixed point of $\mathscr{P}$, $\xi = \mathscr{P}(\xi)$. Since $\mathscr{P}(\xi) \in \mathscr{T}$ for all ξ in the domain of $\mathscr{P}$, we see that $\mathscr{P}(\xi) = \mathscr{P}(\mathscr{P}(\xi))$, briefly $\mathscr{P} = \mathscr{P}^2$, so that $\mathscr{P}$ is a (nonlinear) *projection operator*. Noting that $\mathscr{P}(\xi + \zeta) \approx \mathscr{P}(\xi) + D\mathscr{P}(\xi) \cdot \zeta$, we see that the Fréchet derivative of $\mathscr{P}$ will be the (continuous) linear projection operator

$$D\mathscr{P}(\xi) : \zeta = (\beta(\cdot), \nu(\cdot)) \mapsto \gamma = (z(\cdot), v(\cdot))$$

$$\begin{aligned} \dot{z} &= A(\eta(t))z + B(\eta(t))v, \quad z(0) = 0, \\ v &= \nu(t) + K(t)[\beta(t) - z], \end{aligned}$$

where the system matrices $A(\eta(t)) = f_x(x(t), u(t))$ and $B(\eta(t)) = f_u(x(t), u(t))$ are evaluated on the *trajectory* $\eta = \mathscr{P}(\xi)$. The set of bounded linearized trajectories at a given (nonlinear) trajectory $\xi \in \mathscr{T}$ forms the tangent space to the manifold, denoted $T_\xi \mathscr{T}$. Note that, as in the nonlinear case, tangent trajectories at $\xi \in \mathscr{P}$ are fixed points: $\zeta \in T_\xi \mathscr{T} \iff \zeta = D\mathscr{P}(\xi) \cdot \zeta$. The projection operator $\mathscr{P}$ provides a local representation of the trajectory manifold: given a trajectory $\xi \in \mathscr{T}$, every nearby trajectory $\eta \in \mathscr{T}$ is of the form $\eta = \mathscr{P}(\xi + \zeta)$ for a unique tangent trajectory $\zeta \in T_\xi \mathscr{T}$. That is, using $\mathscr{P}$, tangent trajectories can be used as *local coordinates* for the trajectory manifold. Note that the projection operator depends on the choice of the time-dependent feedback gain $K(\cdot)$, which is typically chosen to provide local stability of the closed-loop system about a given trajectory. For instance, if $\xi \in \mathscr{T}$ and $\dot{z} = [A(\xi(t)) - B(\xi(t))K(t)]z$ is exponentially stable then the domain of the operator $\mathscr{P}$ will include a nice L_∞ neighborhood of ξ. For further insights into the trajectory manifold, and higher derivatives of the projection operator $\mathscr{P}$, see [13].

With a projection operator in hand, one can see [11] that the constrained and unconstrained optimization problems

$$\min_{\xi \in \mathscr{T}} h(\xi) \quad \text{and} \quad \min_\xi h(\mathscr{P}(\xi))$$

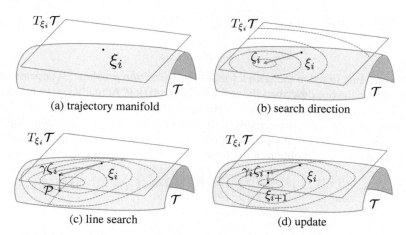

Fig. 1 The Projection Operator approach: at each iteration, **a** the linearization of the control system about the trajectory ξ_i defines the tangent space to the trajectory manifold $\mathscr{T}$ at ξ_i; **b** minimization over the tangent space of a second-order approximation of the extended cost functional $g = h \circ \mathscr{P}$ yields a *search direction* ζ_i; **c** a *step size* is computed via a line search along ζ_i with $\mathscr{P}$ bending that line onto the trajectory manifold $\mathscr{T}$; **d** the search direction ζ_i and step size γ_i are combined and projected to obtain the *updated* trajectory ξ_{i+1}

are essentially equivalent in the sense that a solution to the first *constrained* problem is a solution to the second *unconstrained* problem, while a solution to the second problem is, projected by $\mathscr{P}$, a solution to the first problem. Working with the cost functional $g(\xi) := h(\mathscr{P}(\xi))$ in an essentially unconstrained manner, one may develop effective descent methods for trajectory optimization. For instance, a Newton descent step at $\xi \in \mathscr{T}$ may be obtained by minimizing the quadratic model functional $Dg(\xi) \cdot \zeta + \frac{1}{2} D^2 g(\xi) \cdot (\zeta, \zeta) \approx g(\xi + \zeta) - g(\xi)$ over the linear subspace of tangent trajectories $\zeta \in T_\xi \mathscr{T}$. In fact, one may derive many properties, including first- and second-order optimality conditions, of the constrained optimal control problem though an unconstrained analysis of the functional g.

From an abstract point of view, PRONTO looks pretty much like a standard line search method for unconstrained optimization. The slight differences involved in searching along the tangent space and projecting back onto the trajectory manifold in order to generate a descending sequence of *trajectories* are nicely illustrated by the conceptual diagrams in Fig. 1.

Algorithm (Projection operator Newton method [11])
given initial trajectory $\xi_0 \in \mathscr{T}$
for $i = 0, 1, 2, \ldots$

Redesign feedback $K(\cdot)$ for $\mathscr{P}$, if desired/needed

$$\zeta_i = \arg \min_{\zeta \in T_{\xi_i} \mathscr{T}} Dh(\xi_i) \cdot \zeta + \frac{1}{2} D^2 g(\xi_i) \cdot (\zeta, \zeta) \qquad \textit{(search direction)} \quad (3)$$

$$\gamma_i = \arg \min_{\gamma \in (0,1]} g(\xi_i + \gamma \zeta_i) \qquad \textit{(step size)} \quad (4)$$

$$\xi_{i+1} = \mathscr{P}(\xi_i + \gamma_i \zeta_i) \qquad (update) \quad (5)$$

end

The cost functional in the *search* (or descent) *direction* problem (3) appears to be different than the one described above for the Newton step. Actually, on the manifold, $\xi \in \mathscr{T}$, and tangent space, $\zeta \in T_\xi \mathscr{T}$, we have $Dg(\xi) \cdot \zeta = Dh(\mathscr{P}(\xi)) \cdot D\mathscr{P}(\xi) \cdot \zeta = Dh(\xi) \cdot \zeta$ so that (3) is precisely the problem for computing a valid Newton step, if one exists. This, and related problems, is a time-varying linear quadratic (LQ) optimal control problem of the form

$$\min \int_0^T a^T z + b^T v + \frac{1}{2} \begin{bmatrix} z \\ v \end{bmatrix}^T \begin{bmatrix} Q & S \\ S^T & R \end{bmatrix} \begin{bmatrix} z \\ v \end{bmatrix} d\tau + r_1^T z(T) + \frac{1}{2} z^T(T) P_1 z(T) \quad (6)$$
$$\text{subj to} \quad \dot{z} = A(t)z + B(t)v, \quad z(0) = 0$$

where all terms in the integrand depend on τ: $a(\tau)$, $z(\tau)$, etc. When this problem possesses a unique minimizing trajectory, it can be solved with the help of the solution of a suitable differential Riccati equation (and an associated adjoint system). Since $Q(\tau)$, $R(\tau)$, and $S(\tau)$ arise from second derivatives of the incremental cost and the system dynamics (combined with a particular adjoint solution), they do not necessarily satisfy the typical (positive definiteness) conditions that are usually assumed for LQ. For (6) to possess a unique minimizer, we require that the quadratic portion of the functional be strongly positive definite (in an L_2 sense) on the linear subspace expressed by the dynamic constraint. Fortunately, this occurs precisely [12] when the Riccati equation possesses a bounded solution on $[0, T]$. In the case that $D^2 g(\xi) \cdot (\zeta, \zeta)$ is not strongly positive definite on $T_\xi \mathscr{T}$, one may choose a positive definite approximation $q(\xi) \cdot (\zeta, \zeta)$ such as $D^2 h(\xi) \cdot (\zeta, \zeta)$ (when appropriate) and including, as a last resort, a weighted L_2 inner product $\langle \zeta, \zeta \rangle$.

Once one has an acceptable descent direction $\zeta \in T_\xi \mathscr{T}$, a line search using the actual cost functional $g = h \circ \mathscr{P}$ is required. Step (4) suggests an exact line search, but it is more common and effective to use an Armijo-backtracking line search. Note that, in a sufficiently small neighborhood of a second-order sufficient condition (SSC) minimizer, the Newton step will be accepted (with unit step size) leading to quadratic convergence. Observe that we have used the projection operator to *bend* the line of test points $\xi_i + \gamma \zeta_i$ along the trajectory manifold (see Fig. 1), leading to the natural manifold *update* law (5): $\xi_{i+1} = \mathscr{P}(\xi_i + \gamma_i \zeta_i)$.

2.2 Handling State and Input Inequality Constraints

State and control inequality constraints emerge naturally in the mathematical formulation of motion planning problems representing, e.g., saturation in the actuator forces, velocity and acceleration constraints, position constraints for collision avoidance with the environment and, possibly, other vehicles, as well as geometric constraints to approximately maintain a desired formation for a fleet of vehicles. It is therefore desirable to be able to attack optimal control problems of the form

$$\begin{aligned}
\min \quad & \int_0^T l(x(\tau), u(\tau), \tau) \, d\tau + m(x(T)) \\
\text{subj to} \quad & \dot{x} = f(x, u), \qquad x(0) = x_0 \\
& c_j(x(t), u(t), t) \leq 0, \quad t \in [0, T], \; j = 1, \ldots, k.
\end{aligned} \qquad (7)$$

To address the (state and input) inequality constraints, we take an interior point/barrier function like approach, very much like what is done in finite dimensional, convex optimization [7]. The infinite-dimensional analog is to add a barrier functional to the original cost functional

$$\begin{aligned}
\min \quad & \int_0^T l(x(\tau), u(\tau), \tau) - \varepsilon \sum_{j=1}^k \log(-c_j(x(\tau), u(\tau), \tau)) \, d\tau + m(x(T)) \\
\text{subj to} \quad & \dot{x} = f(x, u), \quad x(0) = x_0
\end{aligned} \qquad (8)$$

and then following the *central path* of minimizers as $\varepsilon \to 0$ we approach the solution of the original problem (7).

In principle, (8) can be directly attacked using PRONTO. Unfortunately, the cost functional is not well defined for curves ξ that are infeasible since the domain of the logarithm is $\mathbb{R}_+$. While this is clearly a problem when starting outside of the feasible set, there is a more subtle situation when doing trajectory optimization with PRONTO. In finite dimensions, once feasibility has been obtained, one may always restrict the step taken to preserve feasibility. Within PRONTO, one may also restrict the step size γ so that $\xi_i + \gamma \zeta_i$ is a feasible curve but this does not ensure that $\mathscr{P}(\xi_i + \gamma \zeta_i)$ is a feasible trajectory! To manage this possibility (and infeasibility), we use $-\log$ when its argument is greater than $\delta > 0$ and a quadratic polynomial (with a C^2 join) otherwise giving the approximate log barrier function [14]

$$\beta_\delta(z) = \begin{cases} -\log z & z > \delta \\ -\log \delta + \frac{1}{2}\left[\left(\frac{z - 2\delta}{\delta}\right)^2 - 1\right] & z \leq \delta \end{cases} \qquad (9)$$

Together with the constraints, we obtain an *approximate* log barrier functional

$$b_\delta(\xi) = \int_0^T \sum_{j=1}^k \beta_\delta(-c_j(\alpha(\tau), \mu(\tau), \tau)) \, d\tau \qquad (10)$$

giving us a family of optimal control problems (parametrized by (ε, δ))

$$\min_{\xi \in \mathscr{T}} h(\xi) + \varepsilon b_\delta(\xi) \qquad (11)$$

without inequality constraints that may easily be attacked with PRONTO. Problem (11) is an approximation of (7) in much the same way that (8) is. Indeed, a trajectory $\xi_\varepsilon^\star$ that is a locally optimal solution of (8), being strictly feasible, is also a locally optimal solution of (11) provided $\delta > 0$ is sufficiently small,

$$\delta < \min_{t \in [0,T]} \min_{j} -c_j(x_\varepsilon^\star(t), u_\varepsilon^\star(t), t). \tag{12}$$

PRONTO may be thus used in the optimization of (11) as part of a continuation (or path-following) method to seek an approximate solution to (7). Our goal is twofold: obtain feasibility and follow the central path (in ε) to a neighborhood of the solution of (7). While there is much room for creativity, it is worth noting that the parameter δ can be helpful in obtaining and retaining feasibility while ε parametrizes *where we are* on the central path (once feasible). A reasonable strategy is to begin with $\varepsilon = 1 = \delta$ and check the feasibility of corresponding (local) minimizer of (11). If not feasible, hold $\varepsilon = 1$ and follow a path of decreasing δ until feasibility is obtained. Once feasibility has been obtained, follow a path of decreasing ε while adjusting δ to ensure that feasibility is retained in computations and such that (12) is satisfied, i.e., each $\xi_\varepsilon^\star$ is δ-feasible (hence, independent of δ). During central path following, it is common to choose the decreasing ε sequence by, e.g., $\varepsilon \leftarrow \varepsilon/10$ or $\varepsilon \leftarrow \varepsilon/6$. For more information on continuation methods, see [2].

Certain kinds of constraints are handled nicely using the approximate log barrier function $\beta_\delta(\cdot)$. For instance, the typical constraint $u^2 \leq 1$ gives rise to the strongly convex function $\beta_\delta(1 - u^2)$ which at $\delta = 1$ is simply $u^2 + u^4/2$. The key thing here is that the *chosen* constraint function $c(x, u, t) = c(u) = u^2 - 1$ is bounded below by -1 ensuring that $\beta_\delta(-c(u)) \geq 0$ for all u and all $\delta \in (0, 1]$.

In contrast, one might choose a constraint of the form $\|p\| \geq 1$ to avoid an obstacle at 0, p as position. The simple constraint function $1 - \|p\|^2$ is not bounded below, thereby rewarding being far from the obstacle when we merely need to avoid it. One way manage this is to *saturate* the argument of $\beta_\delta(\cdot)$ before applying it, replacing $\beta_\delta(-c(x, u, t))$ with $\beta_\delta(\sigma(-c(x, u, t)))$. We have found the *hockey stick* function

$$\sigma(z) = \begin{cases} \tanh(z) & z \geq 0 \\ z & z < 0 \end{cases}$$

to be useful in this regard. The use of the hockey stick on this type of obstacle avoidance function allows us to ignore obstacles that are *far enough* away, see Fig. 2. Also, in spite of the fact the obstacle avoidance constraint functions are definitely *not* convex, they do look convex in the radial direction and we have found them to work well in practice.

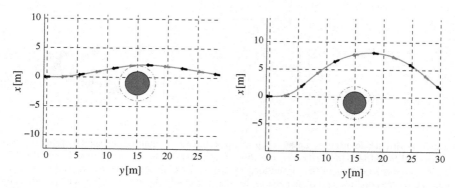

Fig. 2 The barrier function with hockey stick (*left*) does a better job of *just* avoiding the obstacle than without (*right*) during the early stages of optimization (with $\varepsilon \approx 1$)

3 Applications

PRONTO has been employed for trajectory exploration and constrained optimal motion planning for single and, more recently, multiple vehicles. We discuss the single and multiple vehicle cases separately in the following two subsections.

3.1 Trajectory Exploration and Motion Planning for a Single Vehicle

Trajectory exploration for racing motorcycles in the context of virtual prototyping [27, 30] provided a challenging playground and, indeed, one of the first applications [15] of PRONTO. In [31], the computation of feasible trajectories of racing motorcycle has been addressed. The use of PRONTO in [31] was connected with the computation of the minimum time velocity profile for a given vehicle and desired ground path as well as the development of a virtual rider for a multi-body motorcycle model [27, 30, 33]. Figure 3 shows a typical minimum time velocity profile, corresponding roll angle, and closed-loop multibody simulation obtained with the developed optimization and control architecture.

In [31], PRONTO is employed to perform dynamics inversion, i.e., the computation of a (bounded) state and input trajectory pair corresponding to a desired ground trajectory. Dynamics inversion is accomplished by embedding [15] the motorcycle dynamics into an extended control system with extra (artificial) inputs that make the system easy to control.

A motorcycle is an underactuated system where roll dynamics can be influenced by direct control of steering angle and the longitudinal acceleration. In [31], dynamics embedding is achieved by adding an artificial input to the roll dynamics that allows for direct control of the vehicle lean angle. The addition of artificial inputs

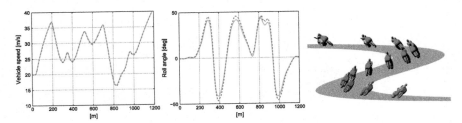

Fig. 3 A virtual rider for racing motorcycles: (*left*) reference minimum-time velocity profile (*solid green*) and closed-loop (*dashed blue*) velocity profile; (*middle*) reference and closed-loop roll angle profile; (*right*) snapshots of the closed-loop multibody simulation related to the handling of the chicane (corresponding, approximately, to the space interval 200–400 m in the velocity and roll angle plots)

virtually renders the mechanical systems fully actuated and allows for the use of standard inverse dynamics (also known as computed torque control, in the robotics literature) to make the embedded system follow any desired velocity-curvature profile. PRONTO is then used to optimize away this artificial control effort, leading to a trajectory of the original underactuated motorcycle model that is still consistent with the specified planar trajectory. By saying "optimized away", we mean that the magnitude of the artificial inputs is reduced to zero by an iterative optimization procedure, by modifying the initial state and input trajectories while keeping the output close to the desired one. For a given desired path and velocity profile, the optimization algorithm is initialized with the so-called "quasi static" trajectory, obtained by computing the equilibrium roll angle that would correspond to assume a constant path curvature pointwise in time. In places when the path curvature varies slowly, one can imagine the feasible, dynamic roll trajectory to be close to the quasi-static one [16]. For more details, see [27, 31].

A similar dynamic inversion technique based on dynamics embedding was employed for computing feasible trajectories for a planar model of a vertical take-off and landing (PVTOL) airplane in [20] with the extra complexity of satisfying state and input inequality constraints. From a computational point of view, the inequality constraints were handled in PRONTO with the approximate log barrier approach reviewed in Sect. 2, based on the seminal contribution [14].

The PVTOL (planar vertical take-off and landing) aircraft, introduced in [17], provides a simplified model that captures the (lateral) non-minimum phase behavior present in many real aircraft. This model has been widely studied in the literature for its characteristic of combining important features of nonlinear systems with a set of *tractable* equations. Furthermore, the dynamics of many other mechanical systems (e.g., the cart-pole system, the pendubot, the nonholonomic bicycle model) can be written in a similar fashion.

Figure 4 provides a graphical representation of the PVTOL as well as an example of the feasible trajectory that results from the application of the optimization strategy. The approach allows one to compute a constrained trajectory that is closest, in an L_2 sense, to a desired input-unconstrained barrel roll airplane trajectory. The figure

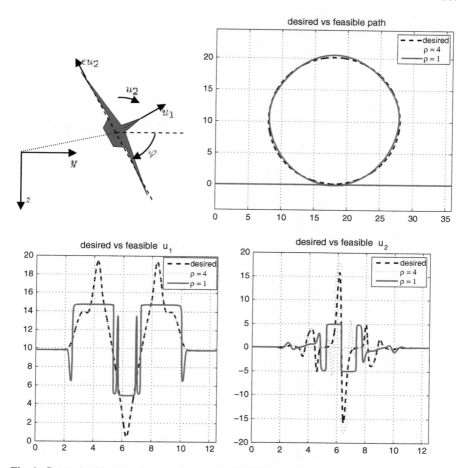

Fig. 4 Constrained barrel roll trajectory: (*top-left*) PVTOL aircraft model, with linear (y and z) and angular (φ) configuration variables, vertical thrust force u_1 and rolling moment u_2 (εu_2 is the cross coupling); (*top-right*) desired, i.e., unconstrained motion of the vehicle (*dashed*), intermediate ($\rho = 4$) and fully constrained ($\rho = 1$) motions; (*bottom*) Unconstrained, intermediate, and constrained control trajectories for the PVTOL barrel roll trajectory: vertical thrust force u_1 rolling moment u_2

furthermore illustrates the trajectories computed during an intermediate step of the optimization (indicated with $\rho = 4$) for which the input constraints are only slightly violated as well as the final optimization result ($\rho = 1$) that fully satisfies the specified inequality constraints. The two control signals, also depicted in Fig. 4, tend to hit the constraint boundary (set to $[0.5, 1.5]g$, being g the gravity acceleration, for the vertical thrust force and $[-5, 5]$ for the rolling moment) for a larger interval of time than the one where the constraints are violated –thus working in a noncausal fashion– in order to compensate the missing availability of input effort in those regions. See [20] for further details.

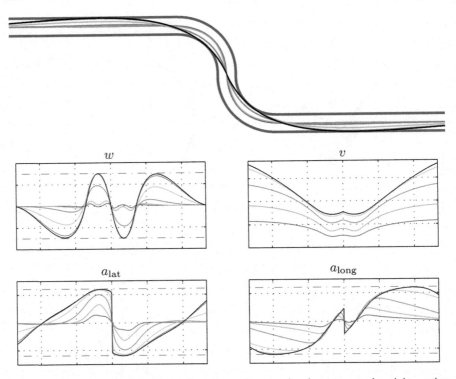

Fig. 5 Point mass racer: minimizing trajectories on the *central path* converge to the minimum time *race line* and velocity profile

PRONTO has also been employed to find the *race line* and corresponding velocity profile for a range of racing vehicles of varying complexity. Figure 5 depicts results from this problem for the case of a point mass vehicle [4], seeking to go through a chicane in minimum time subject to friction ellipse type acceleration constraint. Shown is a sequence of trajectories on the central path as ε is decreased to approach the minimum time trajectory. The lateral displacement w from centerline shows how a racer *flattens* out the curve while actually accelerating and decelerating within the chicane! Working in a good set of maneuver adapted coordinates, described in detail in [3, 4] (where some single track kart type vehicles were also considered), is critical in such applications. Note that the shape of the velocity profile is quite similar to that in [14] wherein the minimum time velocity profile for a given path was computed.

Later on, in [25, 26], the approach is extended to a single-track rigid car and to a two-track car model, respectively. These two car models include a realistic tire model that includes the effect of (longitudinal and lateral) load transfer.

In [4, 25, 26], the minimum-time problem is converted into an equivalent constrained optimal control problem written in terms of longitudinal and transverse coordinates. This change of coordinates allows for the use of a fixed horizon in the optimization where the time-dependent state constraints are converted into space-dependent state constraints. The reformulated constrained optimal control problem is solved using the approximate log barrier function approach reviewed in Sect. 2. One interesting consequence is that the convergence to the optimal trajectory happens in an interior point fashion, somehow resembling the learning process of a real driver in pushing the car toward its limits and allowing to stop the optimization process at any point having the guarantee to possess a feasible trajectory.

As discussed in [26], the performance of this approach (dubbed MTGO) appears extremely promising, taking also into account the performance of other existing algorithms based on multiple shooting methods. In [24], the authors highlight the importance of the trajectory optimization problem formulation for ensuring the *existence* of a solution.

As mentioned, the selection of a suitable set of coordinates used to represent the system state might ease the solution of a given trajectory optimization or motion planning problem. choosing different set of coordinates can significantly influence the ease of coding and obtaining a numerical solution from the optimization algorithm. In [4, 26], the position and orientation of the vehicle are parametrized with respect to the racing track center line by longitudinal and lateral coordinates, in place of absolute longitudinal and lateral inertial coordinates. Longitudinal and transverse coordinates are also encountered in the maneuver regulation controller such as the virtual motorcycle rider in [32].

By nature, a longitudinal and transverse coordinate parametrization is, however, only locally invertible. A key parameter that defines the maximum allowable lateral displacement from the reference path is the desired path curvature. In this case, multiple (even infinite) solutions exist when the vehicle distance from the path is equal to or larger than the maximum absolute curvature: the example of a vehicle at the center of a circle is the prototype example of this situation. If, in some cases, this is not a problem due to the nature of the problem at hand (e.g., the race car example, where the vehicle is confined to be within a tube whose maximum radius is determined by the desired path curvature), in other problems this constitutes a limitation.

To overcome this possible issue, [22] proposes a Virtual target vehicle (VTV) perspective by introducing a virtual target that plays the role of an additional control input. In practice, this approach removes the assumption that the point along the reference path (that is, the position of the VTV) is the point on the path that is closest to the vehicle, because in this setting the motion along the path is determined by the additional input. Removing this assumption leads to an optimization procedure that takes explicitly into account the extra flexibility of where to place the VTV by imposing (during the transient period) a convenient motion of the virtual target with the benefit of also improving the convergence of the solver to obtain the optimal feasible path. The approach has been further elaborated in [23] for a general class of constrained robotic vehicles.

Figure 6 shows an example of the typical error frame employed in the VTV approach, showing the inertial, body-fixed, and Serret–Frenet frames as well as the desired path (typically parametrized using arc length) and the error vector definition.

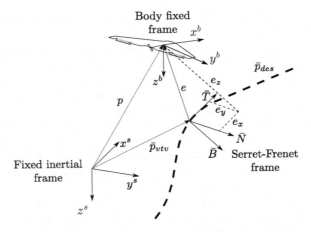

Fig. 6 The error frame for the Virtual Target Approach, illustrated for a UAV

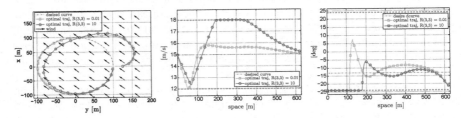

Fig. 7 Loiter maneuver with spatial parametrization: (*left*) The desired path (*dash-dot*) and two optimal paths corresponding to different cost weighting (*solid*) are shown, namely for a lower weight (*light green*) and a higher weight (*dark green*); (*middle*) corresponding air speed profiles; (*right*) roll angle trajectories. State constraints are represented with *red dashed lines*

As previously mentioned, the Serret–Frenet frame origin is not forced to coincide with the point along the desired path having minimum distance from the vehicle, implying the error vector e shown in the figure is not necessary orthogonal to the desired path. Figure 7 shows an example of computation of a feasible trajectory for a UAV asked to perform a circular loiter maneuver at a given constant speed in an area where wind is present. The optimization has a tuning parameter allowing to decide how accurately the VTV should follow the prescribed velocity along the circular path. Depending of this tuning, the resulting optimal trajectory resembles the trajectory that one would obtain by a trajectory-tracking controller (high weight tuning) or by a path-following/maneuver regulation controller (low weigh tuning), both satisfying the state inequality constraints. Numerical results for other testing scenarios and further details regarding the VTV approach are reported in [22, 23]. A receding horizon implementation on this approach to deal with, e.g., time varying wind/current conditions is discussed in the accompanying book chapter [1].

Summarizing, we have shown in this subsection that PRONTO may be used as an effective tool to compute feasible trajectories for accurate dynamic vehicle models in the presence of nontrivial and realistic state and input constraints. In this section, we have mostly emphasized its use as a tool to explore the capabilities of a given vehicle rather than to maneuver in a (known) constrained environment by generating a feasible reference trajectory to be tracked by the low level controllers. This is the subject of the next subsection, where we also document the authors' effort for constrained optimal motion planning in the case of multiple vehicles.

3.2 Constrained Optimal Motion Planning for Multiple Vehicles

The rapid development in the field of robotics in the past decade, going from single robot tasks to missions that require coordination, cooperation, and communication among a number of networked vehicles, makes the availability of versatile motion planners increasingly important. In [10], a numerical algorithm for multiple vehicle motion planning that explicitly takes into account the vehicle dynamics, temporal and spatial specifications, and energy related requirements is introduced (Fig. 9).

As a motivating example, [10] considers the case where a group of vehicles is tasked to reach a number of target points at the same time (simultaneous arrival problem) without colliding among themselves and with obstacles, subject to the requirement that the overall energy required for vehicle motion be minimized. With the theoretical setup adopted, the vehicle dynamics are explicitly taken into account at the planning level. The efficacy of the method is illustrated through examples of the simultaneous arrival problem for the realistic case where the initial guess for the trajectories that is the seed for the optimization procedure is obtained from simple geometrical considerations; for example, by joining the initial and final target points of the vehicles via straight lines. Even though the initial trajectories thus obtained yield inter-vehicle and vehicle/obstacle collisions, in the event that the optimization problem is feasible, PRONTO will generate appropriate trajectories that minimize the overall vehicle energy spent and meet the required temporal and spatial constraints. The method developed applies to a very general class of vehicles; however, for clarity of exposition, [10] adopts as an illustrative example the case of a fleet of wheeled nonholonomic robots. In terms of mathematical formulation, the optimal motion planning algorithm is based on solving a constrained optimal control problem. Indicating with N_v the number of vehicles, with N_o the number of (possibly moving) obstacles, and using the superscripts $[i]$ and $[k]$ to denote quantities related to the ith vehicle and kth obstacle, in its basic formulation the constrained optimal control problem has the following structure:

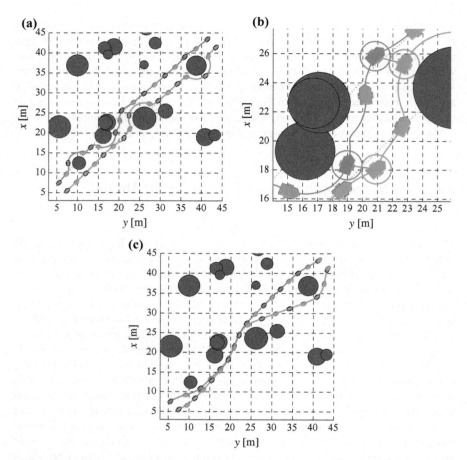

Fig. 8 Maneuvering cooperatively with two vehicles across a random obstacle field. **a** Solution with specified desired motion; **b** Zoomed-in portion of the solution with desired motion, showing collision avoidance with obstacles and other vehicles; **c** Minimum energy solution without specifying a desired motion

$$\text{minimize} \quad \int_0^{t_f} \sum_{i=1}^{N_v} \left(l_{\text{pow}}\big(\mathbf{x}^{[i]}(\tau), \mathbf{u}^{[i]}(\tau)\big) + l_{\text{trj}}\big(\mathbf{x}^{[i]}(\tau), \mathbf{u}^{[i]}(\tau), \tau\big) \right) d\tau + m\big(\mathbf{x}(t_f)\big)$$

$$\text{subject to} \quad \dot{\mathbf{x}}^{[i]}(t) = f(\mathbf{x}^{[i]}, \mathbf{u}^{[i]}, t), \quad \mathbf{x}^{[i]}(0) = \mathbf{x}_0^{[i]} \tag{13}$$
$$c_{\text{col}}\big(\mathbf{x}^{[i]}(t), \mathbf{x}^{[j]}(t)\big) \geq 0, \, i \neq j$$
$$c_{\text{obs}}\big(\mathbf{x}^{[i]}(t), \mathbf{o}^{[k]}\big) \geq 0$$

with $i, j \in \{1, \ldots, N_v\}$ and $k \in \{1, \ldots, N_o\}$, and where l_{pow} denotes the instantaneous power requirement for a single vehicle, l_{trj} an instantaneous cost related to how far a single vehicle is from a desired nominal motion, m the terminal constraint that

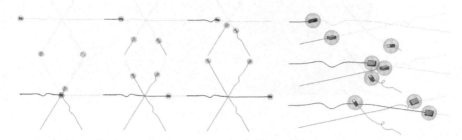

Fig. 9 Constrained optimal motion planning for multiple vehicle where kinematics and computations are carried out in the Lie group SE(3), instead of using an Euler angle like parametrization of the vehicle attitude matrices. (*left*) *top view* of six snapshots of the optimal motion; (*right*) zoomed in and perspective view of three snapshots of the optimal motion illustrating the inter-vehicle collision avoidance

forces the formation to get close to a desired end state, f the kinematics/dynamics model of a single vehicle (possible different for each vehicle), c_{col} a function providing a smooth signed distance function between two vehicles in the formation, and c_{obs} a function providing a smooth signed distance function between a vehicle and an obstacle.

An illustration of the numerical results that can be obtained with this approach is provided in Fig. 8, illustrating the versatility of the method to obtain a trade off between planning a motion where a (possible unfeasible) desired motion is preassigned –moving along straight lines– and a motion, where energy consumption is the only optimization criterion for a given time horizon.

Similar results for a fleet of multiple vehicles are also presented in [28], where the emphasis is to demonstrate the Lie group version of PRONTO [29]. This geometric version of PRONTO allows for singularity free parametrization of the attitude of a rigid body that also results in simpler expression for the gradient and Hessian of the vehicle kinematics with respect to a standard formulation employing Euler angles.

As a last illustration of the versatility of PRONTO, Fig. 10 shows the result of the trajectory optimization process for autonomous marine vehicles going from east to west over the southern rim of the D. João de Castro sea mount (a large submarine volcano located in the central north Atlantic Ocean, in the archipelago of the Azores), where terrain-based information is employed to generate paths with a suitable compromise between the minimization of energy consumption and maximization of useful information for terrain-based navigation as measured by an estimation-theoretic criterion [9].

The above problem was motivated by a simple, often ignored fact: motion planning must, in practical applications, take explicitly into account the constraints introduced by the type of navigation systems that a specific group of vehicles will use - in conjunction with motion control algorithms - to steer them along planned trajectories. Among the methods available for underwater vehicle navigation, terrain-based tech-

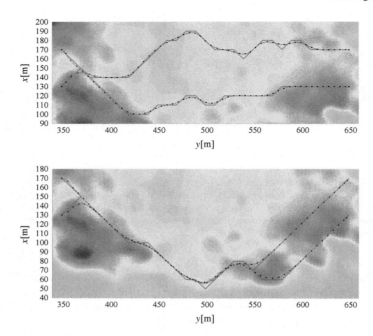

Fig. 10 Constrained motion planning for terrain-based navigation. The colored map is the bathymetric map of a sea mount. Crossing the sea mount, given fixed initial, and final poses: (*top*) More emphasis on energy consumption than feature-rich terrain (for improved terrain-based navigation); (*bottom*) Identical scenario, more emphasis on feature-rich terrain than energy consumption

niques have recently come to the fore. These techniques avoid the use of overly expensive inertial-like motion sensor units and hold considerable promise for the development of a new breed of affordable long range navigation systems. In Fig. 10, the initially specified minimum energy problem would have the vessels go in almost straight, parallel lines (above). Making use of terrain information biases the vehicles towards going over more "information rich" terrain so as to improve navigation (bottom). We refer to [9] for further details.

4 Conclusions and Future Work

This chapter provided an overview of PRONTO, a flexible and effective numerical tool for solving constrained optimal motion planning problems and exploring the trajectory space of single or multiple vehicles.

We have demonstrated PRONTO for two- and four-wheeled vehicles, autonomous underwater vehicles, and unmanned aerial vehicles in several application scenarios. Due to its continuous-time nature that allows to avoid the transcription of the continuous-time optimal control problem into a finite dimensional optimization problem, PRONTO can handle optimization problems with arbitrary fine time discretization of input and state trajectories (up to the desired ODE solver accuracy) and, when second-order information is provided, exhibits second-order convergence rate to a minimizer satisfying second-order sufficient conditions (SSC) for optimality.

Detailed kinematic or dynamics models can be employed, even when the state evolves on a Lie group, e.g., the set of rotation matrices SO(3). State and input inequality constraints are handled with an approximate log barrier approach, allowing to start the iterative process from an unfeasible trajectory.

Current research efforts are directed towards using PRONTO more and more as a reactive planner (i.e., as part of an MPC-type scheme) other than just as an offline planner. To this end, we are experimenting with the use of sensor-data, such as depth camera images, for building an environment description suitable for motion planning in real time. A discrete-time version of PRONTO is also available (see, e.g., [5]) and it is currently being explored in parallel to the standard continuous-time version: the target of this discrete-time implementation is to tackle those applications where approximations in the computations of the dynamics and cost are inevitable in order to allow for real-time optimization, leveraging on the experience in continuous time in the optimal problem formulation.

References

1. Aguiar, A.P., Rucco, A., Alessandretti, A.: A sampled-data model-predictive framework for cooperative path following of multiple robotic vehicles. In: Fossen, T.I., Pettersen, K.Y., Nijmeijer, H. (eds.) Sensing and Control for Autonomous Vehicles: Applications to Land, Water and Air Vehicles. Springer, Heidelberg (2017)
2. Allgower, E.L., Georg, K.: Introduction to Numerical Continuation Methods. Classics in Applied Mathematics. SIAM (2003)
3. Bayer, F.: Time-Optimization of a Raceline: A Projection Operator Approach. Diploma Thesis, Institute for Systems Theory and Automatic Control, University of Stuttgart (2012)
4. Bayer, F., Hauser, J.: Trajectory optimization for vehicles in a constrained environment. In: IEEE Conference on Decision and Control, pp. 5625–5630 (2012)
5. Bayer, F.A., Notarstefano, G., Allgöwer, F.: A projected SQP method for nonlinear optimal control with quadratic convergence. In: IEEE Conference on Decision and Control, pp. 6463–6468 (2013)
6. Betts, J.: Practical Methods for Optimal Control and Estimation Using Nonlinear Programming. Society for Industrial and Applied Mathematics (2010)
7. Boyd, S.P., Vandenberghe, L.: Convex Optimization. Cambridge University Press, Cambridge (2004)
8. Häusler, A.J.: Mission Planning for Multiple Cooperative Robotic Vehicles. Ph.D. thesis, Department of Electrical and Computer Engineering, Instituto Superior Técnico, Lisbon (2015)
9. Häusler, A.J., Saccon, A., Pascoal, A.M., Hauser, J., Aguiar, A.P.: Cooperative AUV motion planning using terrain information. In: MTS/IEEE OCEANS Conference (2013)
10. Häusler, A.J., Saccon, A., Aguiar, A.P., Hauser, J., Pascoal, A.M.: Energy-optimal motion planning for multiple robotic vehicles with collision avoidance. IEEE Trans. Control Sys. Technol. **24**(3), 867–883 (2016)
11. Hauser, J.: A projection operator approach to the optimization of trajectory functionals. In: 15th IFAC World Congress, pp. 377–382 (2002)

12. Hauser, J.: On the computation of optimal state transfers with application to the control of quantum spin systems. In: IEEE American Control Conference, pp. 2169–2174 (2003)
13. Hauser, J., Meyer, D.G.: The trajectory manifold of a nonlinear control system. In: IEEE Conference on Decision and Control, pp. 1034–1039 (1998)
14. Hauser, J., Saccon, A.: A barrier function method for the optimization of trajectory functionals with constraints. In: IEEE Conference on Decision and Control, pp. 864–869 (2006)
15. Hauser, J., Saccon, A., Frezza, R.: Aggressive motorcycle trajectories. In: IFAC Symposium on Nonlinear Control Systems (2004)
16. Hauser, J., Saccon, A., Frezza, R.: Achievable motorcycle trajectories. In: IEEE Conference on Decision and Control, pp. 3944–3949 (2004)
17. Hauser, J., Sastry, S., Meyer, G.: Nonlinear control design for slightly nonminimum phase systems: application to V/STOL aircraft. Automatica **28**(4), 665–679 (1992)
18. Luemberger, D.G.: Optimization by Vector Space Methods. Wiley, New York (1969)
19. Notarnicola, I., Bayer, F.A., Notarstefano, G., Allgwer, F.: Final-state constrained optimal control via a projection operator approach. In: European Control Conference, pp. 148–153 (2016)
20. Notarstefano, G., Hauser, J., Frezza, R.: Computing Feasible Trajectories for Control-Constrained Systems: The PVTOL Example IFAC Symposium on Nonlinear Control Systems, pp. 354–359 (2005)
21. Notarstefano, G., Hauser, J., Frezza, R.: Trajectory manifold exploration for the PVTOL aircraft. In: IEEE Conference on Decision and Control, and the European Control Conference, pp. 5848–5853 (2005)
22. Rucco, A., Aguiar, A.P., Hauser, J.: Trajectory optimization for constrained UAVs: a virtual target vehicle approach. In: International Conference on Unmanned Aircraft Systems (ICUAS), pp. 236–245 (2015)
23. Rucco, A., Aguiar, A.P., Hauser, J.: A virtual target approach for trajectory optimization of a general class of constrained vehicles. In: IEEE Conference on Decision and Control (CDC), pp. 5245–5250 (2015)
24. Rucco, A., Hauser, J., Notarstefano, G.: Optimal control of steer-braking systems: non-existence of minimizing trajectories. Optim. Control Appl. Methods **37**(5), 965–979 (2016)
25. Rucco, A., Notarstefano, G., Hauser, J.: Computing minimum lap-time trajectories for a single-track car with load transfer. In: IEEE Conference on Decision and Control, pp. 6321–6326 (2012)
26. Rucco, A., Notarstefano, G., Hauser, J.: An efficient minimum-time trajectory generation strategy for two-track car vehicles. IEEE Trans. Control Sys. Technol. **23**(4), 1505–1519 (2015)
27. Saccon, A. : Maneuver regulation of nonlinear systems: the challenge of motorcycle control. Ph.D. Dissertation, Department of Information Engineering, University of Padova, Italy (2006)
28. Saccon, A., Aguiar, A.P., Häusler, A.J., Hauser, J., Pascoal, A.M.: Constrained Motion Planning for Multiple Vehicles on SE(3). In: IEEE Conference on Decision and Control, pp. 5637–5642 (2012)
29. Saccon, A., Hauser, J., Aguiar, A.P.: Optimal control on lie groups: the projection operator approach. IEEE Trans. Autom. Control **58**(9), 2230–2245 (2013)
30. Saccon, A., Hauser, J., Beghi, A.: A virtual rider for motorcycles: an approach based on optimal control and maneuver regulation. In: IEEE International Symposium on Communications, Control and Signal Processing, pp. 243–248 (2008)
31. Saccon, A., Hauser, J., Beghi, A.: Trajectory exploration of a rigid motorcycle model. IEEE Trans. Control Sys. Technol. **20**(2), 424–437 (2012)
32. Saccon, A., Hauser, J., Beghi, A.: A virtual rider for motorcycles: maneuver regulation of a multi-body vehicle model. IEEE Trans. Control Sys. Technol. **21**(2), 332–346 (2013)
33. Saccon, A., Hauser, J., Beghi, A.: Virtual rider design: optimal manoeuvre definition and tracking. In: Tanelli, M., Corno, M., Savaresi, S.M. (eds.) Modelling, Simulation and Control of Two-Wheeled Vehicles, pp. 83–117. Wiley, New York (2014)

Part IV
Sensing and Tracking Systems

Observability-Based Sensor Sampling

Kristi A. Morgansen and Natalie Brace

Abstract Systems with a rich array of sensors but limited power and/or processing resources have more potential information available than they can use and are forced to subsample the data. In this work, we build on observability analysis for general nonlinear systems to provide a basis for a framework to investigate dynamic sensor selection to optimize a measure of observability, specifically the condition number of an observability Gramian. This optimization is then applied to a sample system of natural Frenet Frames with sensing allowed to alternate between bearing and range measurements relative to a fixed beacon.

1 Introduction

Small autonomous vehicles can carry an array of simple, lightweight sensors; however, the ability to utilize the sensors is subject both to processing and to power constraints especially for sensors that have intensive power and processing needs. Inertial sensors, acoustic sensors (sonar), compasses, and cameras are commonly used in small aerial and underwater vehicles to provide estimates of the orientation and acceleration of the vehicle as well as local information, such as nearest objects and object recognition. To estimate global position, aerial vehicles typically rely on GPS; in indoor, underwater, or other GPS-denied environments, other methods are required, such as range and bearing measurements to a beacon at a known location. More complex tasks rely on the integration, or 'fusion,' of data from multiple sensors and multiple types of sensors. Given the constraints of power and computation time, the question arises as to how to best select among available sensors to optimize available resources while meeting information requirements.

K.A. Morgansen · N. Brace (✉)
William E. Boeing Department of Aeronautics & Astronautics,
University of Washington, Seattle, WA 98195-2400, USA
e-mail: nbrace@aa.washington.edu

K.A. Morgansen
e-mail: morgansen@aa.washington.edu

To address this question, we draw inspiration from biological sensing. Biological species collect vast quantities of sensory input that is filtered and prioritized based on the potential information that it can supply; for example, one species of bat uses both vision and sonar, but sonar signals become more frequent as the light levels drop (when vision becomes less informative) and during landing (when more precise distance estimates are required) [1]. From this biological example, we develop here a framework for subsampling sensors based on the information they can provide about the state of the system.

The ability of a system to uniquely determine its state from the available sensor readings is characterized by its observability. In contrast to linear systems, observability of a nonlinear system may be affected by its trajectory, and thus by the particular control input history. Further, observability in nonlinear systems may be a local property with some areas of the state space being more or less observable for a given set of sensors and control inputs. Whether or not a system is observable at a given state can be determined by examining whether or not an invertible relationship can be found between the measurement and control histories and the desired states. In many cases, both linear and nonlinear, the question of observability can be addressed locally by generating an appropriate observability matrix (e.g., through linearization) and checking for sufficient rank. Hermann and Krener [2] presented a method for evaluating observability of continuous-time nonlinear systems, and Albertini and D'Alessandro [3] extended the theory to discrete-time nonlinear systems, finding a similar rank condition on the co-distribution of the observation space. Related extensions have also been developed in hybrid systems addressing conditions under which switched systems are observable [4–6].

These rank conditions generally provide a binary measure of observability; one can determine how 'well' a system can be observed by using a metric on a scalar-valued function of the information contained in the observability matrix. A common function used for metric-based observability is the observability Gramian. In order to construct an observability Gramian for a nonlinear system, the choices are to linearize around a viable trajectory and use the linear time-varying observability Gramian [7] or to employ the empirical observability Gramian, which calculates the output energy from perturbations of the initial state [8]. The latter provides a practical method of computing output energy, as in Scherpen's work on balancing nonlinear systems [9], but has more recently found application in studying the observability of nonlinear systems [10] with the distinct advantage that an analytic solution is not required: the method only requires the ability to simulate the system and its outputs.

These tools have been well utilized to study autonomous vehicles, both in proving basic observability and in going a step further to design trajectories or system structure to optimize observability. Discrete-time observability analysis has been performed and observers developed for micro underwater vehicles based on a six degree of freedom (DOF) system model with intermittent beacon range-only measurements augmented by inertial measurements and depth sensors for dead reckoning [11]. The observability of nonholonomic integrators, a class of nonlinear systems, was studied, and the coupling of control and observability was exploited to optimize a measure of the system observability through choice of controls [12]. The sensor placement problem has also been approached using observability analysis to choose optimal sensor locations and types [13].

While sensor placement provides a static answer to observability optimization, a sensor scheduling approach allows for a dynamic subset of available sensors to provide measurements at each time step. Specifically, we are interested in the scheduling problem of selecting from a variety of sensors that provide unique types of sensor data and that are co-located on a moving platform. This problem is studied for a linear time-varying system with linear measurements in [14] and has similar goals and challenges to the more widely studied problem of determining how to best employ a distributed array of networked sensors to measure something in the external environment [15–17] such as tracking a target [18]. One of the larger challenges for these optimal observability problems is the rapidly increasing dimensionality as a function of the number of sensor options (e.g., type, number, possible locations), cost function, number of system states, and number and type of system controls: except for a subset of nicely posed systems (for example, linear systems with quadratic cost functions), finding the exact optimal solution to a particular problem requires an exhaustive tree-search of all possible sensor schedules [15].

In this work, we construct and provide a preliminary study of a framework to address the question of sensor scheduling. We are particularly interested in the application of these methods to nonlinear systems. A comparison is provided for the difference between the results for the Gramian of a linearization of the system dynamics about a valid, nominal trajectory and for the use of the empirical Gramian for the original nonlinear system. To ground the results in likely physical applications, we demonstrate the study for the Frenet frame model of motion of a vehicle in 3D space. The Frenet frame model is a version of the nonholonomic integrator, a canonical nonlinear system model for SE(3) that represents many real systems, such as aircraft, surface water vessels, and underwater vehicles. With this system structure, we incorporate sensors for range and bearing relative to a fixed and known beacon and address the optimal selection of the next sensor measurement. A range of trajectories with different radii of curvature and placement relative to the beacon are considered, and the resulting sensor schedules are compared.

The work here is organized as follows. Section 2 provides an overview of the methods used for determining the observability of general dynamic systems, our optimization problem is introduced in Sect. 3, followed by the introduction, observability analysis, and optimization results for an example system in Sect. 4. Finally, we discuss the results of our initial study of this type of sensor scheduling problem and the future work that we plan to do in Sect. 5.

2 Observability

While all digital systems have continuous-time dynamics and discrete-time measurements, often the update rate of the measurements is fast enough to reasonably use a continuous-time framework for both the dynamics and measurements. Given this assumption, a great deal of the work in nonlinear observability analysis has addressed continuous-time systems. Here, we are specifically interested in switching the sensors on and off, so a discrete-time framework is more relevant. To facilitate the discussion below, both the familiar continuous-time version of observability and the related discrete-time version are presented briefly. In either case, we are concerned with the ability to distinguish the initial state from other states given a set of known controls and measurements.

To begin, we will denote discrete-time system dynamics as

$$\Sigma_d : \begin{array}{l} \mathbf{x}[k+1] = \mathbf{f}(\mathbf{x}[k], \mathbf{u}[k]), \quad k = 0, 1, 2, \ldots \\ \mathbf{y}[k] = \mathbf{h}(\mathbf{x}[k]), \end{array}$$

and continuous-time system dynamics as

$$\Sigma_c : \begin{array}{l} \dot{\mathbf{x}}(t) = \tilde{\mathbf{f}}(\mathbf{x}(t), \mathbf{u}(t)) \\ \mathbf{y}(t) = \mathbf{h}(\mathbf{x}(t)) \end{array}$$

where the states $\mathbf{x}[k], \mathbf{x}(t) \in M$ and measurements $\mathbf{y}[k], \mathbf{y}(t) \in Y$, respectively lie within connected, differentiable manifolds $M \subset \mathbb{R}^n$ and $Y \subset \mathbb{R}^p$, and the controls $\mathbf{u}[k], \mathbf{u}(t) \in U$ lie within connected, differentiable manifold $U \in \mathbb{R}^m$. In cases where a result applies to both Σ_d or Σ_c, the subscript will be dropped for brevity. The following definitions in [3] regarding observability will be employed here.

Definition 1 *Indistinguishable* states $x_1, x_2 \in M$, denoted $x_1 I x_2$, will produce identical outputs for any identical combination of controls. A state $x_1 \in M$ is *observable* if being indistinguishable from state x_2 implies $x_1 = x_2$ for all $x_2 \in M$. The system Σ is observable if $x_1 I x_2 \implies x_1 = x_2 \ \forall \ x_1, x_2 \in M$.

Definition 2 State $x_0 \in M$ is *locally weakly observable* if there exists some neighborhood W of x_0 such that for each $x_1 \in W$, $x_0 I x_1 \implies x_0 = x_1$. A system Σ is locally weakly observable if all states x_0 are locally weakly observable.

Definition 3 State $x_0 \in M$ is *locally strongly observable* if there exists some neighborhood W of x_0 such that for each $x_0, x_1 \in W$, $x_0 I x_1 \implies x_0 = x_1$. A system Σ is locally strongly observable if all states x_0 are locally strongly observable.

2.1 Analytical Observability

The underlying principle for determining analytical observability is the same for continuous and discrete-time systems: check for the existence of a relationship between the outputs, inputs, and states that can be inverted to uniquely determine the initial state from the outputs and inputs. The outcomes of this type of study provide insight into the requirements for control function structure and limitations on allowable trajectories or sensor placement to enable full observability. In either the continuous or discrete scenario, the process can be locally reduced to a rank condition on a matrix composed of either a sequence (discrete) or the derivatives (continuous) of the outputs.

Continuous-Time Systems

Assume that Σ_c is nonlinear in the states, $\mathbf{x}(t)$, linear in the control inputs, $u_i(t)$, and that it can be written in the control affine form

$$\dot{\mathbf{x}}(t) = \tilde{\mathbf{f}}_0(\mathbf{x}(t)) + \sum_{i=1}^{m} \tilde{\mathbf{f}}_i(\mathbf{x}(t))u_i(t) = \tilde{\mathbf{f}}(\mathbf{x}(t), \mathbf{u}(t)),$$

where $\tilde{\mathbf{f}}_0$ is termed the drift, and $\tilde{\mathbf{f}}_i$ are termed the control vector fields. The collection of terms forming the observability space, $\mathscr{O}_c$, are constructed from the time derivatives of the output functions:

$$\frac{d}{dt}\mathbf{h} = \left(\frac{\partial}{\partial x}\mathbf{h}\right)\tilde{\mathbf{f}}(\mathbf{x}(t), \mathbf{u}(t))$$

$$\frac{d}{dt^l}\mathbf{h} = \frac{d}{dt}\left(\frac{d}{dt^{l-1}}\mathbf{h}\right), \; l \in \mathbb{N}.$$

These time derivatives can be equivalently represented using Lie derivatives, defined as

$$L_{\tilde{f}_i}\mathbf{h} = \frac{\partial \mathbf{h}}{\partial \mathbf{x}}\tilde{\mathbf{f}}_i,$$

with repeated and mixed derivatives respectively calculated as

$$L_{\tilde{f}_i}^k = \frac{\partial L_{\tilde{f}_i}^{k-1}\mathbf{h}}{\partial \mathbf{x}}\tilde{\mathbf{f}}_i \quad \text{and} \quad L_{\tilde{f}_j}L_{\tilde{f}_i} = \frac{\partial(L_{\tilde{f}_j}\mathbf{h})}{\partial \mathbf{x}}\tilde{\mathbf{f}}_i,$$

the latter arising from use of switching (area generating) controls captured by the Lie bracket of the vector fields $[\tilde{\mathbf{f}}_i, \tilde{\mathbf{f}}_j] = (\partial \tilde{\mathbf{f}}_j/\partial \mathbf{x})\tilde{\mathbf{f}}_i - (\partial \tilde{\mathbf{f}}_i/\partial \mathbf{x})\tilde{\mathbf{f}}_j$. The observability space of Σ_c, $\mathscr{O}_c$, is thus also the span of the Lie derivatives with respect to the drift and control vector fields, that is $\mathscr{O}_c = \text{span}\{\mathbf{h}, L_{\tilde{f}_i}\mathbf{h}, \ldots\}$, and is alternately termed the observability Lie algebra. Determining whether the relationship between the terms in the observability space, $\mathscr{O}_c$, and the states can be inverted is generally analytically intractable, but one can address the question locally using the inverse function theorem and the local co-distribution of the observability matrix. If $d\mathscr{O}_c$, the co-distribution of $\mathscr{O}_c$ at a point in the state space $\mathbf{x}_0$, is full rank, then the system is locally weakly observable at that point [2].

If the system is linear time-invariant with $\tilde{\mathbf{f}}(\mathbf{x}(t)) = \tilde{A}\mathbf{x}(t) + \tilde{B}\mathbf{u}(t)$ and $\mathbf{y} = \mathbf{h}(\mathbf{x}(t)) = C\mathbf{x}(t)$, then differentiating the output gives

$$\begin{bmatrix} \mathbf{y} \\ \mathbf{y}^{(1)} \\ \vdots \\ \mathbf{y}^{(n)} \end{bmatrix} = \begin{bmatrix} C \\ C\tilde{A} \\ \vdots \\ C\tilde{A}^{n-1} \end{bmatrix} \mathbf{x} + \begin{bmatrix} 0 \\ \tilde{B} \\ \vdots \\ C\tilde{A}^{n-2}\tilde{B} + \cdots + C\tilde{B} \end{bmatrix} \mathbf{u} = \mathscr{O}_{LTI}\mathbf{x} + \mathscr{U}\mathbf{u}. \quad (1)$$

where superscripts in parentheses indicate time derivatives and $\mathscr{O}_{LTI}$ is the continuous-time observability matrix. As all terms other than the state are assumed known, if $\mathscr{O}_{LTI}$ is full rank, and thus has an invertible square submatrix, then the system is observable. Thus, in the linear case, regardless of the control action, observability is completely determined by the observability matrix.

Discrete-Time Systems

The conditions for discrete-time observability are presented for a single input-single output system by Albertini and D'Asessandro in [3]. Given the system Σ_d, let the system dynamics for a choice of control, $\mathbf{u} \in U$, be denoted $f_u(\mathbf{x}) = f(\mathbf{x}, \mathbf{u})$. Then, define the output sequence, $\Theta_1, \Theta_2, \ldots$, at each time step, $k \geq 1$, as a function of the inputs as

$$\Theta_1 = \{\mathbf{h}(\cdot)\}$$
$$\Theta_k = \{\mathbf{h}(f_{u_j} \circ \cdots \circ f_{u_1}(\cdot)) | \forall i = 1, \ldots, j, \mathbf{u}_i \in U, \text{ and } 1 \leq j \leq k-1\}, \quad (2)$$

where $\circ$ denotes composition, and $\mathscr{O}_d = \cup_{k \geq 1} \Theta_k$ is the collection of the measurement functions. As with continuous-time nonlinear systems, the inverse function theorem is generally employed to find a local inversion of this relationship between states, measurements, and controls. For system Σ_d and state $\mathbf{x}_0$, the co-distribution of $\mathscr{O}_d$ at $\mathbf{x}_0$ is given by $d\mathscr{O}_d(\mathbf{x}_0)$. If the co-distribution is full rank, that is, $\text{rank}(d\mathscr{O}_d(\mathbf{x}_0)) = n$, then the system is locally weakly observable at $\mathbf{x}_0$.

This condition naturally holds for linear(ized) systems as well but without the need to employ the inverse function theorem. Assume that Σ_d can be written as the linear time-varying system

$$\Sigma_{d,LTV} : \begin{array}{l} \mathbf{x}[k+1] = A[k]\mathbf{x}[k] + B[k]\mathbf{u}[k], \quad k = 0, 1, 2, \ldots \\ \mathbf{y}[k] = C[k]\mathbf{x}[k]; \end{array}$$

then the measurements in the sequence defined in (2) can be written as

$$\mathbf{y}[k] = C[k]\left(\Phi_d[k, 0]\mathbf{x}[0] + \sum_{i=1}^{k-1} \Phi_d[k, i+1]B[i]u[i]\right),$$

where the discrete state transition matrix is defined as

$$\Phi_d[k_f, 0] = \prod_{k=0}^{k_f-1} A[k]. \tag{3}$$

In the time-sinvariant case, $\Phi_d[k_f, 0]$ simplifies to A^{k_f}, and the sequence of measurements compiled into the discrete-time linear time-invariant observability matrix is the same form as the continuous case, $\mathcal{O}_{LTI}$, defined in (1).

2.2 Observability Gramians

While the above analysis can determine whether or not a system is observable, it does not provide any quantitative characteristics of observability. An alternative approach to observability that admits the use of a metric is that of the observability Gramian. In particular, by incorporating structural or controlled parameters into the system description, these metrics can be used to determine best parameter choice for maximal observability.

Linear Observability Gramian

The traditional form of the observability Gramian applies to linear systems. The discrete-time observability Gramian, $W_{o,d}$, is defined for a linear, time-varying system as

$$W_{o,d}[k_0, k_f] = \sum_{k=k_0}^{k_f} \Phi_d^T[k, k_0] C^T[k] C[k] \Phi_d[k, k_0], \tag{4}$$

where $\Phi_d[k, 0]$ is defined as in (3). The continuous-time observability Gramian, $W_{o,c}$, is constructed using the continuous-time state transition matrix, $\Phi_c(t, t_0)$, as

$$W_{o,c}(t_0, t_f) = \int_{t=t_0}^{t_f} \Phi_c^T(t, t_0) C_c^T(\mathbf{x}^0(t)) C_c(\mathbf{x}^0(t)) \Phi_c(t, t_0) dt, \tag{5}$$

where $\Phi_c(t, t_0)$ satisfies

$$\frac{\partial}{\partial t} \Phi_c(t, t_0) = \tilde{A}(\mathbf{x}(t), \mathbf{u}(t)) \Phi_c(t, t_0).$$

If the observability Gramian, $W_{o,d}[k_0, k_f]$ or $W_{o,c}(t_0, t_f)$, is full rank, n, for some time step k_f or t_f, then the linear time-varying system is observable. Measures of observability based on the Gramian follow in section "Measures of Nonlinear Observability".

The observability Gramian $W_{o,c}$ can be used to study the observability of the nonlinear system Σ_c by linearizing the system about some nominal trajectory, $(\mathbf{x}^0(t), \mathbf{u}^0(t))$, using the first term of a Taylor series expansion of $\tilde{\mathbf{f}}(\mathbf{x}(t), \mathbf{u}(t))$ evaluated along the trajectory to get the linear, time-varying system

$$\Sigma_{c,LTV} : \begin{array}{l} \dot{\bar{\mathbf{x}}} = A_c(\mathbf{x}^0(t), \mathbf{u}^0(t))\bar{\mathbf{x}}(t) + B_c(\mathbf{x}^0(t), \mathbf{u}^0(t))\bar{\mathbf{u}}(t) \\ \bar{\mathbf{y}} = C_c(\mathbf{x}^0(t))\bar{\mathbf{x}}(t), \end{array}$$

where ¯ denotes deviation from the nominal trajectory and

$$A_c(\mathbf{x}(t), \mathbf{u}(t)) = \left. \frac{\partial \tilde{\mathbf{f}}(\mathbf{x}(t), \mathbf{u}(t))}{\partial \mathbf{x}} \right|_{\mathbf{x}(t)=\mathbf{x}^0(t), \mathbf{u}(t)=\mathbf{u}^0(t)}$$

$$B_c(\mathbf{x}(t), \mathbf{u}(t)) = \left. \frac{\partial \tilde{\mathbf{f}}(\mathbf{x}(t), \mathbf{u}(t))}{\partial \mathbf{u}} \right|_{\mathbf{x}(t)=\mathbf{x}^0(t), \mathbf{u}(t)=\mathbf{u}^0(t)}$$

$$C_c(\mathbf{x}(t), \mathbf{u}(t)) = \left. \frac{\partial \mathbf{h}(\mathbf{x}(t))}{\partial \mathbf{x}} \right|_{\mathbf{x}(t)=\mathbf{x}^0(t)}.$$

A linear, time-varying discrete-time approximation of Σ_c can be produced by using the matrix exponential to calculate the discrete state space matrix $A_d[k] = \exp\{A_c(k\Delta_T)\}$, where Δ_T is the size of the time step, which is in turn used to calculate $\Phi_d[k, 0]$ from (3).

Empirical Observability Gramians

The linear Gramians can provide a basis for analysis and optimization of nonlinear systems; however, it is important to remember that the linear approximations do not capture all of the dynamics of the nonlinear system and thus do not always provide accurate results. It is not uncommon that a linearized system would be classified as unobservable when the original nonlinear system in fact observable. To demonstrate this phenomenon, consider the scalar system $\dot{x} = u$ with measurement function $y = \cos(x)$. Assuming nonzero controls, the nonlinear observability codistribution will be $d\mathcal{O} = \text{span}\{\cos(x), \sin(x)\}$ which will always have a dimension of 1 and thus be observable. The linearized measurement function $\bar{y} = -\sin(x)|_{x^0}$ will render the system unobservable when $x^0 = 2\pi k$. Also, in the linearized system, the controls are predetermined along the nominal trajectory, and thus cannot be used for analysis in the same way as in the original nonlinear structure.

To extend the use of observability Gramians to nonlinear systems, Lall, Marsden, and Glavaški introduced the empirical observability Gramian [8], which uses perturbations, $\varepsilon \mathbf{e}_i$, of the initial state, $\mathbf{x}_0 = \mathbf{x}(0)$, to quantify changes in the output measurements. The original version did not include the control explicitly, so instead we use the continuous-time empirical observability Gramian from [19]:

$$W_c^\varepsilon(t_f, \mathbf{x}_0, \mathbf{u}^0(t)) = \frac{1}{4\varepsilon^2} \int_0^{t_f} \Phi^\varepsilon(t, \mathbf{x}_0, \mathbf{u}^0(t))^T \Phi^\varepsilon(t, \mathbf{x}_0, \mathbf{u}^0(t)) dt.$$

Further work by Powel and Morgansen [20] defines the discrete-time version as

$$W_d^\varepsilon[k_f, \mathbf{x}_0, \mathbf{u}^0[k]] = \frac{1}{4\varepsilon^2} \sum_{k=0}^{k_f} \Phi^\varepsilon[k, \mathbf{x}_0, \mathbf{u}^0[k]]^T \Phi^\varepsilon[k, \mathbf{x}_0, \mathbf{u}^0[k]] \qquad (6)$$

where $\Phi^\varepsilon(t, \mathbf{x}_0, \mathbf{u}^0(t))$ and $\Phi^\varepsilon[k, \mathbf{x}_0, \mathbf{u}^0[k]]$ are defined similarly by perturbations of the output in each direction at each time (step). The discrete-time version is given as

$$\Phi^\varepsilon[k, x_0, u] = \begin{bmatrix} y^{+1} - y^{-1} & \cdots & y^{+n} - y^{-n} \end{bmatrix}$$

with $y^{\pm i} = y[k, x_0 \pm \varepsilon \hat{\mathbf{e}}_i, u]$ where $\hat{\mathbf{e}}_i$ denote elements of the standard orthonormal basis in $\mathbb{R}^n$, and the magnitude of the perturbation, ε, is chosen relative to the magnitude of the states; in the subsequent example, $\varepsilon = 0.01$ (a reasonable choice for states with magnitude on the order of one [10]).

Measures of Nonlinear Observability

Eigenvalues of an observability Gramian provide measures of observability of the system [10]. The maximum eigenvalue, $\lambda_1(W)$, gives an idea of how much output energy is in the most observable state. Similarly, the minimum eigenvalue $\lambda_n(W)$, gives an idea of how much output energy is in the least observable state. For a well-conditioned estimate, it is desirable to have similar energy in the most and least observable states so that a small change in one state does not overwhelm changes in another state. This relation is captured by the condition number, $\kappa(W) = \lambda_1(W)/\lambda_n(W)$ that takes a minimum value of unity. It is worthwhile to note that each of these measures can be characterized as convex functions of the observability [21].

One variation on the condition number which will prove useful here is defining it using the largest and smallest singular values instead of eigenvalues:

$$\kappa(W) = \sigma_1(W)/\sigma_3(W). \qquad (7)$$

This change accommodates systems with constraints built into their definitions that cause some of the eigenvalues of the Gramian to have value zero (since the observability Gramian is positive definite by definition, the smallest eigenvalue would be zero, rendering the standard condition number useless).

3 Switched Sensing Optimization

Given a set of sensors with sufficient resources to use only a subset of the available sensors or their data at a time, our goal here is to use measures of observability to determine which sensor(s) should be used at any point in time to maximize observability. We begin by introducing a sensor selection switch into the discrete system description, Σ_d. Specifically, assume that a set of p sensors is available, each of which is respectively described by the measurement function, y_i, $i \in \{1, \ldots, p\}$. For each sensor, incorporate a binary-valued switch function $s_i \in \{0, 1\}$ where $s_i = 1$ if the sensor is active, and $s_i = 0$ if the sensor is inactive. The measurement functions for the system at each time step, k, can then be described by

$$y_i[k] = s_i[k] h_i[k], \quad i \in \{1, \ldots, p\}, \quad s_i[k] \in \{0, 1\}.$$

To restrict the number of active sensors to p_k, the constraint $\sum_{i=1}^{p} s_i[k] = p_k$ would be applied at each time step. Alternatively, p_k could represent some resource budget and c_i could represent the relative resource usage of sensor i; then the constraint would be $\sum_{i=1}^{p} c_i s_i[k] \leq p_k$.

At each time step we then seek to choose $s_i[k]$ to optimize some preferred measure of observability, $J(s)$, subject to the constraints. Stated in the simplest form, this problem is written as

$$\min_{s} \; J(s) \qquad (8)$$
$$\text{subject to} \quad s_i[k] \in \{0, 1\} \quad \forall i, k$$
$$\sum_{i=1}^{p} c_i s_i[k] \leq p_k \quad \forall k.$$

The cost function, $J(s)$, could be one of the convex measures described in the previous section. The constraints require that a sensor be either on or off (unfortunately this constraint is not convex) and that a maximum number of sensors is not exceeded at any time step.

3.1 Finite Look-Ahead Optimization

For the current work, we introduce a model predictivecontrol type scheme to find the exact solution to locally optimize the mixed-integer program for choosing one of two sensors ($p = 2$ and $p_k = 1$ for all k) with a look-ahead of N time steps. We employ this method rather than calculating over an entire trajectory because the computations required to calculate each option N steps for even the simplest case of two sensors grows as 2^N. This optimization becomes infeasible for even moderately large N or for larger numbers of sensors. To determine the optimal sensor selection for time step

$k = K + 1$, the empirical and linear Gramians are calculated with data from $k = 1$ to K with the optimal sensor sequence and then from $K + 1$ to $K + 1 + N$ with each of the potential sensor sequences. The sequence with the minimum condition number at $K + N + 1$ is chosen, and the first sensor (or subset of sensors) in the sequence is selected for time $K + 1$. Thus, (8) becomes

$$\min_{s[k], k=K+1,\ldots,K+N+1} J(W^*[K+N+1], s) \qquad (9)$$

$$\text{subject to} \quad s_i[k] = \{0, 1\}, \quad i = 1, \ldots, p, \text{ and } k = 1, \ldots, K+N+1$$

$$\sum_{i=1}^{p} s_i[k] = 1, \quad k = 1, \ldots, K+N+1.$$

where $W^*[K + N + 1]$ is either the linear Gramian, $W_{o,d}[0, K + N + 1]$, defined in (4) over a nominal trajectory $(\mathbf{x}^0(t), \mathbf{u}^0(t))$ or the empirical Gramian, $W_d^\varepsilon[K + N + 1, \mathbf{x}_0, \mathbf{u}^0[k]]$, defined in (6) calculated from numerical simulation. In other words: at each time step K, the sensor sequence is optimized over a finite look-ahead, the first sensor in that sequence is selected for the next time step $K + 1$, and the process is repeated now incorporating the $K + 1$ measurement, as outlined in Algorithm 1.

Algorithm 1 Algorithm for optimal sensor selection with a finite look-ahead.

Initialize $W^*[0]$ with sufficient measurements to make it sufficient rank
for $K = 1$ to $k_f - N$ **do**
  Calculate $W^*[k]$ with optimal sensor sequence from 1 to K
  for Each sensor sequence S_j **do**
    for Each time step $k = K + 1$ to $K + N + 1$ **do**
      for Each sensor $i = 1$ to p **do**
        Calculate $W^*[k]$ with $s_i[k]h_i[k]$
      end for
    end for
    Calculate $\kappa(W^*[K + N + 1])$ for the sequence S_j
  end for
  Find j corresponding to the minimum condition number κ of S_j
  Set the next measurement $s_i[K + 1]$ as the first measurement in the optimal sequence $S_j(1)$
end for

3.2 Convex Relaxation

To allow for faster solutions, the mixed-integer optimization (8) or (9) could be relaxed to a convex optimization framework by allowing the sensors to be 'partially' on during a given time step; to promote sparsity in s (and thus that sensors are either on or off), an ℓ_1 norm can be added to the cost function [21, 22]. With those modifications, the optimization (8) would become

$$\min_{s} \; J(s) + c\|s\|_1$$
$$\text{subject to } \; 0 \le s_i[k] \le 1 \quad \forall i, k$$
$$\sum_{i=1}^{p} c_i s_i[k] = 1 \quad \forall k.$$

The constant scalar, c, in the cost function allows weighting the importance of sparsity of s versus minimization of $J(s)$.

4 Application to Frenet–Serret Frames

Frenet–Serret and natural Frenet frames both describe the evolution of an orthogonal coordinate system along a trajectory [23]. In previous work, the same authors use natural Frenet frames to design vehicle trajectories for a variety of tasks, such as pursuit, formation flight, and boundary tracking (see references in [23]). The direction of motion along a smooth path is defined as the tangent $\mathbf{T}(s) = \gamma'(s)$, where $\gamma(s)$ is the trajectory normalized along its arc s such that $\gamma'(s)$ is a unit vector ($|\gamma'(s)| = 1$). Frenet–Serret frames define the additional two vectors to make up an orthogonal frame as the normal, $\mathbf{N}(s)$, perpendicular to the motion in the plane of the curve, and the binormal, $\mathbf{B}(s)$, which completes the set. The evolution of the $(\mathbf{T}, \mathbf{N}, \mathbf{B})$ frame is given by the curvature and torsion of the curve. A drawback to this formulation is that the frame is not well defined when the second derivative of $\gamma(s)$ is zero. This constraint motivates the use of a modified version of the Frenet–Serret frames termed the natural Frenet frame, which we will denote by the orthogonal vector set $(\mathbf{t}, \mathbf{n}, \mathbf{b})$. In the natural Frenet structure, the tangent vector is defined as in the Frenet–Serret frames, $\mathbf{t} = \mathbf{T}$, and the remaining vectors are defined to be orthogonal unit vectors that complete the SO(3) group element (see Fig. 1).

To utilize the natural Frenet frames to describe motion of a vehicle, note that the 'particle' model of a vehicle that cannot move directly perpendicular to its current velocity can be written in the following form:

$$\begin{aligned} \mathbf{r}' &= \mathbf{t} \\ \mathbf{t}' &= \mathbf{n}u_1 + \mathbf{b}u_2 \\ \mathbf{n}' &= -\mathbf{t}u_1 \\ \mathbf{b}' &= -\mathbf{t}u_2 \end{aligned} \tag{10}$$

where $'$ indicates differentiation with respect to arc length, s, or with respect to time, t, if we assume unit velocity. The controls, u_1 and u_2, affect the curvature of the path in the $\mathbf{t}(t) - \mathbf{n}(t)$ and $\mathbf{t}(t) - \mathbf{b}(t)$ planes, respectively.

The measurement model considered for this example is a beacon located at the origin, as depicted in Fig. 1, that provides range and/or bearing measurements of the system according to

Fig. 1 Diagram showing the $t-n-b$ Frenet frame along a curve, with position denoted in the global reference frame by $\mathbf{r}$ and the bearing angle β

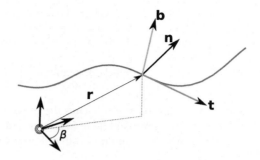

$$h_r(\mathbf{x}) = \frac{1}{2}(r_1^2 + r_2^2 + r_3^2) = \frac{1}{2}\|\mathbf{r}\|_2^2 \qquad (11)$$

$$h_b(\mathbf{x}) = \arctan\left(\frac{r_2}{r_1}\right) \qquad (12)$$

where we take the full state vector to be $\mathbf{x} = \begin{bmatrix} \mathbf{r}^T & \mathbf{t}^T & \mathbf{n}^T & \mathbf{b}^T \end{bmatrix}^T$.

In discrete time, assuming a sufficiently small time step, Δ_T, the dynamics of transitioning from step k to step $k+1$ can be approximated as

$$\mathbf{x}[k+1] = \mathbf{x}[k] + \Delta_T \Big(\mathbf{f}_0(\mathbf{x}[k]) + \mathbf{f}_1(\mathbf{x}[k])u_1[k] + \mathbf{f}_2(\mathbf{x}[k])u_2[k]\Big) \qquad (13)$$

where $\mathbf{f}_1 = \begin{bmatrix} \mathbf{0}^T & \mathbf{n}^T & -\mathbf{t}^T & \mathbf{0}^T \end{bmatrix}^T$ and $\mathbf{f}_2 = \begin{bmatrix} \mathbf{0}^T & \mathbf{b}^T & \mathbf{0}^T & -\mathbf{t}^T \end{bmatrix}^T$, $\mathbf{0}$ denoting a vector of zeros the size of $\mathbf{t}$. It is important to note that while the discrete- and continuous-time systems have a total of twelve variables, the nine variables from the $(\mathbf{t}, \mathbf{n}, \mathbf{b})$ frame have only three degrees of freedom (three dimensional attitude), so the full system has only six independent states. Thus, when performing observability analysis, we do not expect to obtain rank $n = 12$ matrices but rather $n_{DOF} = 6$. When constrained to a plane, the reduced system has six variables with three degrees of freedom and three states.

4.1 Analytical Observability Analysis

Using the discrete approximation of the dynamics (13) and measurement functions (11) and (12), we initially address the observability of the system by considering a drift only scenario with no controls by calculating the first few time steps of the dynamics to construct Θ defined in (2). With $\mathbf{u} = \mathbf{0}$, we have the state and measurement functions

$$\mathbf{x}[k] = \left[(\mathbf{r}_0^T + k\Delta_T \mathbf{t}_0^T) \; \mathbf{t}_0^T \; \mathbf{n}_0^T \; \mathbf{b}_0^T \right]^T$$

$$h_r(\mathbf{x}[k]) = \frac{1}{2} \|\mathbf{r}_0 + k\Delta_T \mathbf{t}_0\|_2^2$$

$$h_b(\mathbf{x}[k]) = \arctan\left(\frac{\mathbf{r}_{0,2} + k\Delta_T}{\mathbf{r}_{0,1} + k\Delta_T}\right)$$

where the zero subscript indicates the initial state, e.g. $\mathbf{r}_0 = \mathbf{r}(0)$. Investigating the range and bearing measurement functions individually, we take the appropriate directional derivatives to build the co-distributions for range-only and bearing-only measurements:

$$d\Theta_r = \begin{bmatrix} \mathbf{r}_0^T & 0 & 0 & 0 \\ \mathbf{r}_0^T + \Delta_T \mathbf{t}_0^T & \Delta_T(\mathbf{r}_0^T + \Delta_T \mathbf{t}^T) & 0 & 0 \\ \vdots & \vdots & \vdots & \vdots \\ \mathbf{r}_0^T + k\Delta_T \mathbf{t}_0^T & k\Delta_T(\mathbf{r}_0^T + k\Delta_T \mathbf{t}^T) & 0 & 0 \end{bmatrix}$$

$$d\Theta_b = \begin{bmatrix} \frac{-\mathbf{r}_{0,2}}{\mathbf{r}_{0,1}^2 + \mathbf{r}_{0,2}^2} & \frac{\mathbf{r}_{0,1}}{\mathbf{r}_{0,1}^2 + \mathbf{r}_{0,2}^2} & 0 & 0 & 0 & 0 & 0 & 0 \\ -\frac{\mathbf{r}_{0,2} + \Delta_T \mathbf{t}_{0,2}}{c(1)} & \frac{\mathbf{r}_{0,1} + \Delta_T \mathbf{t}_{0,1}}{c(1)} & 0 & \frac{\Delta_T(\mathbf{r}_{0,2} + \Delta_T \mathbf{t}_{0,2})}{c(1)} & \frac{\Delta_T(\mathbf{r}_{0,1} + \Delta_T \mathbf{t}_{0,1})}{c(1)} & 0 & 0 & 0 \\ \vdots & \vdots & \vdots & \vdots & \vdots & \vdots & & \\ -\frac{\mathbf{r}_{0,2} + k\Delta_T \mathbf{t}_{0,2}}{c(k)} & \frac{\mathbf{r}_{0,1} + k\Delta_T \mathbf{t}_{0,1}}{c(k)} & 0 & \frac{k\Delta_T(\mathbf{r}_{0,2} + k\Delta_T \mathbf{t}_{0,2})}{c(k)} & \frac{k\Delta_T(\mathbf{r}_{0,1} + k\Delta_T \mathbf{t}_{0,1})}{c(k)} & 0 & 0 & 0 \end{bmatrix}$$

$$c(k) = (\mathbf{r}_{0,1} + k\Delta_T \mathbf{t}_{0,1})^2 + (\mathbf{r}_{0,2} + k\Delta_T \mathbf{t}_{0,2})^2$$

with $\mathbf{0}$ a vector of zeros of the same dimension as $\mathbf{r}_0^T$. The co-distribution generated from range-only measurements, $d\Theta_r$, is not full rank; the first three columns and next three columns are linearly dependent on each other, so the maximum rank is three. Similarly, the rank of the co-distribution from bearing-only measurements, $d\Theta_b$, is at most three. The system has six degrees of freedom, so with either sensor and no input, the system is unobservable; with both bearing and range measurements together, however, the maximum rank is six, and the system is observable except at singular points where the rank drops below six. These results are consistent with an analysis of the continuous-time system which is omitted for brevity.

4.2 Optimization with Observability Gramians

Knowing that the system of evolving natural Frenet frames discussed in the previous section is observable with some combination of measurement functions, we proceed to address the optimization problem (9). For clarity of presentation, we restrict this study to a planar version of the system and note that the results will generalize to the full system. Here, we used the condition number defined in (7). To avoid issues arising from rank deficiency of the Gramian due to insufficient measurement data, we initialized the optimization by providing both range and bearing sensors for $k = 0$.

This approach is not unreasonable, as it is plausible that an autonomous system would collect all possible sensor data before embarking on its mission. In all cases, the look-ahead was $N = 5$ for both the empirical and linear Gramians, and Matlab was used to simulate the system dynamics and perform the optimization.

Some informative results from this initial study are provided here, beginning with the effect of Δ_T on sensor selection. This quantity can be interpreted as the sensor update rate and was held constant for any given trajectory simulation. The optimal sensor selection for a range of Δ_T from 0.25 to 1 for a circular path generated by a constant control of $u_1 = 0.25$ ($u_2 = 0$ for the planar case) is shown in Fig. 2. The optimization results from both the empirical observability Gramian and the linear observability Gramian tended to result in more use of bearing measurements, with range measurements selected more frequently near the initial location and for a couple of outliers elsewhere on the trajectory. The condition numbers calculated for both optimization approaches showed similar trends of an initial decrease in condition number as the trajectory traced the first half of a circle, with values for most test cases stabilizing around the first full revolution.

Based on the geometry of the trajectories in Fig. 2 with respect to the beacon, one might hypothesize that the clustering of range measurements was relative to location within the $r_1 - r_2$ plane. The sensor selection results in Figs. 3 and 4, however, show that the geometry of the trajectory with respect to its own start point plays a larger role. In all cases, there is a cluster of range measurements near the initial location not only at the beginning of the trajectory, but also on subsequent returns to the initial point. The range measurement is more often used for the trajectories with larger radius of curvature. The condition numbers (see Fig. 4) show a general decreasing trend that is somewhat cyclic around the circle.

The sensor selection results for other periodic (but noncircular) trajectories have similar characteristics, as seen in Fig. 5. As the trajectory exhibits more extensive changes in curvature, the range measurements are used more frequently. The trend is more apparent in the empirical Gramian optimization, however, it is present in the linear Gramian optimization as well. The condition number for the empirical observability Gramian maintains a relatively constant mean after an initial decrease. The condition number for the linear observability Gramian is much smoother, and shows that the plots with lower curvature have more poorly conditioned estimates as one would expect based on the work done in [12].

5 Conclusions

We have developed a framework for approaching the sensor selection problem in nonlinear systems and provided some interesting initial results for a Frenet frame motion model with beacon range and bearing measurements. These results indicate that while relative location to the beacon does play a role in sensor selection, the time history of the trajectory has a greater impact on the optimal sensor selection (at least for a look-ahead of $N = 5$). The results support the fact that the choice of

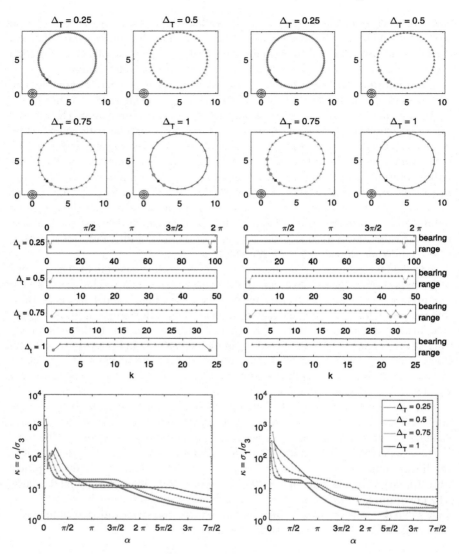

Fig. 2 Trajectories (*top*), sensor selection (*middle*), and condition number with markers for bearing and range measurements (*blue triangles* and *red circles*) to the beacon at the origin (*black circles*), chosen to minimize the condition number of the empirical (*left*), and linear (*right*) observability Gramian. A constant control $u_1 = 0.25$ was applied while varying the time step, $\Delta_t = \{0.25, 0.5, 0.75, 1\}$. The x-axis for the condition number plots is the angle $\alpha[k]$, measured counterclockwise from the center of the circle trajectory between the initial location (*black square*) and $\mathbf{r}[k]$

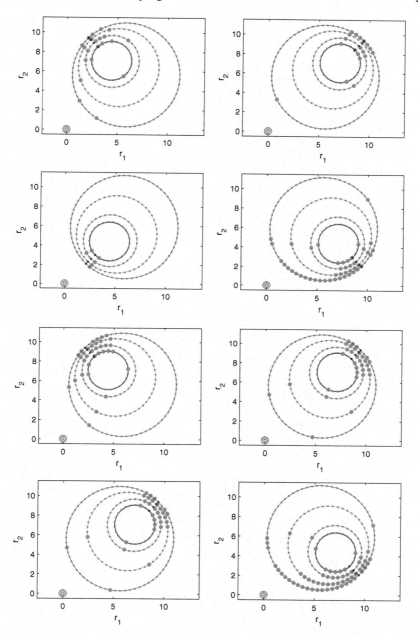

Fig. 3 Trajectories with markers for bearing and range measurements (*blue triangles* and *red circles*) to the beacon at the origin, chosen to minimize the condition number of the empirical, or linear observability Gramian. The time step was $\Delta_T = 0.5$ for all trajectories. Each trajectory represents a constant control from $u = \{0.1875, 0.25, 0.375, 0.5\}$ (largest circle to smallest circle) and the initial state is noted by a *black square*

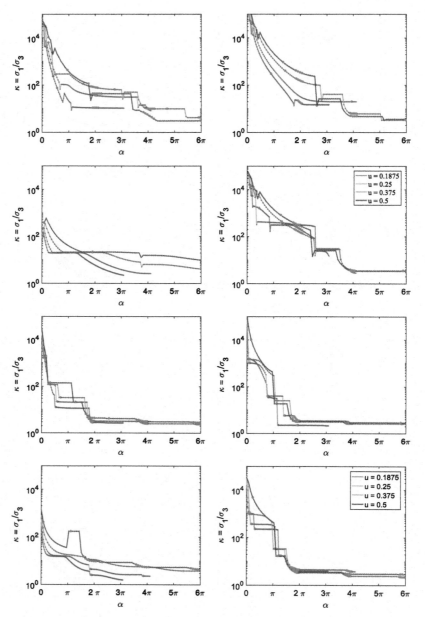

Fig. 4 Condition number of the empirical (*top* four plots) or linear (*bottom* four plots) observability Gramians, corresponding to the trajectories shown in Fig. 3. The time step was $\Delta_t = 0.5$ for all trajectories. Each trajectory represents a single control from $u = \{0.1875, 0.25, 0.375, 0.5\}$, going from small (*blue*) to large (*red*). The x-axis for the condition number plots is the angle $\alpha(k)$, measured counterclockwise from the center of the circle trajectory between the initial location (*black square*) and $\mathbf{r}[k]$

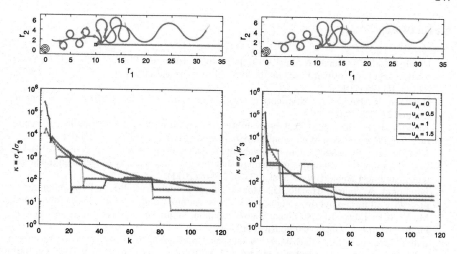

Fig. 5 Trajectories (*top*) and condition number (*bottom*) with markers for bearing and range measurements (*blue triangles* and *red circles*) to the beacon at the origin, chosen to minimize the condition number of the empirical or linear observability Gramian. The time step was $\Delta_t = 0.25$ with control $u = u_A \cos(k\Delta_t/2)$, $u_A = \{0, 0.5, 1, 1.5\}$

trajectory influences the observability and that curvier paths provide better conditioned observability Gramians.

The framework developed here can be used to address some questions about nonlinear observability that will have direct impact on effective design of autonomous vehicles and their controllers. The first set of results (Fig. 2) indicate that sensor sampling rates affect the observability of the system, but further study can be done to investigate how the sampling rate affects the system and then how to use that information to help choose the appropriate sensor. Knowing that control choice, and thus trajectory, influences the observability, we are working to formalize the relationship between path history and optimal sensor choice, particularly with respect to cyclical trajectories and how the history built into the data patterns brings about the repeated selection of certain sensor types. In addition to the history, the effect of longer look-ahead and comparison of local and global results should be performed. Related to these characteristics is the effect of deviating from the planned look-ahead on the conditioning of the estimate. Sensor fidelity and cost of use as well as task-specific requirements will also be incorporated.

Acknowledgements This material is based upon work supported by the National Science Foundation Graduate Research Fellowship Program under grant number DGE-1256082 and by the Air Force Office of Scientific Research grant number FA9550-14-1-0398.

References

1. Danilovich, S., Krishnan, A., Lee, W.-J., Borrisov, I., Eitan, O., Kosa, G., Moss, C.F., Yovel, Y.: Bats regulate biosonar based on the availability of visual information. Curr. Biol. **25**(23), R1124–R1125 (2015)
2. Hermann, R., Krener, A.: Nonlinear controllability and observability. IEEE Trans. Autom. Control **22**, 728–740 (1977)
3. Albertini, F., D'Alessandro, D.: Observability and forward-backward observability of discrete-time nonlinear systems. Math. Control Signals Syst. (MCSS) **15**, 275–290 (2002)
4. Johnson, S.C.: Switched system observability, Master's thesis, Purdue University (2013)
5. Lou, H., Yang, R.: Conditions for distinguishability and observability of switched linear systems. Nonlinear Anal.: Hybrid Syst. **5**(3), 427–445 (2011)
6. Bemporad, A., Ferrari-Trecate, G., Morari, M., et al.: Observability and controllability of piecewise affine and hybrid systems. IEEE Trans. Autom. Control **45**(10), 1864–1876 (2000)
7. Muske, K.R., Edgar, T.F.: Nonlinear State Estimation. Nonlinear Process Control, pp. 311–370. Prentice-Hall Inc., Upper Saddle River (1997)
8. Lall, S., Marsden, J.E., Glavaški, S.: A subspace approach to balanced truncation for model reduction of nonlinear control systems. Int. J. Robust Nonlinear Control **12**(6), 519–535 (2002)
9. Scherpen, J.: Balancing for nonlinear systems. Syst. Control Lett. **21**, 143–153 (1993)
10. Krener, A.J., Ide, K.: Measures of unobservability. In: Proceedings of the 48th IEEE Conference on Decision and Control, pp. 6401–6406 (2009)
11. Wang, T.: An adaptive and integrated multimodal sensing and processing framework for long range moving object detection and classification. Ph.D. thesis, City University of New York (2013)
12. Hinson, B.T., Morgansen, K.A.: Observability optimization for the nonholonomic integrator. In: 2013 American Control Conference, pp. 4257–4262 (2013)
13. Hinson, B.T., Morgansen, K.A.: Gyroscopic sensing in the wings of the hawkmoth manduca sexta: the role of sensor location and directional sensitivity. Bioinspiration Biomim. **10**(5), 056013 (2015)
14. Oshman, Y.: Optimal sensor selection strategy for discrete-time state estimators. IEEE Trans. Aerosp. Electron. Syst. **30**, 307–314 (1994)
15. Gupta, V., Chung, T.H., Hassibi, B., Murray, R.M.: On a stochastic sensor selection algorithm with applications in sensor scheduling and sensor coverage. Automatica **42**(2), 251–260 (2006)
16. Mo, Y., Ambrosino, R., Sinopoli, B.: Sensor selection strategies for state estimation in energy constrained wireless sensor networks. Automatica **47**(7), 1330–1338 (2011)
17. Jawaid, S.T., Smith, S.L.: On the submodularity of sensor scheduling for estimation of linear dynamical systems. In: 2014 American Control Conference, pp. 4139–4144, Portland, Oregon (2014)
18. Li, Y., Jha, D.K., Ray, A., Wettergren, T.A.: Feature level sensor fusion for target detection in dynamic environments. In: 2015 American Control Conference, pp. 2433–2438 (2015)
19. Powel, N.D., Morgansen, K.A.: Empirical observability gramian rank condition for weak observability of nonlinear systems with control. In: 54th IEEE Conference on Decision and Control, pp. 6342–6348 (2015)
20. Powel, N.D., Morgansen, K.A.: Stochastic empirical observability gramian in discrete-time for quadrotor systems. preprint (2016)
21. Boyd, S.P., Vandenberghe, L.: Convex Optimization. Cambridge University Press, Cambridge (2004)
22. Hinson, B.T., Morgansen, K.A.: Observability-based optimal sensor placement for flapping airfoil wake estimation. J. Guid. Control Dyn. **37**, 14771486 (2014)
23. Justh, E.W., Krishnaprasad, P.S.: Optimal natural frames. Commun. Inf. Syst. **11**(1), 17–34 (2011)

Tracking Multiple Ground Objects Using a Team of Unmanned Air Vehicles

Joshua Y. Sakamaki, Randal W. Beard and Michael Rice

Abstract This paper proposes a system architecture for tracking multiple ground-based objects using a team of unmanned air systems (UAS). In the architecture pipeline, video data is processed by each UAS to detect motion in the image frame. The ground-based location of the detected motion is estimated using a geolocation algorithm. The subsequent data points are then process by the recently introduced Recursive RANSAC (R-RANSASC) algorithm to produce a set of tracks. These tracks are then communicated over the network and the error in the coordinate frames between vehicles must be estimated. After the tracks have been placed in the same coordinate frame, a track-to-track association algorithm is used to determine which tracks in each camera correspond to tracks in other cameras. Associated tracks are then fused using a distributed information filter. The proposed method is demonstrated on data collected from two multi-rotors tracking a person walking on the ground.

1 Introduction

The objective of this paper is to describe a new approach to real-time video tracking of multiple ground objects using a team of multi-rotor style unmanned air systems (UAS). Many UAS applications involve tracking objects of interest with an on-board camera. These applications include following vehicles [15], visiting designated sites of interest [12], tracking wildlife [17], monitoring forest fires [8, 13], and inspecting infrastructure [24]. Current approaches to real-time video tracking from UAS can be brittle, and state-of-the-art techniques often require fiducial markings, or extensive human oversight.

There are numerous challenges in developing an object tracking system. For example, object tracking often involves finding image features and tracking those features from frame to frame. However, feature matching has a relatively high error rate, and

J.Y. Sakamaki · R.W. Beard (✉) · M. Rice
Brigham Young University, Provo, UT, USA
e-mail: beard@byu.edu

the errors introduced by incorrect matches do not follow a Gaussian distribution. In addition, each measurement cycle, or image pair, produces many measurements where false measurements occur at a relatively high rate. Another challenge is distinguishing tracks from the background, especially when they stop moving, or have color and features similar to the background. Furthermore, even when objects of interest are correctly identified in each frame, the data association problem, or the problem of consistently associating the measurements with the correct object, can be difficult. Finally, many applications require that the system track many objects in the environment.

In this paper, we introduce a complete solution for tracking multiple ground-based objects using cameras on-board a team of UAS. Our solution draws upon several distinct technologies including geolocation [4, 7, 11], multiple target tracking [6, 28], track-to-track data correlation [1, 2], and distributed sensor fusion [16, 18].

In our framework we assume that there are multiple air vehicles each carrying an on-board camera. The computer vision algorithm on each vehicle returns a set of points in the image frame that may correspond to ground-based objects. In the implementation reported in this paper, we look for moving objects. The set of potential measurements are then processed using a geolocation algorithm and the GPS and IMU measurements on-board each vehicle, to project the measurements onto the ground plane. The geolocation algorithm is described in Sect. 2.2. The data is then processed using a newly introduced multiple target tracker called Recursive RANSAC [15, 21, 22]. The R-RANSAC algorithm produces a set of tracks that are communicated between vehicles on the team. The R-RANSAC algorithm is described in Sect. 2.3. Unfortunately, the geolocation process is imprecise, introducing potential biases between vehicles. For every pair of tracks, the bias must be determined, and our approach to this problem is described in Sect. 2.4. Since each object may not be seen by every UAS, it is necessary to determine whether tracks seen by UAS a, are also seen by UAS b. Our approach to track-to-track association is given in Sect. 2.5. Associated tracks are then fused using an information filter as described in Sect. 2.6. Finally, flight results using the complete system are described in Sect. 3.

2 System Architecture

The architecture for the tracking system that will be described in this paper is shown in Fig. 1. Each UAS instantiates a tracking pipeline that includes six key components. The first component in the pipeline, as shown in Fig. 1, is the UAS and gimbaled camera. We assume that the UAS contains an autopilot system, as well as a pan-tilt camera that can be automatically controlled to point along a desired optical axis. In this paper, we will assume an RGB camera and enough processing power on-board to process images at frame rate, and to implement the other components in the system. The second component shown in Fig. 1 is the Geolocation block. The purpose of the geolocation block is to transform the image coordinates into world coordinates based on the current pose of the UAS. A detailed description of the geolocation block is

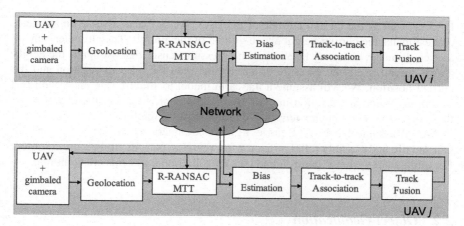

Fig. 1 Architecture for tracking multiple ground-based objects of interest using a team of UAS

given in Sect. 2.2. The next component shown in Fig. 1 is the Recursive RANdom SAmple Consensus Multiple Target Tracking (R-RANSAC MTT) block. This block uses image features in world coordinates to create and manage object tracks. This block performs several key tasks including data association, new track formation, track propagation, track collation, and track deletion. We define a *track* to be the time history of the system state (position, velocity, acceleration, etc.), as well as the associated covariance matrix. A more detailed description of this block will be given in Sect. 2.3. The current tracks maintained by each UAS is shared across the network with other UAS. The current collection of tracks is used in the Bias Estimation block shown in Fig. 1 to estimate the translational and rotational bias between each pair of tracks in the network, and thereby place all tracks in the coordinate system of the ith UAS. Additional details about this process will be described in Sect. 2.4. The collection of tracks are then processed by the Track-to-track Association block shown in Fig. 1. This block uses a statistical test on a past window of the data to determine which tracks maintained by the ith UAS are statistically similar to the tracks maintained by the jth UAS. The details of this block are described in Sect. 2.5. When tracks are determined to be similar, they are fused in the Track Fusion block shown in Fig. 1. Track fusion is accomplished using an information consensus filter, as described in Sect. 2.6.

2.1 UAS and Gimbaled Camera

The techniques that are outlined in this paper are applicable to both multi-rotor systems and fixed wing vehicles. Independent of the type of aircraft used, we will assume that the sensor suite on-board the aircraft consists of a GPS aided IMU and associated filter algorithms that are able to estimate the 3D world position of the

UAS, as well as the inertial attitude of the UAS. We will also assume an altimeter that estimates the current height above ground of the aircraft. The altimeter may be a laser altimeter, or it may be an absolute pressure sensor that estimates the height above ground using the pressure difference between the take-off position and the current position. We will assume a flat earth model to simplify the discussion and equations. When an elevation map of the environment is known, the extension of these ideas to more complex terrain is conceptually straightforward.

We will assume that the UAS carries a gimbaled camera, where the gimbal can both pan and tilt, and possibly roll. For fixed wing vehicles, pan-tilt gimbals are common. For multi-rotor systems, pan-roll and pan-tilt-roll gimbal systems are common.

2.2 Object Geolocation

The UAS and gimbaled camera block shown in Fig. 1 produces a video stream, as well as state estimates obtained by filtering the on-board sensors. We will assume that the video stream is processed to produce a list of pixels, or image coordinates that represent possible measurements of objects on the ground. The task of the Geolocation block shown in Fig. 1 is to transform each image coordinate in the feature list into an inertial position on the ground. Object tracking can be performed in either the camera frame, or in the inertial frame. In order to perform object tracking using a team of UAS, the measurements of the objects need to be in a common reference frame. In this paper, we assume that all UAS on the team have GPS, therefore it makes sense to use GPS to define the common inertial reference frame, and to track the objects in the inertial frame. Transforming image coordinates to inertial coordinates is called geolocation in the literature. Geolocation algorithms for small UAS are described in [4, 5, 7, 10, 11, 23, 29].

Let $\mathscr{I}$ denote the inertial frame, let U_a denote UAS a, and let F_k denote the kth feature of interest. Let $p_{U_a}^{\mathscr{I}}$ denote the inertial position of UAS a, and $p_{F_k}^{\mathscr{I}}$ denote the inertial position of the kth feature. Define the line of sight vector between UAS a and the kth feature, expressed in the camera frame as $\ell_{U_a F_k}^{\mathscr{C}_a} = p_{F_k}^{\mathscr{C}_a} - p_{U_a}^{\mathscr{C}_a}$. If $R_{\mathscr{C}_a}^{\mathscr{G}_a}$ denotes the rotation matrix from the camera frame to the gimbal frame of UAS a, $R_{\mathscr{G}_a}^{\mathscr{B}_a}$ denotes the rotation matrix from the gimbal frame to the body frame of UAS a, and $R_{\mathscr{B}_a}^{\mathscr{I}}$ denotes the rotation of the body frame of UAS a to the inertial frame, then the basic geolocation equation is given by [5]

$$p_{F_k}^{\mathscr{I}} = p_{U_a}^{\mathscr{I}} + R_{\mathscr{B}_a}^{\mathscr{I}} R_{\mathscr{G}_a}^{\mathscr{B}_a} R_{\mathscr{C}_a}^{\mathscr{G}_a} \ell_{U_a F_k}^{\mathscr{C}_a}. \tag{1}$$

The only element that is not available in Eq. (1) is the line of sight vector $\ell_{U_a F_k}^{\mathscr{C}_a}$. If we assume a pin-hole model for the camera, and that the focal length of the camera is f, and that the pixel location of the kth feature is $(\varepsilon_{x_k}, \varepsilon_{y_k})$, then the line of sight vector is given by

$$\ell_{U_aF_k}^{\mathscr{C}_a} = L_{U_aF_k}\lambda_{U_aF_k}^{\mathscr{C}_a}, \tag{2}$$

where $L_{U_aF_k}$ is the unknown length of the line of sight vector, and

$$\lambda_{U_aF_k}^{\mathscr{C}_a} = \frac{1}{\sqrt{\varepsilon_{x_k}^2 + \varepsilon_{y_k}^2 + f^2}} \begin{pmatrix} \varepsilon_{x_k} \\ \varepsilon_{y_k} \\ f \end{pmatrix}$$

is the direction of the line of sight vector expressed in the camera frame. To determine $L_{U_aF_k}$ additional information about the terrain needs to be available. If an elevation map of the terrain is known, then $L_{U_aF_k}$ is determined by tracing the ray given by the right hand side of Eq. (1) to find the first intersection with the terrain. In other words, find the first $\alpha > 0$ such that

$$p_{F_k}^{\mathscr{I}} = p_{U_a}^{\mathscr{I}} + \alpha R_{\mathscr{B}_a}^{\mathscr{I}} R_{\mathscr{G}_a}^{\mathscr{B}_a} R_{\mathscr{C}_a}^{\mathscr{G}_a} \lambda_{U_aF_k}^{\mathscr{C}_a}$$

intersects the terrain model. If the terrain is flat and the altitude h is known, then the equation for the length of the line of sight vector is given by [5]

$$L_{U_aF_k} = \frac{h}{\left(k^{\mathscr{I}}\right)^\top R_{\mathscr{B}_a}^{\mathscr{I}} R_{\mathscr{G}_a}^{\mathscr{B}_a} R_{\mathscr{C}_a}^{\mathscr{G}_a} \lambda_{U_aF_k}^{\mathscr{C}_a}},$$

where $k^{\mathscr{I}} = (0, 0, 1)^\top$ is the unit vector pointing to the center of the earth.

Therefore, the geolocation block projects all 2D features in the image plane, into 3D features in the world frame, and returns a set of features in the inertial frame.

2.3 Multiple Object Tracking

The next step in the tracking pipeline shown in Fig. 1 is Recursive Random Sample Consensus Multiple Target Tracking (R-RANSAC MTT). The function of this block is to process the inertial frame measurements and to produce a set of tracks that correspond with objects on the ground. The R-RANSAC algorithm was recently introduced in [20] for static signals with significant gross measurement error, and extended in [21] to multiple target tracking, and in [15] to video tracking.

A graphic that highlights key elements of the algorithm are shown in Fig. 2. Figure 2a shows a single object on the ground, where the black dots represent current and past measurements. The small box around the object is a decision or measurement gate. As shown in Fig. 2b a set of trajectories that are consistent with the current measurement are created. For a given object, there many be many trajectories that are consistent with the current measurement. A set of these trajectories with the largest number of inlier measurements are retained in memory, and the trajectories that continue to be consistent with the measurements are retained. When other objects

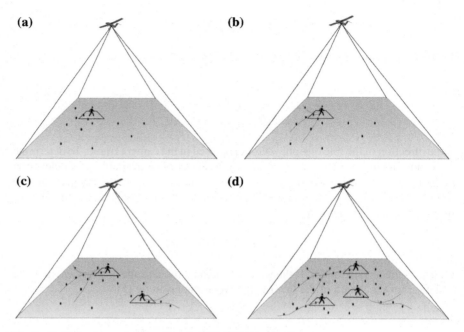

Fig. 2 Multiple object tracking using the R-RANSAC algorithm

appear, as shown in Fig. 2c they will generate measurements that are not in the measurement gate of existing tracks. When that happens, the initialization step is repeated, and a set of trajectories consistent with that measurement are added to memory. The R-RANSAC algorithm will have a bank of M possible trajectories in memory, and so pruning, merging, and spawning operations are key to its operation. In theory, the algorithm is capable of tracking $M - 1$ objects.

The R-RANSAC algorithm assumes a motion model for the objects of the form

$$x[t+1] = Ax[t] + \eta[t] \tag{3}$$
$$y[t] = Cx[t] + \nu[t], \tag{4}$$

where the size of the state is N, and where $\eta[t]$ and $\nu[t]$ are zero mean Gaussian random variables with covariance Q and R, respectively. We have found that for moving objects like pedestrians and vehicles on a road, constant acceleration and constant jerk models tend to work well [14]. The algorithm requires that all past measurements be retained in memory for the past D samples. The R-RANSAC initialization process begins by randomly selecting $N - 2$ time delays in the interval $[1, D - 1]$ denoted as $\{d_1, \ldots, d_{N-1}\}$. At each time delay, one measurement is randomly selected and denoted as $\{y_{d_1}, \ldots, y_{d_{N-1}}\}$. A measurement is also randomly selected at time $t - D$ and denoted $y[t - D]$. The state at time $t - D$ can then be reconstructed from the equation

$$\begin{pmatrix} y[t] \\ y[d_1] \\ \vdots \\ y[d_{N-1}] \\ y[t-D] \end{pmatrix} = \begin{pmatrix} CA^D \\ CA^{D-d_1} \\ \vdots \\ CA^{D-d_{N-1}} \\ C \end{pmatrix} \hat{x}[t-D]. \quad (5)$$

It can be shown that if the system is observable, then there is a unique solution for $\hat{x}[t-D]$ [19]. The state $\hat{x}[t-D]$ is propagated forward to the current time t using the discrete-time steady-state Kalman filter

$$\hat{x}^-[\tau+1] = A\hat{x}[\tau]$$

$$\hat{x}[\tau+1] = \begin{cases} \hat{x}^-[\tau+1] + L(y[\tau] - C\hat{x}[\tau]) & \tau \in \{t, t-d_1, \ldots, t-d_{N-1}, t-D\} \\ \hat{x}^-[\tau+1] & \text{otherwise,} \end{cases} \quad (6)$$

where

$$L = P_p C^\top S^{-1}, \quad (7)$$

is the Kalman gain, $S = (CP_pC^\top + R)$ is the innovation covariance and P_p is the steady state prediction covariance that satisfies the algebraic Riccati equation

$$P_p = AP_pA^\top + Q - AP_pC^\top S^{-1}CP_pA^\top. \quad (8)$$

The quality of the initialized track is then scored by counting the number of measurements that are consistent with that track. Let $\mathcal{Y}_\tau$ be the set of all measurements received at time τ, and let $Y_\tau(z, \gamma)$ be the set of measurements that are a Mahalanobis distance of γ from z at time τ, i.e.,

$$Y_\tau(z, \gamma) = \{y \in \mathcal{Y}_\tau : (z-y)^\top S^{-1}(z-y) \leq \gamma\},$$

then the *consensus set* at time t for the jth track $\{\hat{x}^j[\tau]\}_{\tau=t-D}^t$, is defined to be

$$\chi^j[t] = \bigcup_{\tau=t-D}^{t} Y_\tau(C\hat{x}^j[\tau], \gamma).$$

The *inlier ratio* $\rho^j(t)$ for track j is defined to be the size of the consensus set divided by the total number of measurements, i.e.,

$$\rho^j[t] = \frac{|\chi^j[t]|}{\sum_{\tau=t-D}^{t} |\mathcal{Y}_\tau|}. \quad (9)$$

The inlier ratio is a measure of the quality of the the track.

After a set of tracks have been initialized, the R-RANSAC algorithm processes the set of measurements $\mathcal{Y}[t]$ as follows. Let $G^j[t]$ be a defined gate for the jth track at time t where

$$G^j[t] = \{z \in \mathbb{R}^p : (z - C\hat{x}^j[t])^\top S^{-1}(z - C\hat{x}^j[t]) \leq \gamma\}.$$

The set of measurements $\mathcal{Y}[t] \cap G^j[t]$ are combined using the probabilistic data association (PDA) algorithm [3], and then used to update the associated Kalman filter. Measurements that are outside of the gate for every existing track, are used to spawn new tracks, based on the previously defined initialization method. Two tracks are combined when their outputs are within a certain threshold of each other over a specified window of time.

2.4 Track Alignment

Object tracks produced by the R-RANSAC MTT algorithm are communicated across the network to other team members as shown in Fig. 1. When UAS a receives a track from UAS b, UAS a must determine if the track corresponds to any of its existing tracks. We call this the problem of track-to-track association. However, before testing for track-to-track association, the tracks from UAS a must be aligned with the track from UAS b.

Let $\hat{x}^j_{m|n}[t]$ represent the jth track estimated by UAS m at time t, where the state is represented in the coordinate frame of UAS n. Each UAS will maintain a set of tracks in their own coordinate frame. In other words, UAS a will maintain the track of its jth object as $\hat{x}^j_{a|a}$. The track can be transformed into the coordinate frame of UAS b using

$$\hat{x}^j_{a|b}[t] = R^b_a \left(\hat{x}^j_{a|a}[t] + d^b_a\right),$$

where R^b_a is the transformation matrix that rotates the coordinate frame of UAS a into the coordinate frame of UAS b, and $d^j_{a|b}$ is the associated translation. For example, when the state consists of the 2D ground position, velocity, and acceleration, then

$$R^b_a = I_3 \otimes \begin{pmatrix} \cos\theta & -\sin\theta \\ \sin\theta & \cos\theta \end{pmatrix}$$
$$d^b_a = \begin{pmatrix} \beta_n & \beta_e & 0 & 0 & 0 & 0 \end{pmatrix}^\top$$

where $\otimes$ represents the Kronecker product and θ is defined as the relative rotational bias angle between the two tracks about the negative down axis, and where β_n and β_e are constants representing the north and east translational bias.

Two tracks $\hat{x}^j_{a|a}$ and $\hat{x}^k_{b|b}$ are aligned over a window of length D by solving the optimization problem

$$(R_a^{b*}, b_a^{b*}) = \arg\min_{(R_a^b, b_a^b)} \sum_{\tau=t-D+1}^{t} \left\| \hat{x}_{b|b}^k[\tau] - R_a^b(\hat{x}_{a|a}^j[\tau] + d_a^b) \right\|. \tag{10}$$

It should be noted that obtaining a solution from the optimizer does not guarantee that the tracks are associated. To determine whether the tracks originate from the same object requires solving the track-to-track association problem, which is discussed in the next section.

2.5 Track-to-Track Association

After the tracks have been aligned, the next step shown in Fig. 1 is to test whether the two tracks do in fact originate from the same source. This is the classical track-to-track association problem [1]. The problem is formulated as a hypothesis test, where the two hypotheses are

H_0 : The two tracks originate from the same object.
H_1 : The two tracks do not originate from the same object.

The association problem is solved over the past D measurement. Define the error vector as

$$\tilde{x}^{k_b j_a}[t] = \begin{pmatrix} \hat{x}_{b|b}^k[t-D+1] - R_a^b(\hat{x}_{a|a}^j[t-D+1] + d_a^b) \\ \hat{x}_{b|b}^k[t-D+2] - R_a^b(\hat{x}_{a|a}^j[t-D+2] + d_a^b) \\ \vdots \\ \hat{x}_{b|b}^k[t] - R_a^b(\hat{x}_{a|a}^j[t] + d_a^b) \end{pmatrix}. \tag{11}$$

Under the null hypothesis $\tilde{x}^{k_b j_a}[t]$ is a zero mean Gaussian random variable with covariance P_0, and under the alternative hypothesis $\tilde{x}^{k_b j_a}[t]$ is a zero mean Gaussian random variable with covariance P_1. The covariance matrix P_0 is known and will be discussed below. On the other hand, the covariance matrix P_1 is not known, and depends on the unknown true difference between the two unassociated tracks.

In the ideal case, where both P_0 and P_1 are known, the test statistic that follows from the log-likelihood ratio is [26]

$$L = \tilde{x}^{k_b j_a}[t]^\top \left(P_0^{-1} - P_1^{-1} \right) \tilde{x}^{k_b j_a}[t]. \tag{12}$$

However, because P_1 is unknown, this test statistic is unusable. We instead adopt the test statistic

$$\mathscr{D}[t] = \tilde{x}^{k_a j_b}[t]^\top P_0^{-1} \tilde{x}^{k_a j_b}[t]. \tag{13}$$

Under H_0, $\mathscr{D}[t]$ is a central chi-square random variable with DN degrees of freedom [26]. Under H_1, we use the Cholesky factorization $P_1^{-1} = W^\top W$ to write

$$\mathscr{D}[t] = \tilde{x}^{k_a j_b}[t]^\top W^\top W \tilde{x}^{k_a j_b}[t] = (W\tilde{x}^{k_a j_b}[t])^\top W \tilde{x}^{k_a j_b}[t]. \tag{14}$$

Here, $W\tilde{x}^{k_a j_b}[t]$ is a zero mean Gaussian random variable with covariance $WP_1W^\top$. Depending on the relationship between W and P_1, $\mathscr{D}[t]$ may or may not be a chi-square random variable [26]. In the event that $\mathscr{D}[t]$ is a chi-square random variable, it has less than DN degrees of freedom.

Because the likelihood ratio is an increasing function of $\mathscr{D}[t]$, by the Karlin–Rubin theorem, the following test is a uniformly most powerful test for testing H_0 against H_1 [26]:

$$\phi(\mathscr{D}[t]) = \begin{cases} 1, & \text{if } \mathscr{D}[t] > \mathscr{D}_\alpha \\ 0, & \text{if } \mathscr{D}[t] \leq \mathscr{D}_\alpha \end{cases} \tag{15}$$

where $\phi(\mathscr{D}[t]) = 1$ means H_0 is rejected and $\phi(\mathscr{D}[t]) = 0$ means H_0 is not rejected. The decision threshold is found as follows. For a given false alarm probability

$$\alpha = P(\phi(\mathscr{D}[t]) = 1 \mid H_0) = P(\mathscr{D}[t] > \mathscr{D}_\alpha \mid H_0), \tag{16}$$

$\mathscr{D}_\alpha$ is computed from

$$\begin{aligned}\alpha &= 1 - F_{\mathscr{D}|H_0}(\mathscr{D}_\alpha) \\ &= 1 - \int_0^{\mathscr{D}_\alpha} \frac{1}{\Gamma(Nn_x/2)2^{(Nn_x/2)}} x^{(Nn_x/2)-1} e^{x/2} dx. \end{aligned} \tag{17}$$

Note that under H_0 this produces a probability of detection $P_d = 1 - \alpha$.

Under H_0, $\tilde{x}^{k_a j_b}[t]$ is a zero mean Gaussian random variable with covariance P_0 where the covariance is expressed as

$$P_0 = \lim_{t \to \infty} E\left\{\tilde{x}^{k_a j_b}[t]\tilde{x}^{k_a j_b}[t]^\top\right\}, \tag{18}$$

and where

$$\tilde{x}^{k_a j_b}[t] = \hat{x}^{k_a}[t] - \hat{x}^{j_b}[t],$$

and where $\hat{x}^{k_a}[t] \in \mathscr{R}^{DN \times 1}$ is the stacked vector associated with the past D estimates of the kth track as observed by UAS a. Define the true track to be x^{k_a} and the track estimation error to be $\tilde{x}^{k_a} = \hat{x}^{k_a} - x^{k_a}$. Then P_0 can be written as

$$\begin{aligned} P_0 &= \lim_{t \to \infty} E\{\tilde{x}^{k_a j_b}[t]\tilde{x}^{k_a j_b}[t]^\top\} \\ &= \lim_{t \to \infty} E\{(\hat{x}^{k_a}[t] - \hat{x}^{j_b}[t])(\hat{x}^{k_a}[t] - \hat{x}^{j_b}[t])^\top\} \\ &= \lim_{t \to \infty} E\{(\tilde{x}^{k_a}[t] + x^{k_a}[t] - \tilde{x}^{j_b}[t] - x^{j_b}[t])(\tilde{x}^{k_a}[t] + x^{k_a}[t] - \tilde{x}^{j_b}[t] - x^{j_b}[t])^\top\}. \end{aligned}$$

Under hypothesis H_0, the two tracks originate from the same source, and since they are aligned in the same coordinate frame we have that $x^{k_a} = x^{j_b}$. Therefore

$$P_0 = \lim_{t\to\infty} E\{(\tilde{x}^{k_a}[t] - \tilde{x}^{j_b}[t])(\tilde{x}^{k_a}[t] - \tilde{x}^{j_b}[t])^\top\}$$
$$= \lim_{t\to\infty} E\{\tilde{x}^{k_a}[t]\tilde{x}^{k_a^\top}[t]\} + \lim_{t\to\infty} E\{\tilde{x}^{j_b}[t]\tilde{x}^{j_b^\top}[t]\}$$
$$- \lim_{t\to\infty} E\{\tilde{x}^{k_a}[t]\tilde{x}^{j_b^\top}[t]\} - \lim_{t\to\infty} E\{\tilde{x}^{j_b}[t]\tilde{x}^{k_a^\top}[t]\}.$$

It can be shown using steady-state Kalman filter arguments, that P_0 has the structure

$$P_0 = \begin{pmatrix} P & PG^\top & P(G^2)^\top & \ldots & P(G^{N-1})^\top \\ GP & P & PG^\top & \ldots & P(G^{N-2})^\top \\ G^2 P & GP & \ddots & & \vdots \\ \vdots & \vdots & & \ddots & PG^\top \\ G^{N-1}P & G^{N-2}P & \ldots & GP & P \end{pmatrix}, \quad (19)$$

where $P = 2(P_e - P_c)$, and where P_e is the estimation covariance given by

$$P_e = P_p - P_p C^\top (CP_p C^\top + R)^{-1} CP_p, \quad (20)$$

and P_p is the prediction covariance given by the solution of the Riccati equation in Eq. (8), and where the cross covariance P_c satisfies

$$P_c = (I - LC)(AP_c A^\top + Q)(I - LC)^\top,$$

where L is the Kalman gain given in Eq. (7), and where

$$G = (I - LC)A.$$

The structure of (19) is convenient as it produces an inverse with a tridiagonal block form, as highlighted in the following theorem.

Theorem 1 *Consider the symmetric, positive definite block matrix P_0 defined by (19). The inverse of the matrix P_0 is given by*

$$P_0^{-1} = \begin{pmatrix} U & V & 0 & \ldots & 0 \\ V^T & W & V & \ldots & 0 \\ 0 & V^T & \ddots & & \vdots \\ \vdots & \vdots & & W & V \\ 0 & 0 & \ldots & V^T & Y \end{pmatrix} \quad (21)$$

where

$$U = P^{-1} + G^T Y G$$
$$V = -G^T Y$$
$$W = Y + G^T Y G$$
$$Y = (P - GPG^T)^{-1}.$$

The proof of the theorem is in [25].

The theorem allows for a recursion equation to be developed for $\mathscr{D}$ in Eq. (13), which can be used to update the test statistic for a sliding window of data. Using (21), Eq. (13) can be expanded to

$$\mathscr{D}[t] = \tilde{x}[t]^T P_0^{-1} \tilde{x}[t] \tag{22}$$
$$= \left[\tilde{x}[t-D+1]^T U \tilde{x}[t-D+1] + \tilde{x}[t-D+2]^T V^T \tilde{x}[t-D+1] \right]$$
$$+ \left[\tilde{x}[t-D+1]^T V \tilde{x}[t-D+2] + \tilde{x}[t-D+2]^T W \tilde{x}[t-D+2] + \tilde{x}[t-D+3]^T V^T \tilde{x}[t-D+2] \right]$$
$$+ \left[\tilde{x}[t-D+2]^T V \tilde{x}[t-D+3] + \tilde{x}[t-D+3]^T W \tilde{x}[t-D+3] + \tilde{x}[t-D+4]^T V^T \tilde{x}[t-D+3] \right]$$
$$\vdots$$
$$+ \left[\tilde{x}[t-2]^T V \tilde{x}[t-1] + \tilde{x}[t-1]^T W \tilde{x}[t-1] + \tilde{x}[t]^T V^T \tilde{x}[t-1] \right]$$
$$+ \left[\tilde{x}[t-1]^T V \tilde{x}[t] + \tilde{x}[t]^T Y \tilde{x}[t] \right]. \tag{23}$$

Defining

$$d_1[\tau] \triangleq \tilde{x}[\tau]^T U \tilde{x}[\tau] + \tilde{x}[\tau+1]^T V^T \tilde{x}[\tau]$$
$$d_2[\tau] \triangleq \tilde{x}[\tau-1]^T V \tilde{x}[\tau] + \tilde{x}[\tau]^T W \tilde{x}[\tau] + \tilde{x}[\tau+1]^T V^T \tilde{x}[\tau]$$
$$d_3[\tau] \triangleq \tilde{x}[\tau-1]^T V \tilde{x}[\tau] + \tilde{x}[\tau]^T Y \tilde{x}[\tau]$$

the test statistic in (23) can be expressed as

$$\mathscr{D}[t] = \tilde{x}[t]^T P_0^{-1} \tilde{x}[t]$$
$$= d_1[t-D+1]$$
$$+ d_2[t-D+2] + d_2[t-D+3] + \cdots + d_2[t-2] + d_2[t-1]$$
$$+ d_3[t].$$

At the next time step the test statistic is

$$\mathscr{D}[t+1] = d_1[t-D+2]$$
$$+ d_2[t-D+3] + d_2[t-D+4] + \cdots + d_2[t-1] + d_2[t]$$
$$+ d_3[t+1].$$

$$\mathcal{D}[t] = d_1[t-D+1] + d_2[t-D+2] + d_2[t-D+3] + \cdots + d_2[t-2] + d_2[t-1] + d_3[t]$$

$$\mathcal{D}[t+1] = d_1[t-D+2] + d_2[t-D+3] + d_2[t-D+4] + \cdots + d_2[t-1] + d_2[t] + d_3[t+1]$$

Fig. 3 The test statistic at two subsequent time steps. Notice the values that are carried over to the next time step, as indicated by the *arrows*. The test statistic at time $t+1$ is obtained by taking $D[t]$, subtracting the values in *red*, and adding the values in *blue*

Figure 3 illustrates the manner in which values from the test statistic at t are carried over to $t+1$. From this figure, it is clear to see that the complete recursion is

$$\mathcal{D}[t+1] = \mathcal{D}[t] - (d_1[t-D+1] + d_2[t-D+2] + d_3[t]) \\ + (d_1[t-D+2] + d_2[t] + d_3[t+1]). \quad (24)$$

It should be noted that the original window contains D time steps, however, Eq. (24) requires that a window of $D+1$ time steps be maintained. To avoid this change in the window size the recursion equation can be determined in two steps. First, after the hypothesis test at time t an intermediate value for $\mathcal{D}[t]$ is calculated as

$$\mathcal{D}^+[t] = \mathcal{D}[t] - (d_1[t-D+1] + d_2[t-D+2] + d_3[t]).$$

At time $t+1$ the test statistic can then be updated using

$$\mathcal{D}[t+1] = \mathcal{D}^+[t] + (d_1[t-D+2] + d_2[t] + d_3[t+1]).$$

The method can be extended to the case where during the construction of (11) the time difference between subsequent estimation errors is ℓ time steps. Doing so reduces the observed time correlation in the test statistic (which is theoretically zero, but nonzero in practice), which enhances the power of the test [27]. In that case, the recursion for the test statistic becomes

$$\mathcal{D}[t+\ell] = \mathcal{D}[t] - (d_1[t-\ell(D-1)] + d_2[t-\ell(D-2)] + d_3[t]) \\ + (d_1[t-\ell(D-2)] + d_2[t] + d_3[t+\ell]). \quad (25)$$

Note, the recursion requires that the window is slid by ℓ time steps.

2.6 Track Fusion

The final step in the architectures shown in Fig. 1 is to fuse tracks for which the test statistic exceeds the given threshold. Therefore, if H_0 is accepted for a given pair of tracks, $(\hat{x}^{k_a}, \hat{x}^{j_b})$, the objective is to combine or fuse the estimates. The method presented in this paper takes advantage of the fusion properties of information filters [18]. For the steady-state information filter, the information matrix is given by $J = P_e^{-1}$, where P_e is given in Eq. (20), and the information vector is given by $z^{j_b}[t] = P_p^{-1} x^{j_b}[t]$. When measurements are received locally, the information filter is updated using

$$z^{j_b}[t+1] = JAP_p z^{j_b}[t] + C^\top R^{-1} y_{j_b}[t].$$

When the state $\hat{x}^{k_a}$ is to be fused with z^{j_b} then the fusion equation becomes

$$z^{j_b}[t+1] = JAP_p z^{j_b}[t] + C^\top R^{-1} y_{j_b}[t] + R_a^b J \hat{x}^{k_a} + (J - I) d_a^b.$$

3 Simulation and Flight Results

The complete cooperative estimation system shown in Fig. 1 was tested in a tracking scenario that involved two stationary cameras (simulating a hovering scenario), each viewing the tracking area from a different point of view, as shown in Fig. 4.

For this test the position and orientation of the cameras were calculated by determining the mapping of known inertial coordinates to their corresponding locations in the image frame. A foreground detector based on the KLT method [9] was used to produce pixel measurements for each object of interest, which were geolocated in the inertial frame. Note that the inertial frame was specified using a north-east-down (NED) frame of reference. These measurements in the inertial frame were input to R-RANSAC, producing states for each object.

The tracks produced by each object can be seen in the Fig. 5, where the green tracks are from camera V_1, while the cyan tracks are from camera V_2. This figure illustrates the rotational and translational biases that separate the associated tracks.

A window of $N = 10$ state estimates was stored with $\ell = 10$. This window was used to calculate the rotational and translational biases as in Sect. 2.4. The rotation matrix and bias vector were then used to transform the tracks from one vehicle into the reference frame of the other vehicle. Applying the bias estimation to two associated tracks can be seen in Fig. 6.

Fig. 4 Tracking scenario with two cameras and two ground objects. Each camera views the tracking area from a different angle

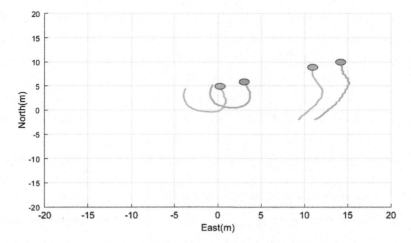

Fig. 5 The two objects from Fig. 4 are geolocated by two cameras. The *circles* represent the object locations at the current time step, while the trails represent the track history. *Green* denotes the tracks from V_1, and *cyan* represents the tracks from V_2. Due to sensor biases, the associated tracks are biased from each other

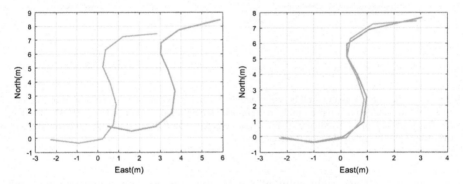

Fig. 6 Two associated tracks, before (*left*) and after (*right*) the bias estimation

It is clear that the bias estimation technique was effective in transforming both tracks into a common reference frame, which is vital for performing the track-to-track association. The application of the bias estimation to two unassociated tracks is shown in Fig. 7. Notice that despite the tracks being unassociated the optimizer still returned a rotation matrix and bias vector that minimized the squared error.

At every time step the window was slid, the bias estimation applied, and the track-to-track association performed. The threshold for the test statistic was based on $\alpha = 0.05$. The results of the track-to-track association over the entire video sequence can be seen in Fig. 8. Note that the track-to-track association was performed between a single track from the first camera with the two tracks from the other camera, which yielded an associated track pair as well as an unassociated track pair. For each

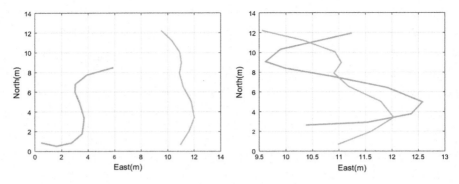

Fig. 7 Two unassociated tracks, before (*left*), and after (*right*) the bias estimation

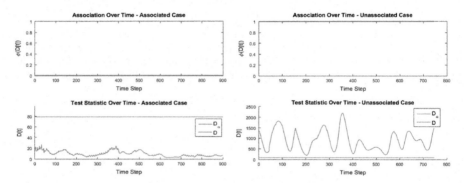

Fig. 8 Track-to-track association between two associated tracks and two unassociated tracks. For each column the *top plots* represent the determined association over time, where a 0 indicates that H_0 was accepted and a 1 indicates that H_0 was rejected. The *bottom plots* show the test statistic over time (*blue*) compared to the threshold (*red*)

column (the left and right columns representing the associated and unassociated cases, respectively) the top plots represent the determined association over time; 0 meaning that H_0 was accepted for the track pair, 1 meaning that H_0 was rejected. The bottom plots show the test statistic over time, compared to the threshold. As seen, over the entire video sequence the track-to-track association algorithm was able to correctly accept and reject H_0 with $P_D = 1$ and $P_R = 1$.

After H_0 was accepted for a given track pair the tracks were fused. The effect of the track fusion can be seen in Fig. 9. The left plot shows two associated tracks that were aligned using the bias estimation technique, with no track fusion. Notice that there were several areas where the two tracks did not fully line up. Over the entire window ($N = 10$, $\ell = 10$) the RMS error of the position states between the two tracks is 0.364 m. On the other hand, the right plot shows the tracks with track fusion applied over the entire window. Here, it can be seen that the fusion caused the two tracks to be more aligned. As a result, the RMS error over the window decreased to 0.037 m.

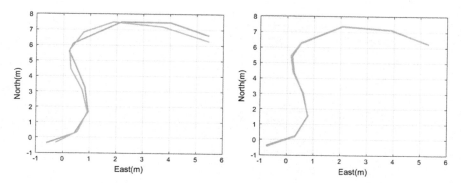

Fig. 9 Results of the track fusion. On the *left* two associated track are aligned, however, no track fusion is performed. It is clear to see that there are areas in which the two tracks did not fully align. On the *right* are the same tracks, however, with track fusion applied over the entire window. The track fusion reduced the differences in the tracks

3.1 Test with Small UAS Platforms

A test was performed with data collected from actual UAS platforms (3DR Y6 multirotor). Again, each UAS viewed the tracking area from a different angle (see Fig. 10). However, unlike the previous test the UAS platforms were not stationary. The vehicle states were provided by the 3DR Pixhawk autopilot. Moreover, each vehicle was equipped with a 3-axis brushless gimbal that was controlled using the BaseCam SimpleBGC 32-bit gimbal controller. Note that the video sequence contained a single object, thus, the data was used to validate the method under the assumption of H_0 only.

The results from the test are summarized in Fig. 11. Overall the algorithm was effective in associating the two tracks from the different vehicles and had a probability of detection $P_D = 1.0$. These results affirm the effectiveness of the method in the presence of actual UAS sensor biases and noise.

Fig. 10 Tracking scenario with two cameras and one ground object of interest. Each camera is mounted to a UAS platform that is maneuvering and views the tracking area from a different angle. The *red circle* indicates the pixel measurement that is used to geolocate the object of interest

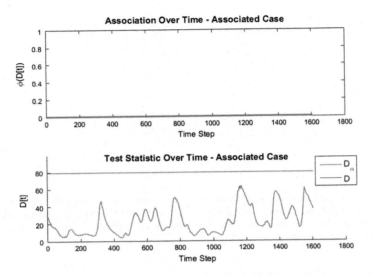

Fig. 11 Track-to-track association between two associated tracks. For each column the *top plots* represent the determined association over time, where a 0 indicates that H_0 was accepted and a 1 indicates that H_0 was rejected. The *bottom plots* show the test statistic over time (*blue*) compared to the threshold (*red*)

4 Conclusions

This paper presents a complete method for cooperative estimation of ground targets using a vision-based object tracking system. The method estimates and accounts for both translational and rotational biases between tracks, and performs a hypothesis test to determine the track-to-track association. The test statistic is calculated using a window of estimates and follows a chi-squared distribution. The correlation between associated tracks, and the correlation in time of the estimation errors, is accounted for in the calculation of the covariance. This paper also presents a track fusion technique, which accounts for the estimated biases.

The complete system is demonstrated in actual tracking scenarios. The results show that the bias estimation is effective in aligning associated tracks from different vehicles. Moreover, the track-to-track association method is able to make the proper assignments with a high probability of detection and rejection. Lastly, the track fusion technique decreases the relative estimation error between associated tracks.

Acknowledgements This work has been funded in part by the Center for Unmanned Aircraft Systems (C-UAS), a National Science Foundation-sponsored industry/university cooperative research center (I/UCRC) under NSF Award No. IIP-1161036 along with significant contributions from C-UAS industry members, and in part by AFRL grant FA8651-13-1-0005.

References

1. Bar-Shalom, Y.: On the track-to-track correlation problem. IEEE Autom. Control **26**(2), 571–572 (1981)
2. Bar-Shalom, Y., Fortmann, T.E.: Tracking and Data Association. Academic Press, New York (1988)
3. Bar-Shalom, Y., Daum, F., Huang, J.: The probabilistic data association filter. IEEE Control Syst. Mag. **26**(9), 82–100 (2009)
4. Barber, D.B., Redding, J.D., McLain, T.W., Beard, R.W., Taylor, C.N.: Vision based target geo-location using a fixed-wing miniature air vehicle. J. Intell. Robot. Syst. **47**(4), 361–382 (2006)
5. Beard, R.W., McLain, T.W.: Small Unmanned Aircraft: Theory and Practice. Princeton University Press, Princeton (2012)
6. Blackman, S.S.: Multiple hypothesis tracking for multiple target tracking. IEEE Aerosp. Electron. Syst. Mag. **19**(1), 5–18 (2004)
7. Campbell, M.E., Wheeler, M.: A vision based geolocation tracking system for UAVs. In: Proceedings of the AIAA Guidance, Navigation, and Control Conference and Exhibit. Keystone, Colorado (2006)
8. Casbeer, D.W., Kingston, D.B., Beard, R.W., McLain, T.W., Li, S.-M., Mehra, R.: Cooperative forest fire surveillance using a team of small unmanned air vehicles. Int. J. Syst. Sci. **37**(6), 351–360 (2006)
9. DeFranco, P.C.: Detecting and Tracking Moving Objects from a Small Unmanned Air Vehicle. Master's thesis, Brigham Young University, Provo, Utah (2015)
10. Dobrokhodov, V.N., Kaminer, I.I., Jones, K.D.: Vision-based tracking and motion estimation for moving targets using small UAVs. In: Proceedings of the AIAA Guidance, Navigation, and Control Conference and Exhibit. Keystone, Colorado (2006)
11. Frew, E.W.: Sensitivity of cooperative target geolocation to orbit coordination. J. Guid. Control Dyn. **31**(4), 1028–1040 (2008)
12. He, Z., Xu, J.-X., Lum, K.-Y.: Targets tracking by UAVs in an urban area. In: IEEE International Conference on Control and Automation (ICCA), pp. 1834–1838. Hangzhou, China (2013)
13. Holt, R.S., Egbert, J.W., Bradley, J.M., Beard, R.W., Taylor, C.N., McLain, T.W.: Forest fire monitoring using multiple unmanned air vehicles. In: Eleventh Biennial USDA Forest Service Remote Sensing Applications Conference. Salt Lake City (2006)
14. Ingersoll, J.K.: Vision Based Multiple Target Tracking Using Recursive RANSAC. Master's thesis, Brigham Young University (2015)
15. Ingersoll, K., Niedfeldt, P.C., Beard, R.W.: Multiple target tracking and stationary object detection using recursive-RANSAC and tracker-sensor feedback. In: Proceedings of the International Conference on Unmanned Air Vehicles. Denver, CO (2015)
16. Kamal, A.T., Bappy, J.H., Farrell, J.A., Roy-Chowdhury, A.K.: Distributed multitarget tracking and data association in vision networks. IEEE Trans. Pattern Anal. Mach. Intell. **38**(7), 1397–1410 (2016)
17. Kumar, R., Sawhney, H., Samarasekera, S., Hsu, S., Tao, H., Guo, Y., Hanna, K., Pope, A., Wildes, R., Hirvonen, D., Hansen, M., Burt, P.: Aerial video surveillance and exploitation. Proc. IEEE **89**(10), 1518–1539 (2001)
18. Mutambara, A.G.O.: Decentralized Estimation and Control for Multisensor Systems. CRC Press, Boca Raton (1998)
19. Niedfeldt, P.C.: Recursive-RANSAC: A Novel Algorithm for Tracking Multiple Targets in Clutter. Ph.D. thesis, Brigham Young University, Provo (2014)
20. Niedfeldt, P.C., Beard, R.W.: Recursive RANSAC: multiple signal estimation with outliers. In: Proceedings of the 9th Symposium on Nonlinear Control Systems, pp. 430-435. Toulouse, France (2013)
21. Niedfeldt, P.C., Beard, R.W.: Multiple target tracking using recursive RANSAC. In: Proceedings of the American Control Conference, pp. 3393–3398. Portland, OR (2014)

22. Niedfeldt, P.C., Beard, R.W.: Convergence and complexity analysis of recursive-RANSAC: a new multiple target tracking algorithm. IEEE Trans. Autom. Control **61**(2), 456–461 (2016)
23. Pachter, M., Ceccarelli, N., Chandler, P.R.: Vision-based target geo-location using camera equipped MAVs. In: Proceedings of the IEEE Conference on Decision and Control. New Orleans, LA (2007)
24. Ruggles, S., Clark, J., Franke, K.W., Wolfe, D., Reimschiissel, B., Martin, R.A., Okeson, T.J., Hedengren, J.D.: Comparison of SfM computer vision point clouds of a landslide derived from multiple small UAV platforms and sensors to a TLS based model. J. Unmanned Veh. Syst. (2016). doi:10.1139/juvs-2015-0043
25. Sakamaki, J.Y.: Cooperative Estimation for a Vision Based Target Tracking System. Master's thesis, Brigham Young University, Provo, Utah (2016)
26. Scharf, L.L.: Statistical Signal Processing. Prentice Hall, Englewood Cliffs (1990)
27. Tian, X., Bar-Shalom, Y.: Sliding window test vs. single time test for track-to-track association. In: Proceedings of the 11th International Conference on Information Fusion, FUSION 2008 (2008). ISBN 9783000248832. doi:10.1109/ICIF.2008.4632281
28. Vo, B.N.: Ma, W.K.: A closed-form solution for the probability hypothesis density filter. In: 2005 7th International Conference on Information Fusion, FUSION 2, pp. 856–863 (2005). ISBN 0780392868. doi:10.1109/ICIF.2005.1591948
29. Whang, I.H., Dobrokhodov, V.N., Kaminer, I.I., Jones, K.D.: On vision-based tracking and range estimation for small UAVs. In: Proceedings of the AIAA Guidance, Navigation, and Control Conference and Exhibit. San Francisco, CA (2005)

A Target Tracking System for ASV Collision Avoidance Based on the PDAF

Erik F. Wilthil, Andreas L. Flåten and Edmund F. Brekke

Abstract Safe navigation and guidance of an autonomous surface vehicle (ASV) depends on automatic sensor fusion methods capable of discovering static and moving obstacles in the vicinity of the ASV. A key component in such a system is a method for target tracking. In this paper, we report a complete radar tracking system based on the classical probabilistic data association filter (PDAF). The tracking system is tested on real radar data recorded in Trondheimsfjorden, Norway.

1 Introduction

Target tracking is a key ingredient in collision avoidance (COLAV) for autonomous vehicles. It is important to discover moving and stationary obstacles in data streams from imaging sensors, such as radar, sonar, lidar, camera etc. Target tracking is required for reliable detection of moving objects under low to moderate signal-to-noise ratio (SNR). Furthermore, the predictions and velocity estimates of a tracking method can enable proactive ownship manoeuvers.

Dozens of tracking methods, or more specifically data association methods, exist [1, 12, 13, 17]. All of these methods attempt to make a judgement regarding the origin of measurements from the imaging sensors, either by hedging over competing association hypotheses, or by minimizing the risk of misassociations. The arguably simplest state-of-the art tracking method is the PDAF [1], which is a single-target tracking method based on hedging. While target tracking has been extensively studied

---

E.F. Wilthil (✉) · A.L. Flåten · E.F. Brekke
Department of Engineering Cybernetics, Centre for Autonomous Marine Operations and Systems, Norwegian University of Science and Technology (NTNU), Trondheim, Norway
e-mail: erik.wilthil@itk.ntnu.no

A.L. Flåten
e-mail: andreas.flaten@itk.ntnu.no

E.F. Brekke
e-mail: edmund.brekke@itk.ntnu.no

in the literature from a military point of view, very few research papers (e.g., [15, 18]) have provided in-depth discuss of sensor fusion for COLAV systems.

In this paper, we report the development of a PDAF-based tracking system intended for usage in future COLAV experiments. This includes the pipeline consisting of plot extraction (detection and segmentation), masking of land returns, integration with the ASV navigation system, and track management (track initiation and termination). We use AIS data to determine suitable values of the process noise covariance. The tracking system is tested on real data recorded by an frequency-modulated continuous-wave (FMCW) X-band radar onboard the Telemetron ASV.

The remainder of document is divided into several sections. In Sect. 2 we describe the radar used and how the raw data from this sensor is preprocessed. Section 3 describes the PDAF, and our particular implementation including track management. Section 4 describes analysis of AIS data recorded as part of the experiments, while Sect. 5 is devoted to the experiments and analysis of the experimental results. In Sect. 6 we discuss some key topics with basis in the results, before a conclusion follows in Sect. 7.

2 Pre-tracking Radar Pipeline

Interpreting radar data in order to track targets is a process that can be broken down into several stages. An overview of the radar processing pipeline that is presented in this section is illustrated in Fig. 1.

2.1 Detection

The tracking system is designed to work with the Navico 4G broadband radar, which has a proprietary detection process optimized for use at sea. The output of the detection stage in the radar processing pipeline is a stream of radar spokes, where each

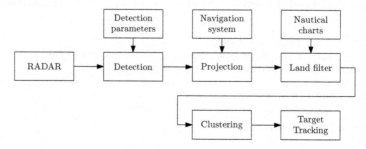

Fig. 1 A block diagram of the proposed radar pre-tracking pipeline

Fig. 2 An illustration of the resolution cells in a full radar scan, and a single radar spoke

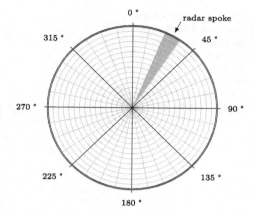

spoke contains detections for the range resolution cells for a given azimuth angle. This is illustrated in Fig. 2.

If a reliable in-built detection procedure is not provided by the radar, standard techniques based on the principle of constant false alarm rate (CFAR) should be considered as the most straightforward approach. See [7].

2.2 Projection

As a next stage in the processing pipeline, we have chosen to transform the detections in to a world fixed reference frame. This mitigates distortion due to the effects of vehicle motion and rotating radar antenna. Since the platform is moving, this step depends on the ability to estimate the pose of the radar relative to a world fixed frame. It will be assumed that this transformation is perfectly known through a navigation system, and that the navigation system is synchronized in time with the radar. It is further assumed that due to the high vertical beam-width of the radar, detections can be interpreted directly in a horizontal reference frame (i.e., parallel to the nominal sea surface). The high beam-width also comes with a cost of distortion when the radar experiences wave-induced motion.

Given a stream of radar spokes with known azimuth angle α, the first step is to transform a spoke from polar to Cartesian coordinates. A single detection can be represented in a horizontal Cartesian reference frame as:

$$p_i^h = \begin{bmatrix} \rho_i \cos \alpha \\ \rho_i \sin \alpha \\ 0 \end{bmatrix} \quad (1)$$

where i is the index of the range resolution cell in the radar spoke, and ρ_i is the range of that resolution cell in meters. Given that the navigation system provides the

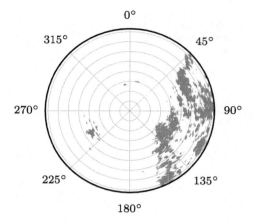

Fig. 3 The radar returns from land. The *small dots* in the *middle* of the plot shows returns from two ships. The points to the *left* shows the island of Munkholmen, while the large clusters to the *right* is Trondheim

translation r_{wh}^w and rotation R_h^w of the horizontal frame, we can express the detection in a world fixed frame as:

$$p_i^w = r_{wh}^w + R_h^w p_i^h \tag{2}$$

If the world fixed frame is a local north-east-down (NED) frame, then this transformation can be approximated by a planar translation and a rotation about one axis.

The spokes are projected when they arrive, and aggregated into a point cloud made from a full rotation of the radar (a complete scan). The output of the projection stage is thus a world fixed two-dimensional point cloud of radar detections.

2.3 Land Masking

When land is within range of the radar, it generates a significant amount of returns, as seen in Fig. 3. If these detections propagates further down the radar pipeline, they will introduce false tracks and extra computational load. To avoid this, we eliminate the detections on land by means of map-based masking. We use data obtained from The Norwegian mapping authority (Kartverket). The data is provided at a vector format describing the land area as a set of polygons. To use this efficiently, the map is preprocessed, and represented as a binary grid in the world-fixed frame which the radar detections are transformed into. If the radar detection hits a cell containing land, the detection is discarded.

The primary error source in the map itself is the accuracy of the data. Kartverket does provide accurate data, but some offsets and reflections close to shore must be expected. We will see some effects of this in Sect. 5. Insufficiently accurate maps can be dealt with in several ways. One simple solution is to inflate the maps such that all detections within some distance to land are discarded as well. More advanced methods includes estimating the mismatch using occupancy grids [16], but no such methods will be considered here.

2.4 Clustering

A standard assumption in target tracking algorithms is that a single target generates at most one measurement per scan. This assumption is useful in order to keep the tracking problem well posed and tractable within the Bayesian framework. If this assumption is relaxed, then the Bayesian framework may need a model for the number of target measurements, which may be difficult to specify.

In practice, however, this assumption will typically be violated when sensor resolution is high relative to the physical target extent. While the PDAF itself does not suffer very much from such violations, other tracking methods are more sensitive to this assumption, and in any case having an abundance of measurements from the same target makes track initialization much more complicated.

Therefore, we attempt to uphold the at-most-one-measurement assumption by means of *clustering* of detections. A clustering algorithm assumes that detections that are in some sense close to each other originate from the same target, and joins these detections together to form a cluster. Once a cluster is established, the detections in the cluster represent a single measurement. The location of the measurement is computed as the centroid of the individual detections, weighted by the strength of the detections if amplitude information is available.

A simple definition of a cluster is a collection of cells where each cell neighbours at least one other cell in the cluster. The definition is not complete without the notion of a neighbouring point. A point p_i^w is defined as a neighbour to another point p_j^w if the Euclidean distance between the points is below some threshold. Mathematically this is defined as the binary function:

$$N_R(p_i^w, p_j^w) = \begin{cases} 1, & \|p_i^w - p_j^w\| \leq R \\ 0, & \text{otherwise.} \end{cases} \quad (3)$$

A set of points together with this function defines a graph $G = (V, E)$, where $V = \{1, 2, \ldots, N\}$ represent the point indices, and there is an edge from vertex i to j whenever (3) is equal to 1:

$$E = \{(i, j) \mid N_R(p_i^w, p_j^w) = 1, \, i, j \in V\} \quad (4)$$

Note that the number of edges in this graph depends on the value of R, and that a larger R implies a denser graph. Clusters can be computed as the *connected components* of G. Finding the connected components of a graph can be done by either a depth-first search (DFS) or breadth-first search (BFS). Constructing this graph has computational complexity $O(N^2)$ since we need to compute the Euclidean distance between each pair of points, while the graph search is $O(V + E)$ for both DFS and BFS. When there are thousands of detections, the $O(N^2)$ time required to build the graph can become prohibitively large. By representing the detected points in a *k-d tree* [6, 11], we can traverse this graph implicitly by performing radius searches at each vertex. Constructing the k-d tree has complexity $O(N \log N)$, and performing a radius search has an *average* complexity of $O(\log N)$. Traversing the graph in either BFS or DFS manner requires a radius search at each vertex, so the total average complexity (including building the tree) will be $O(N \log N)$.

3 PDAF-based Tracking Module

3.1 Motion Model

The state of a target is given as $x = [N, V_N, E, V_E]$, where N, E, V_N, and V_E are the north and east positions and velocities of the target in a stationary NED reference frame. A discrete-time white noise acceleration model [9] is used to model the target motion. This model can be written on the form

$$x_{k+1} = F_T x_k + v_k \qquad p(v_k) = \mathcal{N}(v_k; 0, Q_T) \qquad (5)$$

where v is process noise, $\mathcal{N}$ denotes the normal distribution and the state transition matrix F_T and noise covariance Q_T are given as

$$F_T = \begin{bmatrix} 1 & T & 0 & 0 \\ 0 & 1 & 0 & 0 \\ 0 & 0 & 1 & T \\ 0 & 0 & 0 & 1 \end{bmatrix}, \quad Q_T = \sigma_a^2 \begin{bmatrix} T^4/4 & T^3/2 & 0 & 0 \\ T^3/2 & T^2 & 0 & 0 \\ 0 & 0 & T^4/4 & T^3/2 \\ 0 & 0 & T^3/2 & T^2 \end{bmatrix} \qquad (6)$$

where $T = t_{k+1} - t_k$ is the sample time, in general time varying for a rotating radar. The process noise strength σ_a is selected according to the targets' expected maneuverability (See Sect. 5.3).

The measurements used in the PDAF are of target position only. Thanks to the Projection step described in Sect. 2.2, we can use a linear Cartesian measurement model of the form

$$z_k = H x_k + w_k \qquad p(w_k) = \mathcal{N}(w_k; 0, R) \qquad (7)$$

where w is the measurement noise and H is given by

$$H = \begin{bmatrix} 1 & 0 & 0 & 0 \\ 0 & 0 & 1 & 0 \end{bmatrix}. \qquad (8)$$

We use a constant and diagonal measurement noise covariance matrix, reflecting the fact that target extent generally is larger than the size of the radar resolution cells (See Sect. 5.2).

3.2 Track Initiation

Tracks are formed using 2/2&m/n logic [1]. A measurement that is not associated with any existing targets is a candidate for a new track, also called a *tentative* track. To estimate the state of a target, another measurement is required. This second measurement must fall within a circle-shaped validation gate around the tentative track. The radius of this circle is given by the maximum velocity of the target, the measurement noise statistics, and the probability of missing a target. The expressions are given in [2, p. 247].

3.3 PDAF Tracking

For preliminary and confirmed tracks which have a state estimate and a corresponding covariance, we employ the PDAF to track the targets. Starting from the third scan, the standard Kalman filter prediction equations are used to propagate the tracks

$$\hat{x}_{k|k-1} = F_T \hat{x}_{k-1|k-1} \tag{9}$$

$$P_{k|k-1} = F_T P_{k-1|k-1} F_T^T + Q_T \tag{10}$$

The predicted covariance is used to set up a measurement validation gate in order to reduce the number of measurements we need to consider for each target. Measurements that do not fall within any validation gate are used as candidates for new tracks. The gating process utilizes the predicted measurements and the corresponding covariance

$$\hat{z}_k = H \hat{x}_{k|k-1} \tag{11}$$

$$S_k = H P_{k|k-1} H^T + R \tag{12}$$

from which we can calculate the normalized innovation squared (NIS) as

$$\text{NIS} = (z_k^i - \hat{z}_k)^T S_k (z_k^i - \hat{z}_k) \tag{13}$$

$$= v_k^{iT} S_k^{-1} v_k^i < \gamma_G \tag{14}$$

where γ_G is a threshold used to determine if the measurement should be associated with the target or not. The gating threshold is found from the inverse cumulative distribution function (CDF) of the χ^2-distribution with degrees of freedom corresponding to the dimension of the measurement. The value v_k^i is called the *measurement innovation*.

After each target has been assigned a (possibly empty) set of measurements, the PDAF provides a moment-matched Gaussian approximation of the posterior of the target state by hedging on all the measurements in the association gate, as well as the possibility that no measurements may originate from the target. The probability of measurement z^i being the correct is given as

$$\beta_k^i = \begin{cases} \frac{1}{c} \exp\left(-\frac{1}{2} v_k^{iT} S_k^{-1} v_k^i\right) & i = 1 \ldots m_k \\ \frac{1}{c} \frac{2(1-P_D P_G)}{\gamma} m_k & i = 0 \end{cases} \quad (15)$$

where c is a normalization constant. The moments $\hat{x}_{k|k}$ and $P_{k|k}$ of the posterior are then calculated according to

$$v_k = \sum_{i=1}^{m_k} \beta_k^i v_k^i \quad (16)$$

$$K_k = P_{k|k-1} H^T S_k^{-1} \quad (17)$$

$$\hat{x}_{k|k} = \hat{x}_{k|k-1} + K_k v_k \quad (18)$$

$$P_{k|k} = P_{k|k-1} - (1 - \beta_k^0) K_k H P_{k|k-1} + \tilde{P}_k \quad (19)$$

where

$$\tilde{P}_k = K_k \left(\sum_{i=1}^{m_k} \beta_k^i v_k^i v_k^{iT} - v_k v_k^T \right) K_k^T \quad (20)$$

is called the spread of innovations (SOI). The posterior covariance estimate differs from the regular Kalman filter in two ways. The second term in Eq. (19) is adjusted to account for the possibility of missing measurements, and the SOI inflates the covariance to account for the uncertainty in data association. See [1] for further details.

3.4 Track Management and Track Termination

As validation gates for tentative, preliminary and confirmed tracks can overlap and capture the same measurements, some form of assignment hierarchy is needed. Measurements will be assigned to confirmed, preliminary and tentative tracks in that order, such that the more established tracks take precedence over the newer tracks. For a given category, the measurement will be assigned to all tracks that gate it, for example two confirmed tracks. This hierarchy reduces the number of false tracks, but it does not provide any guarantees against track coalescence, i.e. that multiple tracks start to associate the same set of measurements, and converge to each other.

As a remedy for this, we prune tracks by a similarity test [1]. Denote the distance between two tracks as $d = x^1 - x^2$. The same-target hypothesis H_0 is that $d = 0$, and the two estimates with means $\hat{x}^1$, $\hat{x}^2$ and covariances P^1, P^2 are estimates of the same target. To test this hypothesis, calculate the mean $\hat{d}$ and covariance T of d as

$$\hat{d} = \hat{x}^1 - \hat{x}^2 \qquad (21)$$

$$T = P^1 + P^2 - P^{12} - P^{21} \qquad (22)$$

where P^{ij} is the cross-covariance of track i and j. This is nonzero because of the common process noise in the tracks under the same-target hypothesis. This value is impractical in a real-time tracking system and it has been approximated by the individual covariances P^1 and P^2 as per the technique described in [1]. With these values, we can accept the same-target hypothesis if

$$\gamma = \hat{d}^T T^{-1} \hat{d} < \gamma_\alpha \qquad (23)$$

where γ_α is a threshold such that

$$P(\gamma > \gamma_\alpha | H_0) = \alpha \qquad (24)$$

Furthermore, any track which has failed to gate any measurements over the last five scans is terminated. This criterion terminates all tracks that go outside of the radar range.

4 AIS Filtering

In addition to maritime radar, the automatic identification system (AIS) is a core component of maritime collision avoidance. While most larger ships are required to have AIS onboard, it is not sufficient to rely on AIS information alone for collision avoidance since it is not universally adopted, unreliable, and may even be switched off on purpose. Further information about AIS and its reliability can be found in [8].

Nevertheless, most of the time AIS contains valuable collision avoidance information, and will in many cases compliment radar when detection is limited. In the ideal case, the AIS system can be thought of as a target tracking system without the problems induced by false alarms or misdetections. AIS data come in the form of structured messages consisting of fields, such as maritime mobile service identity (MMSI) transmitter identifier, coordinated universal time (UTC) second time stamp, UTC second time of arrival, World Geodetic System (WGS84) latitude and longitude, course over ground (COG) and speed over ground (SOG) relative true north.

In this section, we discuss how raw AIS messages can be processed in a probabilistic framework, to quantify the uncertainty of AIS messages, and to motivate the use of AIS as a compliment to radar target tracking and as a tool to validate target motion models.

4.1 Out-of-Order Message Arrival

One challenge in state estimation based on AIS is that AIS messages frequently arrive out-of-order, meaning that the UTC second time stamp is not locally monotonously increasing for a single MMSI. Optimal filtering will therefore require out-of-sequence measurements (OOSM) filtering methods [4]. An alternative to OOSM processing is to buffer messages and re-order them on the fly. This would add a time delay to the message arrival, which will propagate to target tracking and collision avoidance algorithms.

A simpler and more convenient solution is to sequentially throw away messages that are defined as "old". While this clearly is a suboptimal solution, the information loss is not necessarily significant in practical terms. This effect can be compared to the information loss induced by reducing the sampling rate by a factor equal to the proportion of messages discarded, which is adequate if the resulting sampling time still is small enough to capture significant target dynamics.

A sequential time stamp filter can be implemented by storing the latest AIS message and comparing it to newly arriving messages. If an arrived message is newer than the currently stored message, it is assigned to be the latest message, else it is discarded. An illustration of this issue and the proposed solution can be seen in Fig. 4, where the UTC time stamp does not always increase.

4.2 AIS Measurement Model

The AIS messages can be transformed into linear measurements of the kinematic state as defined in Sect. 3.1. Given a local stationary NED reference frame defined by a latitude, longitude and height over the WGS84 ellipsoid (equal to the local sea level), the position can be calculated according to formulas found in [5, p. 38]. The velocity in the same NED frame can be calculated as:

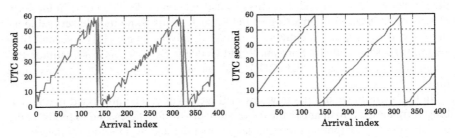

Fig. 4 Sequences of observed UTC second time stamps from the Trondheimsfjord II ferry, before and after filtering (*left* and *right* respectively). The "Arrival index" corresponds to the message arrival time. About 40% of the data is discarded in this case. Note that resulting sampling time is still on the order of a few seconds, implying that the loss of information is minimal

$$\begin{bmatrix} V_N \\ V_E \end{bmatrix} = \begin{bmatrix} V \cos \chi \\ V \sin \chi \end{bmatrix} \qquad (25)$$

where V is the SOG and χ is the COG in the message. Since AIS is usually based on one or several Global Navigation Satellite Systems (GNSSs), a reasonable estimate of the measurement accuracy is conventional GNSS accuracy. Ideally, the sampling time of the AIS messages should be exact, however if the actual sampling time is not an integer multiple of the UTC seconds, this will induce large position errors due to sampling time quantization. The SOG and COG estimates are less affected by this, since the course and speed usually do not change much within a single second. The actual error induced by quantization depends on the SOG and SOG of the vehicle, and how much the time stamp has been rounded off. The worst case rounding error can be assumed to be ± 0.5 s. To approximate this as an additional independent Gaussian error source, we add a position measurement covariance proportional to the speed in each direction (north or east), scaled by a suitable variance related to the quantization error. Since the actual quantization error is uniform in $(-0.5, 0.5)$ s, a reasonable choice is a moment-matched Gaussian distribution with zero mean and standard deviation $\sigma_t = \frac{1}{\sqrt{12}}$. Using these definitions, the measurement matrix for the AIS measurements can be written as:

$$H = I_4 \qquad (26)$$

and measurement covariance:

$$R_{AIS} = R_{GNSS} + \frac{1}{12} R_V, \quad R_V = \text{diag}\left[V_N^2, 0, V_E^2, 0\right] \qquad (27)$$

where the values used for R_{GNSS} are the following:

$$R_{GNSS} = \text{diag}\left[0.5^2, 0.1^2, 0.5^2, 0.1^2\right] \qquad (28)$$

Note that since the measurement noise varies with speed, the measurement noise is time varying. An interesting observation is that with these values for R_{GNSS}, for speeds over approximately 2 m per second, the discretization noise will dominate the position error.

4.3 Missing Measurement Kalman Filter

In addition to arriving out of sequence, AIS messages are recieved at varying rates. This is also the case for messages from a single ship, where the rate may vary significantly over time due to varying ship speed and ship status [8]. It may also be the case that a message was discarded or not recieved at all. However, following the

Fig. 5 The Telemetron ASV is equipped with several sensors such as a radar, AIS reciever, camera and navigation systems. For the purpose of this test, it was also equipped with a Kongsberg Seapath 330+ navigation system. The vehicle can both be used in autonomous mode and with manual control

discussion in Sect. 4.2, it is convenient to run the AIS filter with a fixed sampling time of one second for all ships. When measurements are not available, the AIS filter simply uses Kalman filter prediction instead of Kalman filter update.

5 Experimental Validation

The complete tracking system has been validated on real data recorded during field trials with the Telemetron vehicle (owned by Maritime Robotics) in Trondheimsfjorden December 10 2015 and January 11 2016. The Telemetron vessel is equipped with a Navico 4G broadband radar, a commercial off-the-shelf AIS reciever and a Seapath 330+ high-grade navigation system supplied by Kongsberg Seatex in Trondheim (Fig. 5). The Robot operating system (ROS) [14] software framework was used for data acquisition and algorithm implementation. The experiments involved two scenarios. In the first scenario, referred to as "Trondheimsfjord II", Telemetron followed the catamaran Trondheimsfjord II, while also passing by several other ships. An overview is shown in Fig. 9. In the second scenario, referred to as "Gunnerus", Telemetron approached and encircled the research vessel Gunnerus, which lay stationary in the water east of Munkholmen, north of Trondheim harbor. An overview is shown in Fig. 7.

5.1 Clustering Runtime Experimental Analysis

The only computational bottleneck of significance in the tracking system is the clustering algorithm described in Sect. 2.4. In this section, the runtime of the clustering algorithm is experimentally analyzed with data from the "Gunnerus" scenario, which

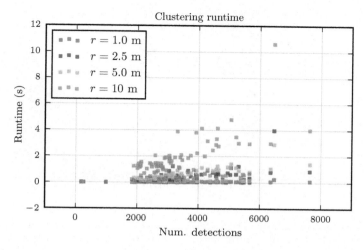

Fig. 6 Scatter plot of number of detection versus empirical runtime of the clustering algorithm described in Sect. 2.4

is further elaborated in Sect. 5.2. The clustering algorithm was implemented in C++ and run on a laptop with a 2.8 GHz Intel Core i7 processor. The results of the runtime experiments can be seen in Fig. 6. We see that the runtime increases both with respect to the number of detections per scan, and the clustering radius r. A lower clustering radius will in general lead to more and smaller clusters. This means that there is a tradeoff between being able to cluster fragmented detections and handling this further down in the processing pipeline. Even if real-time implementation was not the main focus of this study, a critical remark is in order; the maximum rotation period for the radar used in these experiments is 2.5 s, implying that a runtime close to or above 2.5 s will mean that running this algorithm in a real-time scenario will be infeasible. This "time limit" is often violated for a radius of 10 m, and is also dangerously close for the other values of r. The clustering radius used in the rest of the experimental results is $r = 5.0$ m.

5.2 Radar Measurement Noise Covariance Estimation

To determine a reasonable value of the RADAR measurement noise covariance, we analyzed the data from the "Gunnerus" scenario (see Fig. 7). The ownship circled around Gunnerus for one and a half revolution, while recording radar data in which Gunnerus was observed. Since we know that Gunnerus was stationary, the data from these recordings are convenient to use for estimation of the radar measurement noise covariance. Figure 8 shows the AIS position of Gunnerus, and the set of measurements within 100 m of that position. These measurements form the basis of our covariance estimation.

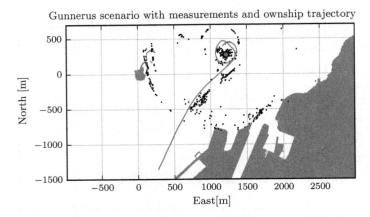

Fig. 7 The "Gunnerus" scenario. The *red* trajectory shows the ownship starting in the lower part of the image, and going north to circle around the NTNU research vessel Gunnerus, whose AIS position is shown as a *green dot*. The *shaded areas* shows the land, and the *black dots* are the cumulated set of measurements from this experiments. In addition to Gunnerus, the detections from a small boat can be seen passing on starboard side of the ownship

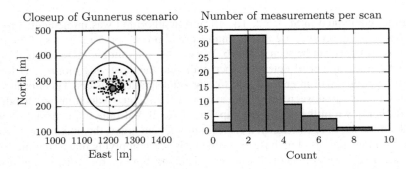

Fig. 8 253 measurements were found within 100 m of Gunnerus at 107 timesteps. The histogram to the right shows the number of measurements per scan

By assuming that at most one detection per scan can originate from Gunnerus, we calculate the sample covariance as

$$\hat{R} = \frac{1}{K-1} \sum_{k=1}^{K} \left(z_{A_{\max}} - \bar{z}_{A_{\max}}\right)\left(z_{A_{\max}} - \bar{z}_{A_{\max}}\right)^T \quad (29)$$

where the measurement $z_{A_{\max}}$ is selected such that the convex hull of its cluster has the *largest area* within the 100 m radius at this timestep. This is justified by assuming that the cluster returned from Gunnerus is larger than sea clutter and other reflections. The difference between the average measurement and the average AIS position of Gunnerus is approximately 6.5 m.

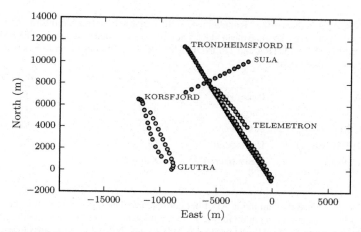

Fig. 9 An overview of the "Trondheimsfjord II" scenario. Only 1 out of 15 samples are shown for each trajectory, for visualization purposes

Notice that due to violations of the maximum-one-measurement assumption, it is not meaningful to use the rest of the measurements to calculate the clutter density of the scenario, as some tracking methods such as the Integrated Probabilistic Data Association (IPDA) [4] would do.

5.3 Determination of Process Noise Covariance

Given that the discrete white noise acceleration model presented in Sect. 3.1 is a realistic model, and that the measurement noise covariances have been determined, the only remaining parameter is the process noise variance σ_a^2. Having process and measurement noise parameters that are close to reality is crucial to performance, both in terms of Root Mean Squared Error (RMSE) and covariance consistency. Since the process noise parameter represents the assumed target maneuverability, the optimal choice of this parameter will likely vary somewhat between ships and even over time, especially with respect to the size of the ship. This section experimentally analyzes the choice of process noise variance for five ships in the vicinity of Telemetron over a 19 min time interval during the "Trondheimsfjord II" scenario, shown in Fig. 9. Relevant characteristics based on AIS information from each vessel in the scenario is shown in Table 1.

The average NIS [2] was calculated as a measure of filter consistency, for a range of process noise variances. Note that the number of samples for each vessel differs, so that the two-sided confidence interval for the NIS varies. Filter bias is also estimated by calculating the Average Innovation (AI) for all states, which should ideally be zero. The results are shown in Table 2. As emphasised by the bold font, it seems that the vessels can be grouped into two categories. Either they show little maneuverability,

Table 1 Vessel parameters in the "Trondheimsfjord II" scenario based on AIS information. Note the large diversity in ship type, size and mean SOG. The data was recorded on board the TELEMETRON vessel. KORSFJORD, GLUTRA and TRONDHEIMSFJORD II are passenger ferries that cross the fjord in Trondheim regularly

Name	Type	Length × Breadth	Mean SOG (m/s)
GLUTRA	Passenger	94.8 m × 16.0 m	5.3
SULA	Cargo	87.9 m × 12.8 m	5.7
KORSFJORD	Passenger	122 m × 16.7 m	5.6
TRONDHEIMSFJORD II	High speed	24.5 m × 8 m	12.8
TELEMETRON	Pleasure craft	8 m × 3 m	13.1

Table 2 Process noise evaluation via AIS filter consistency. The (r_1, r_2) interval is the two-sided 95% probability concentration region for the χ^2 distribution related to the corresponding NIS. This varies with according to the AIS data record length N. The NIS values that are closest to being covariance-consistent, i.e. closest to the 95% probability region, are emphasised in bold

Name	$\sigma_a = 0.05$		$\sigma_a = 0.5$		(r_1, r_2)	N
	NIS	AI	NIS	AI		
GLUTRA	**4.67**	−0.02	0.90	0.01	(3.47, 4.55)	109
SULA	**3.61**	−0.22	0.51	−0.10	(3.49, 4.56)	106
KORSFJORD	71.8	−1.33	**4.31**	−0.44	(3.52, 4.51)	127
TR.FJORD II	11.3	−0.62	**3.24**	−0.16	(3.76, 4.24)	533
TELEMETRON	371	−0.04	**4.45**	−0.01	(3.77, 4.23)	579

which corresponds to $\sigma_a = 0.05$, or they show significantly more maneuverability, where $\sigma_a = 0.5$ gives more appropriate values of the NIS. For simplicity it may be desirable to have a single value for the process noise variance. Choosing $\sigma_a = 0.5$ will thus in some cases give conservative values for the filter covariance, which presumably will inflict less risks of track-loss than the opposite choice. This value has been used in the following radar target tracking results.

5.4 Tracking Performance

We study the PDAF-based tracking system applied to the "Gunnerus" scenario in this section. The resulting tracks are shown in Fig. 10.

First, observe that many of the tracks originate from land. The tracks stay close to or on land as long as the feature is in range of the radar. When the feature is outside of the radar range, due to the movement of Telemetron or radar range adjustments, some tracks start to move out into open sea. Since they are out of range of the radar, no measurements are associated with them, and they are terminated after 5 scans without measurements. The average length of a confirmed track which is eventually

A Target Tracking System for ASV Collision Avoidance Based on the PDAF

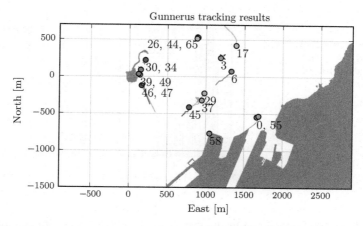

Fig. 10 Resulting tracks from the "Gunnerus" scenario. The circles mark the start of the tracks. The track labels are numbered in the order they are initiated. There are a total of 18 confirmed tracks

terminated is 22.3 scans. When excluding the tracks that are assumed to originate from actual targets, this is reduced to 15.7.

There is also an additional 47 tracks that die at the preliminary stage from the track initiation test. This brings the total number of confirmed and preliminary tracks to 65 for this scenario.

Track 3 is on Gunnerus. It remains stable on Gunnerus during the entire experiment. Tracks 6, 29, 37 and 45 originate from another target, which we refer to as the ToO-boat, for target of opportunity. This target was not equipped with an AIS transponder, and we have not been able to identify its callsign. We notice the open interval between tracks 6 and 29, which could be due to ToO-boat passing through the radar shadow of Gunnerus. Another collection of tracks, 26, 44 and 65, originated from a buoy northwest of Gunnerus. The origin of track 17 is somewhat uncertain, but there is some evidence that it may be due to multi-path returns from Gunnerus, reflected by the mast or other installations on Telemetron.

Although 18 confirmed tracks may seem dramatic in a scenario where only two proper targets were present, it should be noted that most of these tracks originate from land, and stays close to it until they are terminated. This means that although the maps used to mask out detections from land is not perfect, they are sufficient for our purposes.

5.5 Clustering Radius Effect on Number of Measurements

The tracking results can be used to investigate how the assumption of maximum-one measurement per target holds. This is done by considering the tracks of the two known targets in Fig. 10. The number of measurements inside the validation gates

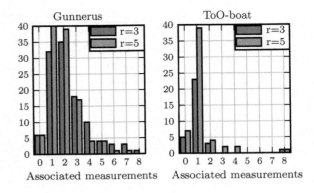

Fig. 11 Number of measurements in the validation gate for clustering radii 3 and 5 m. Gunnerus on the *left* and the ToO-boat on the *right*

are counted, and Fig. 11 shows the number of gated measurements for Gunnerus and the target of opportunity with different clustering radii. There is clearly a difference between the boats, with Gunnerus having a heavier tail of measurements. We believe this to be caused by Gunnerus having both a significantly larger area than the ToO-boat, and being closer on the radar. Both of these effects make Gunnerus occupy more radar cells than the ToO-boat, which can lead to a more fragmented target if the clustering does not capture all the effects. It can also be caused by more subtle radar effects, such as sidelobes from the radar.

6 Discussion and Topics for Future Research

The results presented in the previous section lead to some implications and suggestions for radar-based tracking in the context of maritime COLAV. We summarize these reflections in the four subsequent paragraphs.

Multi-target issues did not apear to be of critical importance in these experiments. Both in the "Trondheimsfjord II" scenario and the "Gunnerus" scenario, we never came across situations that involved confusion between tracks beyond what a parallel bank of PDAFs can handle.

Track continuity appears to be a more important issue. This is also a matter of data association, since deciding to keep or kill a track boils down to a decision about measurement origin hypotheses. The tracking community has a somewhat ambivalent attitude towards the importance of track continuity. While track continuity is a cornerstone of classical tracking methods such as PDAF, more recent methods based on random finite sets [10] do not provide continuous tracks. In any case, it could be argued that the track on ToO-boat should have been maintained for the 22 scans during which it resides in Gunnerus' radar shadow. However, standard existence-based tracking methods such as the IPDA would give the track extremely small existence probabilities under these circumstances, even if poor visibility was accounted for [3].

Target extent also appears to deserve closer scrutiny. The detrimental effect of radar shadows could be mitigated by modeling the radar shadow as part of the measurement model, by means of knowledge about the shadowing target's footprint. The radar footprint of an extended target may also carry other useful information that could aid both data association and state estimation.

The results demonstrate how false tracks inevitably occur in a tracking system. On the one hand, a COLAV system should be designed so that this has a minimal impact. On the other hand, more refined track quality measures can be useful to remove or discriminate false tracks. The track existence concepts of the IPDA [4] and random set methods [10] is a good starting point for this, and additional measures based on "track coherence" [3], consistency of target footprint, etc., may also be explored.

7 Conclusion

In this paper, we have described all the building blocks of a PDAF-based radar tracking system for usage onboard an ASV. Experimental results demonstrate reasonable performance of the tracking system. While the tracking system described in this paper is mature enough to be used as part of experimental work on COLAV, further developments, especially with regard to track continuity, extended targets and track quality measures, should be conducted to make the tracking system truly reliable for COLAV.

Acknowledgements This work was supported by the Research Council of Norway through the projecs 223254 (Centre for Autonomous Marine Operations and Systems at NTNU) and the project 244116/O70 (Sensor Fusion and Collision Avoidance for Autonomous Marine Vehicles). The authors would like to express great gratitude to Kongsberg Maritime and Maritime Robotics for placing high-grade navigation technology and the Telemetron vehicle at our disposal, and especially Thomas Ingebretsen for help with implementing the software interfaces.

References

1. Bar-Shalom, Y., Li, X.R.: Multitarget-Multisensor Tracking: Principles and Techniques. YBS Publishing, England (1995)
2. Bar-Shalom, Y., Li, X.R., Kirubarajan, T.: Estimation with Applications to Tracking and Navigation: Theory Algorithms and Software. Wiley, New York (2001)
3. Brekke, E., Hallingstad, O., Glattetre, J.: The signal-to-noise ratio of human divers. In: Proceedings of OCEANS'10, Sydney, Australia (2010)
4. Challa, S., Morelande, M.R., Mušicki, D., Evans, R.J.: Fundamentals of Object Tracking. Cambridge University Press, Cambridge (2011)
5. Fossen, T.I.: Handbook of Marine Craft Hydrodynamics and Motion Control. Wiley, New York (2011)
6. Friedman, J.H., Bentley, J.L., Finkel, R.A.: An algorithm for finding best matches in logarithmic expected time. ACM Trans. Math. Softw. (TOMS) **3**(3), 209–226 (1977)

7. Gandhi, P.P., Kassam, S.A.: Analysis of CFAR processors in non-homogeneous background. IEEE Trans. Aerosp. Electron. Syst. **24**(4), 427–445 (1988)
8. Harati-Mokhtari, A., Wall, A., Brooks, P., Wang, J.: Automatic identification system (AIS): data reliability and human error implications. J. Navig. **60**(03), 373 (2007)
9. Li, X.R., Jilkov, V.P.: Survey of maneuvering target tracking. Part I: dynamic models. IEEE Trans. Aerosp. Electron. Syst. **39**(4), 1333–1364 (2003)
10. Mahler, R.: Statistical Multisource-Multitarget Information Fusion. Artech House, Norwood (2007)
11. Muja, M., Lowe, D.G.: Scalable nearest neighbor algorithms for high dimensional data. IEEE Trans. Pattern Anal. Mach. Intell. **36**(11), 2227–2240 (2014)
12. Niedfeldt, P.C., Beard, R.W.: Multiple target tracking using recursive RANSAC. In: American Control Conference (ACC), Portland, OR, USA, pp. 3393–3398 (2014)
13. Pulford, G.W.: Taxonomy of multiple target tracking methods. IEE Proc. Radar Sonar Navig. **152**(5), 291–304 (2005)
14. Quigley, M., Conley, K., Gerkey, B., Faust, J., Foote, T., Leibs, J., Wheeler, R., Ng, A.Y.: ROS: an open-source robot operating system. In: ICRA Workshop on Open Source Software. Kobe, Japan (2009)
15. Svec, P., Thakur, A., Raboin, E., Shah, B.C., Gupta, S.K.: Target following with motion prediction for unmanned surface vehicle operating in cluttered environments. Auton. Robots **36**(4), 383–405 (2014)
16. Thrun, S., Burgard, W., Fox, D.: Probabilistic Robotics. MIT Press, Cambridge (2005)
17. Vo, B.-N., Mallick, M., Bar-Shalom, Y., Coraluppi, S., Osborne, R., Mahler, R., Vo, B.-T.: Multitarget tracking. Wiley Encyclopedia of Electrical and Electronics Engineering. Wiley, New York (2015)
18. Wolf, M.T., Assad, C., Kuwata, Y., Howard, A., Aghazarian, H., Zhu, D., Thomas, L., Trebi-Ollennu, A., Huntsberger, T.: 360-degree visual detection and target tracking on an autonomous surface vehicle. J. Field Robot. **27**(6), 819–833 (2010)

Detection and Tracking of Floating Objects Using a UAV with Thermal Camera

Håkon Hagen Helgesen, Frederik Stendahl Leira, Tor Arne Johansen and Thor I. Fossen

Abstract This paper develops a vision-based tracking system in unmanned aerial vehicles based on thermal images. The tracking system are tailored toward objects at sea and consists of three main modules that are independent. The first module is an object detection algorithm that uses image analysis techniques to detect marine vessels in thermal images and extract the center of each object. Moreover, as long as the size of the vessel is known or computed in an image where the whole vessel is visible, the center can be identified in situations where only a part of the object is visible. The pixel position of the center is used in a nonlinear state estimator to estimate the position and velocity in a world-fixed coordinate frame. This is called the filtering part of the tracking system. The state estimator is nonlinear because only two coordinates in the world-frame can be computed with the pixel coordinates. This originates from the fact that cameras are bearing-only sensors that are unable to measure range. The last module in the tracking system is data association, which is used to relate new measurements with existing tracks. The tracking system is evaluated in two different case studies. The first case study investigates three different measures for data association in a Monte Carlo simulation. The second case study concerns tracking of a single object in a field experiment, where the object detection algorithm and the filtering part of the tracking system are evaluated. The results show that the modules in the tracking system are reliable with high precision.

H.H. Helgesen (✉) · F.S. Leira · T.A. Johansen · T.I. Fossen
NTNU AMOS, Department of Engineering Cybernetics, Norwegian University of Science and Technology, O.S. Bragstads plass 2D, 7491 Trondheim, Norway
e-mail: hakon.helgesen@ntnu.no

F.S. Leira
e-mail: frederik.s.leira@ntnu.no

T.A. Johansen
e-mail: tor.arne.johansen@ntnu.no

T.I. Fossen
e-mail: thor.fossen@ntnu.no

© Springer International Publishing AG 2017
T.I. Fossen et al. (eds.), *Sensing and Control for Autonomous Vehicles*,
Lecture Notes in Control and Information Sciences 474,
DOI 10.1007/978-3-319-55372-6_14

1 Introduction

Detection and tracking of objects and structures with visual sensors have been studied thoroughly through the last decades. This is because visual sensors are useful in a vast number of applications and industries, including the maritime sector, car industry, surveillance, monitoring, and obviously also in many other applications. Visual sensors can capture an enormous amount of information through images and that may be why machine and robotic vision are attractive research fields.

Manufacturers have realized that commercial visual sensors are used more and more, which has pushed the price and weight rapidly down in recent years. This has revealed the possibility of utilizing visual sensors in light-weight unmanned aerial vehicles (UAVs). Moreover, the computational capacity of small on-board computers grows every year and permits the use of cameras to perform complex tasks in real-time applications. Visual sensors can be exploited in many ways in UAV operations, including navigation [1, 2], sense and avoid technology [3, 4], inspection [5], object detection and tracking [6, 7], georeferencing [8, 9], and airborne simultaneous localization and mapping (SLAM) [10, 11].

UAVs equipped with a visual sensor can be beneficial at sea because information captured from the air may be very different from the information available at the sea surface. This can for instance be utilized in iceberg detection in the arctic or for detecting floating objects hidden in waves [12]. Moreover, UAVs can be used to monitor and conduct operations on its own or in cooperation with vessels at sea. One particular scenario where a UAV can be useful is as a scouting system for autonomous ship operations [13]. In order to avoid obstacles and plan the most effective path, autonomous ships need the location of floating objects near their desired path. UAVs can be used to detect and track floating objects and assist ships at sea.

Object detection is the problem of identifying objects of interest within an image. Detecting objects in images captured from a moving platform is different from detecting objects in images captured at rest, even though the goal is equal. Moreover, detecting objects in a maritime environment is not similar to identify, e.g., pedestrians at an intersection. Therefore, it is necessary to find viable solutions for a specific application in most cases, especially if the aim is a real-time solution working on an embedded computer with restricted memory and processing capacity. When looking for floating objects at sea, small structures, and vessels are expected to be the majority of the obstacles. It may be challenging to distinguish small objects from the sea surface and detect dark vessels with a regular day camera, particularly during poor lighting conditions. That is why sensors capturing images at another spectrum than the visible may be attractive. Vessels at sea have heat sources and materials that are warmer or with higher emissivity than the sea surface. Therefore, using thermal cameras for object detection at sea is a viable alternative.

Target tracking is the problem of estimating the position and most often also the velocity of an object. Since many objects may be present, multiple target tracking need to be considered. It is much harder than tracking a single target. This is because several objects may be located within the same image and all of these observations must be connected to an existing track (or a new track if the object is detected for the first time). In the literature, this is normally referred to as data association and it is the first part of multiple target tracking. The second part is the filtering problem where the measurements are used to estimate the position and velocity of all targets. The filtering part usually consists of individual state estimators for every object.

1.1 Related Literature

Object detection and tracking are studied thoroughly in the literature and a great number of surveys exist [14–17]. Covering all of the relevant research in one section is impossible. Therefore, this section will mainly be focused toward object detection and tracking with UAVs and for applications at sea.

Object detection and tracking of floating structures in images captured from a UAV operating at high speed are challenging because the velocity of the visual sensor typically exceeds the velocity of the objects significantly. Therefore, motion-based segmentation techniques are not appropriate unless one can compensate for the camera motion accurately. This requires knowledge of the navigation states of the UAV and time-synchronized data with high precision. This is not necessarily available and it is assumed that motion-based segmentation techniques are inappropriate in this paper. Thus, it is necessary to identify algorithms that work well under this assumption.

Feature-based techniques are an alternative and include popular point detectors, such as SIFT [18], SURF [19], and the KLT detector [20]. Point detectors are usually computationally effective and quite robust since they may be scale and rotation invariant. However, they need to be combined with another detection method because features may be located in places where objects are not present. Moreover, several features can be present on the same object. Therefore, it is necessary to identify the ones that belong to objects and not the background. Another issue is that it is easier to associate features between two subsequent images than in a large sequence of images. Therefore, using single features obtained with, e.g., SIFT, is not very robust for tracking purposes where you want to track objects over longer periods of time. This is obviously also a consequence of point detectors ability to find many features because association is harder to do on a large set of features. The large set is, on the other hand, very beneficial when calculating, e.g., optical flow [6].

A template matching approach for visual detection and tracking with UAVs is presented in [21]. However, in order to be robust, a great number of templates need to be available because template matching is neither scale nor rotation invariant. Moreover, the objects of interest need to be known beforehand, which is unavailable information in this scenario. Color segmentation has also been applied in object detection with UAVs [22], but requires tailored parameters toward a specific type of objects. Thus, it is not as robust since floating objects at sea can have many different appearances. Leira et al. [7] demonstrates a stepwise solution for detecting marine vessels in thermal images, which works well as long as the temperature difference between the sea surface and objects of interest is large.

For target tracking, several solutions exist both for the filtering part and the data association part. The greatest variation in object tracking is the distinction between tracking a single and multiple targets (if data association is necessary or not). The filtering part of object tracking consists of state estimators that estimate the position and velocity of the targets. Several alternatives are available, including stochastic methods, such as the Kalman filter and particle filter, and deterministic solutions such as observers. State estimation with applications for tracking is described thoroughly in [23] and includes multiple model approaches. A tracking strategy for vision-based applications on-board UAVs is described in [24], and a correlation method for ship detection with a visual sensor mounted in a UAV is presented in [25]. Tracking strategies for vessels at sea from UAVs equipped with a thermal camera are presented in [6, 7]. Other tracking strategies with UAVs are described in [9, 26, 27].

Data association has also been studied thoroughly and is closely related to classification. A general description of pattern recognition (classification) and machine learning with methods useful for data association is described in [28]. Data association in multiple target tracking with the recursive RANSAC algorithm is described in [27, 29], but requires the knowledge of the number of objects beforehand. Moreover, it can be computationally demanding. Nearest neighbor data association is described in [30] and will be utilized in this paper because of the limited processing capacity on the on-board computer.

1.2 Main Contributions

The robustness of vision-based tracking systems is a challenge because of issues like changes in illumination, occlusion, limited field of view, and real-time requirements. This paper will extend the tracking system in [7] and propose a more robust system for multiple object tracking.

A complete stepwise solution for multiple object detection and tracking is thoroughly described in this paper. A region-based detection algorithm will be used to extract bounding boxes for floating objects at sea. Moreover, the center of the object can be computed, even if just a part of the object is visible as long as the entire object has been available at some point beforehand. In addition, several properties of each object can be extracted in a feature vector for data association purposes. This vector is tailored toward the use of thermal cameras in maritime environments. The distance between the current estimate and a new observation is added to the feature vector to enable the possibility of distinguishing two equal objects. Three different distance measures are evaluated with the Mahalanobis distance to increase the robustness of the data association. A nonlinear tracking filter will be used to estimate the position and velocity of detected objects. The tracking system is divided into modules, which makes it possible to change and expand every part of the system.

The proposed system is evaluated in real-time simulations of data gathered at a flight experiment at sea. A marine vessel is detected and tracked in several image sequences. Furthermore, different distance measures for data association will be evaluated with Monte Carlo simulations in a separate case study.

1.3 Organization

The remainder of this paper is divided into four parts. Section 2 describes how objects are detected in images and the method is tailored toward the use of airborne thermal cameras at sea. Section 3 describes a tracking system that is able to track multiple targets where the number of targets is unknown beforehand. This includes both a data association part, which is necessary to connect existing targets with new observations and a filtering part that estimates the position and velocity of each target. Section 4 presents experiments that have been carried out to evaluate the performance of the detection and tracking system together with the results. The paper is concluded in Sect. 5.

2 Machine Vision

This section describes a machine vision system that utilizes thermal images to find floating objects at sea. The object detection module extracts the center of each object in the image frame and is described thoroughly in the first part of this section. It is more useful to estimate the position of the objects in a world-fixed frame. Therefore, a georeferencing algorithm can be used to transform the image frame coordinates to world-fixed coordinates. This is briefly described in the second part. The latter part of this section describes features that can be extracted in the machine vision system and useful for data association.

2.1 Object Detection

Object detection at sea using thermal images has not been described extensively in the literature even though some examples exist. In this section, an edge detection-based method for locating floating objects will be presented.

In order to automatically detect objects of interest, the thermal image, **I**, is first smoothed. This is conducted to reduce the thermal noise present in the image and was found to make the process of detecting edges in the image more robust than when performed on the raw thermal images. The smoothing is conducted by convolving (denoted by $*$) the image with a Gaussian kernel **g**. **g** is a $n \times n$ kernel approximating a Gaussian distribution with a standard deviation of σ_g, i.e.,

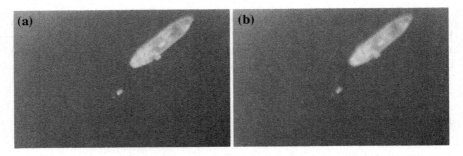

Fig. 1 Before (**a**) and after (**b**) smoothing of the original image. The image is showing a large boat (length of 55 m), a rigid-inflatable boat and a small buoy

$$\mathbf{I_s}[x, y] = (\mathbf{I} * \mathbf{g})[x, y] = \sum_{k=-\frac{n-1}{2}}^{k=\frac{n-1}{2}} \sum_{m=-\frac{n-1}{2}}^{m=\frac{n-1}{2}} \mathbf{I}[x - m, y - k]\mathbf{g}[m, k] \quad (1)$$

$\mathbf{I}$ and $\mathbf{I_s}$ are $w \times h$ matrices, where w and h are the width and height of the original thermal image. $[x, y]$ are integers representing a pixel coordinate in the image, and $[m, k]$ are integers representing a coordinate in the Gaussian kernel approximation $\mathbf{g}$. The result of smoothing an image showing a big boat (length of 55 m), a rigid-hulled inflatable boat (RHIB) and a small buoy can be seen in Fig. 1. Notice that the upper left corner has slightly brighter pixels than the rest of the image, due to inaccurate camera calibration. Now, to detect the edges in the resulting smoothed image $\mathbf{I}_s$, the gradient image, $\mathbf{G}$, of $\mathbf{I}_s$ is calculated. The gradient image of $\mathbf{I}_s$ is found by the following calculation

$$\mathbf{G_{I}}_x[x, y] = (\mathbf{I}_s * \mathbf{P})[x, y] = \sum_{k=-1}^{k=1} \sum_{m=-1}^{m=1} \mathbf{I}_s[x - m, y - k]\mathbf{P}[m, k],$$

$$\mathbf{G_{I}}_y[x, y] = (\mathbf{I}_s * \mathbf{P}^T)[x, y] = \sum_{k=-1}^{k=1} \sum_{m=-1}^{m=1} \mathbf{I}_s[x - m, y - k]\mathbf{P}^T[m, k], \quad (2)$$

$$\mathbf{G_I}[x, y] = \sqrt{\mathbf{G_{I}}_x^2[x, y] + \mathbf{G_{I}}_y^2[x, y]}$$

$\mathbf{P}$, also referred to as the Prewitt operator [31], is defined as the 3×3 matrix

$$\mathbf{P} := \begin{bmatrix} -1 & 0 & 1 \\ -1 & 0 & 1 \\ -1 & 0 & 1 \end{bmatrix} \quad (3)$$

The resulting gradient image $\mathbf{G_I}$ can be seen in Fig. 2a. It is seen that the big boat, the RHIB and the small buoy are clearly visible after these image processing operations. However, it is apparent that the waves and ripples in the ocean in addition to some of

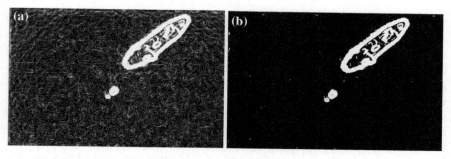

Fig. 2 Before (**a**) and after (**b**) gradient magnitude thresholding

the noise in the image are still visible, albeit smaller in magnitude than the objects of interest. Because of this, removing them can be done by using a threshold value for the magnitude of the gradients. That is, all pixels in the gradient image that have a magnitude less than a certain threshold T_g can be removed. This is achieved by the following operation

$$\mathbf{G_I}(x, y) = \begin{cases} maxValue & \text{if } \mathbf{G_I}(x, y) \geq T_g \\ 0, & \text{otherwise} \end{cases} \quad (4)$$

where $maxValue$ is the maximum brightness value a pixel in $\mathbf{G_I}$ can have. From Fig. 2b it is readily seen that, post processing, it is mostly objects with a distinct heat signature that are left in the image.

Looking at Fig. 3a, it is obvious that some of the blobs clearly do not originate from any object of interest (i.e., the small dots scattered across the image), and therefore have to be filtered out. To filter out the unwanted blobs from the image, a connected component algorithm [32] is used to group and label components together in blobs. Furthermore, the area of each blob is calculated, and blobs with a smaller or larger area than what is expected from an object of interest are removed from the image.

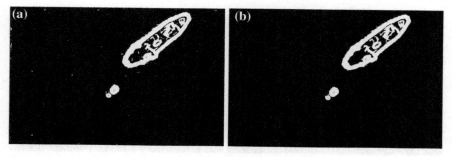

Fig. 3 Before (**a**) and after (**b**) removing blobs which are either too small or too large to be an object of interest

The result of this process is seen in Fig. 3. The resulting image (Fig. 3b) is hereby referred to as the binary image, **B**, of the original image **I**.

After applying the image analysis methods just described and finding the bounding boxes for each detected object, the detected objects can be seen in Fig. 4a. However, it is seen that big objects which have some texture in them can trigger detections within the interior of the actual object. This is because the texture of the object shows up in the edge detector. In order to make every detection of an object only show up as one unique detection, bounding boxes completely contained in a larger bounding box are removed. The result of this process is seen in Fig. 4b. Looking at Fig. 4b, it is apparent that the three detected objects are of further interest, and they are now ready for further evaluation.

The detection step provides the position of objects of interest in the image. However, since the detector is using edge detection, the areas of an image that is highlighted as interesting will often only contain the exterior edges of an object. When performing for instance recognition based on characteristics, such as size, average temperature, and overall form, it is crucial that the whole object is evaluated. To expand the detections to also include the interior of the objects of interest, an algorithm that seeks to fill holes in the binary image [31] is applied. The result of applying this algorithm can be seen in Fig. 5.

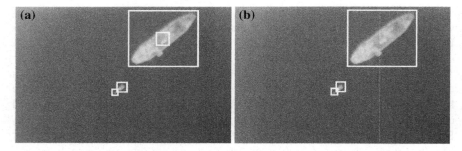

Fig. 4 Before (**a**) and after (**b**) removing detections completely contained in the interior of other detections

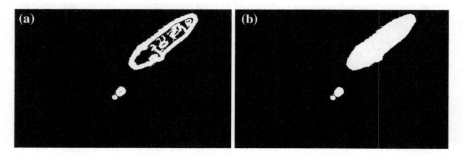

Fig. 5 Before (**a**) and after (**b**) filling the interior holes in the detected objects

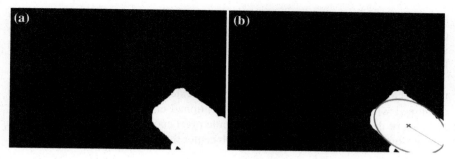

Fig. 6 Partially visible object (**a**), and its corresponding ellipse approximating the object's form (*red circle*), object center (*red cross*), and the orientation of the major axis of the ellipse (*blue line*) illustrated in image (**b**)

Using the location of the bright pixels in the binary image seen in Fig. 5b, the center positions of the remaining blobs are calculated and passed on to the tracking module as measurements.

Note that in some cases, the center of the object blob detected in an image is not necessarily the same as the center of the detected object. Such an example can be seen in Fig. 6. The calculated blob center is illustrated with a red cross in Fig. 6b, and since the boat is only partly visible in the image, the blob center does not coincide with the true object center. In order to adjust the calculated blob center to a pixel coordinate which is closer to the actual object center, the orientation of the detected blob is calculated by approximating the detected blob as an ellipse with a minor and a major axis. This can be done by using the second-order central moments to construct a covariance matrix for the detected blob's pixels. The $(q + p)$th order central moment of a blob or an image region $\mathbf{O}$ is here given as [32]

$$m_{pq} = \sum_{x,y \in \mathbf{O}} x^p y^q \mathbf{B}(x, y), \quad p, q = 0, 1, \ldots \quad (5)$$

The second-order covariance matrix of an image region $\mathbf{O}$ can be calculated as [32]

$$cov[\mathbf{O}(x, y)] = \begin{bmatrix} m_{20} & m_{11} \\ m_{11} & m_{02} \end{bmatrix} \quad (6)$$

Now, the eigenvectors of this matrix correspond to the major and minor axes of the ellipse approximating the detected object blob in the image region $\mathbf{O}$, hence the angle between these two eigenvectors can be used to approximate the orientation of the detected blob. Figure 6b shows the ellipse constructed using the two eigenvectors as major and minor axes for the approximation of the detected blob's form, as well as the calculated blob center marked with a red cross. The blue line shows the direction of the major axis of the blob, and can be used as a measurement of the direction.

Note that in order for this process to yield information about the orientation of an object, the object has to be noncircular, and a sufficient part of the object has to be visible so that the length of the major axis of the blob is larger than the expected full length of the minor axis.

Assuming that the full length of the partially detected object is already known, the expected length of the boat in pixels can be calculated. Using, this information combined with the results of the above mentioned calculations, a more accurate estimate of the detected object's center in the pixel coordinate frame can be found. This is done by moving the calculated blob center the following amount

$$\Delta \mathbf{c} = \frac{\mathbf{L}}{2} - (\mathbf{a} - \mathbf{i}) \tag{7}$$

where $\mathbf{L}$ is the expected object length decomposed in the image plane according to the approximated orientation, i.e., $\|\mathbf{L}\| = L$, where L is the expected length of the boat. Further, $\mathbf{a}$ is the major axis and $\mathbf{i}$ is the vector from the blob's centroid to the closest intersection with the image frame along the direction of the major axis, illustrated as a blue line in Fig. 6. Adding $\Delta \mathbf{c}$ to the calculated blob center will effectively move the detected object's center along the calculated major axis of the blob toward the expected location of the object center. This can sometimes lead to the object center being computed to be located outside the image borders, but this is not an issue since the theoretical camera model is not restricted by the field of view.

2.2 Georeferencing

Georeferencing can be defined as the transformation from a pixel position in the image frame to world-fixed coordinates. It might be very advantageous in UAVs because the pixel position of an object contains a little amount of information when the scene (location of the image frame) changes continuously. This is the main motivation behind georeferencing and is described thoroughly in [6, 8, 9, 33]. However, in this scenario, where we are interested in continuous tracking of floating objects (and not acquire "standalone" measurements of world-fixed coordinates), georeferencing is not necessary because it happens indirectly in the nonlinear measurement model described in Sect. 3.1. Nevertheless, georeferencing has worked well in the aforementioned papers, and therefore, it will be used as a benchmark in this paper. Note that georeferencing without a range measurement is only possible as long as all points in the image are located in the same plane. This is the case at sea and that is why georeferencing can be very useful in maritime environments.

2.3 Extracting Feature Vector for Classification and Data Association

When multiple objects are tracked simultaneously, it is necessary to associate new measurements with objects that are being tracked. In other words, it is necessary to find out which object a new measurement belongs to. This can be achieved by extracting features for detected objects through image analysis and is described extensively in [7], but the main motivation is to use features in the image that can distinguish between several objects. However, complex machine vision routines are challenging to run in real-time on resource-limited embedded computers on-board UAVs. Therefore, it is important to extract features that are computationally effective, but still robust in the sense of data association. Moreover, features should be scale and rotation invariant because the sensor position and field of view change continuously. A feature vector is reliable if it describes an object uniquely with high accuracy. In other words, two different objects should never have feature vectors that are similar. However, this may be the case if features are only gathered from the images. Consider a situation where two circles of identical temperature and size are located in two locations in a thermal image. It is very hard to find unique features for similar objects, just by using image properties. Therefore, it is important to find an additional feature that not directly depends on the image properties. This can for example be some sort of position measure and is described more closely in Sect. 3.2. Note that a position measure in the image plane is not necessarily reliable because the location in the image changes continuously with the UAV. Hence, the distance measure should be based on world-fixed coordinates.

Thermal images consist of pixels where a single channel is used to describe the intensity (temperature) of each pixel. Image features that can be used for data association are very diverse and can include information about intensity, the gradient, edges and of course many others. In [7], the area, average temperature and the first normalized central moment proposed by Hu [34] are features used for data association. They work well in thermal images and can be combined with a distance measure. A global nearest neighbour search can be used to associate the feature vector of detected objects with existing tracks. This is described extensively in [7].

3 Multiple Target Tracking

This section describes a tracking system that is able to estimate the position and velocity of several objects simultaneously. The first part describes the filtering part of the tracking system. The latter part presents how data association is solved in order to track several objects automatically at the same time. In this section, it will be assumed that a set of observations is available from the machine vision system. Moreover, the image frame coordinates of the center are obtained by the object detection algorithm. Note that initialization of new tracks will not be covered in this paper. Leira et al. [7] covers this in more detail.

3.1 The Filtering Part of the Tracking System - State Estimation

In the filtering part of the tracking system, some challenges arise when images are captured from a moving platform with a pose (position and attitude) that changes rapidly. This is because the object position in the image will vary with the attitude and North–East-Down (NED) positions (also called world-frame position) of the UAV. Moreover, it is advantageous to track objects in world-fixed coordinates because it is less informative to track objects in the image plane when the position of the camera changes quickly. Therefore, it is necessary to know the pose of the UAV accurately in order to track objects in a world-fixed coordinate frame. Furthermore, time synchronization between the camera and UAV navigation system becomes crucial since it is necessary to know the pose of the UAV as close as possible to when an image is captured. How navigation uncertainty can be included in the tracking system is described in [35]. Moreover, the concept of simultaneous localization and mapping with moving landmarks is closely related to multiple target tracking so the airborne SLAM literature is also relevant [10, 11, 36]. However, in this paper, the UAV position and attitude are assumed to be known from the autopilot, but the time synchronization is somewhat uncertain. This will obviously affect the performance and robustness of the tracking system in situations, where the navigation data are unreliable, but this is not the focus of this paper.

Mono-cameras are only able to capture measurements of the bearing and not the range. This originates from the fact that only two coordinates in the world-frame can be acquired with two image coordinates when the range is unknown. This is not critical at sea because floating objects are located on the sea surface where the altitude is known. Therefore, all objects can be assumed to be located in the same plane. This is referred to as the flat-earth assumption, which theoretically makes it possible to find the horizontal world coordinates as a function of the UAV attitude, altitude and image plane coordinates. This is most often the basis for georeferencing unless a range measurement is available. However, this assumption is fragile for variations in altitude and errors in the navigation states. The navigation states are known from an extended Kalman filter (EKF), but are not necessarily entirely correct even though the accuracy is expected to be reliable. Moreover, because the navigation states are available from the EKF, the error is most likely not Gaussian and could perhaps be correlated from frame to frame. This might violate the Gaussian assumptions in the Kalman filter. In addition, georeferencing is a nonlinear transformation where the true measurement is the pixel position and not the NE positions. Thus, the noise related to the true measurement (pixel position) theoretically has to go through the same transformation to find the measurement covariance in the Kalman filter and that is very challenging. Therefore, the measurement noise covariance is normally a matter of trial and error and not directly related to the accuracy of the sensors. Considering these issues, it is natural to look for another solution where the georeferencing operation is avoided, but where one still can take advantage of the fact that objects are located on the sea surface. This leads to a nonlinear measurement model, which entails the use of a nonlinear tracking filter, such as the extended Kalman filter (EKF). The goal of the filtering part is to estimate the position and velocity of an object with a state estimator. Since one instant of the state estimator will be created for every object, it is only necessary to consider how this is achieved for a single target.

Target Motion Model

In stochastic state estimation it is necessary to define a motion model for how the states are evolving. How to choose a motion model is described in [37] and depends on the type of targets that are expected to be detected. In this paper, a constant velocity model (white noise acceleration) is chosen because the dynamics of floating objects are assumed to be slow. The goal is to estimate the position and velocity in the NE plane. The discrete-time constant velocity motion model at time step $[k]$ is defined as

$$\mathbf{x}^t[k+1] = \mathbf{F}^t \mathbf{x}^t[k] + \mathbf{E}^t \mathbf{v}^t[k] \tag{8}$$

where $\mathbf{x}^t = [p_t^N, p_t^E, v_t^N, v_t^E]^\top$ is the state vector consisting of the target position and velocity in the NE frame, and $\mathbf{v}^t = [v_v^N, v_v^E]^\top$ is assumed to be zero-mean Gaussian white noise with covariance $\mathbf{Q}$. $\mathbf{F}^t$ and $\mathbf{E}^t$ are the system matrices defined as

$$\mathbf{F}^t = \begin{bmatrix} 1 & 0 & T & 0 \\ 0 & 1 & 0 & T \\ 0 & 0 & 1 & 0 \\ 0 & 0 & 0 & 1 \end{bmatrix}, \quad \mathbf{E}^t = \begin{bmatrix} \frac{1}{2}T^2 & 0 \\ 0 & \frac{1}{2}T^2 \\ T & 0 \\ 0 & T \end{bmatrix}$$

where T is the sampling period of the camera. The down position is zero for surface objects at sea (given that the origin of NED is placed at sea level) and not a part of the state vector. Note that the motion model is linear, which enables the use of the standard equations for time update (prediction) in the Kalman filter.

Tracking System Based on the Extended Kalman Filter

The bearing-only measurement model is based on the pinhole camera model. It relates the image plane coordinates (u, v) to a camera-fixed coordinate frame $\{C\}$ (see Fig. 7):

$$\mathbf{z}_1(k) = \begin{bmatrix} z_{1_1} \\ z_{1_2} \end{bmatrix} = \begin{bmatrix} u \\ v \end{bmatrix} = \underbrace{\frac{f}{z^c} \begin{bmatrix} x^c \\ y^c \end{bmatrix}}_{\mathbf{h}_1}, \quad z^c \neq 0 \tag{9}$$

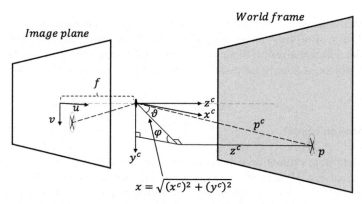

Fig. 7 Illustration of the azimuth (φ) and elevation (ϑ) angles. The camera points straight down from the image plane to the world-frame where z^c is the range from the camera to the plane the pixel is located in. This is simply the altitude of the UAV with zero roll, pitch, and gimbal angles

where f is the focal length of the camera lens. The image plane coordinates (u, v) are measured by the detection module. The camera-fixed coordinates $[x_t^c, y_t^c, z_t^c]^\top$ of the target are used to relate the world-fixed coordinates of the camera ($\mathbf{p}_{uav}^n$) and the target ($\mathbf{p}_t^n$) through the following model

$$\mathbf{p}_t^c = \begin{bmatrix} x_t^c \\ y_t^c \\ z_t^c \end{bmatrix} = \mathbf{R}_n^c(\mathbf{p}_t^n - \mathbf{p}_{uav}^n) \tag{10}$$

where $\mathbf{R}_n^c$ is the rotation matrix from {C} to NED. In the filtering part, $\mathbf{h}_1$ will be used as the measurement model where the covariance can be chosen as a diagonal matrix $\mathbf{R}$ with a pixel uncertainty related to each pixel coordinate. Note that it is necessary to insert (10) in (9) to get a measurement model that depends on the UAV attitude, gimbal orientation, UAV NED positions, and the NED positions of the target. This is beneficial because the measurement model now directly depends on parameters of interest, and not the camera-fixed coordinates of the target.

The bearing-only observation model can also be represented in terms of azimuth (φ) and elevation (ϑ) angles

$$\mathbf{z}_2(k) = \begin{bmatrix} z_{2_1} \\ z_{2_2} \end{bmatrix} = \begin{bmatrix} \varphi \\ \vartheta \end{bmatrix} = \underbrace{\begin{bmatrix} \arctan\left(\frac{y^c}{x^c}\right) \\ \arctan\left(\frac{z^c}{\sqrt{(x^c)^2+(y^c)^2}}\right) \end{bmatrix}}_{\mathbf{h}_2} = \begin{bmatrix} \arctan\left(\frac{\frac{1}{f_v}v}{\frac{1}{f_u}u}\right) \\ \arctan\left(\frac{f_u}{u}\cos(\varphi)\right) \end{bmatrix} \tag{11}$$

where f_u and f_v are camera parameters in the intrinsic camera matrix obtained by the lens specification (or camera calibration). The azimuth and elevation angles are illustrated in Fig. 7 and this representation of the measurement can be used for data association.

The measurement model (9) is nonlinear, and the use of a nonlinear state estimator is necessary. The most common and straightforward solution is to use an Extended Kalman filter (EKF) to estimate the position and velocity of the target. In order to use the EKF, it is necessary to find the Jacobian of $\mathbf{h}_1$ with respect to the states:

$$\frac{\partial \mathbf{h}_1}{\partial \mathbf{x}^t} = \begin{bmatrix} \frac{\partial z_{1_1}}{\partial x_t^n}|_{\hat{\mathbf{x}}_{k|k-1}^t} & \frac{\partial z_{1_1}}{\partial y_t^n}|_{\hat{\mathbf{x}}_{k|k-1}^t} & 0 & 0 \\ \frac{\partial z_{1_2}}{\partial x_t^n}|_{\hat{\mathbf{x}}_{k|k-1}^t} & \frac{\partial z_{1_2}}{\partial y_t^n}|_{\hat{\mathbf{x}}_{k|k-1}^t} & 0 & 0 \end{bmatrix} \qquad (12)$$

where $\hat{\mathbf{x}}_{k|k-1}^t$ is the predicted state $\mathbf{x}^t$ at the current time step. The last two columns of the Jacobian are zero because the measurement model $\mathbf{h}_1$ not depends on the target velocities.

The filtering part proposed here is somewhat more involved than a linear tracking filter based on georeferencing [6, 7] since it is nonlinear. Problems with linearization and initialization follow with the EKF and the measurements (pixel coordinates) might in many cases be less informative than the NE positions (for humans). On the other hand, the uncertainty related to the measurement model is much more properly defined in the nonlinear tracking filter. In the georeferencing approach, the true measurement (pixel position) goes through a complex transformation to obtain the NE positions that are used as a measurement in the tracking filter. This means that the covariance matrix of the measurement noise related to the pixel position should go through the same transformation. This is not necessary in this case.

3.2 Data Association

Data association is the task of assigning newly detected measurements to existing tracks, or creating a new track when an object is entering the image for the first time. Solving the data association problem is most often easy for the human eye and brain, but more difficult for computers. This is especially the case in situations where the scale and rotation of the object change with the movement of the UAV. Moreover, changes in illumination or appearance are challenges that are much harder to deal with for computers than humans. In addition, because of the restricted processing capacity, advanced and complex methods are not suitable for use in real-time solutions on-board UAVs. Therefore, fast algorithms are necessary and this is likely to affect the reliability and robustness of the association. A trade-off between robustness and processing time is most often inevitable. Luckily, a few errors in the association part may not be critical because the filtering part will remove measurements that are outliers.

In this paper, data association will be based on [7] and the global nearest neighbor (GNN) search. GNN is an approach where new measurements are associated with existing objects (tracks) that are closest in distance with respect to a predefined similarity measure. The similarity measure can be based on several different features including distance (from measurement to estimated position), size, intensity, color distribution and so forth. The least squares distance or a similar measure is used to connect new measurements with existing tracks. In situations where a new measurement is far from every existing track with respect to the similarity measure (a threshold must be assigned), a new object is most probably detected. Thus, a new track is added to the tracking system. The search is called global because all measurements are considered together to find the optimal solution for the entire set of measurements at each time instant. If association is not handled globally, the order of which measurements are received will affect the result significantly because the first observation can be assigned to an object the second observation is closer to.

In Sect. 2.3 it was emphasized that data association should not only be based on image properties because there is a possibility of two similar objects being present within the same image. Therefore, it is necessary to add another measure to the association process. In target tracking, a tracking gait (an ellipse based on the mean and covariance of the estimates) is usually created around each object. An observation of an object is expected to fall within the tracking gait. This is usually described in terms of the Mahalanobis distance

$$\gamma = \mathbf{v}^\top \mathbf{S}^{-1} \mathbf{v} \tag{13}$$

where ν (innovation) and $\mathbf{S}$ (innovation covariance) are defined as

$$\begin{aligned} \mathbf{v} &= \mathbf{z}^t - \mathbf{h}^t(\mathbf{p}_{uav}^n, \mathbf{R}_n^c, \mathbf{p}_{target}^n) \\ \mathbf{S} &= \frac{\partial \mathbf{h}^t}{\partial \mathbf{x}^t} \mathbf{P}^t \left(\frac{\partial \mathbf{h}^t}{\partial \mathbf{x}^t}\right)^\top + \mathbf{R} \end{aligned} \tag{14}$$

γ is now a measure of the "inverse likelihood" that a certain observation is related to a specific object. The minimal Mahalanobis distance specifies the object that most likely is the origin of the measurement. Since the Mahalanobis distance is a normalized quantity, it can be combined with the information gathered in the feature vector to do the GNN least squares search for data association. This is described thoroughly in [7] and will not be the main focus here. In this paper, the focus will be directed toward finding the most robust measure for the Mahalanobis distance because the innovation can be computed in different ways. Three models for the innovation are presented and evaluated.

The Linear Case

When the tracking is conducted in a Kalman filter where georeferencing is used to obtain measurements, the innovation in (14) can be calculated directly as the difference between the measurement and the predicted position for the tracked object. The innovation covariance can be extracted from the Kalman filter. The only difference for equation (14) is that the innovation covariance is a linear function of the matrix **H** so linearization is avoided. This is a very straightforward approach and works well in [7]. However, this method can only be used in a linear tracking filter. Therefore, it is important to find an alternative solution with equal or better performance in the nonlinear case. This method is, therefore, used as a benchmark. It will be referred to as the linear association method (or georeferencing) in the rest of this paper.

The Nonlinear Case

In the nonlinear case, two different representations of the innovation can be used to find the Mahalanobis distance. The first method is based on (9) and the second on (11). Both representations have advantages and disadvantages. Bryson and Sukkarieh [11] argues that the azimuth and elevation representation is beneficial because it is more accurate in the border of the image. However, the situation might be different when all objects of interest are located in the same plane and perhaps very close together as well. Therefore, both representations will be evaluated.

The first representation is based on the pinhole measurement model and referred to as the pixel representation. The innovation is calculated as

$$v_1 = \begin{bmatrix} u \\ v \end{bmatrix} - \underbrace{\frac{f}{z^c} \begin{bmatrix} x^c \\ y^c \end{bmatrix}}_{\mathbf{h}_1} \tag{15}$$

where (x^c, y^c, z^c) is calculated from (10). It is obviously important that the pixel coordinates (u, v) are represented in meters and this can be achieved by the camera and lens specifications. The innovation covariance is calculated with (14) where the measurement function $\mathbf{h}_1$ is given by (9) and linearized with respect to the position of the target.

The second representation is based on the azimuth and elevation angles and referred to as the angle representation. The innovation can be calculated as

$$v_2 = \begin{bmatrix} \arctan\left(\frac{\frac{1}{f_v}v}{\frac{1}{f_u}u}\right) \\ \arctan\left(\frac{f_u}{u}\cos(\varphi)\right) \end{bmatrix} - \underbrace{\begin{bmatrix} \arctan\left(\frac{y^c}{x^c}\right) \\ \arctan\left(\frac{z^c}{\sqrt{(x^c)^2+(y^c)^2}}\right) \end{bmatrix}}_{\mathbf{h}_2} \tag{16}$$

The innovation covariance are calculated with (14) where the measurement function $\mathbf{h}_2$ is linearized with respect to the position of the target. Note that the analytical Jacobian is quite complex for both $\mathbf{h}_1$ and $\mathbf{h}_2$. Therefore, it might be clever to evaluate the Jacobian numerically to reduce the computation time.

4 Experiments and Results

This section describes two different case studies that have been conducted to evaluate separate parts of the tracking system. The first case study looks into data association and the methods described in Sect. 3.2. The second case study looks into tracking of a single target.

4.1 Case Study 1 - Evaluation of the Data Association Methods

The main motivation behind the first case study is to compare the different methods for data association in terms of the Mahalanobis distance. Note that only the distance measures described in Sect. 3.2 will be used so image properties are not relevant in this case study. Evaluation of the methods is achieved through Monte Carlo simulations. Two objects at rest that are located 10 (north) and 15 m (east) from each other are simulated (mean values). Gaussian noise with a standard deviation of 10 m is added to the north and east initial positions for both objects. Therefore, the true distance between the objects will vary for each simulation to test the robustness of the data association in different situations. The true pixel positions in the image are calculated with the navigation states of the UAV and the position of the objects for each time step (with backwards georeferencing), before Gaussian noise (properties described in Table 1) is added to the true pixel positions and used as measurements. Moreover, white noise with properties described in Table 1 is added to the navigation states of the UAV so that georeferencing will lead to different north and east positions than the true ones. This ensures a realistic simulation where there are uncertainty related to both the image detection algorithm and the navigation states of the UAV.

One object (referred to as the main object) is tracked with a Kalman filter so that the estimated position and covariance are available. The goal is to find out which of the two measurements that belong to the main object at each time instant. 141 images are processed for each Monte Carlo simulation and the association is either right or

Table 1 Covariance of the Gaussian white noise added to the UAV navigation states and the object detection algorithm in case study 1

State	Covariance Gaussian noise
North and east positions UAV	$(2\,\text{m})^2$
Down position UAV	$(5\,\text{m})^2$
Roll and pitch UAV	$(1°)^2$
Yaw UAV	$(2°)^2$
Gimbal pan and tilt angles	$(1°)^2$
Uncertainty detection algorithm (for the image plane coordinates)	$(2\,\text{pixels})^2$

Table 2 Initial conditions and covariance of the Gaussian white noise related to the measurements for the different association methods in discrete-time

Property	Covariance
North and east initial positions main object	$(10\,\text{m})^2$
North and east initial velocities main object	$(\sqrt{20}\,\text{m/s})^2$
Uncertainty georeferencing north and east positions	$(10\,\text{m})^2$
Pixel uncertainty horizontal coordinate (measurement model (9))	$360\,\text{pixels}^2$
Pixel uncertainty vertical coordinate (measurement model (9))	$160\,\text{pixels}^2$
Uncertainty azimuth angle	$(6°)^2$
Uncertainty elevation angle	$(3°)^2$

wrong for every image. The success rate (number of right associations divided by the number of images) will be the most important measure for the performance. The states for the main object go through 50 iterations in the Kalman filter initially (before association begins) to ensure that the estimates are somewhat consistent before data association is conducted. The measurements for these iterations are generated as the true position where Gaussian noise with standard deviation of 10 m in both the north and east positions are added. However, the estimated position for the main object is not more accurate than what realistically can be expected (as will be shown in the results) in a real situation. 10000 Monte Carlo simulations are conducted.

In order to conduct the Monte Carlo simulations, some design properties need to be stated. The frame rate of the camera is 0.1 s and, thus, each simulation consists of a time period of 14 s (first image starts at zero). The navigation states of the UAV are sampled at the same frequency. Image capture and sampling of the navigation states are not synchronized and, therefore, navigation states at the exact time instant when an image is captured are in general not available. The initial conditions and measurement noise related to each association method are summarized in Table 2. Note that all states are initially uncorrelated (only zeros on the off-diagonal in the covariance matrices). The navigation states are sampled at a field experiment and the objects have been simulated in locations which are visible in the image plane. The covariance of the measurement noise (the bottom four elements in Table 2) is quite large, but that is mainly because the accuracy of the navigation states is limited and thus the measurements are quite poor at times. In other words, the magnitude of the covariance is caused by the navigation states and not the accuracy of the object detection algorithm. The payload used in the field experiment is described more closely in [8].

Results - Evaluation of Data Association Methods

Table 3 presents the main result for the first case study. The success rate for the different methods is very similar, which means that all of the representations are applicable

Table 3 Success rate for the data association methods in 10000 Monte Carlo Simulations

Data association method	Success rate (%)
The linear case (georeferencing)	96.64
The nonlinear case with EKF (15)	96.83
The nonlinear case with azimuth and elevation angles (16)	95.99

in practice. More importantly, the nonlinear methods have the same performance as the linear method. Therefore, a nonlinear tracking filter can be used for multiple target tracking. Note, that the azimuth and elevation approach is the least accurate. However, the performance is so similar that one cannot rule out the possibility of the design parameters being the reason for the difference. Nevertheless, different simulation parameters (covariances) have been tried and the results seem to be reliable in different situations. Note that the nonlinear association in pixel form is the most accurate one overall.

Figures 8, 9 and 10 show results of a single Monte Carlo simulation. The success rate for the angle representation is 92.2% in this particular simulation. The success rate for the nonlinear approach in pixel form and the linear representation is 91.5%. Hence, in this simulation, the angle representation is able to associate one more image correctly compared to the other representations. One interesting consideration in Fig. 8a is even though the pixel and angles form have almost the same success rate, the mistakes in the association are not happening at the exact same set of images. Therefore, it seems to be situations where the pixel representations have advantages and other situations where the angle form is better. Thus, it could be possible to get

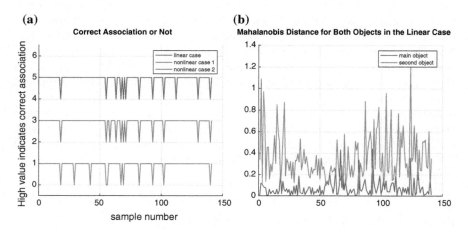

Fig. 8 Results for a single simulation in Case Study 1. **a** Correct association or not for the three methods. High value indicates successful association. The nonlinear case 1 and 2 are the pixel form and angle form, respectively. **b** Mahalanobis distance in the linear case

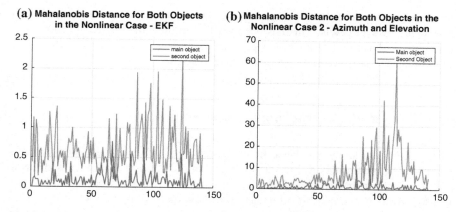

Fig. 9 Results for a single simulation in Case Study 1. **a** Mahalanobis distance in the nonlinear case with measurement model (9). **b** Mahalanobis distance in the nonlinear case with azimuth and elevation angles (11)

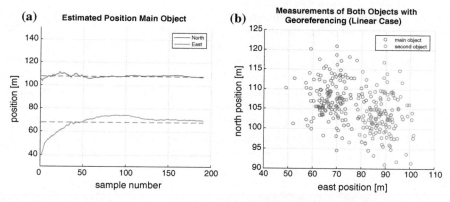

Fig. 10 Results for a single simulation in Case Study 1. **a** Estimated north and east positions for the main object. **b** Measurements of the main object and the second object (with georeferencing)

an even more robust association by combining the methods in the nonlinear case. However, this is not investigated further in this paper.

Figures 8b and 9 show the Mahalanobis distance for the three methods. The Mahalanobis distance for the measurement from the main object is obviously much smaller, but some outliers exist. Figure 10 shows the estimated north and east positions for the main object and the measurements for the main and second object with georeferencing. Notice that the measurements are distributed on a large area in the north–east plane and that is mainly because of the noise related to the UAV navigation states. Noise in the image detection algorithm has a much smaller influence than errors in the attitude. This is because georeferencing uses the attitude of the UAV to calculate the ray from the camera to the sea surface. An error of two degrees in roll will move the ray much further on the sea surface (when the altitude is large) compared to a small error in pixels. This has been verified in the simulations and shows the necessity of having reliable knowledge about the navigation states to get the good performance in target tracking from a moving platform.

4.2 Case Study 2 - Tracking a Single Target

The second case study evaluates the performance of the tracking system when a single target is considered. The target is the ship displayed in Fig. 11 and has a length of approximately 70 m. Data association is not necessary because all of the measurements are known to originate from the target. Thus, the object detection algorithm and the nonlinear filtering part of the tracking system are the focus of this case study. This experiment is based on navigation data and images captured at a flight experiment with the X8 Skywalker fixed-wing UAV and the payload described in [8]. The target is visible in 441 images over a period of 45 seconds. Note that the images are based on five different segments in time that have been merged into one segment. This ensures that the ship is visible in all images in this case study. The true position of the target is measured with GPS and used as a benchmark (ground truth). Moreover, estimates with and without the centroid adjustment will be shown to illustrate why it can be beneficial to calculate the center of the ship in situations where only a part of the vessel is visible. Only a part of the vessel (as illustrated in Fig. 11) is visible in the majority of the images.

The design parameters that need to be stated in this case study include initial conditions for the states and covariance of the measurement ($\mathbf{R}$) and process ($\mathbf{Q}$) noise. $\mathbf{R}$, $\mathbf{Q}$ and the initial covariance of the states ($\mathbf{P}(0)$) are given as

Fig. 11 Thermal images of the ship that is tracked in the second case study. Many images only contain a small part of the ship (as illustrated in the *right* image). That is why it is beneficial to use the size of the ship to get a better measure of the center in situations like this one

$$\mathbf{P}(0) = \begin{bmatrix} (10\,\text{m})^2 & 0 & 0 & 0 \\ 0 & (10\,\text{m})^2 & 0 & 0 \\ 0 & 0 & (\sqrt{20}\,\text{m/s})^2 & 0 \\ 0 & 0 & 0 & (\sqrt{20}\,\text{m/s})^2 \end{bmatrix} \quad (17)$$

$$\mathbf{R} = \frac{1}{h}\begin{bmatrix} (100\,\text{pixels})^2 & 0 \\ 0 & (80\,\text{pixels})^2 \end{bmatrix} \quad \mathbf{Q} = \frac{1}{h}\begin{bmatrix} (0.5\,\text{m/s}^2)^2 & 0 \\ 0 & (0.5\,\text{m/s}^2)^2 \end{bmatrix}$$

where h is the step length (frame rate of camera). $\mathbf{R}$ and $\mathbf{Q}$ are given in discrete time. The initial position is given by georeferencing of the first available measurement and the initial velocity is set to zero. Moreover, an approximation of the size of the ship (in pixels) is assumed to be known. In practice, the size is determined the first time the whole vessel is visible in an image. The ship is almost at rest during the tracking period, but the system has no knowledge about the behavior beforehand. The process noise covariance is chosen as white noise acceleration with standard deviation of $0.5\,\text{m/s}^2$ in continuous time (in both north and east directions). This is assumed to be realistic because large vessels have slow dynamics. Therefore, it is expected that this covariance would work for moving targets as well. The camera captures 10 images each second with a resolution of 640×512 pixels and the navigation states of the UAV are stored at the same rate.

Results - Tracking a Single Target

441 images have been processed during the second case study. The object detection algorithm is able to detect the vessel in 438 of the images, which gives a success rate of 99.3%. The high success rate is explained by the high signal-to-noise ratio in the images (see Fig. 11). There is clearly a large difference in temperature between the vessel and the background. Hence, it is possible to separate the vessel from the background.

Figure 12 shows the NE positions obtained with georeferencing with and without the centroid adjustment. A larger part of the measurements are distributed near the center of the vessel with the centroid adjustment, even though the difference is quite small. Some important measures are summarized in Table 4. The mean georeferenced position is somewhat closer to the GPS reference with the centroid adjustment. Therefore, the benefit of finding the center is clearly illustrated. The variance increases a bit with the centroid adjustment. That is perhaps because the adjustment only is performed in situations where the major axis is significantly larger than the semi-major axis. Therefore, the center is not adjusted in every image and the distribution is likely to have a larger variance. This is because the difference between measurements with and without the adjustment is expected to be larger than the difference between two measurements without adjustment. Note that the variance is mainly related to sensor noise in the navigation states of the UAV and not uncertainty in the object detection

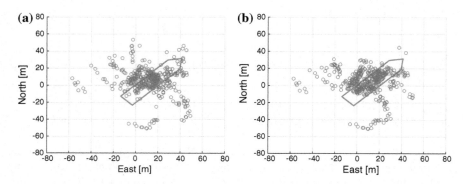

Fig. 12 Georeferenced north and east positions with **a** and without **b** the centroid adjustment

Table 4 Results of georeferencing with and without the centroid adjustment during object detection

Measure	GPS - Reference	With adjustment	Without adjustment
Mean north position	6.6	4.8	3.59
Mean east position	16.4	12.16	9.66
Variance north position	x	267.48	231.67
Variance east position	x	329.49	307.33

Fig. 13 Total distance between the reference (GPS measured position) and the estimated position in the tracking period with (*red*) and without (*blue*) the centroid adjustment

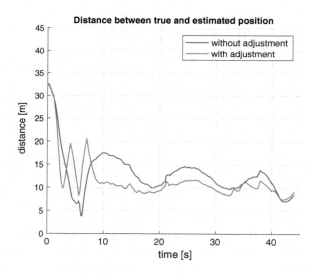

algorithm. Hence, the variance would be reduced significantly if the navigation states are perfectly known.

Figure 13 shows the total distance between the GPS measured position and the estimated position with and without the centroid adjustment. After the initialization period, the estimate with the adjustment is more accurate, even though the difference is below 5 m for most of the time. The mean distance between the true and estimated position is 11.54 and 13.07 m with and without the centroid adjustment, respectively. However, if the initialization period is neglected (the first 10 s), the mean is 9.96 and 12.21 m, respectively. The accuracy increases with the centroid adjustment and the adjustment seems to be beneficial. Nevertheless, it is important to emphasize the fact that the centroid adjustment only is necessary for targets of a certain length. The difference would be negligible for small vessels.

The estimated position is convincing, especially because the accuracy of the navigation states is limited. This is also supported by Fig. 10b where it can be seen that the distribution of georeferenced points is large even though the noise related to the object detection algorithm (Table 1) is quite low. The GPS on the ship was not mounted exactly in the center of the ship and GPS is also affected by noise. Therefore, one cannot rule out that the distance between the estimates and the GPS position is slightly smaller or larger. The estimated position is more accurate than the estimates obtained with georeferencing in [6]. Thus, the nonlinear filtering part seems to be a viable alternative to a linear tracking filter.

Figure 14 shows the estimated speed for the vessel. The vessel is nearly at rest during the simulation (dynamic positioning) and the estimated speed is close to zero after the first period. It is interesting to see that the speed is closer to zero with the centroid adjustment. A speed of exactly zero is hard to achieve because it is impossible to compensate perfectly for the motion of the camera. Nevertheless, the estimated speed is very small so the performance of the filtering part seems to be reliable.

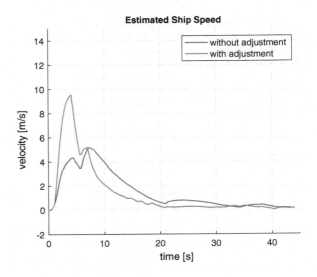

Fig. 14 Estimated speed with (*red*) and without (*blue*) the centroid adjustment

5 Concluding Remarks

This paper presented a vision-based tracking system for unmanned aerial vehicles. The tracking system is tailored toward floating objects and structures at sea. The tracking system consists of an object detection algorithm that is able to find marine vessels in thermal images. Moreover, if the size of the vessel is known (or computed the first time the whole structure is visible), the detection algorithm is able to compute the center of the ship, even when only a part of the vessel is visible in the image. This is obviously very beneficial for applications that require estimates with high precision. The tracking system also contains a filtering part that estimates the position and velocity of each target. In this paper, a nonlinear state estimator has been used to estimate the states based on the pixel position of the center of the object. Moreover, the tracking system includes a data association part that is used to find the origin of each measurement when multiple targets are tracked simultaneously.

Every part of the tracking system have been evaluated in this paper. The data association part was studied in the first case study where it was shown that data association is equally accurate with a nonlinear tracking filter as in the linear case. This is an important result as it permits the use of the nonlinear tracking filter. The second case study evaluated the detection algorithm and the nonlinear filtering part. A single target was considered and it was shown that the detection algorithm is very reliable when the temperature difference between the vessel and the sea surface is large. Moreover, the estimated position and speed were very close to the true position and speed. Therefore, all parts of the tracking system work well individually and multiple targets can be tracked.

Acknowledgements The authors would like to thank Lars Semb and Krzysztof Cisek for their technical support and flawless execution of the practical aspects of the field experiments. They would also like to thank Laboratório de Sistemas e Tecnologia Subaquática (LSTS) at the University of Porto, João Sousa and Kanna Rajan for inviting us to participate in their yearly Rapid Environment Picture (REP) exercise in the Azores. This work was partly supported by the Norwegian Research Council (grant numbers 221666 and 223254) through the Center of Autonomous Marine Operations and Systems at Norwegian University of Science and Technology (NTNU AMOS).

References

1. Hosen, J., Helgesen, H.H., Fusini, L., Fossen, T.I., Johansen, T.A.: Vision-aided nonlinear observer for fixed-wing unmanned aerial vehicle navigation. J. Guid. Control Dyn. **39**(8), 1777–1789 (2016)
2. Fusini, L., Hosen, J., Helgesen, H.H., Johansen, T.A., Fossen, T.I.: Experimental validation of a uniformly semi-globally exponentially stable non-linear observer for gnss- and camera-aided inertial navigation for fixed-wing uavs. In: Proceedings of the International Conference on Unmanned Aircraft Systems, pp. 851–860 (2015)
3. Yu, X., Zhang, Y.: Sense and avoid technologies with applications to unmanned aircraft systems: review and prospects. Progr. Aerosp. Sci. **74**, 152–166 (2015)

4. Yu, H., Beard, R.: A vision-based collision avoidance technique for micro air vehicles using local-level frame mapping and path planning. Auton. Robots **34**(1–2), 93–109 (2013)
5. Ortiz, A., Bonnin-Pascual, F., Garcia-Fidalgo, E.: Vessel inspection: a micro-aerial vehicle-based approach. J. Intell. Robot. Syst. **76**(1), 151–167 (2014)
6. Helgesen, H.H., Leira, F.S., Johansen, T.A., Fossen, T.I.: Tracking of marine surface objects from unmanned aerial vehicles with a pan/tilt unit using a thermal camera and optical flow. In: Proceedings of the International Conference on Unmanned Aircraft Systems, pp. 107–117 (2016)
7. Leira, F.S., Johansen, T.A., Fossen, T.I.: Automatic detection, classification and tracking of objects in the ocean surface from uavs using a thermal camera. In: Proceedings of the IEEE Aerospace Conference, Big Sky, US (2015)
8. Leira, F.S., Trnka, K., Fossen, T.I., Johansen, T.A.: A ligth-weight thermal camera payload with georeferencing capabilities for small fixed-wing UAVs. In: Proceedings of the International Conference on Unmanned Aircraft Systems, pp. 485–494 (2015)
9. Barber, D.B., Redding, J.D., McLain, T.W., Beard, R.W., Taylor, C.N.: Vision-based target geo-location using a fixed-wing miniature air vehicle. J. Intell. Robot. Syst. **47**(4), 361–382 (2006)
10. Kim, J.-H., Sukkarieh, S.: Airborne simultaneous localisation and map building. In: Proceedings of the IEEE International Conference on Robotics and Automation, vol. 1 (2003)
11. Bryson, M., Sukkarieh, S.: Building a robust implementation of bearing-only inertial slam for a uav. J. Field Robot. **24**(1–2), 113–143 (2007)
12. Haugen, J., Imsland, L., Løset, S., Skjetne, R.: Ice observer system for ice management operations. In: Proceedings of the International Offshore and Polar Engineering Conference, pp. 1120–1127 (2011)
13. Johansen, T.A., Perez, T.: Unmanned aerial surveillance system for hazard collision avoidance in autonomous shipping. In: Proceedings of the International Conference on Unmanned Aircraft Systems (2016)
14. Hu, W., Tan, T., Wang, L., Maybank, S.: A survey on visual surveillance of object motion and behaviors. IEEE Trans. Syst. Man Cybern. Part C (Appl. Rev.) **34**(3), 334–352 (2004)
15. Yilmaz, A., Javed, O., Shah, M.: Object tracking: a survey. ACM Comput. Surv. **38**(4) (2006)
16. Kumar, R., Sawhney, H., Samarasekera, S., Hsu, S., Tao, H., Guo, Y., Hanna, K., Pope, A., Wildes, R., Hirvonen, D., Hansen, M., Burt, P.: Aerial video surveillance and exploitation. Proc. IEEE **89**(10), 1518–1539 (2001)
17. Yang, H., Shao, L., Zheng, F., Wang, L., Song, Z.: Recent advances and trends in visual tracking: a review. Neurocomputing **74**(18), 3823–3831 (2011)
18. Lowe, D.: Object recognition from local scale-invariant features. In: Proceedings of the International Conference on Computer Vision, pp. 1150–1157 (1999)
19. Bay, H., Ess, A., Tuytelaars, T., Van Gool, L.: Speeded-up robust features (surf). Comput. Vis. Image Underst. **110**(3), 346–359 (2008)
20. Shi, J., Tomasi, C.: Good features to track. In: Proceedings of the IEEE Computer Society Conference on Computer Vision and Pattern Recognition, pp. 593–600 (1994)
21. Qadir, A., Neubert, J., Semke, W.: On-board visual tracking with unmanned aircraft system (UAs). In: AIAA Infotech at Aerospace Conference and Exhibit 2011 (2011)
22. Teulire, C., Eck, L., Marchand, E.: Chasing a moving target from a flying UAV. In: IEEE/RSJ International Conference on Intelligent Robots and Systems, pp. 4929–4934 (2011)
23. Bar-Shalom, Y., Li, X.R., Kirubarajan, T.: Estimation with Applications to Tracking and Navigation: Theory Algorithms and Software. Wiley, New Jersey (2004)
24. Martínez, C., Mondragón, I.F., Campoy, P., Sánchez-López, J.L., Olivares-Méndez, M.A.: A hierarchical tracking strategy for vision-based applications on-board UAVs. J. Intell. Robot. Syst. **72**(3), 517–539 (2013)
25. Kadyrov, A., Yu, H., Liu, H.: Ship detection and segmentation using image correlation. In: Proceedings of the IEEE International Conference on Systems, Man, and Cybernetics, pp. 3119–3126 (2013)

26. Prevost, C.G., Desbiens, A., Gagnon, E.: Extended kalman filter for state estimation and trajectory prediction of a moving object detected by an unmanned aerial vehicle. In: 2007 American Control Conference, pp. 1805–1810 (2007)
27. Niedfeldt, P.C., Beard, R.W.: Multiple target tracking using recursive RANSAC. In: 2014 American Control Conference, pp. 3393–3398 (2014)
28. Bishop, C.: Pattern Recognition and Machine Learning. Springer, Heidelberg (2006)
29. Niedfeldt, P.C., Beard, R.W.: Recursive RANSAC: multiple signal estimation with outliers. IFAC Proc. **46**(23), 430–435 (2013)
30. Konstantinova, P., Udvarev, A., Semerdjiev, T.: A study of a target tracking algorithm using global nearest neighbor approach. In: Proceedings of the International Conference on Computer Systems and Technologies (CompSysTech03), pp. 290–295 (2003)
31. Soille, P.: Morphological Image Analysis: Principles and Applications, 2nd edn. Springer Inc, New York, Secaucus, NJ, USA (2003)
32. Suzuki, S., Abe, K.: Topological structural analysis of digitized binary images by border following. Comput. Vis. Graph. Image Process. **30**(1), 32–46 (1985)
33. Hemerly, E.M.: Automatic georeferencing of images acquired by UAV's. Int. J. Autom. Comput. **11**(4), 347–352 (2014)
34. Hu, M.-K.: Visual pattern recognition by moment invariants. IRE Trans. Inf. Theory **8**(2), 179–187 (1962)
35. Wilthil, E.F., Brekke, E.F.: Compensation of navigation uncertainty for target tracking on a moving platform. In: Proceedings of the 19th International Conference on Information Fusion, pp. 1616–1621 (2016)
36. Bryson, M., Sukkarieh, S.: Bearing-only slam for an airborne vehicle. In: Proceedings of the Australasian Conference on Robotics and Automation, vol. 4 (2005)
37. Li, X.R., Jilkov, V.P.: Survey of maneuvering target tracking: dynamic models. Proc. SPIE **4048**, 212–235 (2000)

Part V
Identification and Motion Control of Robotic Vehicles

Experimental Identification of Three Degree-of-Freedom Coupled Dynamic Plant Models for Underwater Vehicles

Stephen C. Martin and Louis L. Whitcomb

Abstract This paper addresses the modeling and experimental identification of five different three degree-of-freedom (DOF) coupled nonlinear second-order plant models for low-speed maneuvering of fully actuated open-frame underwater vehicles for the surge, sway, and yaw DOFs. A comparative experimental evaluation of five different candidate plant models, whose unknown plant parameters are estimated from data obtained in free-motion vehicle trials, is reported. Model performance is evaluated for each of the five different 3-DOF coupled nonlinear finite-dimensional second-order plant models as identified by ordinary least squares (OLS) and total least squares (TLS), respectively, by comparing the mean absolute error between the experimentally observed vehicle velocities and the velocities obtained by a numerical simulation of the identified plant models. A cross-validation is reported which evaluates the performance of a plant model to accurately reproduce observed plant velocities for experimental trials differing from the trials from which the plant model parameters were estimated.

1 Introduction

Underwater vehicle control research is presently limited by the paucity of explicit experimentally validated plant models. Although the generic form of approximate lumped parameter finite-dimensional plant models for underwater vehicles is known, actual plant parameters (e.g., hydrodynamic added mass, buoyancy/bias, and drag

---

S.C. Martin (✉)
Maritime Systems Division, U.S. Navy Space and Naval Warfare Systems Center Pacific,
San Diego, CA, USA
e-mail: stephen.c.martin1@navy.mil

L.L. Whitcomb
Department of Mechanical Engineering, G.W.C. Whiting School of Engineering,
Johns Hopkins University, Baltimore, MD, USA
e-mail: llw@jhu.edu

© Springer International Publishing AG 2017
T.I. Fossen et al. (eds.), *Sensing and Control for Autonomous Vehicles*,
Lecture Notes in Control and Information Sciences 474,
DOI 10.1007/978-3-319-55372-6_15

(a) JHU ROV with major sensors identified. (b) JHU ROV with coordinate axes identified.

Fig. 1 Johns Hopkins University remotely operated underwater vehicle

parameters) must be determined experimentally [11–15, 20]. Improved experimentally validated plant dynamical models for underwater vehicles employing doppler-based or inertial navigation systems will enable the development of model-based controllers and also enable high-fidelity numerical simulations to predict the performance of underwater vehicles under a great variety of disparate missions, vehicle control laws, and operating conditions.

This paper addresses the modeling, experimental identification, and performance evaluation of three degree-of-freedom (DOF) coupled nonlinear second-order plant models for low-speed maneuvering of fully actuated open-frame underwater vehicles, for the surge, sway, and yaw degrees-of-freedom. An example vehicle of this class of vehicles is depicted in Fig. 1. A comparative experimental evaluation of five different candidate plant models, whose unknown plant parameters are estimated experimentally from data obtained in free-motion vehicle trials, is reported. A preliminary version of this paper appeared in [25].

Few previously reported studies of the dynamics of this class of vehicles have reported experimentally validated plant models. Most studies that have reported experimentally identified models have addressed decoupled one degree-of-freedom plant models [2–4, 6–8, 21, 23, 30]. Fewer reported studies have addressed coupled multi-degree-of-freedom plant models [10, 16, 17].

This paper provides the following novel contributions that have not previously been reported:

1. This study reports the first 3-DOF fully coupled plant model identification of a low-speed, fully actuated, and neutrally buoyant underwater vehicle. This result differs from previously reported studies which primarily focus on decoupled plants or a single form of the coupled dynamical plant.
2. This study reports a comparative performance analysis of five different experimentally identified 3-DOF plant models (both decoupled and fully coupled).
3. This study compares the performance of parameter identification using ordinary least squares (OLS) and total least squares (TLS) techniques.

4. This study reports a cross-validation performance of all five plant models. This cross-validation evaluates whether the models accurately reproduce observed plant velocities for experimental trials differing from the trials from which the plant model parameters were estimated.

This paper is organized as follows: Sect. 1.1 reviews the previously reported results on finite-dimensional modeling of underwater vehicles. Section 2 defines the plant model equations of motion employed in this paper. Section 3 describes the experimental methodology used in these experiments. Section 4 reports a comparative analysis of experimentally identified plant models of the 3-DOF coupled plant of the Johns Hopkins University Remotely Operated Vehicle (JHU ROV) using OLS and TLS. Section 5 reports a comparative cross-validation of the experimentally identified plant models reported in Sect. 4. Section 6 provides conclusion and summary.

1.1 Related Literature

The most commonly accepted finite-dimensional dynamical models for submarine vehicles trace their lineage to studies beginning in the 1950s at the U.S. Navy's David Taylor Model Basin [11, 14, 15, 20]. Experimental hydrodynamic model identification is often performed using planar motion mechanism (PMM) systems. These systems require a substantial amount of infrastructure to be constructed and maintained [2, 4, 10]. In [10], the author reports the identification of a high speed nonlinear coupled 6-DOF dynamical plant model of a 1/3 scale model of the Deep Submergence Rescue Vehicle (DSRV Scheme A) using the David Taylor Model Basin PMM System. This report identified most of the hydrodynamic mass parameters, all dominant hydrodynamic drag terms, and a few quadratic drag terms relating the coupling between the different degrees-of-freedom. This report does not include a numerical simulation using the identified parameters to evaluate the accuracy of the model. In [2, 4], these authors report the use of a PMM to identify a decoupled dynamical plant of the surge direction of the LAURS vehicle using ordinary least squares and weighted least squares methods. The identified model was validated by first computing an estimated force vector computed from the identified model and experimentally measured velocities and then comparing the estimated force vector to the actual forces measured by the PMM system.

Numerous studies have reported the identification of decoupled hydrodynamic dynamical models, [3, 6–8, 29], of underwater vehicles during free-flight experiments using least squares techniques. The authors employed the identified plant models and experimentally commanded control force and moment data as inputs to numerical plant simulations that compute a simulated plant velocity. The quality of the identified model typically was evaluated by comparing the experimentally measured velocity to the simulated plant velocity predicated by the identified plant model. In [17], the authors report the identification of a surge, heave, and yaw coupled dynamical plant model for the Hugin 4500 Autonomous Underwater Vehicle

(AUV), a torpedo-shaped forward-flying vehicle, using OLS. They identified a partial set of hydrodynamic mass parameters, five linear drag terms, and five quadratic drag terms relating the coupling between the lateral and yaw directions. The authors compared the coupled dynamical plant with a 1-DOF decoupled plant model. The authors found that the error between the numerical simulation's velocity data for the coupled plant model and experimentally observed velocity data is smaller than the error between the simulation's velocity data and experimentally observed velocity data for the decoupled plant model. In [3], the authors cross-validate the identified 1-DOF yaw direction model by comparing the performance of the predicted vehicle velocity with the experimentally measured velocity on experimental trial data different from the one used to identify the plant model. Cross-validation is a technique commonly employed which evaluates the performance of a plant model to accurately reproduce observed plant velocities for experimental trials differing from the trials from which the plant model parameters were estimated. The authors report that the parameter set with the lowest tracking error in self-validation may not be the parameter set that performs best in cross-validation.

Least squares methods for identifying the dynamical plant model parameters of underwater vehicles typically require position, velocity, and acceleration measurements. Acceleration is often difficult to measure directly. Differentiating velocity or position to estimate acceleration may increase the noise in the data, which can degrade the quality of the identified model. In [30], the authors report a comparative analysis of the adaptive and OLS techniques when applied to decoupled underwater robotic vehicle plant identification of the surge, sway, heave, and yaw DOFs. The reported adaptive identification technique does not require acceleration state measurements to estimate the unknown plant parameters of an underwater robotic vehicle. The experimentally identified plant models were evaluated by computing the difference between the experimentally observed vehicle velocities and the velocities obtained by a numerical simulation of the identified plant models for a commanded thruster force and moment.

Several previous computational studies have examined computational fluid dynamics (CFD) approaches to investigate the dynamics of underwater vehicles. While these computational approaches have been shown to accurately predict the hydrodynamic parameters for axisymmetric underwater vehicle hull forms with and without appendages at constant forward speed, they have not yet been successfully applied to the more complex case of low-speed 3-DOF unsteady motion of ROVs with complex hull form geometry. In [28], the authors report a CFD study of four different axisymmetric underwater vehicle hull forms without appendages at constant forward speed. The CFD predictions for pressure distribution profiles, velocity profiles, and drag coefficients are compared to those observed in experimental model trials. In [1], the authors report an evaluation of three computational methods—large eddy simulation, detached eddy simulation, and Reynolds averaged Navier Stokes (RANS) simulation—to predict longitudinal hydrodynamic coefficients of axially symmetric submarine vehicle with and without appendages in uniform axial (surge) motion. The study compares the computed longitudinal hydrodynamic coefficients, including the static pressure coefficient and skin friction coefficient, with that observed experimentally in the Defense Advanced Research Projects Agency SUBOFF experimental model tests of this hull form [18]. In [26], the authors report the use of RANS and CFD techniques to predict the longitudinal and transverse added mass of axisymmetric underwater vehicle hull forms without appendages at constant forward speed. The authors compare the added mass predicted by end-member-case simulations of a sphere and an infinitely long cylinder to analytical solutions for added mass for these special cases, but no comparison to experimental data is reported. In [19], the authors report the determination of the effects of various plant parameters on the motion of an underwater robotic vehicle. The authors also report a comparison of geometrically computed hydrodynamic parameters with a set of previously reported experimentally identified hydrodynamic parameters based on their performance using a numerical simulation. It was concluded that the geometrically computed hydrodynamic parameters matched the experimentally identified parameter if the parameter contributes significantly to the dynamics of the underwater vehicle. In [27], the authors report a comparison of the hydrodynamic parameters of the C-Scout AUV using the *Estimate Submarine Added Mass* computer program. The goal of this study was to gain insight into the effect of the geometry of the body and appendages on the calculated hydrodynamic coefficients of the system. A problem with applying geometric methods is that these methods are applicable to only a limited class of axisymmetric vehicle shapes that does not include the class of open-frame underwater vehicles addressed in the present paper.

2 3-DOF Plant Model for Underwater Vehicles

The generic form of finite-dimensional plant models for the 3-DOF plant representing the surge, sway, and yaw DOFs take the following general form [12, 13]

$$\tau_{3\times 1} = M\dot{v} + C(v)v + DQ(v)v + DLv + b, \tag{1}$$

where $\tau_{3\times 1} \in \mathbb{R}^{3\times 1}$ is the vector of body-frame forces and moments applied to the vehicle in the surge, sway, and yaw degrees-of-freedom and $\tau_{3\times 1} = [f_1; f_2; t_6]$, $v \in \mathbb{R}^{3\times 1}$ is the vehicle body-velocity and $v = [v_1; v_2; v_6]$, $\dot{v} \in \mathbb{R}^{3\times 1}$ is the vehicle body acceleration and $\dot{v} = [\dot{v}_1; \dot{v}_2; \dot{v}_6]$, $M \in \mathbb{R}^{3\times 3}$ is the positive definite vehicle mass matrix, $C(v) \in \mathbb{R}^{3\times 3}$ is the vehicle Coriolis matrix, $DQ(v) \in \mathbb{R}^{3\times 3}$ is the positive semi-definite quadratic drag matrix valued function, $DL \in \mathbb{R}^{3\times 3}$ is the positive definite linear drag matrix, and $b \in \mathbb{R}^{3\times 1}$ represents the sum of any buoyancy or bias terms (e.g., due to the force of the tether acting on the vehicle, or bias forces due to small zero-bias offsets in the thrusters). In 3-DOF model discussed herein, with only surge, sway, and yaw, clearly these terms will be due to any systematic biases and not due to buoyancy. If the model included heave, roll, or pitch, then true buoyancy terms would arise.

The following subscript notation was adopted

$$1 = Surge \text{ or } X$$
$$2 = Sway \text{ or } Y$$
$$6 = Yaw \text{ or } Heading. \quad (2)$$

Although the form of these equations is known, the actual plant parameter (e.g., hydrodynamic added mass and drag parameters) generally must be determined experimentally.

The rationale for addressing the 3-DOF plant model of vehicle dynamics for the surge, sway, and yaw degrees-of-freedom is as follows: Remotely operated vehicles of the type investigated experimentally herein typically have significant vertical separation of their center of buoyancy and center of gravity. In consequence, these vehicles exhibit a significant natural righting moment rendering them passively stable in roll and pitch. If thrusters are positioned and controlled so as to not apply roll and pitch moments to the vehicle, then the roll and pitch dynamics are not excited and can be neglected, and the dynamics of the surge, sway, and yaw degrees-of-freedom are largely decoupled from the heave dynamics.

The remainder of this section defines an explicit form of the matrices defined in (1) for the case of motion in the surge, sway, and yaw degrees-of-freedom.

2.1 3-DOF Coupled Mass and Coriolis Terms

The general 3-DOF coupled symmetric mass matrix is

$$M = \begin{bmatrix} m_{1,1} & m_{1,2} & m_{1,6} \\ m_{1,2} & m_{2,2} & m_{2,6} \\ m_{1,6} & m_{2,6} & m_{6,6} \end{bmatrix}. \quad (3)$$

This can be equivalently expressed as

$$M = \begin{bmatrix} {}^{b}M_T & M_l \\ M_l^T & m_{6,6} \end{bmatrix} \quad (4)$$

where ${}^{b}M_T \in \mathbb{R}^{2 \times 2}$

$${}^{b}M_T = \begin{bmatrix} m_{1,1} & m_{1,2} \\ m_{1,2} & m_{2,2} \end{bmatrix} \quad (5)$$

and $M_\ell \in \mathbb{R}^{2 \times 1}$

$$M_\ell = \begin{bmatrix} m_{1,6} \\ m_{2,6} \end{bmatrix}. \quad (6)$$

Experimental Identification ...

The coupled 3-DOF system is

$$C(v) = \begin{bmatrix} J(v_6)^b M_T & J(v_6) M_\ell \\ v_{1,2}^T J(1)^b M_T & v_{1,2}^T J(1) M_\ell \end{bmatrix} \quad (7)$$

where $C(v) \in \mathbb{R}^{3\times 3}$, $v_{1,2} \in \mathbb{R}^{2\times 1}$

$$v_{1,2} = \begin{bmatrix} v_1 \\ v_2 \end{bmatrix}, \quad (8)$$

and $J : \mathbb{R}^1 \to \mathbb{R}^{2\times 2}$ is the operator given by

$$J(x) = \begin{bmatrix} 0 & -x \\ x & 0 \end{bmatrix}. \quad (9)$$

2.2 Hydrodynamic Drag Terms

The coupled 3-DOF linear drag matrix is

$$DL = \begin{bmatrix} DL_{1,1} & DL_{1,2} & DL_{1,6} \\ DL_{2,1} & DL_{2,2} & DL_{2,6} \\ DL_{6,1} & DL_{6,2} & DL_{6,6} \end{bmatrix} \quad (10)$$

and the coupled 3-DOF quadratic matrix valued function of v is

$$DQ(v) = |v_1| \begin{bmatrix} DQ1_{1,1} & DQ1_{1,2} & DQ1_{1,6} \\ DQ1_{2,1} & DQ1_{2,2} & DQ1_{2,6} \\ DQ1_{6,1} & DQ1_{6,2} & DQ1_{6,6} \end{bmatrix}$$
$$+ |v_2| \begin{bmatrix} DQ2_{1,1} & DQ2_{1,2} & DQ2_{1,6} \\ DQ2_{2,1} & DQ2_{2,2} & DQ2_{2,6} \\ DQ2_{6,1} & DQ2_{6,2} & DQ2_{6,6} \end{bmatrix} \quad (11)$$
$$+ |v_6| \begin{bmatrix} DQ6_{1,1} & DQ6_{1,2} & DQ6_{1,6} \\ DQ6_{2,1} & DQ6_{2,2} & DQ6_{2,6} \\ DQ6_{6,1} & DQ6_{6,2} & DQ6_{6,6} \end{bmatrix}.$$

2.3 Buoyancy and Bias Terms

The term $b \in \mathbb{R}^{3\times 1}$ represents the sum of any buoyancy or bias terms (for example, due to the force of the tether acting on the vehicle, or bias forces due to small zero-bias offsets in the thrusters). In 3-DOF model discussed herein, with only surge, sway, and yaw, clearly these terms will be due to any systematic biases and not due to buoyancy. If the model included heave, roll, or pitch, then true buoyancy terms would arise. A b vector of the form was employed

$$b = \begin{bmatrix} 0 \\ 0 \\ b_6 \end{bmatrix} \qquad (12)$$

where b_6 is a yaw bias term. This parametric form was chosen because pilot experimental studies with the vehicle indicated that the first two terms of the bias vector were negligible, and that the third term was small but not negligible. This parameter was surmised to be principally the consequence of the vehicle tether being attached to the aft-port vehicle-frame strut.

3 Experimental Setup

The experiments reported in this paper employed the Johns Hopkins University Remotely Operated Vehicle (JHU ROV)—a research testbed vehicle developed at Johns Hopkins University. The vehicle displacement is 240 kg. It measures 1.37 m long, 0.85 m wide, and 0.61 m high. Figure 1a depicts the JHU ROV and its sensor suite. Figure 1b depicts the JHU ROV's vehicle coordinate frame axes.

The JHU ROV is equipped for full 6-DOF position measurement. Vehicle roll, pitch, yaw, and angular body-velocities are instrumented with a three-axis KVH ADGC gyro-stabilized magnetic compass (KVH Industries, Middletown, Rhode Island, USA), Microstrain 3DM-GX1 gyro-stabilized magnetic compass (Microstrain, Inc., Williston, Vermont, USA), and an IXSEA Phins North-seeking three-axis fiber-optic gyrocompass (IXSea SAS, Marly-le-Roi, France). Depth is instrumented with a Paroscientific 10 m range depth sensor Model Number 8CDP010-1 (Paroscientific, Inc., Redmond, Washington, USA). A 1,200 kHz Doppler Sonar Model Number WHN1200/6K (Teledyne RD Instruments, Poway, California, USA) provides XYZ body-velocity measurements and, in combination with the gyrocompasses and depth sensor, enables Doppler navigation via the DVLNAV software program [22]. Vehicle XYZ position is instrumented with a SHARPS 300 kHz time-of-flight acoustic navigation system (Marine Sonic Technology, Ltd., White Marsh, Virginia, USA) and the Paroscientific depth sensor. Vehicle XYZ body acceleration was obtained by both numerically differentiating the XYZ velocities measured by the Doppler sonar and directly as measured by the IXSEA Phins.

For these experiments, the vehicle was configured with six thrusters that produce force and torque to enable active control authority in all 6-DOF. The vehicle has two longitudinal thrusters, two lateral thrusters, and two vertical thrusters. The vehicle coordinate frame was defined to be co-located and aligned with the IXSEA Phins coordinate frame.

The force and moment applied to the vehicle is a linear function of the vector of thruster thrust forces

$$\tau_{6\times 1} = Af \qquad (13)$$

where $f = [f_1; f_2; \cdots f_n]$ is the $n \times 1$ vector of thruster forces, $\tau_{6\times 1} = [f_1; f_2; f_3; t_4; t_5; t_6]$ is the 6×1 vector of vehicle-frame forces and moments acting on the vehicle, and the thruster allocation matrix $A \in \mathbb{R}^{6\times n}$ maps the $n \times 1$ vector of thruster forces to the 6×1 vector of vehicle-frame forces and moments. The thruster allocation matrix A, defined in (15) on Page 9, is computed from the 3×1 unit vector (u_i) representing the direction of the thruster axis and measuring the XYZ position (p_i) of each thruster relative to the vehicle frame of reference. The 3×1 force moment vector $\tau_{3\times 1}$ of the 3-DOF model (1) is linearly related to the 6×1 force moment vector $\tau_{6\times 1}$ by the matrix $B \in \mathbb{R}^{3\times 6}$ as follows:

$$\tau_{3\times 1} = B\, \tau_{6\times 1} \text{ where } B = \begin{bmatrix} 1 & 0 & 0 & 0 & 0 & 0 \\ 0 & 1 & 0 & 0 & 0 & 0 \\ 0 & 0 & 0 & 0 & 0 & 1 \end{bmatrix}. \qquad (14)$$

$$A = \begin{bmatrix} u_1 & u_2 & u_3 & u_4 & u_5 & u_6 \\ u_1 \times p_1 & u_2 \times p_2 & u_3 \times p_3 & u_4 \times p_4 & u_5 \times p_5 & u_6 \times p_6 \end{bmatrix} \qquad (15)$$

A static thrust model was employed to model the thrust produced by each thruster. Thrust is linearly proportional to motor torque at steady state, and motor torque is linearly proportional to motor current [5, 32]. The motor current was controlled using calibrated current amplifiers. Although the steady-state thruster model (i.e., thrust linearly proportional to torque) ignores the thruster's transient dynamics, [5, 32], in the reported experiments low-frequency to moderate-frequency sinusoidal reference trajectories and open-loop thrust commands were employed to minimize the effects of thruster transient dynamics. In the next section the resulting identified plant models agree very well with the experimentally observed results, thus supporting the validity of this approach.

A median filter was applied to the velocity data from the Doppler sonar velocity data to reject outliers resulting from the experimental setup. A median filter was applied to the velocity data from the IXSEA Phins differentiated angular velocity data to compensate for outliers generated in the numerical differentiation of angular velocity data. A median filter of length 9 and width 0.03 m/s was applied to the velocity data from the Doppler sonar velocity data, where length defines the number of data points used in the median filter and width defines the maximum absolute difference between the measured value of the sensor and the median value of the datum points used in the median filter. To reject outliers in the estimated angular acceleration, a median filter of length 9 and width $0.3 \deg/\text{s}^2$ was applied to differentiated angular velocity data from the IXSEA Phins gyrocompass. These median filter parameter was selected empirically to perform reasonable outlier rejection.

4 Comparison of Experimentally Identified Plant Models

This section compares the dynamic performance of five different experimentally identified plant models to experimentally observed dynamic performance of the underwater vehicle in three degrees-of-freedom motion. This section also reports a comparative analysis of the ordinary least squares (OLS) and total least squares (TLS), [9, 31], algorithms when applied to these experimental data. The following five plant models were investigated:

1. **DIAG ALL**: This model employs a 3-parameter diagonal mass matrix, 3 diagonal quadratic drag parameters, and 1 buoyancy/bias term.
2. **DIAG QD**: This model employs a full 6 mass parameters, 3 diagonal quadratic drag parameters, and 1 buoyancy/bias term.
3. **FULL LD**: This model employs a full 6 mass parameters, 9 linear drag parameters, and 1 buoyancy/bias term.
4. **DIAG MASS**: This model employs a 3-parameter diagonal mass matrix, a full 27 quadratic drag parameters, and 1 buoyancy/bias term.
5. **FULL QD**: This model employs a full 6 mass parameters, a full 27 quadratic drag parameters, and 1 buoyancy/bias term.

The five different plant models are summarized in Table 1.

The experimental methodology employed was as follows: First, the JHU ROV's two horizontal translational DOFs were commanded to follow a sinusoidal position and velocity world reference trajectory under closed-loop control and the rotational DOF was given an open-loop sinusoidal force trajectory. Table 2 reports the magnitude, frequency, and reference trajectory type of sinusoidal profile for each DOF. This excitation resulted in coupled motion in the surge, sway, and yaw degrees-of-freedom, with relatively little excitation of the heave, roll, and pitch degrees-of-freedom. Table 3 reports the update rate and precision of the sensors used in this identification. Second, the least squares identification algorithm was employed to identify an average set of the plant model parameters from the experimental data.

Table 1 Five different plant models evaluated in this study

Parameter set label	Parameters	Description
DIAG ALL	7	3-parameter diagonal mass matrix, 3 diagonal quadratic drag parameters, and 1 buoyancy/bias term
DIAG QD	10	Full 6 mass parameters, 3 diagonal quadratic drag parameters, and 1 buoyancy/bias term
FULL LD	16	Full 6 mass parameters, a 9 full linear drag parameters, and 1 buoyancy/bias term
DIAG MASS	31	3-parameter diagonal mass matrix, a full 27 quadratic drag parameters, and 1 buoyancy/bias term
FULL QD	34	Full 6 mass parameters, a full 27 quadratic drag parameters, and 1 buoyancy/bias term

Table 2 Magnitude and frequency of reference signal for parameter identification for the v_1 (Surge), v_2 (Sway), and v_6 (Yaw) directions

Degree-of-Freedom	Trajectory type	Frequency (rad/s)	Magnitude
Surge (v_1)	Closed-Loop	0.25	1 m
Sway (v_2)	Closed-Loop	0.35	1 m
Yaw (v_6)	Open-Loop	0.3	70 Nm

Table 3 Navigation sensors used for 3-DOF parameter identification experiments

Variable	Sensor	Precision	Update rate (Hz)
$v_{1,2}$	1200 kHz RD Instruments acoustic doppler current profiler	$\sigma = 3$ mm/s single ping standard deviation	1–10
v_6	IXSEA Phins Inertial Navigation system	0.01 deg/s	10

Third, the identified model parameters for each plant model and the commanded force data from the identification experiment were employed in a numerical simulation, the output of which is a simulated velocity The model simulation velocity for each DOF was then compared with the corresponding experimentally observed plant velocity. The mean absolute error between the experimental velocity and the simulated velocity was computed to evaluate how closely each plant model behavior agrees with the experimentally observed plant. The mean absolute error for each DOF i was calculated as $e_i = mean(|v_{\text{model}i} - v_i|)$ (Table 4).

This section reports a comparative analysis of the five identified plant models for 3-DOF excitations of the Surge (v_1), Sway (v_2), and Yaw (v_6) directions. Figures 2a–3c are representative plots comparing the body-velocity data of the numerical simulation with the actual experimentally observed velocity. The five sets of plant

Table 4 Table comparing mean absolute error between simulation velocity and measured velocity using the identified parameters for the OLS and TLS parameter estimate for the JHU ROV's Surge (v_1), Sway (v_2), and Yaw (v_6) directions. The mean absolute error is in units of cm/s for the v_1 and v_2 directions and deg/s for the v_6-direction

Method	Surge (v_1)		Sway (v_2)		Yaw (v_6)	
	TLS	OLS	TLS	OLS	TLS	OLS
DIAG ALL	6.57	7.47	5.25	6.22	5.69	5.81
DIAG QD	9.67	7.37	5.85	6.31	6.10	6.15
FULL LD	6.68	6.23	4.26	5.05	4.52	4.21
DIAG MASS	6.49	5.84	4.13	4.67	4.60	4.09
FULL QD	6.20	5.99	4.29	4.74	4.88	4.40

Fig. 2 Comparison of the performance of the OLS FULL QD identified plant model and the experimentally observed body-velocities. Body-velocities shown are **a** Surge (v_1); **b** Sway (v_2); and **c** Yaw (v_6). The plots show numerical simulation velocity and actual velocity versus time (A) and the velocity error between simulated and experimentally observed velocities (B)

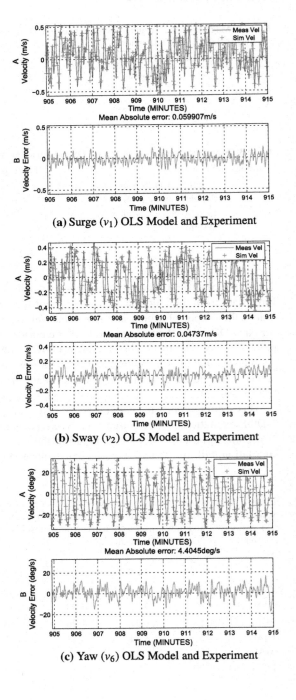

Fig. 3 Comparison of the performance of the TLS FULL QD identified plant model and the experimentally observed body-velocities. Body-velocities shown from the experimental data are **a** Surge (v_1); **b** Sway (v_2); and **c** Yaw (v_6). The plots show numerical simulation velocity and actual velocity versus time (A) and the velocity error between simulated and experimentally observed velocities (B)

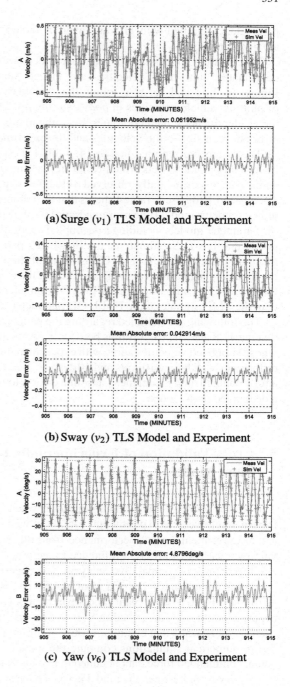

(a) Surge (v_1) TLS Model and Experiment

(b) Sway (v_2) TLS Model and Experiment

(c) Yaw (v_6) TLS Model and Experiment

parameters identified with TLS are given in Table 5, and those identified with OLS are given in Table 6.

Note that the sinusoidal XY reference trajectories are specified in inertial world coordinates, thus their velocities are indeed sinusoidal in world coordinates. Figures 2a, b, c and 3a, b, c follow the customary naval architecture convention of plotting vehicle velocities in *body coordinates*. The parameter models and parameter identification methodologies are compared by computing the mean absolute error between simulation velocities and the experimental velocities.

The results are reported in Table 4. Table 4 shows the TLS and OLS algorithms both yield parameter sets which result in plant models whose simulated velocities agree with experimentally observed velocities with mean differences well below 10 cm/s and 10 deg/sec. Table 4 shows that the parameter models including a coupled drag parameter model (linear or quadratic) have a smaller mean absolute error than the parameter models including a decoupled quadratic drag model. Finally, this table shows that the TLS estimate of the FULL LD parameter model to have a smaller mean absolute error than the TLS estimate of the FULL QD parameter model for the Sway (v_2) and Yaw (v_6) DOFs, but higher mean absolute error for the Surge (v_1) DOF. The TLS estimate of the FULL LD parameter model performed better overall than the TLS estimate of the FULL QD parameter model.

The following is concluded regarding the parameter identification methods: First, that the TLS and OLS algorithms both yield parameters which result in plant models whose simulated velocities agree with experimentally observed velocities with mean differences well below 10 cm/s and 10 deg/sec. Second, that plant models employing a fully coupled drag model mean absolute error is lower then plant models employing a decoupled drag model.

5 Cross-Validation of Experimentally Identified Plant Models

This section addresses the following question: when using model parameters obtained from a single vehicle experimental trial, how well do the plant models predict the actual experimental plant performance from other vehicle experimental trials that employ a variety of different reference trajectories? The term *cross-validation* was employed for this comparative analysis. The experimental methodology for cross-validating a particular form of a plant model (i.e., decoupled, fully coupled, etc.) is as follows: First, a set of plant model parameters is identified from a single vehicle experimental trial as described in Sect. 4. Second, multiple additional experimental trials are performed that exhibit a variety of different vehicle forces and trajectories. Third, numerical simulations compute the model's predicted state in response the actual commanded force data obtained in the multiple additional experimental trials. Finally, the error between the model performance and the experimental performance was computed as the normalized mean absolute error between the actual experimental

Table 5 JHU ROV TLS identification of v_1 (Surge), v_2 (Sway), and v_6 (Yaw) directions identified model parameters

Parameter	FULL QD	DIAG QD	DIAG ALL	DIAG MASS	FULL LD
$m_{1,1}$	768.02	773.5	789.44	775.69	857.28
$m_{2,2}$	948.24	897.5	873.15	904.48	1030.48
$m_{6,6}$	136.27	127.0	124.13	131.94	137.33
$m_{1,6}$	12.36	27.2	0.0	0.0	16.920
$m_{2,6}$	−95.55	−75.0	0.0	0.0	−104.75
$m_{1,2}$	−7.23	−2.88	0.0	0.0	−13.25
$DQ1_{1,1}$	720.32	959.5	951.19	574.63	0.0
$DQ1_{1,2}$	−116.2	0.0	0.0	−52.92	0.0
$DQ1_{1,6}$	44.4	0.0	0.0	88.33	0.0
$DQ1_{2,1}$	−73.4	0.0	0.0	−67.02	0.0
$DQ1_{2,2}$	476.36	0.0	0.0	474.06	0.0
$DQ1_{2,6}$	−170.46	0.0	0.0	−257.97	0.0
$DQ1_{6,1}$	56.2	0.0	0.0	76.72	0.0
$DQ1_{6,2}$	7.9	0.0	0.0	−11.81	0.0
$DQ1_{6,6}$	183.0	0.0	0.0	202.14	0.0
$DQ2_{1,1}$	233.9	0.0	0.0	250.08	0.0
$DQ2_{1,2}$	−56.3	0.0	0.0	13.51	0.0
$DQ2_{1,6}$	217.65	0.0	0.0	−154.51	0.0
$DQ2_{2,1}$	57.1	0.0	0.0	181.90	0.0
$DQ2_{2,2}$	1019.1	1947.0	1946.64	1156.01	0.0
$DQ2_{2,6}$	126.89	0.0	0.0	138.00	0.0
$DQ2_{6,1}$	109.9	0.0	0.0	31.35	0.0
$DQ2_{6,2}$	−101.86	0.0	0.0	−38.31	0.0
$DQ2_{6,6}$	157.66	0.0	0.0	90.98	0.0
$DQ3_{1,1}$	281.8	0.0	0.0	517.52	0.0
$DQ3_{1,2}$	−0.3	0.0	0.0	−84.38	0.0
$DQ3_{1,6}$	−32.2	0.0	0.0	82.40	0.0
$DQ3_{2,1}$	29.4	0.0	0.0	44.01	0.0
$DQ3_{2,2}$	688.2	0.0	0.0	527.01	0.0
$DQ3_{2,6}$	−72.5	0.0	0.0	−40.90	0.0
$DQ3_{6,1}$	80.3	0.0	0.0	40.02	0.0
$DQ3_{6,2}$	−36.19	0.0	0.0	−96.51	0.0
$DQ3_{6,6}$	197.49	313.13	323.92	226.75	0.0
$DL_{1,1}$	0.0	0.0	0.0	0.0	345.53
$DL_{2,2}$	0.0	0.0	0.0	0.0	496.32
$DL_{6,6}$	0.0	0.0	0.0	0.0	136.93

(continued)

Table 5 (continued)

Parameter	FULL QD	DIAG QD	DIAG ALL	DIAG MASS	FULL LD
$DL_{1,2}$	0.0	0.0	0.0	0.0	−22.74
$DL_{1,6}$	0.0	0.0	0.0	0.0	15.89
$DL_{6,1}$	0.0	0.0	0.0	0.0	2.6
$DL_{2,6}$	0.0	0.0	0.0	0.0	−20.94
$DL_{6,1}$	0.0	0.0	0.0	0.0	38.53
$DL_{6,2}$	0.0	0.0	0.0	0.0	−27.58
b_6	0.36	3.17	4.11	1.95	2.09

plant velocity and the simulated plant velocity. The normalized error, for a particular experiment, is computed by dividing the mean absolute error of each DOF by the mean absolute measured velocity for that experiment. For example, the normalized error for the Surge (v_1) DOF of an individual experiment, FULL QD parameter set is $ne_1 = mean(|v_{\text{model1}} - v_{\text{measured1}}|)/mean(|v_{\text{measured1}}|)$.

Table 7 tabulates the results of cross-validation experiments where the v_1 (Surge), v_2 (Sway), and v_6 (Yaw) DOFs are simultaneously excited. Table 8 reports the magnitude, frequency, of the sinusoidal profile for each DOF for cross-validation. Entries 1–4 used the closed-loop profile for each DOF, and entries 5–10 use the closed-loop profile for DOF 1 and 2, and open-loop for DOF 6. Table 7 shows that for every parameter set that only identifies quadratic drag, the TLS estimates mean absolute error is significantly smaller than its OLS counterpart. The OLS estimate of the FULL LD parameter model mean absolute error was insignificantly smaller than the TLS estimate of the FULL LD parameter model. This table shows that the parameter models identifying a fully parametrized drag model mean absolute error is significantly smaller than parameter models that identify only linear drag or a decoupled quadratic drag model. It also shows that the TLS estimate of both models identifying a fully parametrized drag model mean absolute error is significantly smaller than the OLS estimate. When comparing the two parameter estimates that include a fully parametrized drag matrix, the DIAG MASS parameter estimate has a smaller mean absolute error for the Surge (v_1) and Sway (v_2) DOF in the case of all three DOF are being excited, but the normalized mean absolute error is more evenly distributed for the FULL QD estimate. The difference between the FULL LD and FULL QD model performance is negligible for the yaw DOF, but the FULL QD model is superior to the FULL LD model for the surge and sway DOF. Thus the FULL QD model was observed to perform best in cross-validation.

The following is concluded for cross-validation: First, the DIAG MASS and FULL QD models yielded the smallest error for TLS and OLS algorithms. Second, the TLS estimate that best matches the dynamics of the system's mean absolute error is smaller than or equal to the best OLS estimate. Third, the TLS estimate of the FULL QD

Table 6 JHU ROV OLS identification of v_1 (Surge), v_2 (Sway), and v_6 (Yaw) directions identified model parameters

Parameter	FULL QD	DIAG QD	DIAG ALL	DIAG MASS	FULL LD
$m_{1,1}$	660.76	678.9	693.95	678.05	667.50
$m_{2,2}$	779.55	763.4	740.00	756.70	772.89
$m_{6,6}$	116.0	111.8	109.59	109.99	111.6
$m_{1,6}$	24.0	18.5	0.0	0.0	15.30
$m_{2,6}$	−63.3	−72.6	0.0	0.0	−79.29
$m_{1,2}$	29.4	−4.57	0.0	0.0	−7.00
$DQ1_{1,1}$	452.73	837.26	831.93	444.67	0.0
$DQ1_{1,2}$	−125.9	0.0	0.0	−31.07	0.0
$DQ1_{1,6}$	82.2	0.0	0.0	88.33	0.0
$DQ1_{2,1}$	−15.5	0.0	0.0	20.077	0.0
$DQ1_{2,2}$	507.2	0.0	0.0	513.54	0.0
$DQ1_{2,6}$	−101.3	0.0	0.0	−120.14	0.0
$DQ1_{6,1}$	53.4	0.0	0.0	39.17	0.0
$DQ1_{6,2}$	−76.0	0.0	0.0	−77.72	0.0
$DQ1_{6,6}$	170.3	0.0	0.0	192.49	0.0
$DQ2_{1,1}$	902.9	0.0	0.0	915.91	0.0
$DQ2_{1,2}$	−67.4	0.0	0.0	−101.54	0.0
$DQ2_{1,6}$	87.8	0.0	0.0	109.64	0.0
$DQ2_{2,1}$	37.5	0.0	0.0	35.53	0.0
$DQ2_{2,2}$	1001.7	1513.3	1510.50	985.42	0.0
$DQ2_{2,6}$	−57.4	0.0	0.0	−141.90	0.0
$DQ2_{6,1}$	136.0	0.0	0.0	110.57	0.0
$DQ2_{6,2}$	−50.3	0.0	0.0	−55.50	0.0
$DQ2_{6,6}$	175.8	0.0	0.0	172.72	0.0
$DQ1_{1,1}$	192.1	0.0	0.0	174.95	0.0
$DQ1_{1,2}$	−20.6	0.0	0.0	−8.78	0.0
$DQ1_{1,6}$	−18.8	0.0	0.0	−58.12	0.0
$DQ3_{2,1}$	62.0	0.0	0.0	46.99	0.0
$DQ3_{2,2}$	368.3	0.0	0.0	417.93	0.0
$DQ3_{2,6}$	65.8	0.0	0.0	99.12	0.0
$DQ3_{6,1}$	44.5	0.0	0.0	40.35	0.0
$DQ3_{6,2}$	−4.13	0.0	0.0	−3.94	0.0
$DQ3_{6,6}$	166.8	293.4	301.76	160.01	0.0
$DL_{1,1}$	0.0	0.0	0.0	0.0	331.73
$DL_{2,2}$	0.0	0.0	0.0	0.0	479.14
$DL_{6,6}$	0.0	0.0	0.0	0.0	134.39
$DL_{1,2}$	0.0	0.0	0.0	0.0	−24.20

(continued)

Table 6 (continued)

Parameter	FULL QD	DIAG QD	DIAG ALL	DIAG MASS	FULL LD
$DL_{1,6}$	0.0	0.0	0.0	0.0	20.24
$DL_{2,1}$	0.0	0.0	0.0	0.0	11.40
$DL_{2,6}$	0.0	0.0	0.0	0.0	−7.78
$DL_{6,1}$	0.0	0.0	0.0	0.0	37.54
$DL_{6,2}$	0.0	0.0	0.0	0.0	−24.97
b_6	1.38	3.19	4.10	2.27	2.37

Table 7 Cross-validation results: Table comparing normalized average of the mean absolute error between simulation velocity and measured velocity using the OLS and TLS parameter estimate for the JHU ROV on trajectories with significant motion in all 3-DOF for the v_1 (Surge), v_2 (Sway), and v_6 (Yaw) directions

Method	Surge (v_1)		Sway (v_2)		Yaw (v_6)	
	TLS	OLS	TLS	OLS	TLS	OLS
DIAG ALL	0.434	0.495	0.390	0.483	0.480	0.492
DIAG QD	0.491	0.561	0.386	0.471	0.474	0.486
FULL LD	0.458	0.455	0.356	0.349	0.308	0.300
DIAG MASS	0.312	0.378	0.303	0.370	0.360	0.398
FULL QD	0.334	0.416	0.330	0.364	0.314	0.363

parameter model has the lowest overall mean absolute error and as a result the best overall parameter model.

5.1 Physical Interpretation of Identified Models

The experimentally identified plant parameters are reasonable given the size and physical structure of the vehicle (shown in Fig. 1). The vehicle experiments were conducted with highly *unsteady* motion corresponding to a wide variation on velocity and Reynolds number. The experimentally identified plant parameters thus represent an average value over the ensemble of unsteady motions. The cross-validation experiments reported in this section show that the plant models employing the identified plant parameters accurately predict vehicle dynamic behavior over a wide variety of unsteady motion. The plant mode parameters are diagonally dominant, with the largest mass and drag terms appearing on the diagonal of the mass and drag matrices, as one would expect for a vehicle that is roughly symmetric with respect to the coordinate center of the vehicle along each of its three principal axes.

Table 8 Magnitude and frequency of reference signals used to evaluate the performance of the parameter model. The frequency is in units of rad/s. The magnitude is in units of m for the x_1 and x_2 directions and radians for the x_6 direction entry 1 through 4, and N-m for entries 5–10

Entry No.	Frequency			Magnitude		
	x_1	x_2	x_6	x_1	x_2	x_6
1	0.35	0.35	0.35	1.0	1.0	1.0
2	0.2	0.2	0.2	1.0	1.0	1.0
3	0.15	0.15	0.15	1.0	1.0	1.0
4	0.1	0.1	0.1	1.0	1.0	1.0
5	0.15	0.25	0.2	1.0	1.0	70
6	0.1	0.2	0.15	1.0	1.0	70
7	0.1	0.15	0.25	1.0	1.0	50
8	0.3	0.2	0.25	1.0	1.0	70
9	0.25	0.15	0.1	1.0	1.0	70
10	0.2	0.15	0.1	1.0	1.0	50

The parameters $m_{1,1}$ and $m_{2,2}$ each represent the **sum** of the rigid-body mass and the hydrodynamic added mass in the, respectively, x and y (surge and sway) body coordinates. The rigid-body mass is the same in all dimensions, but the hydrodynamic added mass will differ. The vehicle displacement (i.e., its dry mass) is 240 kg. The vehicle's cross section is smallest in the x (surge) degree-of-freedom (DOF), thus we expect its added mass to be smallest in this direction. The vehicle's cross section is larger in the y (sway) degree-of-freedom (DOF), thus we expect its added mass to be larger in this direction. The experimentally identified mass parameters confirm this. In the x_1-direction, the vehicle has the smallest cross-sectional area and the smallest hydrodynamic added mass of 528 kg, which is computed subtracting the dry mass of the JHU ROV (240 kg) from the FULL QD TLS estimate of $m_{1,1}$ = 768 kg. The x_2-direction has a larger cross-sectional area and has a comparatively larger hydrodynamic added mass of 708 kg, which is computed subtracting the dry mass of the JHU ROV (240 kg) from the FULL QD TLS estimate of $m_{2,2}$ = 948 kg. The value of the off-diagonal mass term $m_{2,6}$ shows that the vehicle center of mass is aft the Doppler. Moreover, the first two diagonal total mass terms (representing the sum of the vehicle dry mass and hydrodynamic added mass) are roughly equal to the approximately 700 kg mass of a volume of water of the same overall dimensions as the vehicle itself.

The identified quadratic drag terms were found to be diagonally dominant (as expected) and, moreover, the symmetric part of the experimentally identified drag parameter matrices are positive semi-definite, in agreement with the physical principal that drag is *dissipative*. The symmetric part of the experimentally identified quadratic drag matrices are

Table 9 Summary of the eigenvalues of the component drag matrices DQ_{v1}^S, DQ_{v2}^S, and DQ_{v6}^S appearing in (16)

DQ_{v1}	160.8	456.42	762.45
DQ_{v2}	27.54	363.83	1019.29
DQ_{v6}	184.83	288.18	694.49

$$DQ = |v_1|DQ_{v1}^S + |v_2|DQ_{v2}^S + |v_6|DQ_{v6}^S$$

$$= |v_1| \begin{bmatrix} 720.32 & -94.8 & 50.3 \\ -94.8 & 476.36 & -81.28 \\ 50.3 & -81.28 & 183.00 \end{bmatrix}$$

$$+ |v_2| \begin{bmatrix} 233.91 & 0.4 & 163.78 \\ 0.4 & 1018.10 & 12.52 \\ 163.78 & 12.52 & 157.66 \end{bmatrix} \quad (16)$$

$$+ |v_6| \begin{bmatrix} 281.8 & 14.55 & 24.05 \\ 14.55 & 688.2 & -54.34 \\ 24.05 & -54.34 & 197.50 \end{bmatrix}.$$

The eigenvalues of the symmetric part of the component matrices DQ_{v1}^S, DQ_{v2}^S, and DQ_{v6}^S appearing in (16) are reported in Table 9. Recall that the eigenvalues of a symmetric real matrix are real. Table 9 shows the eigenvalues of DQ_{v1}, DQ_{v2}, and DQ_{v6} are real and positive; therefore, $DQ(v(t))$ is positive semi-definite when $v(t) \neq 0$ and $DQ(v(t)) = 0_{3 \times 3}$ when $v(t) = 0$. Thus, the identified drag term is indeed dissipative. It is straightforward to show that the positive-definiteness of the DQ term guarantees the open-loop stability of this system. A proof is given in Sect. 7.3.5 of [24].

The buoyancy/bias term represents a combination of actual buoyant force, any possible thruster zero-calibration offset error, possibly the result of the interaction of the JHU ROV's tether with the dynamics of the overall vehicle, and the tether being attached to the JHU ROV to its aft port support strut. It could mean that the vehicle is fairly evenly ballasted because the TLS estimate of B_6 is relatively small.

6 Conclusion

This paper reported the modeling and experimental identification of five different three degree-of-freedom (DOF) coupled nonlinear second-order plant models for low-speed maneuvering of fully actuated open-frame underwater vehicles for the surge, sway, and yaw degrees-of-freedom. A comparative experimental evaluation of five different candidate plant models whose unknown plant parameters are estimated experimentally from data obtained in free-motion vehicle trials was reported. The following was concluded:

1. Free-flight experiments with coupled motion in the surge, sway, and yaw degrees-of-freedom, described in Sect. 4, provide data from which both the OLS and TLS algorithms yield model parameters that accurately model the 3-DOF planar dynamics of an underwater robotic vehicle.
2. Models employing a coupled drag model perform best on the trajectory from which that set of model parameters were obtained.
3. Fully coupled linear and quadratic drag models perform better than corresponding decoupled drag models.
4. In cross-validation, the TLS estimate of the parameter model that best matches the dynamics of the system performed as good or better than the best OLS estimate.
5. The model employing a fully coupled quadratic drag model (FULL QD) performed best overall in cross-validation when the Surge (v_1), Sway (v_2), and Yaw (v_6) DOFs are simultaneously excited.
6. The TLS estimate of the FULL QD parameter model identifying the Surge (v_1), Sway (v_2), and Yaw (v_6) directions identified quadratic damping parameters for which each individual symmetric part of the quadratic damping matrix has positive eigenvalues; therefore, each matrix is positive semi-definite when $v(t) \neq 0$ and $DQ(v(t)) = 0_{3\times 3}$ when $v(t) = 0$, and the assumption that the hydrodynamic drag associated with these directions is dissipative has been validated.

Improved experimentally validated plant dynamical models for underwater vehicles may enable the development of model-based controllers and may also enable high-fidelity numerical simulations to predict the performance of underwater vehicles under a great variety of disparate missions, vehicle control laws, and operating conditions.

Acknowledgements The authors would like to thank Dr. D. R. Yoerger who graciously provided the use of his 300-kHz SHARPS navigation system. The authors gratefully acknowledge the support of the National Science Foundation under Award #0812138. Stephen C. Martin was supported by a National Defense Science and Engineering Graduate Fellowship, a Link Foundation Doctoral Research Fellowship in Ocean Engineering and Instrumentation, and an Achievement Rewards for College Scientists Foundation Scholarship.

References

1. Alin, N., Bensow, R.E., Fureby, C., Huuva, T., Svennberg, U.: Current capabilities of DES and LES for submarines at straight course. J. Ship Res. **54**(3), 184–196 (2010)
2. Avila, J., Adamowski, J.: Experimental evaluation of the hydrodynamic coefficients of a ROV through Morison's equation. Ocean Eng. **38**(17–18), 2162–2170 (2011)
3. Avila, J., Adamowski, J., Maruyama, N., Takase, F., Saito, M.: Modeling and identification of an open-frame underwater vehicle: the yaw motion dynamics. J. Intell. Robot. Sys. **64**, 1–20 (2011)

4. Avila, J., Nishimoto, K., Sampaio, C., Adamowski, J.: Experimental investigation of the hydrodynamic coefficients of a remotely operated vehicle using a planar motion mechanism. J. Offshore Mech. Arct. Eng. **134**, 021,601(1–6) (2012)
5. Bachmayer, R., Whitcomb, L.L., Grosenbaugh, M.: An accurate finite-dimensional dynamical model for the unsteady dynamics of marine thrusters. IEEE J. Ocean. Eng. **25**(1), 146–159 (2000)
6. Caccia, M., Indiveri, G., Veruggio, G.: Modeling and identification of open-frame variable configuration underwater vehicles. IEEE J. Ocean. Eng. **25**(2), 227–240 (2000)
7. Choi, S.K., Yuh, J., Takashi, G.Y.: Development of the omni-directional intelligent navigator. IEEE Robot. Autom. Mag. **2**(1), 44–53 (1995)
8. Conte, G., Zanoli, S.M., Scaradozzi, D., Conti, A.: Evaluation of hydrodynamics parameters of a UUV: a preliminary study. In: First International Symposium on Control, Communications and Signal Processing, 2004, pp. 545–548 (2004)
9. Elbert, T.F.: Estimation and Control of Systems. Van Nostrand Reinhold Company Inc., New York, NY (1984)
10. Feldman, J.: Model investigation of stability and control characteristics of a preliminary design for the deep submergence rescue vessel (DSRV scheme A). Technical report AD637884, Hydromechanics Laboratory, David Taylor Model Basin, Washington D.C. (1966)
11. Feldman, J.: DTNSDC revised standard submarine equations of motion. Technical report, US Department of Defense (1979)
12. Fossen, T.I.: Guidance and Control of Ocean Vehicles. Wiley, New York (1994)
13. Fossen, T.I.: Marine Control Systems: Guidance, Navigation and Control of Ships and Underwater Vehicles. Marine Cybernetics, Trondheim (2002)
14. Gertler, M., Hagen., G.: Standard equations of motion for submarine simulation. Technical report, US Department of Defense (1967)
15. Goodman, A.: Experimental techniques and methods of analysis used in submerged body research. Technical report, US Department of Defense (1960)
16. Healey, A.J., Papoulias, F.A., Cristi, R.: Design and experimental verification of a model based compensator for rapid AUV depth control. In: Proceedings of the Sixth International Symposium on Unmanned Untethered Submersible Technology, pp. 458–474 (1989)
17. Hegrenaes, Ø., Hallingstad, O., Jalving, B.: Comparison of mathematical models for the HUGIN 4500 AUV based on experimental data. In: International Symposium on Underwater Technology. UT 2007 - International Workshop on Scientific Use of Submarine Cables and Related Technologies 2007, pp. 558–567. Tokyo, Japan (2007)
18. Huang, T., Liu, H., Groves, N., Forlini, T., Blanton, J., Gowing, S.: Measurements of flows over an axisymmetric body with various appendages (DARPA SUBOFF Experiments). In: Proceedings of the 19th Symposium on Naval Hydrodynamics. Seoul, South Korea (1992)
19. Humphreys, D., Watkinson, K.: Prediction of acceleration hydrodynamic coefficients for underwater vehicles from geometric parameters. Technical report, Naval Coastal Systems Laboratory (1978). (NCSL TR 327-78)
20. Imlay, F.: The complete expressions for added mass of a rigid body moving in an ideal fluid. Technical report, US Department of Defense (1961)
21. Kim, J., Kim, K., Choi, H., Seong, W., Lee, K.: Estimation of hydrodynamic coefficients for an AUV using nonlinear observers. IEEE J. Ocean. Eng. **27**(4), 830–840 (2002)
22. Kinsey, J., Whitcomb, L.: Preliminary field experience with the DVLNAV integrated navigation system for manned and unmanned submersibles. In: Proceedings of the 1st IFAC Workshop on Guidance and Control of Underwater Vehicles. GCUV '03, pp. 82–89. New Port, South Wales, UK (2003)
23. Marco, D., Healey, A.: Surge motion parameter identification for the NPS Phoenix AUV. In: Proceedings International Advanced Robotics Program IARP 98, pp. 197–210. University of South Louisiana (1998)
24. Martin, S.: Advances in six-degree-of-freedom dynamics and control of underwater vehicle. Ph.D. thesis, The Johns Hopkins University, Department of Mechanical Engineering, Baltimore, MD (2008)

25. Martin, S.C., Whitcomb, L.L.: Preliminary results in experimental identification of 3-DOF coupled dynamical plant for underwater vehicles. In: Proceedings of IEEE/MTS Oceans'2008, pp. 1–9. Quebec City, Canada (2008)
26. Mishra, V., Vengadesan, S., Bhattacharyya, S.: Translational added mass of axisymmetric underwater vehicles with forward speed using computational fluid dynamics. J. Ship Res. **55**(3), 185–195 (2011)
27. Perrault, D., Bose, N., O'Young, S., Williams, C.D.: Sensitivity of AUV added mass coefficients to variations in hull and control plane geometry. Ocean Eng. **30**(5), 645–671 (2003)
28. Sarkar, T., Sayer, P., Fraser, S.: A study of autonomous underwater vehicle hull forms using computational fluid dynamics. Int. J. Numer. Methods Fluids **25**(11), 1301–1313 (1997)
29. Smallwood, D., Whitcomb, L.L.: Preliminary identification of a dynamical plant model for the jason 2 underwater robotic vehicle. In: Proceedings of IEEE/MTS Oceans'2003, pp. 688–695. San Diego, CA (2003)
30. Smallwood, D.A., Whitcomb, L.L.: Adaptive identification of dynamically positioned underwater robotic vehicles. IEEE Trans. Control Sys. Technol. **11**(4), 505–515 (2003)
31. Van Huffel, S., Vandewalle, J.: The Total Least Squares Problem: Computational Aspects and Analysis. Society for Industrial and Applied Mathematics, Philadelphia (1991)
32. Whitcomb, L., Yoerger, D.: Development, comparison, and preliminary experimental validation of nonlinear dynamic thruster models. IEEE J. Ocean. Eng. **24**(4), 481–494 (1999)

Model-Based LOS Path-Following Control of Planar Underwater Snake Robots

Anna M. Kohl, Eleni Kelasidi, Kristin Y. Pettersen and Jan Tommy Gravdahl

Abstract This chapter presents a model-based control system for straight-line path-following of neutrally buoyant underwater snake robots that move with a planar sinusoidal gait in the presence of an unknown, constant and irrotational ocean current. The control system is based on a cascaded design, where a line-of-sight guidance law is employed in the outer control loop in order to provide a heading reference for the robot. In the presence of currents, the guidance scheme is augmented with integral action in order to compensate for the steady-state error. This work reviews the theoretical control concept and provides experimental test results with a swimming snake robot that demonstrate the concept of the control system and validate the theoretical analysis.

1 Introduction

Modelling, implementation and control of underwater snake robots is a growing field in the intersection of biomimetics and marine robotics. Underwater snake robots have emerged from the more established land-based snake robots. Research on land-based

---

A.M. Kohl (✉) · E. Kelasidi · K.Y. Pettersen
Centre for Autonomous Marine Operations and Systems (NTNU-AMOS),
Department of Engineering Cybernetics, NTNU, Norwegian University of Science
and Technology, 7491 Trondheim, Norway
e-mail: anna.kohl@ntnu.no

E. Kelasidi
e-mail: eleni.kelasidi@ntnu.no

K.Y. Pettersen
e-mail: kristin.y.pettersen@ntnu.no

J.T. Gravdahl
Department of Engineering Cybernetics, NTNU Norwegian University
of Science and Technology, 7491 Trondheim, Norway
e-mail: jan.tommy.gravdahl@ntnu.no

© Springer International Publishing AG 2017
T.I. Fossen et al. (eds.), *Sensing and Control for Autonomous Vehicles*,
Lecture Notes in Control and Information Sciences 474,
DOI 10.1007/978-3-319-55372-6_16

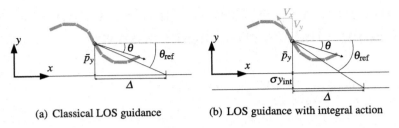

Fig. 1 The line-of-sight guidance scheme. The tuning parameters are the look-ahead distance Δ and the integral gain σ

snake robots dates back to the 1970s [5] but is still an evolving field of research [13]. A survey of the mechanical design of snake robots can be found in [6]. Another literature review is presented in [14], which focuses additionally on the modelling, analysis and control of such robots. Underwater snake robots are closely related with robotic fish and are sometimes even considered a special kind of fish robots [20]. Surveys on fish inspired robots and their control can be found in [2, 20].

One important challenge in marine robotics is the development of autonomous control systems for path-following. A strategy for straight-line path-following is to apply the well-known line-of-sight (LOS) guidance [3], to determine a reference heading for the control system of the robot. The guidance scheme is visualized in Fig. 1a: the robot steers towards a point on the path that is located at the look-ahead distance Δ in front of the robot along the path. In the presence of ocean currents, this strategy will result in a steady-state offset of the path if the currents have a component transverse to the path. This problem can be solved by augmenting the guidance law with integral action, which makes the robot target a point at the look-ahead distance Δ along a displaced path that lies upstream of the desired path, as illustrated in Fig. 1b. A formulation of the integral LOS guidance scheme with a strategy to prevent significant integral windup can be found in [1].

In the context of snake and fish robotics, research on path-following control systems is quite limited. A maneuvering control system for land-based snake robots is proposed in [16] and extended to planar underwater snake robots in [10]. The control system is based on a first-principle model, it considers both velocity and path-following control for generic paths, and is formally shown to be practically stable. A motion planning strategy that is similar to LOS guidance is presented for an eel-like robot in [18]. In [12], trajectory tracking is performed with a fish robot with flow sensors. Another approach is proposed in [4], where a LOS guidance law is employed in order to make a fish-like robot head towards predefined waypoints. The LOS guidance scheme for path-following of snake robots according to Fig. 1 has been used both on land and in water. For land-based snake robots, the guidance strategy is investigated in [13] in combination with two different controllers. First, the strategy is implemented in combination with a proportional controller that steers the robot towards the path. Secondly, the heading is controlled with a model-based strategy, which enables an analysis that formally shows stability. For swimming snake robots, the LOS guidance scheme in combination with a proportional heading controller is experimentally investigated in [7]. The augmented integral LOS strategy with the same heading controller is successfully tested in [8]. However, a formal stability analysis of the path-following control system based on this heading controller is challenging, since a simple proportional controller is combined with a highly non-linear model, and therefore a simulation based Poincaré map analysis is provided instead. The model-based LOS path-following control system that was proposed for land-based snake robots in [13] was recently extended to include integral action and thus be suitable for planar underwater snake robots that are affected by ocean currents in [11], where a formal stability analysis shows uniform semiglobal exponential stability of the control system.

In this chapter, we present experimental results with a swimming snake robot for the model-based heading controller in combination with the LOS guidance law without and with integral action. In particular, the experimental results show that the model-based LOS path-following controller for land-based snake robots from [13] also works for swimming robots. Furthermore, the integral LOS path-following controller from [11] is experimentally validated. The chapter is structured as follows. Section 2 presents the control-oriented model of an underwater snake robot that has been used for the design of the control systems that are the subject of Sects. 3 and 4. The model-based control scheme for snake robots that employs LOS guidance is reviewed and experimentally validated with a swimming snake robot in Sect. 3. The extension of the model-based control strategy using integral LOS guidance is reviewed in Sect. 4, where experimental results for this control approach are presented for the first time. Some concluding remarks are made in Sect. 5.

2 The Control-Oriented Model of the Underwater Snake Robot

The control-oriented model of a snake robot moving in a two-dimensional plane that has been used to design the control systems, which are the subject of this work, is based on several simplifying assumptions. To begin with, the robot is assumed to be neutrally buoyant and conduct slow planar motion with a sinusoidal gait such that the angles between adjacent links remain small. For such limited joint angles, the motion of the links with respect to each other can be approximated by linear displacements. It has been shown that the control-oriented model based on these simplifying assumptions captures the behaviour of the robot very well for angles smaller than 30° [9, 13]. The approximation gets gradually less accurate for larger angles, but the qualitative behaviour is still similar.

In essence, the rotational joints of the snake robot are modelled as translational joints, which can be seen in Fig. 2. The robot consists of N links of equal length and mass m that are connected by $N-1$ actuated joints. The joint coordinates ϕ_i, $i = 1, \ldots, N-1$ are assembled in the vector $\boldsymbol{\phi} \in \mathbb{R}^{N-1}$. Since the model disregards

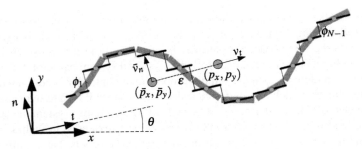

Fig. 2 The control-oriented model of a snake robot is based on approximating the rotational joints by translational joints when the robot conducts a sinusoidal gait

the rotational motion of the links with respect to each other, they all have the same orientation θ, which also defines the rotation of the robot with respect to the global frame. The robot turns about a virtual point $[\bar{p}_x, \bar{p}_y]^T$ that is located at a distance ε behind its centre of mass (CM) at the position $[p_x, p_y]^T$, which is also indicated in Fig. 2. The absolute velocity of the robot is defined in the $t-n$ coordinate frame that is aligned with the robot, and thus given by the tangential velocity v_t, and the normal velocity $\bar{v}_n$. Since the robot is affected by a current $[V_x, V_y]^T$ in the global coordinate frame, the relative velocities have to be taken into account when modelling the hydrodynamic forces that act on the robot. The relative velocities are given by $v_{t,\mathrm{rel}} = v_t - V_x \cos\theta - V_y \sin\theta$ and $\bar{v}_{n,\mathrm{rel}} = \bar{v}_n + V_x \sin\theta - V_y \cos\theta$. Based on the analysis in [9], the hydrodynamic effects are modelled as linear drag forces, which results in the following model intended for control design:

$$\dot{\boldsymbol{\phi}} = \mathbf{v}_\phi,$$
$$\dot{\theta} = v_\theta,$$
$$\dot{p}_y = v_{t,\mathrm{rel}} \sin\theta + \bar{v}_{n,\mathrm{rel}} \cos\theta + V_y,$$
$$\dot{\mathbf{v}}_\phi = -\frac{c_n}{m}\mathbf{v}_\phi + \frac{c_p}{m}v_{t,\mathrm{rel}}\mathbf{A}\mathbf{D}^T\boldsymbol{\phi} + \frac{1}{m}\mathbf{D}\mathbf{D}^T\mathbf{u},$$
$$\dot{v}_\theta = -\lambda_1 v_\theta + \frac{\lambda_2}{N-1}v_{t,\mathrm{rel}}\bar{\mathbf{e}}^T\boldsymbol{\phi},$$
$$\dot{\bar{v}}_{n,\mathrm{rel}} = (X + V_x\cos\theta + V_y\sin\theta)v_\theta + Y\bar{v}_{n,\mathrm{rel}}.$$

The joints of the robot are actuated by the control input $\mathbf{u} \in \mathbb{R}^{N-1}$. The parameter c_n is the drag parameter of a single link in the normal direction, c_p is a propulsion coefficient, and λ_1, λ_2 are empirical constants that characterize the turning motion. Furthermore, X and Y are defined as $X = \varepsilon(\frac{c_n}{m} - \lambda_1)$ and $Y = -\frac{c_n}{m}$. The vector operators $\mathbf{A}$, $\mathbf{D}$ and $\bar{\mathbf{e}}$ are defined in [13]. More details on the modelling can be found in [9] and on the transformation to a purely relative velocity representation in [11].

Remark 1 This model does not include the dynamics of the relative forward velocity $v_{t,\mathrm{rel}}$. This is because the purpose of the model is to design a path-following control system where the forward velocity is not feedback controlled. Instead, the robot propels itself forward by using a biologically inspired gait, which results in some positive forward velocity $v_{t,\text{rel}} \in [V_{\min}, V_{\max}]$. In the control design process, $v_{t,\text{rel}}$ is therefore treated as a positive model parameter.

3 The Model-Based Control System for LOS Path-Following

The structure of the relative velocity model presented in Sect. 2 is the same as that of a land-based snake robot presented in [13]. In fact, if the underwater robot is not exposed to any currents, the relative velocities and the absolute velocities are the same, and the models become identical when replacing the hydrodynamic drag parameters by the ground friction coefficients. It is therefore possible to achieve straight-line path-following with a swimming snake robot by using the LOS path-following controller that was presented for land-based snake robots in [13], as long as there is no ocean current.

This section reviews the control system from [13], explains how it was implemented on the amphibious snake robot Mamba, and finally presents an experimental study that validates the conjecture that the control system is also applicable for swimming robots.

3.1 The Control System

In the following, the model-based LOS path-following control system presented in [13] will be shortly reviewed.

> The control objective is to make a snake robot converge to a straight line, align with it, and subsequently travel along it. Without loss of generality, the global coordinate frame is defined such that the path and the global x-axis coincide, and the control objective is formulated as
>
> $$\lim_{t \to \infty} \bar{p}_y(t) = 0, \quad (1)$$
> $$\lim_{t \to \infty} \theta(t) = 0. \quad (2)$$

The control problem is solved by a cascaded approach, where the robot achieves a forward velocity with the well-known gait lateral undulation, a sinusoidal wave that travels through the snake from head to tail. This way of propulsion is inspired by the motion of biological snakes and can be achieved by controlling each joint i to track the reference signal

$$\phi_{i,\text{ref}} = \alpha \sin\left(\omega t + (i-1)\delta\right) + \phi_0, \quad i = 1, \ldots, N-1. \tag{3}$$

In (3), α is the amplitude of the joint motion, ω is the frequency of the body undulation, δ is the phase shift between adjacent joints, which makes the wave propagate, and ϕ_0 is a constant offset that induces turning motion to the robot. The controller that enforces the reference (3) closes the inner control loop of the cascaded control system and is given by the feedback linearizing control law

$$\mathbf{u} = m(\mathbf{D}\mathbf{D}^T)^{-1}\left[\bar{\mathbf{u}} + \frac{c_n}{m}\dot{\boldsymbol{\phi}} - \frac{c_p}{m}v_t\mathbf{A}\mathbf{D}^T\boldsymbol{\phi}\right], \tag{4}$$

$$\bar{\mathbf{u}} = \ddot{\boldsymbol{\phi}}_{\text{ref}} + k_{v_\phi}(\dot{\boldsymbol{\phi}}_{\text{ref}} - \dot{\boldsymbol{\phi}}) + k_\phi(\boldsymbol{\phi}_{\text{ref}} - \boldsymbol{\phi}), \tag{5}$$

with the positive control gains k_ϕ and k_{v_ϕ}. It was proven in [13] that this control law exponentially stabilizes the joint coordinate errors $\tilde{\phi}_i = \phi_i - \phi_{i,\text{ref}}$ to zero. After closing the inner control loop according to the above equations, ϕ_0 can be interpreted as a new control input that induces turning motion to the inner cascade.

In the outer control loop, the robot is steered towards the path and thus forced to fulfil the control objectives (1), (2) by enforcing the heading reference

$$\theta_{\text{ref}} = -\arctan\left(\frac{\bar{p}_y}{\Delta}\right) \tag{6}$$

given by the LOS guidance law. It was shown in [13] that the heading controller

$$\phi_0 = \frac{1}{\lambda_2 v_t}\left[\ddot{\theta}_{\text{ref}} + \lambda_1 \dot{\theta}_{\text{ref}} - k_\theta(\theta - \theta_{\text{ref}})\right] - \frac{1}{N-1}\sum_{i=1}^{N-1}\alpha\sin\left(\omega t + (i-1)\delta\right) \tag{7}$$

with the positive control gain k_θ makes the equilibrium of the heading error $\tilde{\theta} = \theta - \theta_{\text{ref}} = 0$ uniformly globally exponentially stable.

The structure of the control system is summarized in the block diagram in Fig. 3. In [13], the control system was proven to κ-exponentially stabilize the error dynamics to zero, corresponding to satisfying the control objectives (1), (2) under the assump-

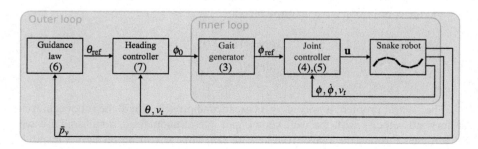

Fig. 3 The structure of the LOS path-following controller for snake robots

tion that the robot moves at a velocity $v_t \in [V_{\min}, V_{\max}]$ and the following sufficient condition:

> **Theorem 1** (see Theorem 8.2 in [13]) *If the look-ahead distance Δ of the LOS guidance law (6) is chosen such that*
>
> $$\Delta > \frac{|X|}{|Y|}\left(1 + \frac{V_{\max}}{V_{\min}}\right), \qquad (8)$$
>
> *the control objectives (1), (2) are enforced by the path-following control system in Fig. 3.*

Remark 2 The model-based heading controller (7) has to be implemented with care in order to avoid singularity issues when the forward velocity of the robot is zero, $v_t = 0$. Since the robot gains a positive forward velocity by moving with a sinusoidal gait, this issue will only occur when starting up the system and can be avoided by adding a saturation function to ϕ_0 or setting the control input ϕ_0 to zero if the forward velocity v_t is smaller than a certain bound.

3.2 Implementation of the Control System

The control system presented in Sect. 3.1 was implemented on a laptop that runs LabVIEW 2013. The proportional controllers that are implemented in the microcontrollers of each joint of the test robot replaced the low level control law (4), (5), because the theoretical feedback linearizing control law (4) requires torque control while the joints of the test robot are position controlled. This does not invalidate the theoretical control structure, because the cascaded analysis just requires that the joint error dynamics is exponentially stabilized, regardless which controller is used. The control input ϕ_0, which is used to induce turning motion, is a linear displacement in the control-oriented model. However, since it has been shown previously [9, 13] that the control-oriented model presented in Sect. 2 still captures the qualitative behaviour of the robot with its revolute joints, (7) was implemented as the heading controller. The model parameters of the control-oriented model λ_1 and λ_2 that show up in the heading controller (7) were treated as control gains analogously to the implementation in [13], where the control system was tested with a land-based snake robot. In order to implement the heading controller (7), the forward velocity v_t needed to be approximated from the data of an external motion capture system. It was estimated as the displacement of the CM divided by a sampling interval of 2 s. In order to obtain smooth time derivatives of the heading reference θ_{ref}, the commanded angle θ_{ref} was passed through a third-order low-pass filtering reference model. The parameters of the reference model were $T = \frac{1}{2\pi}$ and $\zeta = 1$. Details on the reference model can be found in [13]. Finally, in order to avoid the singularity in (7) or self-collision of

Table 1 The control gains of the LOS path-following control system

α	ω	δ	k_θ	λ_1	λ_2	Δ
30°	90°	40°	0.4	0.5	0.2	1.8 m

the physical robot, the heading control input ϕ_0 was saturated at $\phi_{0,\max} = \pm 20°$, and filtered with a first-order low-pass filter with the cut-off frequency 1.25 Hz. Since the exact model parameters that are required for calculating the offset ε are unknown, it was assumed that the robot turns about its CM, i.e. the parameter was set to $\varepsilon = 0$. In order to close the feedback control loop, reflective markers were attached to the robot tail, and the angle and position measurements of these markers were obtained from an external motion capture system. On the second laptop, both Qualisys Track Manager (QTM) and Labview 2013 were installed and communicated with each other. The obtained position data were sent from the second laptop in Labview 2013 via UDP in real time at a sample frequency of 10 Hz. The angles of each single link and the position of the CM p_x, p_y were then obtained from the angle and position measurements of the reflective markers in combination with the single joint angles, analogously to the implementation in [7]. The orientation of the robot was estimated by the average of the link angles, $\bar{\theta}$. The control gains of the system were tuned according to Table 1. Since neither the exact parameters of the control-oriented model in Sect. 2 were known, nor did we know the velocity of the robot in advance, it was difficult to find a numeric value for the lower bound on the look-ahead distance Δ according to (8). Consequently, we interpreted this condition as the requirement to choose Δ sufficiently large.

3.3 Experimental Validation of the Control System

The LOS path-following control system that was reviewed in Sect. 3.1 was experimentally tested in the Marine cybernetics laboratory (MC-lab) at NTNU [17]. The MC-lab is a small wave basin that is suitable for the testing of marine control systems. The tank dimensions are 40 m in length, 6.45 m in width and 1.5 m in depth. For our tests, the wave maker was deactivated. In order to obtain accurate position measurements for the control system, six cameras of the underwater motion capture system Qualisys [19] were mounted on the walls of the basin, three on each side. The snake robot Mamba [15] was used to test the LOS control system. Figure 4 shows the robot in the MC-lab. Mamba is a modular snake robot that was designed and built at NTNU and is suitable for operations both on land and in water. It consists of nine horizontal joints and nine vertical joints and is connected to a power source and communication unit with a slender, positively buoyant cable. During all our tests an angle of zero was enforced on the vertical joints in order to test the two-dimensional control scheme. The robot joints are waterproof down to 5 m and equipped with a servo motor, a microcontroller card, and several sensors. The joint angles are controlled by a proportional controller, which is implemented on the microcontroller

Fig. 4 The snake robot Mamba in the MC-lab. The cameras of the motion capture system are mounted on both sides of the tank

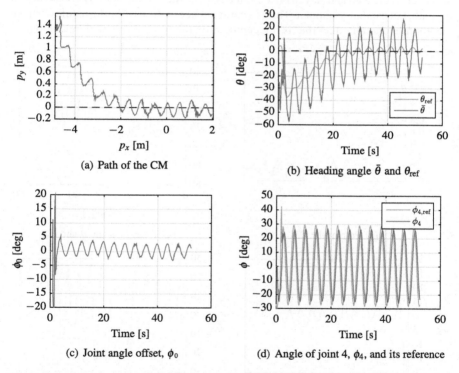

(a) Path of the CM

(b) Heading angle $\bar{\theta}$ and θ_{ref}

(c) Joint angle offset, ϕ_0

(d) Angle of joint 4, ϕ_4, and its reference

Fig. 5 Experimental results of the first scenario: Model-based LOS path-following with the robot initially headed towards the path

card that communicates over a CAN bus. More details about the physical robot can be found in [15]. For our tests, reflective markers for the Qualisys motion capture system were attached to the tail of the robot and the robot was put into a synthetic skin for additional waterproofing. Detailed information about the skin is provided in [7]. The amount of air that is contained inside the skin can be varied manually by a pneumatic valve, which changes the buoyancy of the robot. For the two-dimensional control system that was tested, a slightly positive buoyancy was enforced in order to make the robot stay close to the surface and thus not require depth control. The control system from Sect. 3.1 was successfully tested in two different scenarios. In the first case, the robot was initially headed towards the path, and in the second case it was initially headed approximately parallel to the path. In both cases, the robot was initially straight and kept in a fixed position.

The results of the different test scenarios are presented in Figs. 5 and 6. The path of the CM of the robot is plotted in Figs. 5a and 6a. It can be seen that the robot approached the path and stayed on it. In steady state, the CM did not stay constantly on the path, but oscillated about it. This is a consequence of the oscillating nature of snake locomotion and an effect that is not captured by the control-oriented model, where

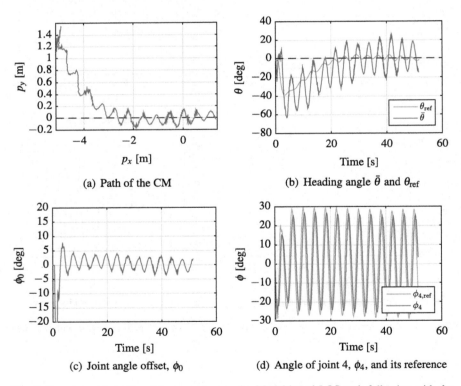

Fig. 6 Experimental results of the second scenario: Model-based LOS path-following with the robot initially headed parallel to the path

the rotational motion of the links is disregarded. The same applies to the heading of the robot, $\bar{\theta}$, as can be observed in Figs. 5b and 6b. Instead of converging to zero, the measured signal kept oscillating about zero as a consequence of the simplifications in the control-oriented model. However, after reaching the reference, the measured signal clearly stayed as close to it as the oscillating motion allowed. The reference signal θ_{ref} itself kept oscillating about zero as a consequence of the deviations of the CM from the path. The control input for turning motion, ϕ_0, is displayed in Figs. 5c and 6c. In the first seconds, before the robot had reached sufficient forward velocity, the signal was saturated. After the robot had reached the path on the other hand, ϕ_0 was oscillating about zero, thus allowing the robot to follow the oscillating heading reference θ_{ref}. Finally, the reference signal and the measured signal of the arbitrarily chosen joint number four is shown in Figs. 5d and 6d. It is obvious that the joints were tracking their references.

It can be concluded from the experimental results that the model-based LOS path-following controller works for swimming snake robots without any modification, despite the simplifications in the control-oriented model. The control objectives (1), (2) are approximately satisfied. The remaining oscillations after convergence are an inherent element of snake robots conducting sinusoidal gaits, and they are not captured by the control-oriented model, which explains why they are not cancelled by the model-based control approach.

4 The Model-Based Control System for Integral LOS Path-Following

The LOS path-following controller for snake robots, which was reviewed and tested in the previous section, works for underwater snake robots only under the condition that there are no ocean currents acting on the robot. In the presence of currents, the LOS guidance law will not enable the robot to converge to and stay on the path, so a different guidance law has to be employed. In addition, the model of the robot is a different one, since the current effect has to be accounted for in the model and the hydrodynamic forces that act on the robot now no longer depend on the absolute velocities, but rather on the relative velocities. In this section, the model-based integral LOS path-following control system that was presented in [11] will therefore be used for snake robot control in the presence of ocean currents. At first, the control system will be reviewed; secondly, the implementation of the control laws on the physical robot will be discussed, and finally an experimental validation of the control system will be presented.

4.1 The Control System

In the following the model-based integral LOS path-following control system presented in [11] will be shortly reviewed.

> The control objective is to make an underwater snake robot converge to a straight line, and subsequently travel along it at some heading θ^{eq}. Without loss of generality, the global coordinate frame is defined such that the path and the global x-axis coincide and the control objective is formulated as
>
> $$\lim_{t \to \infty} \bar{p}_y(t) = 0, \quad (9)$$
>
> $$\lim_{t \to \infty} \theta(t) = \theta^{\text{eq}}. \quad (10)$$
>
> The steady-state heading θ^{eq} is in general non-zero, and is a crab angle that enables the robot to compensate for the sideways component of the current. Its value depends on the speed of the robot and the magnitude of the ocean current.

The structure of the control system is the same as of the LOS path-following controller in Sect. 3.1 and visualized in the block diagram in Fig. 7. The reference for the joint controller in the inner control loop is augmented to the more general

$$\phi_{i,\text{ref}} = \alpha g(i) \sin(\omega t + (i-1)\delta) + \phi_0, \quad i = 1, \ldots, N-1, \quad (11)$$

where the amplitude of the undulation can be varied along the body with the scaling function $g(i)$. For instance, choosing $g(i) = \frac{N-i}{N+1}$ will mimic the swimming motion of eels [7]. The feedback linearizing joint controller and the heading controller in the outer control loop are based on the complete underwater model in Sect. 2, which takes the relative velocity into account, and the new gait reference (11). They read as

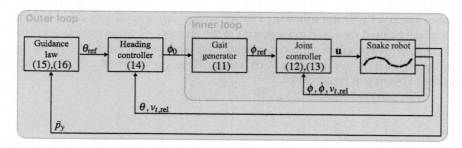

Fig. 7 The structure of the integral LOS path-following controller for snake robots

$$\mathbf{u} = m(\mathbf{DD}^T)^{-1}\left[\bar{\mathbf{u}} + \frac{c_n}{m}\dot{\boldsymbol{\phi}} - \frac{c_p}{m}v_{t,\text{rel}}\mathbf{AD}^T\boldsymbol{\phi}\right], \tag{12}$$

$$\bar{\mathbf{u}} = \ddot{\boldsymbol{\phi}}_{\text{ref}} + k_{v_\phi}(\dot{\boldsymbol{\phi}}_{\text{ref}} - \dot{\boldsymbol{\phi}}) + k_\phi(\boldsymbol{\phi}_{\text{ref}} - \boldsymbol{\phi}) \tag{13}$$

and

$$\phi_0 = \frac{1}{\lambda_2 v_{t,\text{rel}}}\left[\ddot{\theta}_{\text{ref}} + \lambda_1 \dot{\theta}_{\text{ref}} - k_\theta(\theta - \theta_{\text{ref}})\right] \tag{14}$$
$$- \frac{1}{N-1}\sum_{i=1}^{N-1}\alpha g(i)\sin\left(\omega t + (i-1)\delta\right).$$

The guidance law in the outer loop controller needs to be changed in order to account for the ocean current. Instead of the LOS guidance in Sect. 3.1, the augmented integral LOS guidance scheme is employed now:

$$\theta_{\text{ref}} = -\arctan\left(\frac{\bar{p}_y + \sigma y_{\text{int}}}{\Delta}\right), \tag{15}$$

$$\dot{y}_{\text{int}} = \frac{\Delta \bar{p}_y}{(\bar{p}_y + \sigma y_{\text{int}})^2 + \Delta^2}. \tag{16}$$

The integral action in (16) enables the robot to compensate for the steady-state error that would result from applying the LOS guidance in Sect. 3.1 in the presence of ocean currents. Due to the integral action, when $\bar{p}_y$ converges to zero, the reference angle will converge to a non-zero crab angle θ^{eq}, which is necessary to counteract currents that are transverse to the path.

In [11], the control system in Fig. 7 was transformed into error coordinates that form a cascaded system. The assumption was made that the robot moves at a relative velocity $v_{t,\text{rel}} \in [V_{\min}, V_{\max}]$ that is large enough to counteract the current. Furthermore, the ocean current was assumed to be constant, irrotational, and of a magnitude smaller than the bound $V_{c,\max}$. It was shown in [11] that the origin of the system is uniformly semi-globally exponentially stable under the following sufficient condition:

Theorem 2 (see Theorem 2 in [11]) *If the look-ahead distance Δ and the integral gain σ of the integral LOS guidance law (15), (16) are chosen such that*

$$\Delta > \frac{|X| + 2V_{c,\max}}{|Y|}\left[\frac{5}{4}\frac{V_{\max} + V_{c,\max} + \sigma}{V_{\min} - V_{c,\max} - \sigma} + 1\right], \tag{17}$$
$$0 < \sigma < V_{\min} - V_{c,\max}, \tag{18}$$

the control objectives (9), (10) are enforced by the path-following control system in Fig. 7, and the steady-state heading is

$$\theta^{\text{eq}} = -\arctan\left(\frac{V_y}{\sqrt{v_{t,\text{rel}}^2 - V_y^2}}\right). \tag{19}$$

Remark 3 For implementing the heading controller (14) on a robotic system, measurements of the relative velocity are needed. In addition, the first two derivatives of the heading reference θ_{ref} are required. These depend on the absolute velocity of the robot, but for a practical system it is much preferred to obtain them by passing the reference signal θ_{ref} through a low-pass filtering reference model in order to obtain smooth signals. It is therefore not necessary to equip the system with sensors for the absolute velocity, and the knowledge of the relative velocity is sufficient for the implementation.

4.2 Implementation of the Control System

The integral LOS control system presented in Sect. 4.1 was implemented in LabVIEW 2013 analogously to the LOS control system, the implementation details of which have been presented in Sect. 3.2. In addition to the implementation there, (16) was numerically integrated in LabVIEW in order to account for the effect of the current. The heading controller (14) requires the relative velocity $v_{t,\text{rel}}$ instead of the absolute velocity v_t that was used in Sect. 3.2. The relative velocity was approximated by subtracting the current speed from the absolute velocity v_t. This way of approximating $v_{t,\text{rel}}$ is only accurate when the robot is heading against the current and will become less accurate when the robot turns. However, this was a reasonable approximation since the current was mainly opposing the forward motion of the robot, and the calculations were significantly simplified. In order to avoid the singularity in the heading controller and thus achieve a smoother transient in the beginning, the heading controller ϕ_0 was started only after 2 s. The control gains of the system can be found in Table 2. Since a lower bound on the velocity of the robot was not known a priori, it was not feasible to use the theoretical condition (18) to determine a value for σ, and consequently (17) could not be employed to find a bound on Δ. In this light, and following the same line of thoughts as when choosing the look-ahead distance Δ for the classical LOS system in Sect. 3.2, we made sure to choose a sufficiently large Δ, and a sufficiently small σ in order to converge to the path.

Table 2 The control gains of the integral LOS path-following control system

	α	$g(i)$	ω	δ	k_θ	λ_1	λ_2	Δ	σ
Lateral undulation	30°	1	110°	40°	0.8	0.5	0.2	2 m	0.5
Eel-like motion	50°	$\frac{N-i}{N+1}$	130°	40°	0.9	0.5	0.2	2 m	0.3

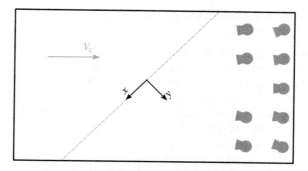

Fig. 8 The coordinate transformation: the global coordinate frame is rotated by 45° with respect to the walls of the tank, such that the current has a negative x-component and a positive y-component. The cameras of the motion capture system (displayed in *grey*) are mounted on one end of the tank. The path is indicated in *orange*

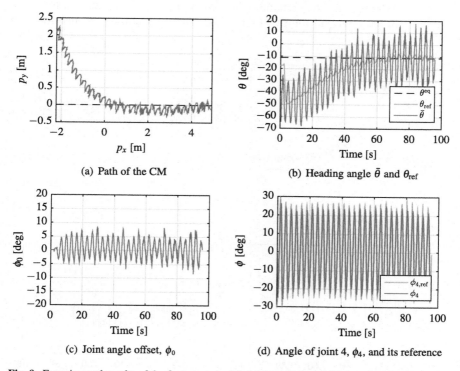

(a) Path of the CM

(b) Heading angle $\bar{\theta}$ and θ_{ref}

(c) Joint angle offset, ϕ_0

(d) Angle of joint 4, ϕ_4, and its reference

Fig. 9 Experimental results of the first scenario: Model-based integral LOS path-following with a flow speed $V_c = 0.07$ m/s, lateral undulation, and the robot initially headed towards the path

4.3 Experimental Validation of the Control System

The experimental tests of the integral LOS control system were performed in the North Sea Centre Flume Tank [21] operated by SINTEF Fisheries and Aquaculture in Hirtshals, Denmark. The flume tank is 30 m long, 8 m wide and 6 m deep and is equipped with four propellers that can generate a circulating flow of up to 1 m/s. For our tests, nine cameras of the Qualisys motion capture system were mounted on one end of the tank. During the experiments, the global coordinate frame was rotated by 45° with respect to the basin, such that the generated current, which is aligned with the long side of the tank, had both an x- and a y-component. The coordinate transformation is sketched in Fig. 8. The snake robot Mamba that was introduced in Sect. 3.2 served as the test platform also for these experiments. The control system from Sect. 4.1 was tested experimentally in four different scenarios. In the first two scenarios, the robot was moving with the gait lateral undulation and was exposed to a constant current of $V_c = 0.07$ m/s. In the first case, it was initially headed towards the path; in the second, it was initially headed approximately parallel to the path. Similarly, in the third and fourth scenario, the robot was initially headed towards

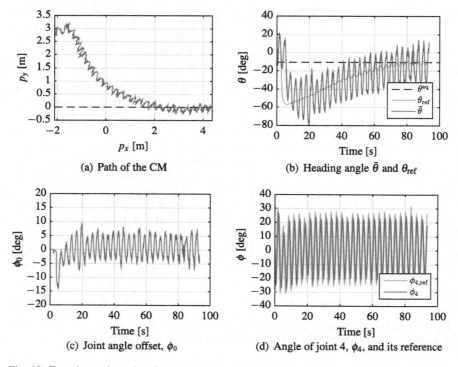

Fig. 10 Experimental results of the second scenario: Model-based integral LOS path-following with a flow speed $V_c = 0.07$ m/s, lateral undulation, and the robot initially headed parallel to the path

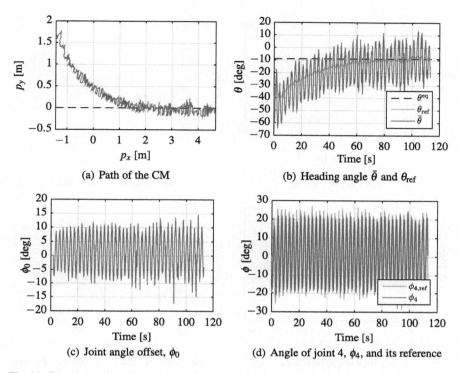

Fig. 11 Experimental results of the third scenario: Model-based integral LOS path-following with a flow speed $V_c = 0.05$ m/s, eel-like motion, and the robot initially headed towards the path

the path and approximately parallel to the path respectively, and propelled with eel-like motion against a current of $V_c = 0.05$ m/s. In all cases, the robot was initially straightened and kept in a fixed position.

The results of the four different test scenarios are presented in Figs. 9, 10, 11 and 12. The path of the CM of the robot is presented in Figs. 9a, 10a, 11a and 12a, respectively. In the first and fourth scenario, there was a small overshoot, but nevertheless, the robot approached the path and stayed on it in all four scenarios. Just like for the examples in Sect. 3.3, the CM did not stay constantly on the path after convergence, but oscillated about it. This was expected, since the oscillations of the CM are a consequence of the sinusoidal gait and merely not captured by the simplifications in the control-oriented model. The same effect can also be observed for the heading of the robot, $\bar{\theta}$, which is plotted in Figs. 9b, 10b, 11b and 12b. The small overshoots in the first and fourth scenario can also be seen in these plots. Unlike in Sect. 3.3, where the robot was more or less aligned with the path after convergence, the reference signal θ_{ref} instead oscillated about the steady-state angle θ^{eq}. This allowed the robot to side-slip along the path and thus compensate for the sideways component of the current. The oscillating nature of the reference signal θ_{ref} is again a result of the oscillations of the CM. The steady-state crab angle θ^{eq}

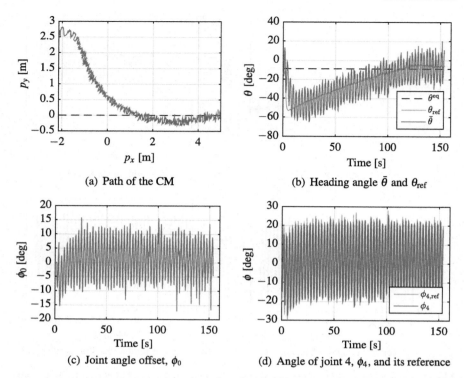

Fig. 12 Experimental results of the fourth scenario: Model-based integral LOS path-following with a flow speed $V_c = 0.05$ m/s, eel-like motion, and the robot initially headed parallel to the path

that is depicted in Figs. 9b, 10b, 11b and 12b was determined from (19) a posteriori. To this end, we needed the current component V_y and the relative velocity $v_{t,\text{rel}}$. The current component V_y was directly calculated as $V_y = V_c \cos(45°)$ since both the direction and magnitude V_c of the current were known, and the relative velocity $v_{t,\text{rel}}$ was approximated by the average speed $\bar{v}_{\text{rel}}$. The average speed $\bar{v}_{\text{rel}}$ was calculated from the norm of the relative velocity, and the relative velocity components in x and y were obtained by subtracting the current components V_x and V_y from the velocity components of the CM, $\dot{p}_x$ and $\dot{p}_y$. These had been extracted from the position measurements that were obtained during the experiments by using finite differences with a time step of 0.2 s. In order to calculate the average speed for lateral undulation, the measurements of the first scenario were evaluated, and for the average speed for eel-like motion, those of the third scenario were used. It can be seen from the figures that the theoretical result (19) predicted θ^{eq} correctly. The control input for the turning motion, ϕ_0, is shown in Figs. 9c, 10c, 11c and 12c, respectively. Compared to the tests in Sect. 3.3, the heading controller was started later, and thus the saturation of ϕ_0 was prevented. After convergence of the robot, ϕ_0 was oscillating about zero, because the robot stayed approximately at the constant angle θ^{eq}. The reference signal and the

Fig. 13 The snake robot Mamba during a test run of scenario two. The *yellow line* indicates the path and the buoy visualizes the current effect. The robot is initially approximately parallel to the path (1), turns towards it (2), approaches the path (3) and (4) and subsequently follows it (5) while side-slipping (6)

measured signal of joint number four are displayed in Figs. 9d, 10d, 11d and 12d. The measured joint angle clearly also tracked its reference. Pictures of the physical robot during the second scenario are presented in Fig. 13.

The experimental results validate the model-based integral LOS path-following controller in the presence of constant irrotational currents. The steady-state crab angle θ^{eq} was predicted correctly by the analytical relation in (19). The control objectives (9), (10) are satisfied in the sense that the robot oscillates about the desired values after convergence. As already pointed out in Sect. 3.3, these oscillations are a result of the sinusoidal motion that the robot conducts and that are not captured by the model in Sect. 2, which was used for the control design. It was therefore not expected from the theory that these oscillations would be suppressed by the model-based path-following controller.

5 Conclusions

This chapter reviewed and validated a model-based control system for straight-line path-following of neutrally buoyant underwater snake robots that move with a planar sinusoidal gait in the presence of unknown, constant and irrotational ocean currents. The control design was based on a simplified control-oriented snake robot model, which disregards the rotational motion of the single links but captures the overall behaviour of the system. The guidance method that was used in the path-following control system was a LOS guidance scheme, which was augmented with integral action in the presence of currents in order to eliminate the steady-state error that the original LOS guidance would give. This was achieved by allowing the robot to head towards a look-ahead point upstream of the path and thus travel at a non-zero crab angle. Experimental results were presented that verify the concept of the control system. Furthermore, the experimental results showed that the crab angle was correctly predicted by the theoretical analysis. Future work will focus on extending the control approach to a three-dimensional control strategy for underwater snake robots.

Acknowledgements The authors gratefully acknowledge the engineers at the Department of Engineering Cybernetics, Glenn Angell and Daniel Bogen, for the technical support before and during the experimental tests, Stefano Bertelli and Terje Haugen for preparing the necessary components for the experimental setup, the team at the SINTEF Fisheries and Aquaculture flume tank, Kurt Hansen, Nina A. H. Madsen, and Anders Nielsen for the support during the tests there and Martin Holmberg from Qualisys for setting up the motion capture system.

This work was supported by the Research Council of Norway through its Centres of Excellence funding scheme, project no. 223254-NTNU AMOS, and by VISTA - a basic research programme in collaboration between The Norwegian Academy of Science and Letters, and Statoil.

References

1. Børhaug, E., Pavlov, A., Pettersen, K.Y.: Integral LOS control for path following of underactuated marine surface vessels in the presence of constant ocean currents. In: Proceedings of the 47th IEEE Conference on Decision and Control, Cancun, Mexico, pp. 4984–4991 (2008)
2. Colgate, J.E., Lynch, K.M.: Mechanics and control of swimming: a review. IEEE J. Ocean. Eng. **29**(3), 660–673 (2004)
3. Fossen, T.I.: Handbook of Marine Craft Hydrodynamics and Motion Control. Wiley, London (2001)
4. Guo, J.: A waypoint-tracking controller for a biomimetic autonomous underwater vehicle. Ocean Eng. **33**, 2369–2380 (2006)
5. Hirose, S.: Biologically Inspired Robots: Snake-Like Locomotors and Manipulators. Oxford University Press, Oxford (1993)
6. Hopkins, J.K., Spranklin, B.W., Gupta, S.K.: A survey of snake-inspired robot designs. Bioinspiration Biomim. **4**, 021001 (2009)
7. Kelasidi, E., Liljebäck, P., Pettersen, K.Y., Gravdahl, J.T.: Innovation in underwater robots: biologically inspired swimming snake robots. IEEE Robot. Autom. Mag. **23**(1), 44–62 (2016)

8. Kelasidi, E., Liljebäck, P., Pettersen, K.Y., Gravdahl, J.T.: Integral line-of-sight guidance for path following control of underwater snake robots: theory and experiments. IEEE Trans. Robot. **1–19**, 99 (2017). doi:10.1109/TRO.2017.2651119
9. Kohl, A.M., Pettersen, K.Y., Kelasidi, E., Gravdahl, J.T.: Analysis of underwater snake robot locomotion based on a control-oriented model. In: Proceedings of the IEEE International Conference on Robotics and Biomimetics, Zhuhai, China, pp. 1930–1937 (2015)
10. Kohl, A.M., Kelasidi, E., Mohammadi, A., et al.: Planar maneuvering control of underwater snake robots using virtual holonomic constraints. Bioinspiration Biomim. **11**(6), 065005 (2016). doi:10.1088/1748-3190/11/6/065005
11. Kohl, A.M., Pettersen, K.Y., Kelasidi, E., Gravdahl, J.T.: Planar path following of underwater snake robots in the presence of ocean currents. IEEE Robot. Autom. Lett. **1**(1), 383–390 (2016)
12. Kruusmaa, M., Fiorini, P., Megill, W., et al.: FILOSE for svenning: a flow sensing bioinspired robot. IEEE Robot. Autom. Mag. **21**(3), 51–62 (2014)
13. Liljebäck, P., Pettersen, K.Y., Stavdahl, Ø., Gravdahl, J.T.: Snake Robots: Modelling, Mechatronics, and Control. Springer, London (2012)
14. Liljebäck, P., Pettersen, K.Y., Stavdahl, Ø., Gravdahl, J.T.: A review on modelling, implementation, and control of snake robots. Robot. Auton. Syst. **60**, 29–40 (2012)
15. Liljebäck, P., Stavdahl, Ø., Pettersen, K.Y., Gravdahl, J.T.: Mamba – a waterproof snake robot with tactile sensing. In: Proceedings of the IEEE/RSJ International Conference on Intelligent Robots and Systems, Chicago, IL, USA, pp. 294–301 (2014)
16. Mohammadi, A., Rezapour, E., Maggiore, M., Pettersen, K.Y.: Maneuvering control of planar snake robots using virtual holonomic constraints. IEEE Trans. Control Syst. Technol. **24**(3), 884–899 (2015)
17. Marine cybernetics laboratory (MC-lab) – operated by the Department of Marine Technology, Trondheim, Norway
18. McIsaac, K., Ostrowski, J.: Motion planning for anguilliform locomotion. IEEE Trans. Robot. Autom. **19**(4), 637–652 (2003)
19. Qualisys – Motion Capture Systems
20. Raj, A., Thakur, A.: Fish-inspired robots: design, sensing, actuation, and autonomy - a review of research. Bioinspiration Biomim. **11**, 031001 (2009)
21. The North Sea Centre Flume Tank – Managed and operated by SINTEF Fisheries and Aquaculture, Hirtshals, Denmark

Robotized Underwater Interventions

Giuseppe Casalino, Enrico Simetti and Francesco Wanderlingh

Abstract Working in underwater environments poses many challenges for robotic systems. One of them is the low bandwidth and high latency of underwater acoustic communications, which limits the possibility of interaction with submerged robots. One solution is to have a tether cable to enable high speed and low latency communications, but that requires a support vessel and increases costs. For that reason, autonomous underwater robots are a very interesting solution. Several research projects have demonstrated autonomy capabilities of Underwater Vehicle Manipulator Systems (UVMS) in performing basic manipulation tasks, and, moving a step further, this chapter will present a unifying architecture for the control of an UVMS, comprehensive of all the control objectives that an UVMS should take into account, their different priorities and the typical mission phases that an UVMS has to tackle. The proposed strategy is supported both by a complete simulated execution of a test-case mission and experimental results.

1 Introduction

Underwater operations are costly and demanding. Remotely Operated Vehicles (ROVs) are usually employed in lieu of professional divers to perform these tasks. The advantages of this solution are clearly an inherently higher safety (no human involved) and an higher operational time. On the downside, ROVs require a support vessel equipped with a dynamic positioning system and a tether management system

---

G. Casalino · E. Simetti (✉) · F. Wanderlingh
DIBRIS, Interuniversity Research Center on Integrated Systems
for Marine Environment (ISME), University of Genova, Genova, Italy
e-mail: enrico.simetti@unige.it

G. Casalino
e-mail: giuseppe.casalino@unige.it

F. Wanderlingh
e-mail: francesco.wanderlingh@dibris.unige.it

for handling the tether cable connecting the vessel to the ROV. In an effort to reduce costs of underwater operations, marine robotic research has taken two main paths: on the one hand improving ROVs autonomy, on the other hand paving the way for the use of autonomous UVMSs (Underwater Vehicle Manipulator Systems).

The DexROV project [1] is a recent EU Horizon 2020 (H2020) project that has the goal of reducing the costs of ROVs operations by increasing their autonomy. In this way, crew numbers onboard the ship can be reduced and can be transferred to an onshore control facility, from where they can control the ROVs operations. Naturally, controlling the ROV from a remote station through a satellite communication channel introduces heavy latency that must be properly taken into account. DexROV plans to solve this problem by using an advanced environment simulator, integrating a physics engine and a haptics engine. This environment will support a model-mediated teleoperation approach. Furthermore, as a robust approach to address communication latencies and disruptions, a cognitive engine relying on probabilistic movement/feedback manipulation primitives will be developed to analyze, interpret, and anticipate user interactions with the simulated environment and translate them into consistent *high level* commands that the ROV can execute autonomously in the real environment.

Another relevant project on the topic of underwater manipulation task is the recently started EU H2020 ROBUST project [2], whose main objective is to develop an autonomous, reliable, cost effective technology to map vast terrains, in terms of mineral and raw material contents, which will aid in reducing the cost of mineral exploration, currently performed by ROVs and dedicated support vessels and crew. Another objective is to identify, in an efficient and nonintrusive manner (minimum impact to the environment), the most rich mineral sites. To tackle these objectives, the ROBUST project proposes to use a fully autonomous UVMS. The robot should dive, identify the resources to be scanned and autonomously perform qualitative and quantitative in situ analyses, by positioning a laser, mounted on the end-effector of a manipulator, in close proximity of the rocks to be analyzed.

Research in the field of autonomous or semi-autonomous intervention is of course not limited to H2020 projects, but has roots that go back to early 90s, with the works of Woods Hole Oceanographic Institute on design and control of compliant underwater manipulators [3] and coordinated vehicle/arm control for teleoperation [4]. During that decade, two milestones were achieved. The AMADEUS project [5] demonstrated underwater dual arm autonomous manipulation in water tank experiments [6]. The second milestone was instead achieved within the UNION project [7], where for the first time the mechatronic assembly of an autonomous UVMS took place.

The following decade marked two further important milestones in underwater manipulation. The SAUVIM project [8, 9] tackled, for the first time, the problem of autonomous floating operations. However, considering the mass of the vehicle was 6 tons, while the arm weighted only 65 kg, the two subsystems were practically decoupled from a dynamic point of view. The second milestone was instead achieved within the EU project ALIVE [10, 11], which demonstrated how an UVMS could autonomously dock at a ROV-friendly panel, and successively perform fixed-base manipulation tasks.

The current decade has seen another important milestone, represented by the EU project TRIDENT [12, 13], where autonomous floating manipulation has been performed like in the SAUVIM project, but with the important difference that the arm and vehicle were much closer in terms of masses. The final demonstration took place in a harbor environment and the UVMS successfully recovered a mock-up of a black box from the seafloor [14, 15].

Finally, the authors have been involved in an Italian research project named MARIS [16], whose main objective was the development of control algorithms capable of integrating single [17], dual arm [18] and cooperative UVMSs [19] within a common control framework [20].

In this chapter, the kinematic control layer that has been used for the MARIS project, and is currently used and enhanced within the DexROV [21] and ROBUST projects will be presented. The focus is the kinematic control layer, because most ROV and arm commercial systems can only be controlled through velocity commands. A dynamic control layer (DCL) based on independent PI (proportional integrative) loops will be assumed, as this is the kind of control laws implemented by the commercial systems of the aforementioned projects.

The chapter is structured as follows. Basic definitions are reported in Sect. 2. Section 3 presents the common phases and control objectives in typical intervention missions. The successive Sect. 4 introduces the kinematic control layer and its core concepts, such as the phase definition, the activation functions, and the task hierarchy resolution. Section 5 reports a complete simulated execution of an intervention mission, as well as one free floating experiment of the MARIS project. Finally, the last section draws some conclusions and future line of developments.

2 Definitions and Preliminaries

2.1 Notation

Vectors and matrices are expressed with a bold face character, such as **M** for matrices or *v* for vectors, whereas scalar values are represented with a normal font such as γ. Given a matrix **M** and a vector *v*:

- $M_{(i,j)}$ indicates the element of **M** at the i-th row and j-th column;
- $v_{(k)}$ refers to the k-th element of *v*;
- $\mathbf{M}^{\#}$ is the exact generalized pseudoinverse (see [22] for a review on pseudoinverses and their properties), i.e., the pseudoinverse of **M** performed without any regularizations.

Further, less used notation will be introduced as needed.

Fig. 1 The UVMS and its relevant frames: $\langle v \rangle$ vehicle frame, $\langle c \rangle$ camera frame, $\langle e \rangle$ end-effector frame, $\langle t \rangle$ tool frame, $\langle g \rangle$ goal frame, $\langle o \rangle$ object frame

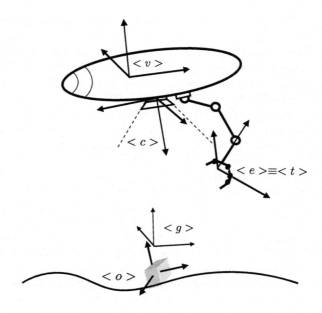

2.2 Definitions

Let us consider a free floating UVMS, such as the one depicted in Fig. 1, and let us first introduce some basic definitions, often used throughout the chapter:

- the system configuration vector $c \in \mathbb{R}^n$ of the UVMS as

$$c \triangleq \begin{bmatrix} q \\ \eta \end{bmatrix}, \qquad (1)$$

where $q \in \mathbb{R}^l$ is the arm configuration vector and $\eta \in \mathbb{R}^6$ is the vehicle *generalized coordinate position vector*, which is the stacked vector of the position vector η_1, with components on the inertial frame $\langle 0 \rangle$, and the orientation vector η_2, the latter expressed in terms of the three angles yaw, pitch and roll (applied in this sequence) [23]. From the above definitions it results $n = l + 6$;
- the system velocity vector $\dot{y} \in \mathbb{R}^n$ of the UVMS as

$$\dot{y} \triangleq \begin{bmatrix} \dot{q} \\ v \end{bmatrix}, \qquad (2)$$

where $\dot{q} \in \mathbb{R}^l$ is the joint velocities and $v \in \mathbb{R}^6$ is the stacked vector of the vehicle linear and angular velocity vectors, with components on the vehicle frame $\langle v \rangle$. To simplify the discussion, this chapter will assume the vehicle fully actuated, and

the system velocity will be used as control vector. Details on how the proposed algorithm can be adapted for under-actuated vehicles are given in the Appendix;
- a configuration dependent scalar variable $x(c)$ is said to correspond to an *equality control objective* when it is required to satisfy

$$x(c) = x_0, \qquad (3)$$

or to an *inequality control objective* when it is required to satisfy

$$x(c) \geq x_m \quad \text{or} \quad x(c) \leq x_M, \qquad (4)$$

where the m and M subscripts indicate a minimum and maximum value respectively.

Note that if m different variables $x_i(c)$ are considered, each of them corresponding to the i-th component of a vector $p \in \mathbb{R}^m$, then it is possible to control the vector to any desired value. Thus, limiting the discussion to scalar objectives does not influence the generality of the approach. Furthermore, if $x(c)$ is the modulus of a certain vector p, then it can be used to require a particular value for the norm of p (e.g., to nullify it), or to be below or above a given threshold. To ease the notation, the dependency of x on c is dropped from now on;
- for such variables, we also consider the existing Jacobian relationship between x and the system velocity vector $\dot{y}$ as

$$\dot{x} = g^T(c)\dot{y}, \qquad (5)$$

where $g \in \mathbb{R}^n$ is a vector. To simplify the notation, the dependency of g on c will not be repeated in the following.
- a *task* is defined as tracking a given feedback reference rate $\dot{\bar{x}}$ (see the remarks below), capable of driving the associated variable x toward the corresponding objective. Thus, for instance, a task is tracking at best a velocity reference rate generated to bring the arm's end-effector in the required Cartesian position; The control objectives may have different *priorities* and the same holds for their associated tasks. The achievement of a task with lower priority should not interfere with the achievement of an *active* task (see Sect. 4.2) with higher priority, and tasks with the same priority should be achieved simultaneously, if possible. A set of tasks with different priorities is also called a *hierarchy of tasks*.

Remark 1: If our objective is zeroing the norm of a vector $p = [p_1, \ldots, p_m]$, since the following equivalence holds

$$\|p\| = 0 \iff p_i = 0, \forall i \quad i = 1, \ldots, m, \qquad (6)$$

instead of imposing its norm to be zero with a single equality objective, we can also consider separately all of its m components to be zero. The main difference between these two approaches is that in the first case we have to employ only one degree of freedom (d.o.f.) in the Cartesian space, while in the latter use m d.o.f. since we are separately controlling each of its components.

Remark 2: For equality or inequality control objectives, a suitable feedback reference rate $\dot{\bar{x}}$ that drives our x toward any arbitrary point x^*, where the objective is satisfied, is

$$\dot{\bar{x}} \triangleq \gamma(x^* - x), \; \gamma > 0, \tag{7}$$

where γ is a positive gain to control the convergence speed. For inequality control objectives, Sect. 4.2 will explain how to disregard the reference rate whenever the variable lies within the desired range, in order to avoid overconstraining the system.

3 Underwater Intervention Missions and Related Control Objectives

3.1 Reference Mission and Relevant Phases

A typical reference mission carried out by an UVMS, involving manipulation and transportation, can be decomposed into the following sequential phases:

1. navigation: the vehicle should get in close proximity with the target object to be manipulated;
2. grasping: the UVMS must perform the grasping of the object;
3. transportation: the UVMS must transport the object to the target area;
4. deployment: whenever in close proximity with the target area, the object must be deployed in the required position.

Another mission, involving the inspection of a certain subsea structure, can be decomposed as follows:

1. navigation: the vehicle should get in close proximity with the target object to be inspected;
2. docking: the UVMS might need to dock to the underwater structure using an auxiliary arm;
3. inspection: the UVMS should perform the interaction with the structure (e.g., turn valve, plug connector, move a sensor on a surface).

At the end of the above phases, the UVMS can leave the target area and each of them can be assigned to a new mission.

In general, the execution of complex missions leads to identify different control phases, each one with its own control objectives. Obtaining a smooth transition between two consecutive phases will constitute a particular challenge to be solved.

3.2 Control Objectives Categories

The control objectives of the UVMS can be divided in five broad categories, which are listed in their natural descending order of priority:

- physical constraints objectives, i.e., interacting with the environment;
- system safety objectives, e.g., avoiding joint limits or obstacles;
- objectives that are a prerequisite for accomplishing the mission, e.g., maintaining the manipulated object in the camera angle of view;
- mission-oriented objectives, i.e., what the system really needs to execute to accomplish the user defined mission;
- optimization objectives, i.e., objectives that do not influence the mission, but allow to choose between multiple solutions, if they exists.

3.3 Typical Underwater Control Objectives

This section reports some of the control objectives that a UVMS has to tackle in a typical mission, without a particular ordering. For each objective, the type (equality/inequality) and its category are reported in brackets:

- Joint Limits (inequality, safety): the arm must operate within its joint limits, which means having the following inequality control objectives fulfilled:

$$q_{i,m} \leq q_i \leq q_{i,M} \quad i = 1, \ldots, l \tag{8}$$

where q_i is the i-th joint variable, $q_{i,m}$ and $q_{i,M}$ are the lower and higher joint bounds, and l is the total number of joints of the manipulator.
- Manipulability (inequality, prerequisite): the arm must operate as far as possible from kinematic singularities, which means to keep the manipulability measure μ [24] above a minimum threshold, i.e., $\mu > \mu_m$.
- Arm Reference Shape: maintain the arm in a preferred shape, which allows to perform repetitive tasks minimizing the internal motions and can also be used as an alternative method to ensure good manipulability.
- Camera Occlusion (inequality, prerequisite): keep the arm links away from the camera system's cone of vision to avoid unnecessary occlusions of the target object frame $\langle o \rangle$, e.g., to avoid arm elbow interferences with the camera well before the grasping phases.
- Force Regulation (equality, physical constraint): regulate to a constant value the force exerted by the end-effector. Let us define λ^* as the desired force that the end-effector must exert on the environment and λ as the actual force, then the objective is to have $\lambda = \lambda^*$. Note that this regulation has to be done at kinematic level, since the underlying DCL is assumed as given.

- Camera Centering (inequality, prerequisite): a stereo vision system is often used to obtain the position of the objected to be manipulated. To help in guaranteeing continuous visual contact with the target while approaching it, the norm of the misalignment vector between the camera and object frames $\boldsymbol{\xi}$ should be maintained lower than a given bound, e.g., $\|\boldsymbol{\xi}\| \leq \xi_M$.
- Horizontal Attitude (inequality, safety): avoid vehicle overturning. The norm of the misalignment vector $\boldsymbol{\varphi}$ between the absolute world frame z axis and the vehicle's one should be lower than a give value, i.e., $\|\boldsymbol{\varphi}\| \leq \varphi_M$. For vehicles and arms with comparable masses, such thresholds should be big enough to avoid excessive energy consumption whenever the arm moves its joints away from the center of buoyancy, tilting the vehicle. This objective should be considered only for fully actuated vehicles since for under-actuated ROVs or AUVs passively stable in roll and pitch these d.o.f. are not controllable.
- Vehicle Position (inequality, mission execution): has the vehicle frame $\langle v \rangle$ roughly aligned with a particular goal frame $\langle g_v \rangle$. This could be required in order to bring the vehicle close to the area where the manipulation needs to be carried out. This goal requires the achievement of the following inequality conditions:

$$\|\boldsymbol{r}_v\| \leq r_{v,M}, \quad \|\boldsymbol{\vartheta}_v\| \leq \vartheta_{v,M}, \tag{9}$$

where $\boldsymbol{r}_v$ is the position error and $\boldsymbol{\vartheta}_v$ the orientation error.
- End-effector/Tool-frame Position (equality, mission execution): this objective requires that the tool frame $\langle t \rangle$, rigidly attached to the end-effector space, converges to a given goal frame $\langle g \rangle$. In other words, the following two equality objectives must be eventually satisfied

$$r_{(i)} = 0, \ i = 1, 2, 3; \quad \vartheta_{(i)} = 0, \ i = 1, 2, 3; \tag{10}$$

where $\boldsymbol{r}$ is the position error and $\boldsymbol{\vartheta}$ the orientation error between the tool and goal frames. Note that the component by component zeroing has been used, in lieu of the norm zeroing, for achieving a straight convergence to the target (especially important for grasping tasks).

Moreover there can also be tasks directly specified at velocity level, which are useful for better controlling the overall behavior of the system, i.e.:

- Arm motion minimization (equality, optimization): in some cases (e.g., during transportation) it might be better if the arm minimizes its movements, leaving to the vehicle most of the work to move the end-effector. This allows to use the arm d.o.f. to compensate for the vehicle controller errors in tracking its reference velocity, as explained in Sect. 4.4;
- Vehicle motion minimization (equality, optimization): it might be preferred to have the vehicle roughly stationary (e.g., during grasping and manipulation), since usually the arm exhibits better control performances than the vehicle.

In order to prioritize one of the two subsystems these two task should be mutually exclusive. Let us also remark that these tasks exploit any residual arbitrariness on their relevant control variables so it is a good practice to place them at the bottom of the task hierarchy, since all the higher priority tasks are related to the safety of the system (e.g., joint limits, force regulation, vehicle attitude) or a prerequisite to complete the mission (e.g., manipulability, camera centering).

4 A Unifying Control Framework

The control of an UVMS is achieved by solving a sequence of optimization problems, following the assigned priority of each control objective. This mechanism descends from the original task priority framework [25] that has been extended to also encompass scalar tasks corresponding to inequality control objectives [15], where each of them is assigned with a different priority, and successively to include clusters of control objectives with equal priorities [18, 26]. We will now recall the basic steps behind the algorithmic structure of the task priority based control layer.

4.1 Phase Control Priorities and Unified Task List

Each phase of a mission is characterized by its control objectives, and consequently by a hierarchy of tasks. Two phases may have multiple tasks in common, despite they might have a different priority order within each list. For instance, consider the following two lists of tasks (now abstractly labeled with alphabetic letters) for two different phases, where $A \prec B$ denotes that A has higher priority than B:

$$P1: \quad A \prec B, C, D$$
$$P2: \quad A \prec D \prec C, E$$

where A, C, D are in common, but with D at a different priority ordering w.r.t. C. Now consider the following merged list:

$$P1, P2: \quad A \prec D \prec B, C, D, E;$$

It is clear that, through insertion/deletion of some of the entries, the two original lists can be reconstructed. To do so, the mechanism of activation functions will be exploited, as explained in the next section.

Before concluding, from now on tasks at the same priority level will be stacked together forming a so-called multidimensional task $\dot{\bar{x}}_i$, where i indicates the priority level. For generality, we will consider scalar tasks as a particular case of the multidimensional ones, and consequently we shall indicate a prioritized task list simply as $\dot{\bar{x}}_1, \ldots, \dot{\bar{x}}_N$. and the Jacobians relevant to the actual task velocities $\dot{x}_1, \ldots, \dot{x}_N$ as $J_1, \ldots, J_N$.

4.2 Activation Functions

Let us consider a multidimensional task, and let us consider an activation function associated to each j-th of its components, called $a_{(j)}$, to be then organized in a diagonal activation matrix $\mathbf{A}$, whose meaning is the following:

- if $a_{(j)} = 1$, the associated scalar task is called *active* and the corresponding actual $\dot{x}_{(j)}$ should therefore track $\dot{\bar{x}}_{(j)}$ as close as possible;
- if $a_{(j)} = 0$, the scalar task is termed *inactive* and the actual $\dot{x}_{(j)}$ should be unconstrained;
- if $0 < a_{(j)} < 1$ the scalar task is termed *in transition* and the actual $\dot{x}_{(j)}$ should smoothly evolve between the two previous cases.

In particular, the overall activation function $a_{(j)}$ is the product of two functions

$$a_{(j)} \triangleq a_{(j)}^{p} a_{(j)}^{i}, \tag{11}$$

which have the following specific purposes:

- $a_{(j)}^{i}$ is a function of the control objective variable $x_{(j)}$, and its purpose is to deactivate the task whenever the inequality objective is satisfied, to avoid overconstraining the system. Note that for equality control objectives it clearly holds that $a_{(j)}^{i} = 1$;
- $a_{(j)}^{p}$ is a function of the mission phases, and is used to activate or deactivate a task during a phase transition. For example, it can be a function of the time elapsed within the current phase, allowing the proper activation of new tasks, and deactivation of objectives that are not anymore relevant.

More specifically, $a_{(j)}^{i}$ is defined as follows for objectives of the type $x_{(j)} \leq x_{(j),M}$ (a similar function can be constructed for objectives $x_{(j)} \geq x_{(j),m}$):

$$a_{(j)}^{i} \triangleq \begin{cases} 1, & x_{(j)} > x_{(j),M} \\ s_j(x), & x_{(j),M} - \Delta_{(j)} \leq x_{(j)} \leq x_{(j),M} \\ 0, & x_{(j)} < x_{(j),M} - \Delta_{(j)} \end{cases} \tag{12}$$

where $s_j(x)$ is any sigmoid function exhibiting a continuous behavior from 0 to 1. The $\Delta_{(j)}$ value allows to create a buffer zone, where the inequality is already satisfied, but the activation value is still greater than zero, to prevent any chattering problem around the inequality control objective threshold. An example of such a function is reported in Fig. 2.

Let us briefly clarify how the activation function mechanism works with inequality control objectives through an example. Suppose that $x_{(j),M} = 0.1$ and $\Delta_{(j)} = 0.05$ (this is the case depicted in Fig. 2). Then if we consider x^* in (7) equal to $x_{(j),M} - \Delta_{(j)} = 0.05$ (note that 0.05 is inside the validity region of the inequality objective) we have the following behavior:

Fig. 2 Example of activation function corresponding to a control objective $x_{(j)} \leq 0.1$, with $\Delta = 0.05$

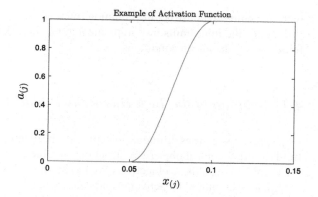

- When $x_{(j)} > 0.1$, the task is active $a_{(j)}^i = 1$ and the feedback reference rate defined in (7) will drive $x_{(j)}$ toward the region where $x_{(j),M} < 0.1$, in particular toward the point $x^* = 0.05$.
- When $0.05 < x_{(j)} < 0.1$, the task is in the transition zone $0 < a_{(j)}^i < 1$. The feedback reference rate will still try to drive $x_{(j)}$ towards the point $x^* = 0.05$. However, the priority of this task is not any longer fully enforced. Indeed, as $a_{(j)}^i$ decreases from 1 to 0, lower priority tasks will have the opportunity to influence the current task, possibly pushing it to either its lower bound (0.05, not in conflict with the current task) or its upper bound (0.1, in conflict with the current task).
  Let us for a moment consider a continuous time, purely kinematic system. In the former case, the task will be simply completely deactivated. In the latter case $x_{(j)}$ will decrease until the point where lower priority tasks' influence on the current task matches the feedback reference rate, achieving a stable condition within the transition zone, where by definition the inequality control objective is satisfied. This behavior is enforced by the specific regularization mechanism that will be briefly presented in Sect. 4.3 and that can be found in [26].
  In practice, discrete time control system has to be used both at kinematic and dynamic velocity control level, where the latter is usually characterized by higher control rates. For given dynamic performances, which will depend on the control rate, gains and specific dynamic controller structure, a tuning of both kinematic gains and activation functions parameters is actually required to ensure that chattering phenomena do not appear. Since this may sometime result into a tedious trial-and-error process, research effort are currently devoted toward the possible formalization of specific procedures for the problem in hand.
- Finally, when $x_{(j)} < 0.05$, the task is inactive and the task velocity is determined only by the rest of the task hierarchy.

Remark: We have implicitly considered objectives of the type $x_m < x < x_M$ as two separate ones. Note that if x_m and x_M are sufficiently spaced, i.e., $x_m + \Delta < x_M - \Delta$, then they can be considered together by using as activation function the sum of the two activation functions, and by choosing an arbitrary point inside the validity of both inequality to construct the common reference rate in (7). This is actually what is done for the joint limits task implementation, since the minimum and maximum limits satisfy the above conditions.

4.3 Solution of the Task Hierarchy Problem

In the previous sections different concepts have been introduced. The UVMS needs to fulfil different control objectives. Tracking suitable feedback reference rates allows the UVMS to meet these objectives. We have termed this need as task. We have also seen that tasks should be activated and deactivated for two main reasons: either the current control phase does not require that specific objective and its related task, or the task corresponds to an inequality objective which is currently satisfied. For the reasons outlined above, whenever that situation occurs we do not want the control to be overconstrained. To comply with this need, activation functions have been introduced, which describe whether a given task should or not be fulfilled.

Having said that, the problem becomes that of tracking the given reference velocities, following the required priority order, and taking into account their corresponding activation values. The solution of this problem can be found solving the following sequence of minimization problems:

$$S_k \triangleq \left\{ \arg \text{R-} \min_{\dot{y} \in S_{k-1}} \left\| \mathbf{A}_k(\dot{\bar{x}}_k - \mathbf{J}_k \dot{y}) \right\|^2 \right\}, \quad k = 1, 2, \ldots, N, \quad (13)$$

where S_{k-1} is the manifold of solutions of all the previous tasks in the hierarchy.

The solution of each problem in (13) is not straightforward. It is well known that minimization problems can be invariant to weights (such is the role of $\mathbf{A}_k$) [27, 28]. For that reason, we have used the notation R- min, to highlight the fact that we employ a special regularized pseudoinverse solution of that problem, as defined in [26]. In fact, the solution of the regularized minimization problem exploits the following definition of pseudoinverse operator $\mathbf{X}^{\#,\mathbf{A},\mathbf{Q}}$ for given non-negative definite matrices $\mathbf{A}$, $\mathbf{Q}$, with dimensions equal to the rows and columns of $\mathbf{X}$ respectively:

$$\mathbf{X}^{\#,\mathbf{A},\mathbf{Q}} \triangleq \left(\mathbf{X}^T \mathbf{A} \mathbf{X} + (\mathbf{I} - \mathbf{Q})^T (\mathbf{I} - \mathbf{Q}) + \mathbf{V}^T \mathbf{H} \mathbf{V} \right)^{\#} \mathbf{X}^T \mathbf{A} \mathbf{A}, \quad (14)$$

where $\mathbf{V}$ is the right orthonormal matrix of the SVD decomposition of $\mathbf{X}^T \mathbf{A} \mathbf{X} + (\mathbf{I} - \mathbf{Q})^T (\mathbf{I} - \mathbf{Q})$. The matrix $\mathbf{H}$ is a diagonal (singular value-oriented regularization) matrix, whose elements $h_{(i,i)}$ are bell-shaped, finite support functions (similar to the activation functions) of the corresponding singular value of the same mentioned SVD decomposition. As a brief insight, let us remark how the pseudoinverse operator *explicitly* depends on the weights $\mathbf{A}$ (the activation function) and $\mathbf{Q}$ (the nonorthogonal projection matrix). This fact allows the operator to avoid the problems of discontinuity arising from the invariance of the minimization w.r.t. the weights. The interested reader can find more details in [26].

Using the above operator for solving a hierarchy of tasks, a methodology termed *iCAT task priority framework*, results in the following algorithm:

$$\boldsymbol{\rho}_0 = \mathbf{0}, \quad \mathbf{Q}_0 = \mathbf{I}, \tag{15}$$

then for $k = 1, \ldots, N$

$$\begin{aligned}
\mathbf{W}_k &= \mathbf{J}_k \mathbf{Q}_{k-1} (\mathbf{J}_k \mathbf{Q}_{k-1})^{\#, \mathbf{A}_k, \mathbf{Q}_{k-1}}, \\
\mathbf{Q}_k &= \mathbf{Q}_{k-1} (\mathbf{I} - (\mathbf{J}_k \mathbf{Q}_{k-1})^{\#, \mathbf{A}_k, \mathbf{I}} \mathbf{J}_k \mathbf{Q}_{k-1}), \\
\boldsymbol{\rho}_k &= \boldsymbol{\rho}_{k-1} + \mathbf{Q}_{k-1} (\mathbf{J}_k \mathbf{Q}_{k-1})^{\#, \mathbf{A}_k, \mathbf{I}} \mathbf{W}_k \left(\dot{\bar{\mathbf{x}}}_k - \mathbf{J}_k \boldsymbol{\rho}_{k-1} \right),
\end{aligned} \tag{16}$$

where

- $\boldsymbol{\rho}_k$ is the control vector, which is computed in an iterative manner by descending the various priority levels;
- $\left(\dot{\bar{\mathbf{x}}}_k - \mathbf{J}_k \boldsymbol{\rho}_{k-1} \right)$ is the modified task reference that takes into account the contribution of the control vector $\boldsymbol{\rho}_{k-1}$ established at the previous iteration;
- $\mathbf{Q}_{k-1}$ is the projection matrix that is used to take into account the control direction (totally or partially) used by the higher priority tasks;
- $\mathbf{W}_k$ is a $m \times m$ matrix, where m is the row-dimension of the task at the current priority level, whose effect is to modify the task reference $\left(\dot{\bar{\mathbf{x}}}_k - \mathbf{J}_k \boldsymbol{\rho}_{k-1} \right)$ to avoid discontinuities between priority levels.

A further improvement that can be applied to this technique is to take into account velocity saturations at each priority level, following the methodology first proposed in [29]. To do so, the update equation of $\boldsymbol{\rho}_k$ is redefined as follows:

$$\boldsymbol{\rho}_k = \boldsymbol{\rho}_{k-1} + \text{Sat}\left(\mathbf{Q}_{k-1} (\mathbf{J}_k \mathbf{Q}_{k-1})^{\#, \mathbf{A}_k, \mathbf{I}} \mathbf{W}_k \left(\dot{\bar{\mathbf{x}}}_k - \mathbf{J}_k \boldsymbol{\rho}_{k-1} \right) \right), \tag{17}$$

where the function $\text{Sat}(\cdot)$ implements the saturation proposed in [29].

The above task hierarchy resolution ends with the N-th manifold $S_N = \{\dot{\mathbf{y}} = \boldsymbol{\rho}_N + \mathbf{Q}_N \dot{\mathbf{z}}_N ; \forall \dot{\mathbf{z}}_N\}$. To ensure the continuity, a final minimization on the control vector needs to be performed [26], leading to the following final velocity control vector:

$$\dot{\mathbf{y}} = \arg \text{R-} \min_{\dot{\mathbf{y}} \in S_N} \| \dot{\mathbf{y}} \|^2 = \boldsymbol{\rho}_{N+1}. \tag{18}$$

4.4 Vehicle Velocity Tracking Error Compensation Scheme

An important consideration when dealing with UVMS is that the two subsystem are characterized by very different dynamic performances. Indeed, it is well known that thrusters have nonlinear properties [30–32]. This nonlinearity, coupled with the higher mass of the vehicle compared to the manipulator's one, makes the velocity control of the vehicle far less accurate than the arm's one.

The proposed algorithm (16) solves the task hierarchy considering the vehicle and arm velocity together in the stacked vector $\dot{\mathbf{y}}$. However, for reasons states at the start of this section, inevitable vehicle velocity tracking errors will occur. To overcome this problem, the idea is to add, in parallel to (16), another task hierarchy resolution, where only the arm variables are subject to optimization. The *actual* vehicle velocity, as measured by onboard sensors, is used as a parameter to solve the task hierarchy.

This procedure leads to the following set of equations:

$$\begin{aligned}
\mathbf{W}_k &= \mathbf{J}_k^a \mathbf{Q}_{k-1} (\mathbf{J}_k^a \mathbf{Q}_{k-1})^{\#, \mathbf{A}_k, \mathbf{Q}_{k-1}}, \\
\mathbf{Q}_k &= \mathbf{Q}_{k-1} (\mathbf{I} - (\mathbf{J}_k^a \mathbf{Q}_{k-1})^{\#, \mathbf{A}_k, \mathbf{I}} \mathbf{J}_k^a \mathbf{Q}_{k-1}), \\
\boldsymbol{\rho}_k &= \boldsymbol{\rho}_{k-1} + \text{Sat}\left(\mathbf{Q}_{k-1} (\mathbf{J}_k \mathbf{Q}_{k-1})^{\#, \mathbf{A}_k, \mathbf{I}} \mathbf{W}_k \left(\dot{\bar{\mathbf{x}}}_k - \mathbf{J}_k^a \boldsymbol{\rho}_{k-1} - \mathbf{J}_k^v \boldsymbol{\nu} \right) \right),
\end{aligned} \tag{19}$$

where now $\boldsymbol{\rho}_k \in \mathbb{R}^l$ contains only the arm joint velocities and each Jacobian $\mathbf{J}_k$ has now been considered as its two separate vehicle contributions, i.e., $\mathbf{J}_k \triangleq \begin{bmatrix} \mathbf{J}_k^a & \mathbf{J}_k^v \end{bmatrix}$. The resulting arm control law (19), since parametrized by the vehicle velocity $\boldsymbol{\nu}$, results to be the optimal one in correspondence of *any* vehicle velocity.

5 Simulation and Experimental Results

A series of experiments performed using the proposed task priority based kinematic controller are presented in this section. Experimental results on underwater grasping taken from the MARIS project are reported in the first subsection. The successive one presents a simulation of underwater pipe inspection with force regulation taken from the DeXROV project.

5.1 Approaching and Grasping

The following section presents the data collected during the MARIS project. The setup consisted in a UVMS composed of an under-actuated vehicle with controllable d.o.f. x, y and yaw angle ψ, and a 7 d.o.f. manipulator, with a 3-fingered gripper attached as shown in Fig. 3. Note that the z axis of the vehicle was disabled for safety reasons, since the trials were done in a pool with only 3 m of water. The vehicle was controlled at 10 Hz rate, while the arm was controlled at 33 Hz. The reference mission was approaching and grasping an underwater pipe, whose position was estimated using an onboard stereo camera [33] at an approximate frequency of 3 Hz.

The first trial, depicted in Fig. 4 shows the generated joint velocities using the approach presented in Sect. 4, with the vehicle velocity compensation of Sect. 4.4. It can be noted that the vehicle feedback was particularly noisy on the angular velocity

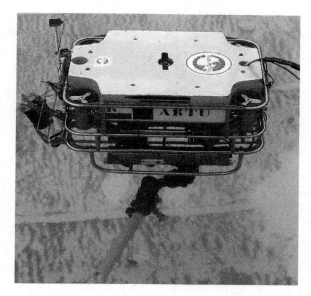

Fig. 3 The UVMS R2, used throughout the on-field experiments, while approaching an underwater pipe to grasp it (photo courtesy of the MARIS consortium)

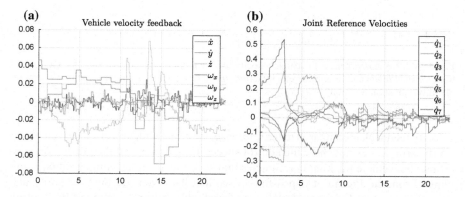

Fig. 4 MARIS experiment without vehicle velocity feedback filtering: **a** vehicle velocity feedback and **b** reference arm joint velocities

components. In order to compensate for the noisy feedback, the arm moved quite quickly, inducing actual oscillations in the system due to the dynamic coupling with the floating base.

To cope with the above problem, a simple first-order filter on the vehicle feedback was introduced, with a cut-off frequency at angular frequency of 50 rad/s. New trials were then performed; one of these is depicted in Fig. 5a, where now the filtered

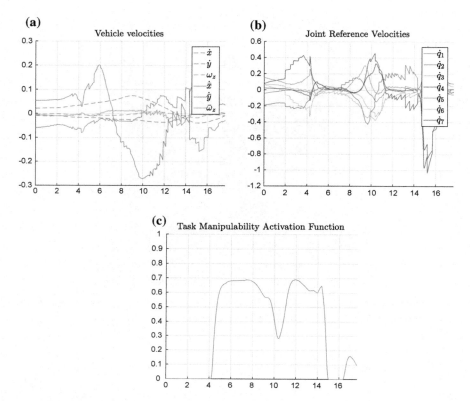

Fig. 5 MARIS experiment with vehicle velocity feedback filtering: **a** comparison of vehicle feedback (*dashed lines*) and reference velocity (*solid lines*), **b** reference arm joint velocities, **c** manipulability activation function

velocity feedback is compared to the reference one along with the reference velocities for the arm. Let us note how the control successfully completed the grasp sequence despite the vehicle tracking inaccuracies thanks to the proposed compensation technique. From the Fig. 5b, it is possible to see that the generated joint velocities are smoother compared to the first trial, even though the multi-rate nature of the control and feedback can be appreciated. Figure 5c shows the time behavior of the activation function of the manipulability task. A video showing one of the experiments can be seen at the URL https://youtu.be/b3jJZUoeFTo.

Finally, it must be noted how the experiments were necessarily preceded by a general tuning of all the parameters, ranging from the control gains of each task reference rate, to the size of the activation functions buffers Δ.

5.2 Force Regulation

A series of simulations were carried out to test the effectiveness of the proposed algorithm including a force regulation task. The reference mission is the inspection of a pipeline weld. To carry out the inspection, a sensor attached to the end-effector must be put in contact with the pipe along all the weld. The shape of the inspected object is not known, although the knowledge of a two-dimensional reference path defined on an known underlying surface was assumed. The UVMS must regulate the force to a given reference value to maintain contact with the pipe without damaging the sensor's probe and adapting to the unknown surface. We recall that this regulation is performed at kinematic level, with a dedicated task at the *top* of the task hierarchy. This is done by measuring the force normal, and generating a velocity along that direction proportional to the force error. This reference velocity, at dynamic level, will become an actual force. This further ensures that lower priority task do not generate velocity references along the direction of the physical constraint.

Compared to previous results [34], this simulation includes realistic saturation values for the vehicle generalized force/moments resulting at the vehicle frame and arm joint torques, other than an hydrodynamic model of the UVMS. The thruster dynamics have not been modeled; the vehicle tracking accuracy are attributed only to the inertia effects and dynamic control performances. In any case, the proposed velocity compensation technique takes into account the resulting effect and would not be different if the thruster dynamic was included in the simulation (Fig. 6).

In the simulation, the underlying DCL implements a PI control law. This case reproduces the actual implementation of DexROV, where the arm and vehicle DCLs are provided by the respective manufacturers and implementing global dynamic controllers, such as a computed torque, are not possible. The regulation of the force

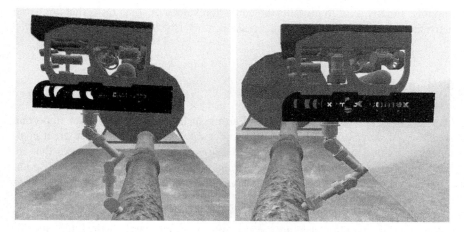

Fig. 6 Two snapshots of the force regulation experiments during the inspection of an underwater pipeline weld

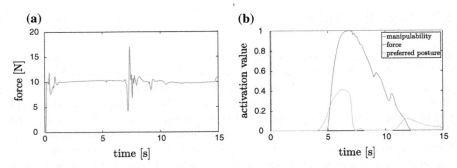

Fig. 7 Pipeline inspection simulation: **a** force exerted by the end-effector on the pipe (desired value 10 N) and **b** activation values of the relevant tasks

is done through a proportional term, since the underlying dynamic level already contains an integral part. The simulation shows that the regulation is accomplished despite different tasks are being activated and deactivated during the trial, e.g., the joint limits and manipulability ones. The proposed controller implements the saturation of reference joint and vehicle velocities as proposed in (17), to avoid generating unrealistic values. Finally, the simulation implements a multi-rate control, where the kinematic control is run at 100 Hz, while the DCL and the dynamic simulation runs at 1 kHz.

Figure 7a shows the force exerted on the pipe, which is very close to the desired value of 10 N. The spikes are in correspondence of the end of the legs, where the vehicle needs to move in a difference direction and where most of the vehicle velocity tracking inaccuracies occur. Those spikes were present also in the simulations presented in [34]; however, the saturation on the actuators increase this effect. Further investigations on how to mitigate these spikes despite saturation are currently ongoing. Finally, Fig. 7b presents the activation values of some tasks that are being activated and deactivated during the mission.

6 Concluding Remarks

Oceans cover approximately 70% of Earth's surface and represent an important source of resources. Unfortunately, the underwater environment poses great challenges to robotic systems, due to high pressures and hydrodynamic forces that are both nonlinear. A further important constraint is the low bandwidth of acoustic communications, which are the de facto standard since electromagnetic communications only work at very short ranges.

Despite such premises, underwater robotic systems capable of performing autonomous intervention tasks are a very active topic of research. In the future, the difference between ROV systems with advanced autonomy capabilities and completely autonomous UVMSs is expected to be less prominent. In fact, one of the trends is to imagine hybrid solutions where the robotic system autonomously navigates toward a target area, docks with a subsea structure and a tether cable, and then performs as a semi-autonomous ROV system. Such a solution is very interesting to reduce the maintenance costs of permanent underwater structures, especially for offshore platforms.

The employment of UVMSs can be of course predicted also for applications that do not target the offshore market. For example, exploration for resources with the possibility of in situ measurements, such as what is proposed in the ROBUST project, is very interesting and promising. Other applications might include the use of UVMSs for deep-sea archaeological sites.

This chapter has presented the overall kinematic control strategy of an UVMS, supported both by experimental results of the TRIDENT and MARIS projects, which is now currently refined in the scope of the DexROV and ROBUST projects. More investigation efforts are still necessary for what concerns interaction tasks with the environment, especially under the presence of saturation effects.

The chapter covered only control issues. However, perception, navigation, communication, and dynamic modeling are other fundamental topics for the development of autonomous underwater robotic systems. The interested reader can find further readings on these topics in the following survey papers and books [35–40].

Acknowledgements This work has been supported by the MIUR (Italian Ministry of Education, University and Research) through the MARIS prot. 2010FBLHRJ project and by the European Commission through the H2020-BG-06-2014-635491 DexROV project and the H2020-SC5-2015-690416 ROBUST project.

Appendix

Under-Actuated Vehicles

This chapter has presented UVMS control algorithms under the assumption of fully actuated vehicle. However, in many cases the vehicles are passively stable in some d.o.f. (typically roll and/or pitch). The algorithm (16) can easily cope with this situation, by using a slightly different initialization. As an example, in lieu of (15) consider the following initial values for a vehicle with roll and pitch not actuated:

$$\boldsymbol{\rho}_0 = \begin{bmatrix} \mathbf{0}_{l \times 1} \\ \mathbf{0}_{3 \times 1} \\ \omega_x \\ \omega_y \\ 0 \end{bmatrix} \quad \mathbf{Q}_0 = \begin{bmatrix} \mathbf{I}_{l \times l} & \mathbf{0}_{l \times 3} & \mathbf{0}_{l \times 1} & \mathbf{0}_{l \times 1} & \mathbf{0}_{l \times 1} \\ \mathbf{0}_{3 \times l} & \mathbf{I}_{3 \times 3} & \mathbf{0}_{3 \times 1} & \mathbf{0}_{3 \times 1} & \mathbf{0}_{3 \times 1} \\ \mathbf{0}_{1 \times l} & \mathbf{0}_{1 \times 3} & 0 & 0 & 0 \\ \mathbf{0}_{1 \times l} & \mathbf{0}_{1 \times 3} & 0 & 0 & 0 \\ \mathbf{0}_{1 \times l} & \mathbf{0}_{1 \times 3} & 0 & 0 & 1 \end{bmatrix}. \quad (20)$$

The idea is that the solution $\boldsymbol{\rho}$ is initialized with the actual angular velocities of the vehicle, as measured by onboard sensors. At the same time, to force the task hierarchy resolution to avoid changing these values, the corresponding diagonal values of the matrix $\mathbf{Q}$ are set to zero. This effectively inhibits the algorithm from changing the initial values. Note that all the tasks will properly take into account the nonactuated d.o.f. velocities due to the term $\left(\dot{\bar{x}}_k - \mathbf{J}_k \boldsymbol{\rho}_{k-1}\right)$.

References

1. Gancet, J., Weiss, P., Antonelli, G., Pfingsthorn, M.F., Calinon, S., Turetta, A., Walen, C., Urbina, D., Govindaraj, S., Letier, P., Martinez, X., Salini, J., Chemisky, B., Indiveri, G., Casalino, G., Di Lillo, P., Simetti, E., De Palma, D., Birk, A., Fromm, T., Mueller, C., Tanwani, A., Havoutis, I., Caffaz, A., L, Guilpain: Dexterous undersea interventions with far distance onshore supervision: the dexrov project. In: 10th IFAC Conference on Control Applications in Marine Systems. IFAC, vol. 49, no. 23, pp. 414–419. Elsevier, Trondheim, Norway (2016)
2. ROBUST website. http://eu-robust.eu (2016). Accessed 25 Oct 2016
3. Yoerger, D.R., Schempf, H., DiPietro, D.M.: Design and performance evaluation of an actively compliant underwater manipulator for full-ocean depth. J. Robot. Syst. **8**(3), 371–392 (1991)
4. Schempf, H., Yoerger, D.: Coordinated vehicle/manipulator design and control issues for underwater telemanipulation. In: Applications in Marine Systems (CAMS 92), Genova, Italy (1992)
5. Lane, D.M., Davies, J.B.C., Casalino, G., Bartolini, G., Cannata, G., Veruggio, G., Canals, M., Smith, C., O'Brien, D.J., Pickett, M., Robinson, G., Jones, D., Scott, E., Ferrara, A., Angelleti, D., Coccoli, M., Bono, R., Virgili, P., Pallas, R., Gracia, E.: Amadeus: advanced manipulation for deep underwater sampling. IEEE Robot. Autom. Mag. **4**(4), 34–45 (1997)
6. Casalino, G., Angeletti, D., Bozzo, T., Marani, G.: Dexterous underwater object manipulation via multi-robot cooperating systems. In: IEEE International Conference on Robotics and Automation, 2001. Proceedings 2001 ICRA, vol. 4, pp. 3220–3225. IEEE (2001)
7. Rigaud, V., Coste-Manière, È., Aldon, M.-J., Probert, P., Perrier, M., Rives, P., Simon, D., Lang, D., Kiener, J., Casal, A., et al.: Union: underwater intelligent operation and navigation. Robot. Autom. Mag. IEEE **5**(1), 25–35 (1998)
8. Yuh, J., Choi, S., Ikehara, C., Kim, G., McMurty, G., Ghasemi-Nejhad, M., Sarkar, N., Sugihara, K., Design of a semi-autonomous underwater vehicle for intervention missions (SAUVIM). In: Proceedings of the 1998 International Symposium on Underwater Technology, pp. 63–68. IEEE, Tokyo, Japan (1998)
9. Marani, G., Choi, S.K., Yuh, J.: Underwater autonomous manipulation for intervention missions AUVs. Ocean Eng. **36**, 15–23 (2008)
10. Evans, J., Redmond, P., Plakas, C., Hamilton, K., Lane, D.: Autonomous docking for Intervention-AUVs using sonar and video-based real-time 3D pose estimation. In: Oceans 2003, vol. 4, pp. 2201–2210. IEEE (2003)
11. Marty, P., et al.: Alive: An autonomous light intervention vehicle. In: Advances in Technology for Underwater Vehicles Conference. Oceanology International, vol. 2004 (2004)
12. Casalino, G., Zereik, E., Simetti, E., Torelli, S., Sperindé, A., Turetta, A.: A task and subsystem priority based control strategy for underwater floating manipulators. In: IFAC Workshop on Navigation, Guidance and Control of Underwater Vehicles (NGCUV 2012), pp. 170–177, Porto, Portugal (2012)
13. Casalino, G., Zereik, E., Simetti, E., Torelli, S., Sperindé, A., Turetta, A.: Agility for underwater floating manipulation task and subsystem priority based control strategy. In: International Conference on Intelligent Robots and Systems (IROS 2012), pp. 1772–1779, Vilamoura, Portugal (2012)

14. Simetti, E., Casalino, G., Torelli, S., Sperindé, A., Turetta, A.: Experimental results on task priority and dynamic programming based approach to underwater floating manipulation. In: OCEANS 2013. Bergen, Norway (2013)
15. Simetti, E., Casalino, G., Torelli, S., Sperindé, A., Turetta, A.: Floating underwater manipulation: developed control methodology and experimental validation within the trident project. J. Field Robot. **31**(3), 364–385 (2014). May
16. Casalino, G., Caccia, M., Caiti, A., Antonelli, G., Indiveri, G., Melchiorri, C., Caselli, S.: Maris: A national project on marine robotics for interventions. In: 22nd Mediterranean Conference of Control and Automation (MED). IEEE, 864–869 (2014)
17. Casalino, G., Caccia, M., Caselli, S., Melchiorri, C., Antonelli, G., Caiti, A., Indiveri, G., Cannata, G., Simetti, E., Torelli, S., Sperind, A., Wanderlingh, F., Muscolo, G., Bibuli, M., Bruzzone, G., Zereik, E., Odetti, A., Spirandelli, E., Ranieri, A., Aleotti, J., Lodi Rizzini, D., Oleari, F., Kallasi, F., Palli, G., Moriello, L., Cataldi, E.: Underwater intervention robotics: an outline of the italian national project MARIS. Mar. Technol. Soc. J. **50**(4), 98–107 (2016)
18. Simetti, E., Casalino, G.: Whole body control of a dual arm underwater vehicle manipulator system. Annu. Rev. Control **40**, 191–200 (2015)
19. Manerikar, N., Casalino, G., Simetti, E., Torelli, S., Sperindé, A.: On autonomous cooperative underwater floating manipulation systems. In: International Conference on Robotics and Automation (ICRA 15), pp. 523–528. IEEE, Seattle, WA (2015)
20. Simetti, E., Casalino, G.: Manipulation and transportation with cooperative underwater vehicle manipulator systems. IEEE J. Ocean. Eng. (2016)
21. Di Lillo, P.A., Simetti, E., De Palma, D., Cataldi, E., Indiveri, G., Antonelli, G., Casalino, G.: Advanced rov autonomy for efficient remote control in the DexROV project. Mar. Technol. Soc. J. **50**(4), 67–80 (2016)
22. Ben-Israel, A., Greville, T., Generalized Inverses: Theory and Applications, vol. 15. Springer, Berlin (2003)
23. Perez, T., Fossen, T.I.: Kinematic models for manoeuvring and seakeeping of marine vessels. Model. Identif. Control **28**(1), 19–30 (2007)
24. Yoshikawa, T.: Manipulability of robotic mechanisms. Int. J. Robot. Res. **4**(1), 3–9 (1985)
25. Siciliano, B., Slotine, J.-J.E.: A general framework for managing multiple tasks in highly redundant robotic systems. In: Proceedings of the Fifth International Advanced Robotics 'Robots in Unstructured Environments', 91 ICAR. Conference, pp. 1211–1216. IEEE, Pisa, Italy (1991)
26. Simetti, E., Casalino, G.: A novel practical technique to integrate inequality control objectives and task transitions in priority based control. J. Intell. Robot. Syst. **84**(1), 877–902 (2016)
27. Doty, K.L., Melchiorri, C., Bonivento, C.: A theory of generalized inverses applied to robotics. Int. J. Robot. Res. **12**(1), 1–19 (1993)
28. Mansard, N., Remazeilles, A., Chaumette, F.: Continuity of varying-feature-set control laws. IEEE Trans. Autom. Control **54**(11), 2493–2505 (2009)
29. Antonelli, G., Indiveri, G., Chiaverini, S.: Prioritized closed-loop inverse kinematic algorithms for redundant robotic systems with velocity saturations. In: IEEE/RSJ International Conference on Intelligent Robots and Systems: IROS 2009, pp. 5892–5897. IEEE (2009)
30. Whitcomb, L.L., Yoerger, D.R.: Comparative experiments in the dynamics and model-based control of marine thrusters. In: OCEANS, vol. 2, pp. 1019–1028. IEEE (1995)
31. Whitcomb, L.L., Yoerger, D.R.: Preliminary experiments in model-based thruster control for underwater vehicle positioning. IEEE J. Ocean. Eng. **24**(4), 495–506 (1999)
32. Bachmayer, R., Whitcomb, L.L., Grosenbaugh, M.A.: An accurate four-quadrant nonlinear dynamical model for marine thrusters: theory and experimental validation. IEEE J. Ocean. Eng. **25**(1), 146–159 (2000)
33. Rizzini, D.L., Kallasi, F., Oleari, F., Caselli, S.: Investigation of vision-based underwater object detection with multiple datasets. Int. J. Adv. Robot. Syst. **12**(77), 1–13 (2015)
34. Simetti, E., Galeano, S., Casalino, G.: Underwater vehicle manipulator systems: control methodologies for inspection and maintenance tasks. In: OCEANS 16. IEEE, Shanghai, China (2016)

35. Yuh, J.: Design and control of autonomous underwater robots: a survey. Auton. Robots **8**(1), 7–24 (2000)
36. Yuh, J., West, M.: Underwater robotics. Adv. Robot. **15**(5), 609–639 (2001)
37. Antonelli, G.: Underwater Robots. Springer Tracts in Advanced Robotics, vol. 96. Springer, Berlin (2014)
38. Kinsey, J.C., Eustice, R.M., Whitcomb, L.L.: A survey of underwater vehicle navigation: recent advances and new challenges. In: IFAC Conference of Manoeuvering and Control of Marine Craft, vol. 88 (2006)
39. Bonin-Font, F., Ortiz, A., Oliver, G.: Visual navigation for mobile robots: a survey. J. Intell. Robot. Syst. **53**(3), 263–296 (2008)
40. Partan, J., Kurose, J., Levine, B.N.: A survey of practical issues in underwater networks. ACM SIGMOBILE Mob. Comput. Commun. Rev. **11**(4), 23–33 (2007)

Adaptive Training of Neural Networks for Control of Autonomous Mobile Robots

Erik Steur, Thijs Vromen and Henk Nijmeijer

Abstract We present an adaptive training procedure for a spiking neural network, which is used for control of a mobile robot. Because of manufacturing tolerances, any hardware implementation of a spiking neural network has non-identical nodes, which limit the performance of the controller. The adaptive training procedure renders the input-output maps of these non-identical nodes practically identical, therewith recovering the controller performance. The key idea is to replace the nodes of the spiking neural network by small networks of synchronizing neurons that we call clusters. The networks (and interaction weights) are generated adaptively by minimizing the errors in the collective input-output behavior of the cluster relative to that of a known reference. By means of numerical simulations we show that our adaptive training procedure yields the desired results and, moreover, the generated networks are consistent over trials. Thus, our adaptive training procedure generates optimal network structures with desired collective input-output behavior.

1 Introduction

Autonomous mobile robots (wheeled robots, legged robots, unmanned arial vehicles, unmanned underwater vehicles, etc.) are everywhere nowadays. Domestic robots mow our lawns and vacuum-clean our houses [7], wheeled robots are used for

---

E. Steur (✉)
Department of Mechanical Engineering, Institute for Complex Molecular Systems,
Eindhoven University of Technology,
PO Box 513, 5600 Eindhoven, MB, The Netherlands
e-mail: e.steur@tue.nl

T. Vromen · H. Nijmeijer
Department of Mechanical Engineering, Eindhoven University of Technology,
PO Box 513, 5600 Eindhoven, MB, The Netherlands
e-mail: t.g.m.vromen@tue.nl

H. Nijmeijer
e-mail: h.nijmeijer@tue.nl

T. Vromen
Océ–Technologies B.V., Venlo, The Netherlands

transportation of goods in factory environments [1], and unmanned vehicles explore unknown (and dangerous) terrain [11]. In many applications, the environment in which the mobile robot operates is unknown and constantly changing, which demands the robot to be able to recognize and avoid obstacles and to do self-localization. Some applications even require the mobile robots to cooperate, cf. [1, 18]. Thus, the tasks for mobile robots become increasingly demanding, which suggests that mobile robots need to be more and more "intelligent" (and even cognitive).

Inspired by the stunning navigation and information processing capabilities of the biological brain, several studies have investigated the possibility of using "artificial brains" for control of mobile robots. Natural candidates for such artificial brains are neural networks (NN) and, in particular, spiking neural networks (SNN) [4, 8, 13, 14, 17, 23]. An SNN consists of (a large number of) spiking neurons that interact via synaptic connections, therewith mimicking the architecture of the biological brain and nervous system. Like in the real biological brain, SNNs generate and process information in spatio-temporal spike patterns. SNNs are a powerful computational paradigm that, in combination with leaning techniques, has been proven successful in solving complicated (optimization) problems, cf. [9] (and the references therein).

The focus of this chapter is however not on the design of a SNN controller neither on learning mechanisms that should result in, e.g., optimal path planning. We rather focus on a particular problem that arises when one equips a mobile robot with an electronic brain, i.e., a hardware implementation of an SNN. In such hardware implementations, the neurons are typically made of analog electronic components while the (synaptic) connections between the neurons are softwired, cf. [3, 16]. Note that it is necessary to have the synaptic connections softwired because learning mechanisms involve changing the connections (rather than the intrinsic dynamics of the neurons). Because of hardware imperfections, the neurons in an electronic brain will never be completely identical whereas in design of the SNN controller one does assume the neurons to be identical. Nonidentical neurons in (a hardware implementation of) the SNN may limit the performance of the control scheme; For instance, in [22], an experiment is presented which shows that nonidentical neurons cause a mobile robot having a bias for turning either left or right.

A solution to eliminate the unwanted effects of nonidentical neurons in an SNN is proposed in [22]. The key idea is to replace the neurons of the SNN by small networks of interacting neurons, which are called *clusters*. Each cluster is exposed to an adaptive training procedure that redefines the connections between the neurons in a way such that

1. all neurons within a cluster synchronize, and
2. the collective responses of all clusters to a common input are identical.

In other words, the adaptive training procedure makes the clusters of neurons to behave like a single neuron (as the neurons are synchronized) and, importantly, the variability between the responses of the neurons that exist because of hardware imperfections are eliminated. Here the (collective) response of interest is the (collective) firing rate of the neurons.

The adaptive training procedure of [22] is based on the observation that the collective firing rate of a cluster of two nonidentical neurons, one neuron firing at a higher rate than the other, can be made faster by increasing of the coupling from the fast neuron to the slow neuron by some amount and decreasing the coupling from the slow neuron to the fast neuron by the same amount. (In the extreme case that the slow neuron will be enslaved to the fast neuron, i.e., a master-slave coupling scheme, and the collective firing rate will be identical to that of the fast neuron). The collective firing rate can be decreased analogously.

These observations suggest that the collective firing rate of the cluster can be adaptively controlled using the following procedure. Given a reference neuron that defines the desired (collective) firing rate, select two neurons from the cluster and establish coupling between the neurons such that they synchronize. Subsequently make the collective firing rate as close as possible to the desired firing rate by changing the coupling according to the procedure described above. Fix the coupling parameters and consider the two synchronized neurons from now on as a single neuron. Next select an other neuron from the cluster and establish coupling between the added neuron and the two synchronized neurons. First, synchronize the three neurons and then change the coupling to make the collective firing rate as identical as possible to the desired firing rate set by the reference neuron. Repeat this process until all neurons in the cluster are synchronized and the collective firing rate is as desired.

The adaptive training procedure of [22] is experimentally validated; After adaptive training the collective firing rates of the clusters to a common input are practically identical and the mobile robot the trained SNN has no bias for turning left or right anymore. However, the incremental nature of that training procedure (i.e., start with two neurons and perform adaptation, add the third neuron to the cluster and perform adaptation, add a fourth neuron and perform adaptation, and so on) is cumbersome and time-consuming.

In this chapter, we present a new adaptive training procedure that synchronizes all neurons in a cluster and renders the collective response of the clusters identical. The main advantage of the new adaptive training procedure, compared to that of [22], is that it is nonincremental and less time-consuming.

In Sect. 2, we introduce the mobile robot and its electronic brain. Furthermore, we discuss the experiment conducted in [22] and we illustrate the issue of degradation of performance due to nonidentical neurons. Next, in Sect. 3 we introduce the clustered SNN and in Sect. 4 we introduce our adaptive training procedure. The results of adaptive training (which are in this chapter obtained only using numerical simulation) are presented in Sect. 5. Section 6 provides conclusions.

2 Experiment: Control of a Mobile Robot by an Electronic Brain

This section introduces the experimental setup as used in [22], in which a wheeled mobile robot is controlled by an electronic brain with the task to move autonomously in an unknown environment therewith avoiding any obstacle. We discuss briefly the

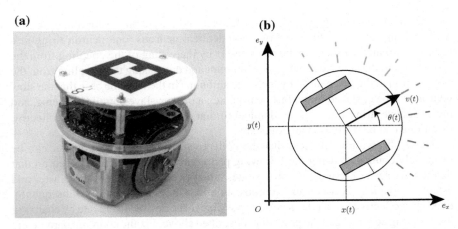

Fig. 1 a The e-puck mobile robot. b Schematic top view of the mobile robot with the nine sensors

electronic brain (SNN controller) of the mobile robot and we comment on an issue that is encountered in real experiments.

The Mobile Robot

The mobile robot that is controlled by the electronic brain is an *e-puck* [15], which is the two-wheeled robot shown in Fig. 1a. The kinematics of the e-puck mobile robot in the Cartesian plane with origin O are given by the system of equations

$$\begin{cases} \dot{x}(t) = v(t)\cos\theta(t), \\ \dot{y}(t) = v(t)\sin\theta(t), \\ \dot{\theta}(t) = \omega(t), \end{cases} \quad (1)$$

where $v(t)$ and $\omega(t)$ are the forward velocity, respectively, rotational velocity of the mobile robot. (See Fig. 1b.) The velocities $v(t)$ and $\omega(t)$ relate to the rotational velocities of the left wheel and right wheel, denoted by $\omega_l(t)$ and $\omega_r(t)$ respectively, via the equations

$$v(t) = \frac{r}{2}(\omega_l(t) + \omega_r(t)),$$
$$\omega(t) = \frac{r}{2b}(\omega_r(t) - \omega_l(t)),$$

where r is the radius of the wheels and b is half the distance between the two wheels. The rotational velocities of the wheels $\omega_l(t)$ and $\omega_r(t)$ are the control inputs for the mobile robot.

The mobile robot is equipped with nine sensors that measure distance to objects. The sensor layout is depicted schematically in Fig. 1b by the dashed lines. Three sensor groups are defined, which are denoted by S_{g1}, S_{g2}, and S_{g3}; S_{g1} is the group of the four sensors at the left of the robot, S_{g3} contains the four sensors on the right

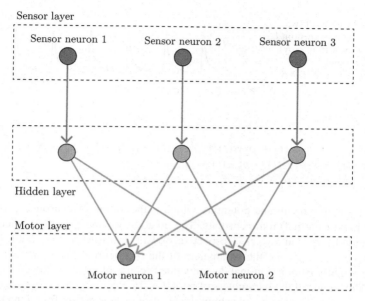

Fig. 2 SNN controller topology for control of the e-puck

of the robot and S_{g2} is the sensor in the middle. (The sensors in groups S_{g1}, S_{g2}, and S_{g3} are colored respectively green, red and blue in Fig. 1b.)

The SNN Controller

The electronic brain of the mobile robot used in [22] is a simplified version of the SNN controller introduced in [23]. Figure 2 shows the simplified SNN, which consists of three layers; a sensor layer, a hidden layer and a motor layer.

The neurons in the sensor layer process the information provided by the distance sensors of the robots. The ℓ-th sensor neuron fires at a rate E_ℓ that is strictly increasing with the distance d_ℓ measured by the sensor group $S_{g\ell}$:

$$d_\ell = \min_{j \in S_{g\ell}} d_j^*, \quad \ell = 1, 2, 3.$$

Here $d_j^* \in [d_{\min}, d_{\max}]$ is the distance measured by sensor $j \in S_{g\ell}$ and $d_{\min}$ and $d_{\max}$ are the minimal distance, respectively, maximal distance that can be measured by the sensors. Consequently, $d_\ell \in [d_{\min}, d_{\max}]$ for each $\ell \in \{1, 2, 3\}$ and the firing rate E_ℓ of sensor neuron ℓ is limited between $E_{\min}$ and $E_{\max}$, $0 < E_{\min} < E_{\max}$, with $E_\ell = E_{\min}$ if and only if $d_\ell = d_{\min}$ and $E_\ell = E_{\max}$ if and only if $d_\ell = d_{\max}$.

The hidden layer consists of three electronic Hindmarsh–Rose (HR) model neurons. These electronic HR model neurons are circuit board realizations of the HR model neuron [10], whose dynamics are governed by the system of equations

Fig. 3 Spike trains of two HR model neurons (neuron 1 in *black*, neuron 15 in *gray*) for constant input $E_1(t) = E_{15}(t) \equiv E = 4.5$

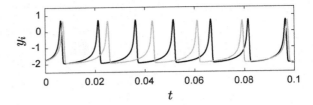

$$\begin{cases} y_i'(t) = -p_1^i y_i^3(t) + p_2^i y_i(t) + p_3^i z_{i,1}(t) - p_4^i z_{i,2}(t) - p_5^i + p_6^i E_i(t), \\ z_{i,1}'(t) = -p_7^i y_i^2(t) - p_8^i y_i(t) - p_9^i z_{i,1}(t), \\ z_{i,2}'(t) = p_{10}^i \left(p_{11}^i y_i(t) + p_{12}^i - z_{i,1}(t) \right). \end{cases} \quad (2)$$

Here $y_i(t)$ is the membrane potential, which is the output of neuron i, $z_{i,1}(t)$, and $z_{i,2}$ represent (internal) ionic currents. The notation y_i' (and $z_{i,1}'$, $z_{i,2}'$) indicates differentiation of y_i (and $z_{i,1}$, $z_{i,2}$, respectively) w.r.t. fast time $t_{\text{fast}} = 1000t$. Positive constants $p_1^i, \ldots, p_{12}^i$ are the parameters of the i-th neuron. Each parameter value deviates slightly from it nominal design value because of, e.g., manufacturing tolerances. The values of the parameters of a total of 16 electronic neurons are provided in the appendix. (Further details about the electronic neurons are found in [16, 21].)

The i-th electronic HR model neuron receives as input the firing rate $E_i(t)$ provided by the i-th sensor neuron. Depending on the value of $E_i(t)$ the neurons in the hidden layer fire at a rate $f_i(t)$, which is strictly increasing with $E_i(t)$. Figure 3 shows the spikes trains of two HR model neurons for the same constant input $E = 4.5$. Figure 4 shows the firing rates of the HR model neurons (16 in total) for constant E_i with $E_{\min} = 4.5$ and $E_{\max} = 10$. Note the difference in intrinsic firing rates of the different neurons. The intrinsic firing rates are determined using numerical simulations with the model (2) and the parameters given in the appendix. Details about computation of the firing rates are provided in the next section.

The motor layer contains two motor neurons, which control the rotational velocities of the wheel of the mobile robot. The rotational velocities of the left wheel and right wheel of the mobile robot are proportional to the firing rates of the left and right motor neurons, respectively. The firing rates of the two motor neurons are controlled by the neurons in the hidden layer. The central neuron in the hidden layer excites both motor neurons (by a same amount). The left hidden layer neuron excites the right motor neuron and inhibits the left motor neuron whereas the right hidden layer neuron excites the left motor neuron and inhibits the right motor neuron. Thus if the mobile robot approaches an obstacle at its left, d_1 is decreasing and the firing rate of the left sensor neuron is decreasing. Consequently, the left hidden layer neuron fires at a lower rate then the right hidden layer neuron. This results in an increase of the input to the left motor neuron and an decrease of the input to the right motor neurons such that the robot makes a turn to the right. As such the mobile robot avoids the obstacle at its left. Because the amount of excitation and inhibition is balanced the robot moves in a straight line in case the distance to a left obstacle equals the distance to a right obstacle. The presence of the central sensor and the central hidden layer

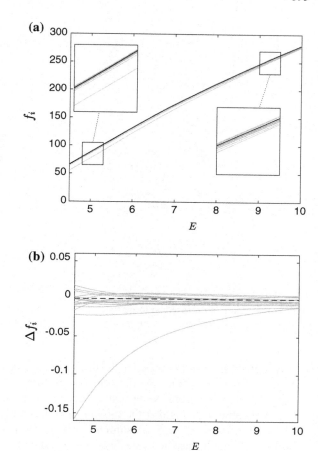

Fig. 4 **a** Intrinsic firing rates f_i of fifteen HR model neurons (in *gray*) and the reference neuron (in *black*) as function of constant sensor input $E_i(t) \equiv E$. **b** Firing rate error relative to the firing rate of the reference neuron, $\Delta f_i = \frac{f_i - f_{ref}}{f_{ref}}$ for constant input $E_i(t) \equiv E$

neuron ensures that the robot comes to a stand-still in case it is moving straight upon an obstacle.

For further details about this SNN controller the reader is referred to [22].

Experimental Results: Issues

The SNN controller described above will function properly in case the neurons respond identically to the same input. Unfortunately, as already remarked in the introduction, for any hardware implementation of such a SNN one cannot expect the neurons to respond identically to identical inputs. In the experiment reported in [22], the sensor neurons and the motor neurons are implemented in software, hence they can be considered identical. However, the neurons in the hidden layer are electronic HR model neurons and the unavoidable mismatches in parameters caused by manufacturing tolerances on the electronic components cause these neurons to fire at different frequencies for identical sensor inputs. See, again, Figs. 3 and 4.

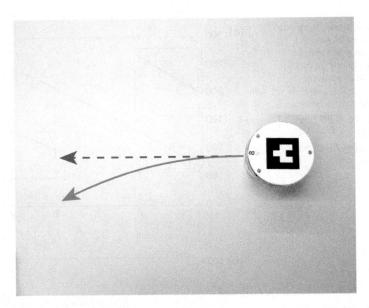

Fig. 5 Illustration of degraded performance due to nonidentical neurons. The desired trajectory is indicated by the *dashed (blue) arrow*, the *solid (red) arrow* shows the actual trajectory

Experimental results reported in [22] show that the unavoidable differences in the HR model neurons degrade the performance of the control scheme; A consequence of the HR model neurons being nonidentical (in particular, having different firing rate at the same input) is that the mobile robot has a bias for turning left or right, which is illustrated in Fig. 5. (A similar figure with experimental results is found in [22].) Recall that in case no obstacle is detected the robot should move in a straight line rather than turning left or right.

3 A Clustered SNN

There are multiple ways to compensate for the effect of nonidentical neurons on the control performance. For instance, assuming that the intrinsic firing rates are known and as such the deviations from a desired firing rate are known, one may redefine the inputs E_i to the HR model neurons by modifying the input–output map of the sensor neurons. However, such an approach may only work if each neuron receives inputs from just a single sensor neuron, which is a restrictive assumption that is hard to fulfill in more sophisticated SNN topologies. Instead, we propose to replace the single neurons in the SNN by clusters of interacting neurons, see Fig. 6.

We define three clusters, which we denote by C_1, C_2, and C_3, and we let the set $\mathscr{C}_\ell$ be the index-set of the neurons belonging to cluster C_ℓ, $\ell = 1, 2, 3$. Every neuron in cluster C_ℓ is stimulated by the ℓ-th sensor neuron. The idea is to define coupling between the neurons within each cluster in a way such that

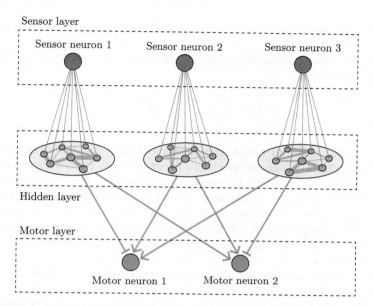

Fig. 6 SNN controller topology with clusters

1. all neurons in the cluster synchronize;
2. the collective firing rate of each cluster C_ℓ at stimulus E_ℓ is identical to the firing rate of a *reference neuron* at that same stimulus.

In other words, we require each cluster to behave as a single neuron and the collective responses of the different clusters should be as identical as possible to that of the reference neuron, which in turn implies that the collective responses of different clusters are practically identical.

Each clusters will consist of the HR model neurons introduced in the previous section. We have a total of 16 HR model neurons and we assign to each cluster five neurons,

$$\mathscr{C}_1 = \{1, 2, 3, 4, 5\},$$
$$\mathscr{C}_2 = \{6, 7, 8, 9, 10\},$$
$$\mathscr{C}_3 = \{11, 12, 13, 14, 15\},$$

and we let the remaining HR model neuron be our reference neuron. The dynamics of the reference neurons are governed by (2) whereas the HR model neurons in the clusters have extra input $u_i(t)$ appearing in the first equation:

$$\begin{cases} \dot{y}_i(t) = -p_1^i y_i^3(t) + p_2^i y_i(t) + p_3^i z_{i,1}(t) - p_4^i z_{i,2}(t) - p_5^i + p_6^i E_i(t) + u_i(t), \\ \dot{z}_{i,1}(t) = -p_7^i y_i^2(t) - p_8^i y_i(t) - p_9^i z_{i,1}(t), \\ \dot{z}_{i,2}(t) = p_{10}^i \left(p_{11}^i y_i(t) + p_{12}^i - z_{i,1}(t) \right), \end{cases} \quad (3)$$

with $i \in \mathscr{C}_\ell$. This extra input $u_i(t)$ is used to establish coupling between the neurons within a cluster.

The interaction between the HR model neurons within a cluster is according to the linear coupling law

$$u_i(t) = \sum_{j \in \mathscr{C}_\ell} w_j a_{ij}(y_j(t) - y_i(t)), \tag{4}$$

where w_j and a_{ij} are positive parameters, respectively, non-negative parameters. (The use of the product $w_j a_{ij}$ instead of a single parameter looks cumbersome; the reason for using the product of two parameters becomes clear at a later stage.) We remark that we have assumed the "input gain" of $u_i(t)$ to be equal to one for each neuron (3). This assumption is without loss of generality as any deviations from the value of one can be compensated for in the parameters $w_j a_{ij}$. It is worth noting that the coupled neurons (3), (4) will synchronize provided that the coupling is sufficiently strong [5, 20, 22].[1]

Because the neurons are not identical we cannot expect perfect synchrony; We may only expect the neurons to synchronize practically. By saying that the neurons practically synchronize, we mean that the outputs of the neurons are practically indistinguishable. Following [5], we measure practical synchronization of two neurons i and j using the synchronization measure

$$s(y_i[t_1, t_2], y_j[t_1, t_2]) = 1 - \rho(y_i[t_1, t_2], y_j[t_1, t_2])$$

where $y_i[t_1, t_2]$, $t_2 > t_1$ denotes the sequence

$$\left\{ y_i(t_1), y_i(t_1 + \tfrac{1}{10000}), y_i(t_1 + \tfrac{2}{10000}), \ldots, y_i(t_2 - \tfrac{1}{10000}), y_i(t_2) \right\},$$

($y_j[t_1, t_2]$ is defined analogously),[2] and $\rho(\cdot, \cdot)$ is Pearson's correlation coefficient,

$$\rho(y_i[t_1, t_2], y_j[t_1, t_2]) = \frac{\text{cov}(y_i[t_1, t_2], y_j[t_1, t_2])}{\sigma_{y_i[t_1, t_2]} \sigma_{y_j[t_1, t_2]}}.$$

Here $\text{cov}(\cdot, \cdot)$ is the covariance and σ denotes the standard deviation. Note that ρ takes values in the interval $[-1, 1]$ with 1 indicating positive linear correlation, -1 being negative linear correlation and 0 indicating no correlation. Thus, the synchronization measure s takes values in the interval $[0, 2]$ and the value $s = 0$ indicates that the outputs of the neurons are perfectly synchronized (as $s = 0$ if and only if $\rho = 1$). We say that the neurons practically synchronize if $s < \varepsilon_s$ for some sufficiently small positive constant ε_s.

---

[1]By saying that the coupling is sufficiently strong we mean the values of the coupling parameters $w_j a_{ij}$ are sufficiently large. A more accurate description is found in [21].

[2]We set a "sample rate" of 10 kHz, which is also used in [22].

The firing rate of neuron i (for constant sensor input E_i) equals

$$f_i = \frac{1}{T_i}$$

with T_i being the (average) time between subsequent spikes. The number T_i is estimated by counting the number of spikes in a time-window of predefined length. Here, we have taken the length of this window to be 0.5 s. The number of spikes in the window is determined by counting the number of local maxima of $y_i(t)$ with the constraint $y_i(t) > 0$. (See Fig. 3.)

The collective output of a cluster $y_{C_\ell}(t)$ is defined as the average of the outputs of the neurons $y_i(t)$ in that cluster:

$$y_{C_\ell}(t) = \frac{1}{5} \sum_{i \in \mathscr{C}_\ell} y_i(t).$$

Thus if the neurons in cluster C_ℓ are practically synchronized, then the output of each neuron in that cluster is practically synchronized to the collective output $y_{C_\ell}(t)$. The collective firing rate of cluster C_ℓ as

$$f_{C_\ell} = \frac{1}{T_{C_\ell}}$$

with T_{C_ℓ} being the time between subsequent spikes in the collective output $y_{C_\ell}(t)$. We say that the collective firing rate f_{C_ℓ} of each cluster C_ℓ at a stimulus E_ℓ is identical to the firing rate of the reference neuron f_{ref} at that same stimulus if the relative error

$$\frac{|f_{C_\ell} - f_{ref}|}{f_{ref}} < \varepsilon_f$$

for some sufficiently small constant ε_f.

4 The Adaptive Training Procedure

For the clustered SNN to work properly (and outperform the nonclustered SNN), we need to find coupling parameters $w_j a_{ij}$ such that the neurons in a cluster practically synchronize and fire collectively at the same rate as the reference neuron. Note that we can only change the (softwired) interactions between the (hardware) neurons. Define the matrix

$$A_\ell = (w_j a_{ij}), \quad i, j \in \mathscr{C}_\ell,$$

and note that A is the *adjacency matrix* of a (weighted and directed) graph. Let L_ℓ be the Laplacian matrix associated to A_ℓ,

$$L_\ell = D_\ell - A_\ell$$

with D_ℓ the diagonal matrix with the row-sums of A_ℓ as entries. Then we can write

$$\begin{pmatrix} u_{\mathscr{C}_\ell(1)}(t) \\ \vdots \\ u_{\mathscr{C}_\ell(5)}(t) \end{pmatrix} = -L_\ell \begin{pmatrix} y_{\mathscr{C}_\ell(1)}(t) \\ \vdots \\ y_{\mathscr{C}_\ell(5)}(t) \end{pmatrix}.$$

Description of the Adaptive Training Procedure

We assume that for each value of E the intrinsic firing rates f_i of the HR model neurons to be known. The adaptive training procedure for a fixed value of E is as follows:

S0. Initialize the training procedure with

$$w_j = 1, \quad a_{ij} = 0, \quad i, j \in \mathscr{C}_\ell,$$

and continue at S1;

S1. Determine the trajectories $y_i(t)$, $i \in \mathscr{C}_\ell$, for $t \in \{0, \frac{1}{10000}, \ldots, 5\}$ and go to S2;

S2. Compute the synchronization measures $s_{ij} := s(y_i[2, 5], y_j[2, 5])$ for all $i, j \in \mathscr{C}_\ell$, $i \neq j$, and check if all neurons in the cluster are practically synchronized:

- Go to S3 if all neurons are practically synchronized;
- Otherwise increase a_{ij} by the amount $\Delta a_{ij} = \gamma s_{ij}$ (with γ a pre-defined positive constant)

$$a_{ij} \leftarrow a_{ij} + \Delta a_{ij},$$

and go back to S1;

S3. Determine the collective firing rate f_{C_ℓ} and compute the firing rate error $\frac{|f_{C_\ell} - f_{ref}|}{f_{ref}}$:

- Stop if the firing rate error is smaller than ε_f;
- Otherwise compute for each $j \in \mathscr{C}_\ell$,

$$\Delta w_j = \eta w_j \left(\frac{f_j - f_{ref}}{f_{ref}}\right) \left(\frac{f_{ref} - f_{C_\ell}}{f_{ref}}\right),$$

(with η a pre-defined positive constant), increase w_j by Δw_j,

$$w_j \leftarrow w_j + \Delta w_j,$$

and go back to S1.

Let us briefly explain the rationale of this adaptive training procedure. In the first place, we require the neurons in each cluster to practically synchronize. As remarked before, we already know that practical synchronization will be achieved for sufficiently large coupling. In S2 we increase the strength of the coupling between neurons i and j proportional to s_{ij}. As such the coupling between nonsynchronous neurons is made stronger until all neurons in the cluster practically synchronize. Here, the neurons are called practically synchronized if $s(y_i[2, 5], y_j[2, 5]) < \varepsilon_s$ for all $i, j \in \mathscr{C}_\ell$. Note that the synchronization errors are determined on the interval [2, 5]. Thus, the signals on the interval [0, 2), which include transient behavior, are ignored in the computation of the synchronization errors.

If the neurons are all practically synchronized, then (in S3) the collective firing rate is determined and compared with that of the reference neuron. In case the collective firing rate deviates to much from the firing rate of the reference neuron the numbers w_j will be updated by an amount Δw_j. The update Δw_j contains the term

$$\frac{f_j - f_{ref}}{f_{ref}}, \qquad (5)$$

which is the difference between the intrinsic firing rate of neuron j and the firing rate of the reference neuron (relative to the firing rate of the reference neuron), and the term

$$\frac{f_{ref} - f_{C_\ell}}{f_{ref}}, \qquad (6)$$

which is the difference between the firing rate of the reference neuron and the collective firing rate (relative to the firing rate of the reference neuron). Suppose that $f_{ref} > f_{C_\ell}$, i.e., the firing rate of the reference is larger than that of cluster C_ℓ. Note that (6) is positive. Then in case $f_j > f_{ref}$, the term (5) is positive such that Δw_j is positive. On the other hand, if $f_j < f_{ref}$ the term (5) is negative which results in a negative Δw_j. This means that the outgoing connections of the neurons with an intrinsic firing rate *larger* than the reference neuron are getting *stronger*, whereas the outgoing connections of the neurons with an intrinsic firing rate *lower* than the firing rate of the reference neuron become *weaker*. As explained in the introduction (and in much more detail in [22]), strengthening the outgoing connections of the "fast neurons" and weakening the strength of outgoing connection of "slow neurons" results in an increase of the collective firing rate f_{C_ℓ}. Likewise, if $f_{ref} < f_{C_\ell}$ such that (6) is negative, the outgoing connections of the neurons with an intrinsic firing rate *larger* than the reference are *weakened* and the outgoing connections of the neurons with an intrinsic firing rate *lower* than the reference are *enhanced*, which results in an decrease of the collective firing rate. The term w_j that appears in the update Δw_j ensures that w_j does not become negative (of course, provided that the adaptation gain η is not too large).

Recall that the objective is that the neurons in a cluster are practically synchronized with a collective firing rate practically identical to that of the reference neuron for every $E \in [E_{\min}, E_{\max}]$. To meet the objective, we choose a set of training points $\mathscr{E} \subset [E_{\min}, E_{\max}]$ and we subsequently require that the convergence criteria (practical synchronization and matching of the firing rates) are satisfied at each point $E \in \mathscr{E}$. We perform the training according to the diagram shown in Fig. 7. Here we choose

Fig. 7 Scheme of the adaptive training procedure

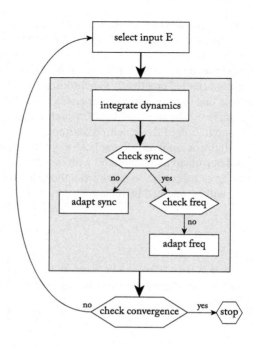

$E \in \mathscr{E}$ randomly from a negative exponential distribution. The negative exponential distribution is used such that lower values of E are visited more frequently (in probability), which is desired as the lower the value of E the more difficult it is to synchronize the network, cf. [5]. The reason for letting E be a random variable rather than, e.g., completing the adaptive training procedure at one training point, then performing the whole adaptive training procedure at another training point (and so on) is that we do not expect the "best" network structure at one training point also to be the "best" network structure at the other training points; Using random sampling, we expect to find a network structure that is (close to) optimal for all $E \in \mathscr{E}$.

5 Results of Training

We have performed the adaptive training procedure for the three cluster with

- $\varepsilon_s = 0.01$ (a value taken from [5], which describes synchronization experiments with the same HR model neurons);
- $\varepsilon_f = 0.003$;
- $\gamma = 0.1$;
- $\eta = 50$;
- $\mathscr{E} = \{4.5, 5, 5.5, \ldots, 9.5, 10\}$;
- $\lambda = 0.25$ (which is the parameter of the negative exponential distribution).

Fig. 8 Relative error in firing rate after adaptation training. *Black dots* show the relative error in collective firing rate for **a** cluster 1, **b** cluster 2, **c** cluster 3. *Gray lines* show the relative error in intrinsic firing rates of the HR model neurons. *Solid black line* indicate the firing rate tolerance ε_f

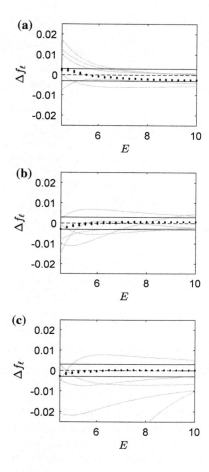

The dynamics (3), (4) are numerically integrated in *Matlab* using the *ode45* solver. Initial conditions are chosen randomly from a normal distribution with zero mean and unit variance. For each cluster we have performed the adaptive training procedure ten times. Each (of the total of 30) adaptive training procedure has converged.

Figure 8 shows the collective firing rate errors of the cluster relative to the reference firing rate,

$$\Delta f_\ell = \frac{f_{C_\ell} - f_{ref}}{f_{ref}}, \quad \ell = 1, 2, 3,$$

as function of E for ten instances of adaptive training. Note that this figure also shows results at point E in between the training points. It can be seen that at each point E the adaptive training reduces the variability in firing rate significantly, in particular for the lower values of E.

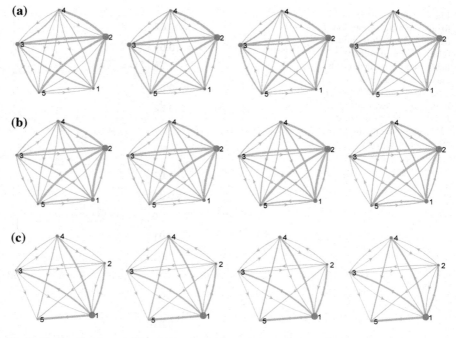

Fig. 9 Four realized network structures for **a** cluster 1, **b** cluster 2, and **c** cluster 3

Figure 9 shows for each of the three clusters four (out of ten) network structures that have emerged out of the adaptive training procedure. In this figure the size of the nodes (red dots) is proportional to the values of w_j after adaptation and the thickness of the edges is proportional to $w_j a_{ij}$. For illustrative purposes we normalized the coupling strengths $w_j a_{ij}$ such that $\max_{i,j} w_j a_{ij} = 1$ and subsequently we binned the normalized coupling strengths with a resolution of 20. It is interesting to note that in this figure the fifth node in cluster 3 does not have any outgoing connection. The reason for this is that the corresponding neuron (neuron 15) has an intrinsic firing which is much lower than all other neurons including the reference. (The intrinsic firing rate of neuron 15 is the bottom gray line in Fig. 4a.) Furthermore, we see that for each cluster the realized network structures are consistent[3]; This provides evidence that our adaptive training procedure renders globally optimal networks, i.e., network structures that are optimal for all $E \in \mathscr{E}$.

---

[3] Actually, for each cluster all ten realizations of the networks are consistent.

6 Concluding Remarks

We have presented a strategy to eliminate unwanted effects in hardware implementations of SNNs that arise because of hardware imperfections. We have suggested to use trained clusters of interaction neurons instead of single neurons in an SNN. An adaptive training procedure is presented that

1. synchronizes all neurons within a cluster (in a practical sense), and
2. makes the collective firing rate of each clusters practically identical to that of a reference neuron.

As a result, the variability in the collective firing rates of different clusters for the same input is practically eliminated, which ensures that the SNN with clusters works properly.

We have shown using numerical simulations that our proposed adaptive training procedure yields the desired results. The network structures that are obtained after adaptation are strikingly similar, which provides evidence that our adaptive training procedure determines globally optimal networks.

It is expected that the new adaptive training procedure is more robust for output/measurement noise than the training procedure of [22]. (It is mentioned in [22] that in presence of measurement noise the number of neurons in a cluster can not be too large because practical synchronization is not possible.) A stability analysis of the synchronized state like the one presented in the appendix of [22] suggests that the bound for practical synchronization increases with the product of the intensity of measurement noise and the largest singular value of the Laplacian matrix. We have observed that the largest singular values of the Laplacian matrices generated by new adaptive training procedure are consistently lower than those generated using the adaptive training procedure of [22]. (These results are not included in the chapter because of space limitations.) In the near future, we will investigate the robustness of the two adaptive training procedures w.r.t. measurement noise in more detail. Furthermore, we will perform real experiments with the new adaptive training procedure. Additionally, it would be interesting to find mathematical proofs that guarantee convergence (to a global optimum) of the adaptive training procedure and to provide practical methods for determining (tight) a priori bounds on synchronization tolerance and the firing rate tolerance.

Probably the most interesting direction for future research is to replace the coupling functions (4), which model so-called gap junctions [12], by synaptic coupling. Synaptic coupling, which is only active when the pre-synaptic neurons are emitting their spikes, is a more natural type of coupling in SNNs as it facilitates *spike synchronization*, i.e., synchronization of timing of spikes without requiring the spikes to have the same shape (e.g., amplitude). Note that we actually require the neurons in each cluster to synchronize only the timing of their spikes. Synaptic coupling (instead of the coupling we considered) would also allow for the use of biologically inspired learning mechanisms [9]. Models of synaptic coupling are found in, for instance, [2, 19]. However, contrary to the case of gap-junction coupling, a comprehensive theory about synchronization of synoptically coupled (heterogeneous) model neurons is lacking; Only a few results concerning local synchronization in networks of identical model neurons are available, cf. [2, 6].

Appendix: Parameters of the HR Model Neurons

Parameters of the electronic HR model neurons, taken from [16], which are estimated using extended Kalman filtering. (Neuron indices have changed compared to [16].)

Neuron	p_1	p_2	p_3	p_4	p_5	p_6	p_7	p_8	p_9	$p_{10} \cdot 10^3$	p_{11}	p_{12}
1	0.9946	2.9925	4.9564	0.9880	7.8380	0.9874	1.0138	2.0271	1.0110	5.0279	3.9897	4.5348
2	0.9902	2.9861	4.9495	0.9829	7.8420	0.9846	1.0083	2.0305	1.0074	4.9914	4.0188	4.6579
3	1.0036	2.9826	4.9312	0.9946	8.0198	1.0009	1.0119	2.0174	0.9977	4.8884	4.1030	4.6043
4	0.9989	2.9737	4.9014	0.9941	7.9129	0.9989	1.0116	2.0161	0.9959	4.8677	4.0143	4.5275
5	0.9982	2.9915	4.9340	0.9860	7.8960	0.9884	1.0132	2.0216	1.0072	4.9686	4.0084	4.4629
6	1.0063	2.9905	4.9246	0.9888	7.9346	0.9949	1.0196	2.0255	1.0050	4.8458	4.0646	4.5447
7	1.0112	2.9898	4.9491	0.9841	7.8532	0.9916	1.0161	2.0262	1.0044	4.8643	4.0427	4.6245
8	0.9913	2.9581	4.9191	0.9981	7.9737	1.0031	0.9958	2.0039	0.9906	4.8605	4.0235	4.6363
9	1.0061	2.9999	4.9819	0.9908	7.8814	0.9935	1.0074	2.0152	1.0034	4.8442	4.0317	4.6456
10	1.0351	3.0026	4.9587	1.0023	7.9654	1.0029	1.0295	2.0119	1.0046	4.8333	4.0277	4.3949
11	1.0137	2.9841	4.9374	0.9889	7.9595	1.0015	1.0183	2.0326	1.0013	4.8833	4.0925	4.6788
12	0.9993	2.9829	4.9317	0.9914	7.9626	1.0036	1.0108	2.0138	1.0003	4.8000	4.0749	4.6086
13	0.9802	2.9825	4.9388	1.0024	8.0299	1.0059	1.0003	2.0100	1.0017	4.8252	4.0299	4.5863
14	1.0061	2.9891	4.9247	0.9867	7.8698	0.9972	1.0136	2.0191	0.9969	4.9159	4.1138	4.7313
15	1.0020	2.9797	4.8917	0.9930	7.9973	1.0313	1.0201	2.0316	0.9992	4.9206	4.0559	4.9206
Ref	1.0080	2.9734	4.9190	0.9872	7.9038	0.9994	1.0142	2.0159	0.9954	4.9425	4.0702	4.6286

References

1. Adinandra, S.: Hierarchical Coordination Control of Mobile Robots. Ph.D. thesis, Eindhoven University of Technology (2012)
2. Belykh, I., de Lange, E., Hasler, M.: Synchronization of bursting neurons: what matters in the network topology. Phys. Rev. Lett. **94**(188101) (2005)
3. Benjamin, B.V., Gao, P., McQuinn, E., Choudhary, S., Chandrasekaran, A.R., Bussat, J.M., Alvarez-Icaza, R., Arthur, J.V., Merolla, P.A., Boahen, K.: Neurogrid: a mixed-analog-digital multichip system for large-scale neural simulations. Proc. IEEE **102**(5), 699–716 (2014)
4. Cao, Z., Cheng, L., Zhou, C., Gu, N., Wang, X., Tan, M.: Spiking neural network-based target tracking control for autonomous mobile robots. Neural Comput. Appl. **26**, 1839–1847 (2015)
5. Castanedo-Guerra, I.T., Steur, E., Nijmeijer, H.: Synchronization of coupled Hindmarsh-Rose neurons: effects of an exogenous parameter. In: Proceedings of the 6th IFAC Workshop on Periodic Control Systems PSYCO 2016, Eindhoven, The Netherlands, vol. 49 of 14, pp. 84 – 89 (2016)
6. Checco, P., Righero, M., Biey, M., Kocarev, L.: Synchronization in networks of hindmarsh-rose neurons. IEEE Trans. Circuits Syst. II **55**(12), 1274–1278 (2008)

7. de Almeida, A.T., Fong, J.: Domestic service robots. IEEE Robot. Autom. Mag. **18**(3), 18–20 (2011)
8. Floreano, D., Epars, Y., Zuffery, J.-C., Mattiussi, C.: Evolution of spiking neural circuits in autonomous mobile robots. Int. J. Intell. Syst. **21**, 1005–1024 (2006)
9. Ghosh-Dastidar, S., Adeli, H.: Spiking neural networks. Int. J. Neural Syst. **19**(04), 295–308 (2009)
10. Hindmarsh, J.L., Rose, R.M.: A model for neuronal bursting using three coupled differential equations. Proc. R. Soc. Lond. B **221**, 87–102 (1984)
11. Kelasidi, E., Liljeback, P., Pettersen, K.Y., Gravdahl, J.T.: Innovation in underwater robots: biologically inspired swimming snake robots. IEEE Robot. Autom. Mag. **23**(1), 44–62 (2016)
12. Koch, C.: Biophysics of Computation, 1st edn. Oxford University Press, Oxford (1999)
13. Manoonpong, P., Pasemann, F., Fischer, J.: Modular neural control for a reactive behavior of walking machines. In: Proceedings of the IEEE International Symposium on Computational Intelligence in Robotics and Automation, pp. 403–408 (2005)
14. Menon, S., Fok, S., Neckar, A., Khatib, O., Boahen, K.: Controlling articulated robots in task-space with spiking silicon neurons. In: 5th IEEE RAS/EMBS International Conference on Biomedical Robotics and Biomechatronics, pp. 181–186 (2014)
15. Mondada, F., Bonani, M., Raemy, X., Pugh, J., Cianci, C., Klaptocz, A., Magnenat, S., Zuffery, J.-C., Floreano, D., Martinoli, A.: The e-puck, a robot designed for education in engineering. Proc. Conf. Auton. Robot Syst. Compet. **1**, 59–65 (2009)
16. Neefs, P.J.: Experimental synchronisation of Hindmarsh-Rose neurons in complex networks. Master's thesis (DCT 2009.107), Eindhoven University of Technology, the Netherlands (2009)
17. Osella Massa, G.L., Vinuesa, H., Lanzarini, L.: Modular creation of neuronal networks for autonomous robot control. In: Proceedings of the IEEE Latin American Robotics Symposium, pp. 66–73 (2006)
18. Pettersen, K.Y., Gravdahl, J.T., Nijmeijer, H.: Group Coordination and Cooperative Control. Springer, Berlin (2006)
19. Somers, D., Kopell, N.: Rapid synchronization through fast threshold modulation. Biol. Cybern. **68**, 393–407 (1993)
20. Steur, E., Tyukin, I., Nijmeijer, H.: Semi-passivity and synchronization of diffusively coupled neuronal oscillators. Phys. D **238**(21), 2119–2128 (2009)
21. Steur, E., Murguia Rendon, C.G., Fey, R.H.B., Nijmeijer, H.: Synchronization and partial synchronization experiments with networks of time-delay coupled Hindmarsh-Rose neurons. Int. J. Bifurc. Chaos **26**(7) (2016)
22. Vromen, T.G.M., Steur, E., Nijmeijer, H.: Training a network of electronic neurons for control of a mobile robot. Int. J. Bifurc. Chaos **26**(12) (2016) (In press)
23. Wang, X., Hou, Z.-G., Zou, A., Tan, M., Cheng, L.: A behavior controller based on spiking neural networks for mobile robots. Neurocomputing **71**, 655–666 (2008)

Modeling, Identification and Control of High-Speed ASVs: Theory and Experiments

Bjørn-Olav Holtung Eriksen and Morten Breivik

Abstract This paper considers a powerful approach to modeling, identification, and control of high-speed autonomous surface vehicles (ASVs) operating in the displacement, semi-displacement, and planing regions. The approach is successfully applied to an 8.45 m long ASV capable of speeds up to 18 m/s, resulting in a high-quality control-oriented model. The identified model is used to design four different controllers for the vessel speed and yaw rate, which have been tested through full-scale experiments in the Trondheimsfjord. The controllers are compared using various performance metrics, and two controllers utilizing a model-based feedforward term are shown to achieve outstanding performance.

1 Introduction

The development of autonomous vehicles is moving rapidly forward. The automotive industry is particularly leading this trend. At sea, there is also a great potential for such vehicles, which are typically referred to as autonomous surface vehicles (ASVs). The use of such vehicles has scientific, commercial, and military applications, and can result in reduced costs, increased operational persistence and precision, widened weather window of operations, improved personnel safety, and more environmentally friendly operations. In [1], an early overview of unmanned surface vehicles is given, while a more recent survey is presented in [8].

In this paper, we focus on modeling and control of small, agile ASVs which can operate at high speeds with aggressive maneuvers. These vehicles typically cover the whole range of speed regions for a surface vehicle, namely the displacement,

---

B.-O.H. Eriksen (✉) · M. Breivik
NTNU AMOS Department of Engineering Cybernetics,
Norwegian University of Science and Technology (NTNU),
Trondheim, Norway
e-mail: bjorn-olav.h.eriksen@ieee.org

M. Breivik
e-mail: morten.breivik@ieee.org

© Springer International Publishing AG 2017
T.I. Fossen et al. (eds.), *Sensing and Control for Autonomous Vehicles*,
Lecture Notes in Control and Information Sciences 474,
DOI 10.1007/978-3-319-55372-6_19

Fig. 1 The Telemetron ASV, which is a Polarcirkel Sport 8.45 m long dual-use ASV capable of speeds up to 18 m/s. Courtesy of Maritime Robotics

semi-displacement, and planing regions. Hence, they are challenging to model and control, and it therefore becomes challenging to develop a robust and precise motion control system which allows the vehicles to utilize their full potential. Specifically, this paper revisits the modeling and control approach originally suggested in [4] and further developed and reported in [3]. The method represents a control-oriented modeling approach and underlines the importance of developing and using good feedforward terms in the control law. The high-quality performance of the resulting motion control system was validated through several full-scale experiments with ASVs in the Trondheimsfjord in 2008 and 2009, both for target tracking and formation control applications. In this paper, we further develop this approach and go into greater details concerning the modeling and identification procedure and results.

Full-scale identification experiments based on the suggested modeling approach are conducted with a dual-use (manned/unmanned) ASV named Telemetron, see Fig. 1, which is owned and operated by the company Maritime Robotics. The resulting identified model is shown to be very precise and cover the entire operational envelope of the ASV. This model subsequently forms the basis for a detailed performance comparison between four qualitatively different controllers, which are implemented and experimentally tested to control the speed and yaw rate of the Telemetron ASV. In particular, the controllers are: A PI feedback (FB) controller; a pure model-based feedforward controller based on the identified model (FF); a controller which is a combination of model-based feedforward and PI feedback (FF-FB); and a controller using feedback signals in the model-based feedforward term in combination with PI feedback, which can be characterized as a feedback linearization (FBL) controller. Relevant performance metrics are defined and used to compare these controllers to determine which is most precise and energy efficient.

Other relevant work concerning a control-oriented modeling approach can be found in, e.g., [10, 11].

The rest of the paper is structured as follows: Chap. 2 presents the main characteristics of the control-oriented modeling approach. Chapter 3 describes the model identification in detail, from experimental design to parameter identification. Chapter 4 describes the four controllers which are considered in the paper, while Chap. 5 presents the results from the motion control experiments. Finally, Chap. 6 concludes the paper.

2 2DOF Control-Oriented Vessel Model

The vast majority of surface vessel models are based on the 3DOF model [7]:

$$\dot{\boldsymbol{\eta}} = \boldsymbol{R}(\psi)\boldsymbol{v} \qquad (1a)$$

$$\boldsymbol{M}\dot{\boldsymbol{v}} + \boldsymbol{C}_{RB}(\boldsymbol{v}) + \boldsymbol{C}_A(\boldsymbol{v}_r)\boldsymbol{v}_r + \boldsymbol{D}(\boldsymbol{v}_r)\boldsymbol{v}_r = \boldsymbol{\tau} + \boldsymbol{\tau}_{\text{wind}} + \boldsymbol{\tau}_{\text{wave}}, \qquad (1b)$$

where $\boldsymbol{\eta} = \begin{bmatrix} N & E & \psi \end{bmatrix}^T \in \mathbb{R}^2 \times S^1$ is the vessel pose, $\boldsymbol{v} = \begin{bmatrix} u & v & r \end{bmatrix}^T \in \mathbb{R}^3$ is the vessel velocity and $\boldsymbol{v}_r$ denotes the relative velocity between the vessel and the water, see Fig. 2. The terms $\boldsymbol{\tau}, \boldsymbol{\tau}_{\text{wind}}, \boldsymbol{\tau}_{\text{wave}} \in \mathbb{R}^3$ represent the control input, wind, and wave environmental disturbances, respectively. The matrix $\boldsymbol{R}(\psi)$ is the rotation matrix about the z-axis, the inertia matrix is $\boldsymbol{M} = \boldsymbol{M}_{RB} + \boldsymbol{M}_A$ where $\boldsymbol{M}_{RB}$ is the rigid-body mass and $\boldsymbol{M}_A$ is the added mass caused by the moving mass of water. The matrices $\boldsymbol{C}_{RB}(\boldsymbol{v})$ and $\boldsymbol{C}_A(\boldsymbol{v}_r)$ represent the rigid-body and hydrodynamic Coriolis and centripetal effects, respectively, while $\boldsymbol{D}(\boldsymbol{v}_r)$ captures the hydrodynamic damping of the vessel. An important limitation of (1b) is that it can be challenging to use for vessels operating outside of the displacement region. For approximating the operating region of a surface vessel, it is common to use the Froude number, defined as [6]:

$$Fn = \frac{U_r}{\sqrt{Lg}}, \qquad (2)$$

where U_r is the vessel speed through water, L is the submerged vessel lengths, and g is the acceleration of gravity. For Fn less than approximately 0.4, the hydrostatic pressure mainly carries the weight of the vessel, and we operate in the displacement region. When Fn is higher than 1.0–1.2, the hydrodynamic force mainly carries the weight of the vessel, and we operate in the planing region. For Fn between these values, we are in the semi-displacement region [6].

Typical ASVs have vessel lengths of up to 10 m, submerged lengths of up to 8 m and operating speeds up to 18 m/s. From Table 1, we see that an ASV with

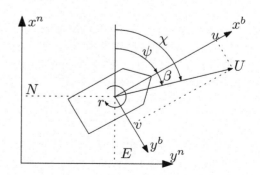

Fig. 2 Vessel variables. The superscripts $(\cdot)^n$ and $(\cdot)^b$ denote the NED and body-frames [7], respectively. The variables N, E, and ψ are the vessel pose, u, v, and r are the vessel velocity and U is the vessel speed over ground. The course χ is the sum of the heading ψ and the sideslip β

Table 1 Operating speeds for displacement and planing regions. *Supply ships typically operate with speeds up to 7 m/s. It is therefore clear that supply ships generally operate in the displacement region

Vessel type	Submerged length	Maximum speed in displacement ($Fn = 0.4$)	Minimum speed in planing ($Fn = 1.0$)
Small ASV	4 m	2.51 m/s	6.26 m/s
Large ASV	8 m	3.54 m/s	8.86 m/s
Small supply ship	50 m	8.86 m/s*	22.1 m/s*
Large supply ship	100 m	12.5 m/s*	31.3 m/s*

a submerged length of 8 m exits the displacement region already at 3.54 m/s, and enters the planing region at 8.86 m/s. Hence, (1b) is typically only suited for a small part of the ASV operating region, which motivates for an alternative model.

ASVs are generally underactuated, hence it not possible to independently control surge, sway and yaw. We therefore choose to reduce the model to the 2DOF which we want to control, namely the speed over ground (SOG) $U = \sqrt{u^2 + v^2}$ and yaw rate (rate of turn, ROT). The kinematic equation (1a) is therefore modified to:

$$\dot{\eta} = \begin{bmatrix} \cos(\chi) & 0 \\ \sin(\chi) & 0 \\ 0 & 1 \end{bmatrix} \begin{bmatrix} U \\ r \end{bmatrix} \qquad (3)$$

$$\dot{\chi} = r + \dot{\beta},$$

where $\chi = \psi + \beta$ is the vessel course angle and β is the vessel sideslip. It should be noted that this model implies that:

- Since $U \geq 0$, we assume that the vessel is traveling forward, that is $u \geq 0$.
- The sideslip β enters the kinematic equation. For kinematic control (e.g., path following), this must be addressed by, e.g., controlling course instead of heading.

To relax the limitation of operating in the displacement region implied by (1b), we propose a normalized non-first-principles model. This is inspired by [3, 4] where a steady-state model in a similar form is developed. Since the actual control input of the vessel is not forces, but rather motor throttle and rudder angle, we select these as inputs to the model. As a result, we also implicitly model the actuator dynamics. Let the motor throttle be given as $\tau_m \in [0, 1]$ and the rudder input be given as $\tau_\delta \in [-1, 1]$. Denoting the vessel velocity as $\boldsymbol{x} = \begin{bmatrix} U & r \end{bmatrix}^T \in \mathbb{R}^2$ and the control input as $\boldsymbol{\tau} = \begin{bmatrix} \tau_m & \tau_\delta \end{bmatrix}^T \in \mathbb{R}^2$, we propose the model:

$$\boldsymbol{M}(\boldsymbol{x})\dot{\boldsymbol{x}} + \boldsymbol{\sigma}(\boldsymbol{x}) = \boldsymbol{\tau}, \qquad (4)$$

where the inertia matrix $\boldsymbol{M}(\boldsymbol{x}) = \mathrm{diag}\,(m_U(\boldsymbol{x}), m_r(\boldsymbol{x}))$ is diagonal with elements of quantities $\left[\frac{1}{\text{m/s}^2} \; \frac{1}{1/\text{s}^2}\right]$, and $\boldsymbol{\sigma}(\boldsymbol{x}) = \begin{bmatrix} \sigma_U(\boldsymbol{x}) & \sigma_r(\boldsymbol{x}) \end{bmatrix}^T$ is a unit-less damping term.

Notice that both are functions of x, which allows for a nonlinear model. The reader should also note that centripetal effects are not explicitly included in (4) due to the choice of coordinates.

3 Model Identification

Identifying the parameters of (4) requires a series of experiments to be performed. In this section, we describe the identification experiments, parameterization of the inertia and damping terms, and the methodology used for parameter identification.

3.1 Vessel Platform and Hardware

As already mentioned, the vessel used in this work is the Telemetron ASV. It is a dual-use vessel for both manned and unmanned operations, and is equipped with a number of sensors and a proprietary control system. Some of the specifications are summarized in Table 2.

3.2 Identification Experiment Design

Since we wish to identify damping and inertia terms, both steady-state and transient information is required. We therefore construct a series of step responses:

- Step changes in τ_m given a series of fixed rudder settings τ_δ, illustrated as orange trajectories in Fig. 3.
- Step changes in τ_δ given a series of fixed throttle settings τ_m, illustrated as blue trajectories in Fig. 3.

The steps are performed both for increasing and decreasing values to include the effect of hysteresis, and is designed to sample the U/r-space of the vessel as shown in Fig. 3. It is assumed that the vessel response is symmetric in yaw, such that it is sufficient to perform experiments only for positive rudder settings (which for the Telemetron ASV result in positive yaw rate). The vessel shall reach steady state between the step changes such that the damping terms can be identified from the steady-state response, while the inertia is identified from the transient response. The motor will be kept in forward gear throughout the entire experiment.

The step changes in τ_m are performed as:

1. Start at $\tau_m = 0$. Select $\tau_\delta = 0$.
2. Step τ_m stepwise from 0 to 1 in steps of 0.1, letting the vessel SOG and ROT reach steady state before the next step is applied. Let the vessel do at least one

Table 2 Telemetron ASV specifications

Component	Description
Vessel hull	Polarcirkel Sport 845
Length	8.45 m
Width	2.71 m
Weight	1675 kg
Propulsion system	Yamaha 225 HP outboard engine
Motor control	Electro-mechanical actuation of throttle valve
Rudder control	Hydraulic actuation of outboard engine angle with proportional-derivate (PD) feedback control
Navigation system	
Identification experiments	Kongsberg Seatex Seapath 330+
Control experiments	Hemisphere Vector VS330

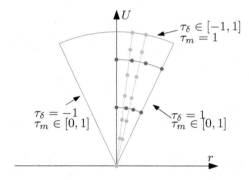

Fig. 3 Expected shape of the vessel velocity space, where the *red line* is the boundary of the velocity space. The *orange* and *blue lines* are examples of step change trajectories for fixed rudder and throttle, respectively. Note that only some trajectories are illustrated. The *dots* on the trajectories illustrate steady-state points

full turn after reaching steady state, to be able to minimize the effect of external disturbances through averaging.
3. Step τ_m stepwise from 1 to 0, in the same fashion as in step 2.
4. Repeat step 2 and 3 with the next rudder setting.

Step changes in τ_δ are performed by interchanging τ_m and τ_δ. Identification experiments were carried out in the Trondheimsfjord 17th and 18th of December 2015.

3.3 Measurement Extraction

To identify parameters for $M(x)$ and $\sigma(x)$, we need measurements of σ_U, σ_r, m_U and m_r for different vessel states x.

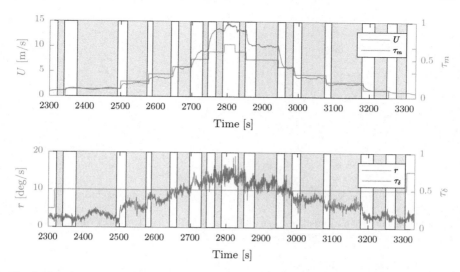

Fig. 4 Vessel response with a fixed rudder setting. The *gray* patches mark steady-state regions

Extraction of Damping Data

When the vessel is in steady state, the model (4) gives the relation:

$$\dot{x} = 0 \rightarrow \sigma(x) = \tau, \qquad (5)$$

hence measurements of the damping term can be taken simply as the control input when the vessel is at steady state. To reduce the influence of external forces, the vessel state is averaged to extract measurements for σ_U and σ_r. This is shown for one of the fixed rudder settings in Fig. 4. We observed that the motor response is greatly reduced for $\tau_m > 0.6$, hence measurements with $\tau_m > 0.6$ are omitted.

By averaging the steady-state regions, we generate a set of N_σ measurements $\mathcal{D}_\sigma = \{\{x_1, x_2, \ldots, x_{N_\sigma}\}, \{\sigma_1, \sigma_2, \ldots, \sigma_{N_\sigma}\}\}$, which can be used for identifying parameters for the damping term. The damping measurements, with mirrored values for negative rudder settings, are shown in Fig. 5.

Extraction of Inertia Data

To extract measurements for m_U and m_r, we have N_m step changes, and we create an estimate of the vessel response using N_m local first-order linear models. We approximate the SOG and ROT dynamics as SISO systems, hence for the i-th step, the linear approximation of the vessel SOG can be written as:

$$m_{U_i} \Delta \dot{U}_i + k_i \Delta U_i = \Delta \tau_{m_i}, \qquad (6)$$

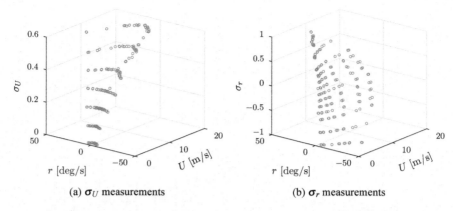

(a) σ_U measurements (b) σ_r measurements

Fig. 5 Damping term measurements from averaging of steady-state responses. Measurements for negative rudder settings are obtained by mirroring the data

Fig. 6 Inertia measurement extraction for a step in the SOG. It is clear that $m_U = 0.1711$ is the best fit

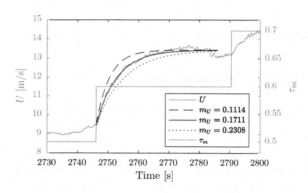

where the inertia m_{U_i} is assumed to be constant during the step, $k_i = \frac{\sigma_{U_i}^+ - \sigma_{U_i}^-}{U_i^+ - U_i^-}$, where $(\cdot)^-$ and $(\cdot)^+$ denotes the value prior to and after the step, is a linearized damping term, $\Delta U_i = U - U_i^-$ and $\Delta \tau_{m_i} = \tau_m - \tau_{m_i}^-$. The only unknown in (6) is the inertia m_{U_i}, hence we can find a suitable inertia m_{U_i} by simulating (6) for a set of possible inertias and selecting the inertia with the smallest squared estimation error, as shown in Fig. 6. The measurement is taken as $(\mathbf{x}, m_U) = \left(\left(\frac{U_i^+ + U_i^-}{2}, \frac{r_i^+ + r_i^-}{2} \right), m_{U_i} \right)$. The same approach is employed for identifying inertia for ROT.

It should be noted that, in contrast to identifying damping, we obtain two sets of measurements $\mathscr{D}_{m_U} = \left\{ \{\mathbf{x}_1, \mathbf{x}_2, \ldots, \mathbf{x}_{N_{m_U}}\}, \{m_{U_1}, m_{U_2}, \ldots, m_{U_{N_{m_U}}}\} \right\}$ and $\mathscr{D}_{m_r} = \left\{ \{\mathbf{x}_1, \mathbf{x}_2, \ldots, \mathbf{x}_{N_{m_r}}\}, \{m_{r_1}, m_{r_2}, \ldots, m_{r_{N_{m_r}}}\} \right\}$ containing N_{m_U} and N_{m_r} measurements, respectively. The inertia measurements are shown in Fig. 7.

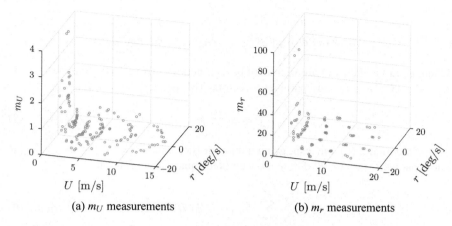

Fig. 7 Inertia term measurements. Measurements for negative rudder settings are obtained by mirroring the data

Data Preprocessing

Before the measurements are used for parameter identification, some preprocessing is required:

- Damping measurements with $\tau_\delta = 0$ should result in zero yaw rate. Even though we average the steady-state response, some offset will be present. Hence, all measurements $(U, r, \sigma_U, \sigma_r) \in \mathscr{D}_\sigma$ with $\sigma_r = \tau_\delta = 0$ should be modified as $(U, r, \sigma_U, \sigma_r) = (U, 0, \sigma_U, \sigma_r)$.
- Since the domains of U and r are different, the measurements should be normalized. We have applied zero-mean and unit variance normalization individually for each measurement set $\mathscr{D}_\sigma$, $\mathscr{D}_{m_U}$ and $\mathscr{D}_{m_r}$.

3.4 Parameter Identification

This section describes identification of the parameters of (4) based on the measurement sets $\mathscr{D}_\sigma$, $\mathscr{D}_{m_U}$ and $\mathscr{D}_{m_r}$.

Linear Regression

For identification of model parameters, we use linear regression [2]. This requires that the terms in (4) are linear in the parameters, e.g., that the damping and inertia terms can be written as:

$$\sigma_U(x) = \phi_\sigma(x)^T \beta_{\sigma_U}, \qquad \sigma_r(x) = \phi_\sigma(x)^T \beta_{\sigma_r}$$
$$m_U(x) = \phi_M(x)^T \beta_{m_U}, \qquad m_r(x) = \phi_M(x)^T \beta_{m_r}, \qquad (7)$$

where $\phi_\sigma(x)$ and $\phi_M(x)$ are vectors of basis functions (also called regressors) while $\beta_{\sigma_U}, \beta_{\sigma_r}, \beta_{m_U}$ and β_{m_r} are parameter vectors. This generalizes as a function:

$$\hat{y} = \phi(x)^T \beta. \qquad (8)$$

For the model (8), one can, given a data set $\{\{x_1, x_2, \ldots, x_N\}, \{y_1, y_2, \ldots, y_N\}\}$ and a parameter vector β, define the weighted square loss function:

$$\varepsilon = \frac{1}{N} \sum_{i=1}^{N} W_{ii} \left(y_i - \phi(x_i)^T \beta\right)^2, \qquad (9)$$

where W_{ii} is a weight for sample i. By defining $Y = \begin{bmatrix} y_1 \, y_2 \, \ldots \, y_N \end{bmatrix}^T$ and $X = \begin{bmatrix} \phi(x_1)^T \, \phi(x_2)^T \, \ldots \, \phi(x_N)^T \end{bmatrix}^T$ one can find the β that minimizes (9) as:

$$\beta = (X^T W X)^{-1} X^T W Y, \qquad (10)$$

where $W = \text{diag}(W_{11}, W_{22}, \ldots, W_{NN})$. This is known as weighted linear least-squares regression.

A well known issue with linear regression, especially with large parameter vectors, is the problem of overfitting. To reduce this problem, one can penalize large parameter values by adding a regularization term to (9) as:

$$\varepsilon = \frac{1}{N} \sum_{i=1}^{N} W_{ii} \left(y_i - \phi(x_i)^T \beta\right)^2 + \lambda R(\beta), \qquad (11)$$

where $\lambda > 0$ is a regularization weight, and the choice of the regularization term $R(\beta)$ is problem dependent. We choose ℓ_1-regularization where $R(\beta) = \|\beta\|_1$, also known as lasso, which has the property of driving parameters to zero for sufficiently high values of λ [2]. This penalizes basis functions with low sensitivities to the loss function, and favors sparsity in the parameter vector.

It should be noted that introducing regularization provides one parameter more to the problem, in form of the regularization weight λ. Additionally, there exist no closed form solution to minimizing (11) with respect to β. However, given a regularization parameter λ, the solution can be found through quadratic programming techniques.

Cross-Validation (CV)

For identifying hyperparameters, such as the regularization weight λ, one can use cross-validation (CV). This involves dividing the available data into a training set and a validation set, where the training set is used for solving the parameter estimation while using the validation set for evaluating the loss. Hyperparameters can then be identified by minimizing the loss with respect to the hyperparameters. There exist different methods for dividing the available data, e.g., κ-fold, leave-p-out, and leave-one-out (which is a special case of leave-p-out). Leave-one-out CV evaluates all possible combinations of leaving one sample for the validation set, hence for a data set of N samples this will result in N combinations of training and validation sets. We chose to use leave-one-out CV based on this property, while the limited data size ensures computational feasibility.

It should be noted that when performing both positive and negative step changes (see Fig. 4), steady-state points with the same τ will have quite similar (U, r) coordinates. Hence, one should handle groups of measurements when dividing the measurements into training and validation data.

Damping Term

From the structure of the damping measurements in Fig. 5, we propose to use polynomial basis functions for the damping term in (4). This is also motivated by [7] where polynomial damping terms are used. The power of the polynomial is chosen as four, which is assumed to be sufficient to capture hydrodynamic damping and actuator dynamics. Hence, the regressor is defined as the 15-element vector:

$$\boldsymbol{\phi}_\sigma(\boldsymbol{x}) = \left[1,\ U,\ r,\ U^2,\ Ur,\ r^2,\ U^3,\ U^2r,\ Ur^2,\ r^3,\ U^4,\ U^3r,\ U^2r^2,\ Ur^3,\ r^4\right]^T. \tag{12}$$

The parameter vectors $\boldsymbol{\beta}_{\sigma_U}$ and $\boldsymbol{\beta}_{\sigma_r}$ are identified by minimizing (11) with respect to $\boldsymbol{\beta}$. The regularization parameter is found using CV. A surface plot of the damping function is shown in Fig. 8.

Inertia Term

From the structure of the inertia measurements in Fig. 7, it is clear that a polynomial model will struggle to fit the data well. We therefore introduce an asymptotic basis function $\tanh(a(U-b))$ in addition to the polynomial terms. The regressor for the inertia terms is hence defined as the 16-element vector:

$$\boldsymbol{\phi}_M(\boldsymbol{x}) = \big[1,\ U,\ r,\ U^2,\ Ur,\ r^2,\ U^3,\ U^2r,\ Ur^2,\ r^3,\ U^4,\ U^3r,$$
$$U^2r^2,\ Ur^3,\ r^4,\ \tanh(a(U-b))\big]^T. \tag{13}$$

Notice that the asymptotic basis function introduces two more hyperparameters in the regression problem, namely a and b. To identify these hyperparameters, we again use leave-one-out CV. Notice that we use regularization when we identify these hyperparameters, individually of the linear regression. The motivation for this is

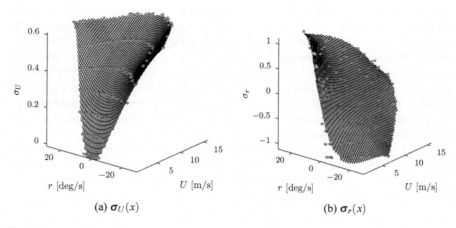

Fig. 8 Polynomial function for the damping term. The scatter points are the measuring points, where *red* points have weight $W = 1$, *blue* points have $W = 0.5$ and *green* points have $W = 0.1$

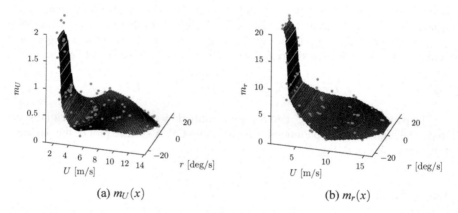

Fig. 9 Function for the inertia term. The scatter points are the measuring points

that the position of the steep asymptote in the inertial measurement will move with changing ocean currents and external forces. Adding regularization when identifying the hyperparameters increases the robustness of the identified inertia term by adding a cost to choosing high parameter values for the asymptotic term and hence limiting the gradient of the asymptotic term.

The parameter vectors β_{m_U}, β_{m_r} are, as for the damping term, identified by minimizing (11) with respect to β. The hyperparameters are identified using CV. It should be noted that we use ℓ_1-regularization when identifying the hyperparameters (a_{m_U}, b_{m_U}) and (a_{m_r}, b_{m_r}). A surface plot of the inertia function is shown in Fig. 9.

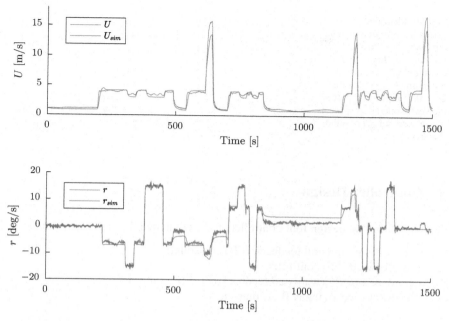

Fig. 10 Real and simulated vessel response. The deviation at high SOG is caused by exiting the valid domain of the identified model

3.5 Model Verification

To qualitatively verify the identified vessel model, we simulate the model with the input sequence from an experiment not used in the model identification and compare the results. The model (4), with damping and inertia parameterization and parameters as identified in Sect. 3.4, is simulated with the recorded input sequence to obtain the response shown in Fig. 10. Based on the comparison, we see that the model captures the dynamics of the vessel, although with slight offsets especially for ROT. The simulated transient response coincides well with the real vessel response.

A design choice for the identification experiments was the assumed shape of the vessel operating space, discussed in Sect. 3.2 and illustrated in Fig. 3. This design choice is verified by estimating the actual vessel operating space. This can be generated by using all the steady-state velocities obtained during the identification, as shown in Fig. 11. By comparing the actual and assumed operating spaces, we see that the shapes are very similar.

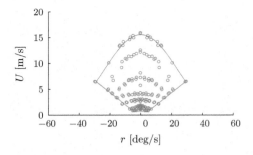

Fig. 11 Identified steady-state velocities. The *red boundary line* is estimated by least-squares curve fitting a fourth order polynomial. Note that that $U < 0.75$ m/s is not part of the vessel operating space as ocean current lower-bound the SOG

4 Controller Design

In this section, we design four controllers to be compared through experiments:

1. A proportional-integral feedback (FB) controller.
2. A feedforward (FF) controller.
3. A combined feedforward and feedback (FF-FB) controller.
4. A feedback-linearizing (FBL) controller.

4.1 Controller Types

This section describes the controller formulations, and the resulting closed-loop dynamics.

Model Uncertainties

The model (4) does not account for modeling uncertainties. For closed-loop analysis, we therefore add an unknown bias term and modify the model as:

$$M(x)\dot{x} + \sigma(x) = \tau + b, \tag{14}$$

where b is assumed to be slowly varying, hence $\dot{b} \approx 0$.

Proportional-Integral Feedback (FB) Controller

The FB controller is a proportional-integral controller with gain scheduling of the proportional gain using the inertia term of the identified model (4):

$$\tau_{FB} = -M(x)K_p\tilde{x} - K_i \int_{t_0}^{t} \tilde{x}(\gamma)d\gamma, \tag{15}$$

where $K_p > 0$ is a diagonal proportional gain matrix, $K_i > 0$ is a diagonal integral gain matrix and $\tilde{x} = x - x_d$. By inserting (15) into (14) we derive the error dynamics:

$$\dot{\tilde{x}} = -K_p \tilde{x} + M(x)^{-1} \left(b - \sigma(x) - K_i \int_{t_0}^{t} \tilde{x}(\gamma) d\gamma \right) + \dot{x}_d, \quad (16)$$

where we see that the integrator must compensate for modeling errors and damping. Even if $K_i \int_{t_0}^{t} \tilde{x}(\gamma) d\gamma = b - \sigma(x)$ and $\tilde{x} = 0$, we will still not be able to track a changing reference since $\sigma(x)$ is changing with x, and $\dot{x}_d \neq 0$ for a changing reference.

Feedforward (FF) Controller

The model-based FF controller feedforwards the desired acceleration and velocity:

$$\tau_{FF} = M(x)\dot{x}_d + \sigma(x_d). \quad (17)$$

Notice that we use the measured state x when computing the inertia term, and the desired state x_d when computing the damping term. The error dynamics becomes:

$$\dot{\tilde{x}} = M(x)^{-1} \left(\sigma(x_d) - \sigma(x) + b \right), \quad (18)$$

which has an equilibrium $x = \sigma^{-1}(\sigma(x_d) + b)$, given that $\sigma^{-1}(\cdot)$ is well defined. Hence, if $b \neq 0$ we will have some tracking and steady-state offset.

Combined Feedforward and Feedback (FF-FB) Controller

The FF-FB controller combines the FF and FB controllers as:

$$\tau_{FF\text{-}FB} = M(x)\dot{x}_d + \sigma(x_d) - M(x)K_p \tilde{x} - K_i \int_{t_0}^{t} \tilde{x}(\gamma) d\gamma. \quad (19)$$

Inserting (19) into (14), we can derive the error dynamics:

$$\dot{\tilde{x}} = -K_p \tilde{x} + M(x)^{-1} \left(\sigma(x_d) - \sigma(x) + b - K_i \int_{t_0}^{t} \tilde{x}(\gamma) d\gamma \right), \quad (20)$$

where one should notice that if $\sigma(x_d)$ would be substituted with $\sigma(x)$ we would have a feedback-linearizing controller. The motivation for using $\sigma(x_d)$ in (17) is to increase the robustness and introduce an extra "driving" term in addition to the proportional feedback driving the error to zero.

Feedback-Linearizing (FBL) Controller

The FBL controller is similar to (19), but computes the damping term for the measured velocity:

$$\tau_{FBL} = M(x)\dot{x}_d + \sigma(x) - M(x)K_p\tilde{x} - K_i \int_{t_0}^{t} \tilde{x}(\gamma)d\gamma, \quad (21)$$

which can cause poor robustness with respect to disturbances and time delays in the control system. The FBL controller is often used for analysis of closed-loop systems due to the simple error dynamics:

$$\dot{\tilde{x}} = -K_p\tilde{x} + M(x)^{-1}\left(b - K_i \int_{t_0}^{t} \tilde{x}(\gamma)d\gamma\right). \quad (22)$$

4.2 Control Architecture

The FB, FF, and FF-FB controllers in Sect. 4.1 are realized by enabling the feedback, feedforward, and both functions shown in Fig. 12, respectively. The FBL controller is realized by combining the feedback and feedforward functions, while computing the feedforward damping as $\sigma(x)$ instead of $\sigma(x_d)$.

To increase robustness in the implementation, saturation elements are placed at each output except for the proportional feedback element. Also to ensure continuous reference signals x_d and $\dot{x}_d$, we employ a second-order reference filter from a possibly discontinuous, user-specified setpoint signal x_{SP}. Additionally, we limit

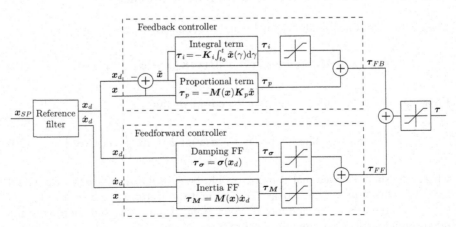

Fig. 12 Control architecture. The different controllers are realized through combinations of the feedback and feedforward functions

the acceleration such that the reference signals are feasible with respect to the vessel capability. The filter is parameterized as [7]:

$$\begin{bmatrix} \dot{x}_d \\ \ddot{x}_d \end{bmatrix} = \begin{bmatrix} 0 & I \\ -\Omega^2 & -2\Delta\Omega \end{bmatrix} \begin{bmatrix} x_d \\ \dot{x}_d \end{bmatrix} + \begin{bmatrix} 0 \\ \Omega^2 \end{bmatrix} x_{SP} \qquad (23)$$

while imposing the acceleration limits:

$$\dot{U}_d \in \left[\dot{U}_{d_{\min}}, \dot{U}_{d_{\max}}\right], \quad \dot{r}_d \in \left[\dot{r}_{d_{\min}}, \dot{r}_{d_{\max}}\right]. \qquad (24)$$

The relative damping ratio matrix $\Delta > 0$ is chosen as identity to achieve a critically damped system, while the diagonal natural frequency matrix $\Omega > 0$ is a tuning parameter.

5 Motion Control Experiments

To evaluate the performance of the controllers described in Sect. 4.1, they were implemented on the Telemetron ASV and tested in the Trondheimsfjord on the 13th and 14th of October 2016. During the first day, the sea state can be characterized as calm, which refer to significant wave heights of 0–0.1 m, while the sea state for the second day can be characterized as slight, which refer to significant wave heights of 0.5–1.25 m [9]. In total, three different scenarios were tested in different sea states.

It should be noted that the time between the model identification and motion control experiments was about 10 months. The top speed of the vessel was reduced from 18 m/s to about 16 m/s, probably caused by algae growth on the hull.

5.1 Tuning Parameters

The reference filter was tuned with a natural frequency of $\Omega = \text{diag}(0.4, 1)$ and acceleration constraints $\dot{U}_{\max} = 0.75 \, \text{m/s}^2$, $\dot{U}_{\min} = -0.75 \, \text{m/s}^2$, $\dot{r}_{\max} = 0.1 \, \text{rad/s}^2$ and $\dot{r}_{\min} = -0.1 \, \text{rad/s}^2$. The feedback tuning parameters were selected as shown in Table 3. Unfortunately, an implementation error resulted in a too high integrator gain for the yaw rate feedback controller during the experiments in calm seas.

5.2 Performance Metrics

To compare controller performance, it is beneficial to define suitable performance metrics. To simplify the analysis, it is also beneficial to combine the control inputs and outputs to one input and one output when calculating the metrics. Since the

Table 3 Feedback tuning parameters

Parameters	Values		
	FB	FF-FB	FBL
Sea state - Calm			
K_p	diag(0.15, 0.75)	diag(0.15, 0.75)	diag(0.15, 0.75)
K_i	diag(0.015, 0.5)	diag(0.015, 0.5)	diag(0.015, 0.5)
Sea state - Slight			
K_p	diag(0.15, 1)	diag(0.1, 0.5)	diag(0.1, 0.5)
K_i	diag(0.01, 0.25)	diag(0.0067, 0.125)	diag(0.0067, 0.125)

outputs have different units, we define the normalized signals $\bar{U}, \bar{U}_d, \bar{r}$ and $\bar{r}_d$ that are in the interval [0, 1] in the expected operation space of the vessel. A combined error and control input can then be computed as:

$$\bar{e}(t) = \sqrt{(\bar{U}(t) - \bar{U}_d(t))^2 + (\bar{r}(t) - \bar{r}_d(t))^2}, \quad \bar{\tau}(t) = \sqrt{\tau_m^2 + \tau_\delta^2}. \quad (25)$$

Given these signals, we can define the integral of absolute error (IAE):

$$IAE(t) = \int_{t_0}^{t} |\bar{e}(\gamma)| d\gamma, \quad (26)$$

which penalizes the error linearly with the magnitude and serves as a measure of control precision. A similar metric is the integral of square error (ISE), which penalizes large errors more than small errors.

The integral of absolute differentiated control (IADC) has been used earlier in a combined performance metric in [13], and is defined as:

$$IADC(t) = \int_{t_0}^{t} |\dot{\bar{\tau}}(\gamma)| d\gamma, \quad (27)$$

which penalizes actuator changes and serves as a measure of actuator wear and tear.

The integral of absolute error times the integral of absolute differentiated control (IAE-ADC) is a combination of IAE and IADC:

$$IAE\text{-}ADC(t) = \int_{t_0}^{t} |\bar{e}(\gamma)| d\gamma \int_{t_0}^{t} |\dot{\bar{\tau}}(\gamma)| d\gamma, \quad (28)$$

which serves as a measure of control precision versus wear and tear.

The integral of absolute error times work (IAEW) scales IAE with energy consumption [12]:

$$IAEW(t) = \int_{t_0}^{t} |\bar{e}(\gamma)| d\gamma \int_{t_0}^{t} P(\gamma) d\gamma, \quad (29)$$

where $P(t)$ is the mechanical power applied by the engine. IAEW measures control precision versus energy consumption, hence it quantifies the energy efficiency. It is common to model the applied propeller force F as proportional to the square of the propeller speed, hence, $F \propto |n|n$ [7]. The mechanical energy can then be written as:

$$P(t) \propto U(t)|n(t)|n(t). \tag{30}$$

Since we use the metric in a relative comparison, we do not care about any scaling constant and set $P(t) = U(t)|n(t)|n(t)$.

5.3 Experiments in Slight Seas

All the scenarios were tested in slight seas, and here we present two of the scenarios.

Test 1 - High-Speed Trajectory Tracking with Steady States

The first test was intended to test a large portion of the vessel operating space while measuring both steady-state and transient performance. The test is symmetric in U_d and anti-symmetric in r_d. The vessel response in slight seas is shown in Fig. 13.

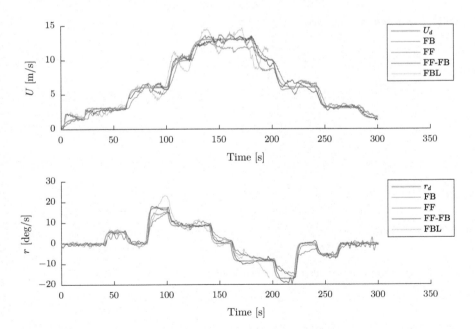

Fig. 13 Test 1 - High-speed trajectory tracking with steady states in slight seas. The feedback-linearizing (FBL) controller fails at $t \approx 194\ s$ due to sensor dropout

Immediately, we observe that the FBL controller suffers from instability, caused by the feedback term $\sigma(x)$ in (21). The oscillatory vessel state causes a dropout of the navigation system, stopping the experiment at $t \approx 194\ s$. In general, using sensor measurements in model-based feedforward terms reduces the robustness with respect to time delays, sensor dropouts, and noise. Using the reference in the feedforward terms avoids these problems. The FBL controller is not used in the other tests. The FF controller achieves good tracking, but naturally with some steady-state offset. The FB controller achieves poor tracking, while also being largely influenced by disturbances. The FF-FB controller has similar (or better) tracking performance than the FF controller while avoiding steady-state offsets, and at the same time better disturbance rejection than the FB controller. The FF, FF-FB, and FBL controllers fail in tracking the first transient due to the control system time delay which causes problems with capturing the steep transient in the inertia term. It might be beneficial to limit the gradient $\nabla_x M(x)$ or saturating the inertia $M(x)$ to avoid this behavior. The IAE (Fig. 14a) shows that the FF-FB controller has the best control precision, while the FF controller is somewhat better than the FB and FBL controllers. From the IADC (Fig. 14b), it is clear that the FB controller is tough on the actuators, while the FBL and FF-FB controllers are comparable. The FF controller is, as expected, the best with respect to wear and tear. From the IAEW (Fig. 14c), the FF-FB controller has the best energy efficiency, the FF controller is second best and the FB controller places third. The FBL controller has a bad IAEW due to the oscillatory behavior. The IAE-ADC (Fig. 14d) shows the same tendencies as the IADC, but the FF-FB controller performs better than the FBL controller, and the gap between the FF-FB and FF controllers is smaller.

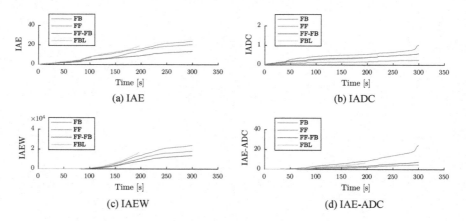

Fig. 14 Performance metrics for Test 1 in slight seas

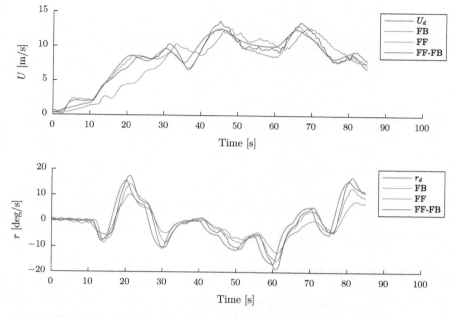

Fig. 15 Test 2 - High speed trajectory tracking without steady states in slight seas. The FF and FF-FB controllers far outperform the FB controller

Test 2 - High-Speed Trajectory Tracking Without Steady States

The second test was intended to investigate the tracking performance of the controllers. The reference is constantly changing without reaching steady state, and both moderate and high velocities are tested. This test was performed only in slight seas. From Fig. 15, we observe that the FB controller again suffers from poor tracking and largely fails this test. The FF controller performs remarkably well and, from the time plot, the FF and FF-FB controllers seem to have equal performance.

From the performance metrics in Fig. 16, the FB controller has the lowest performance, while the FF and FF-FB controllers are quite equal. The FF-FB controller has slightly better control precision (IAE) than the FF controller, at the cost of increased actuator wear and tear (IADC and IAE-ADC).

5.4 Experiments in Calm Seas

Two of the scenarios were tested in calm seas, and here we present one of them.

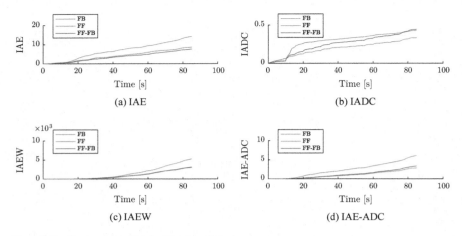

Fig. 16 Performance metrics for Test 2 in slight seas

Test 3 - Lower-Speed Trajectory Tracking with Steady States

The third test was intended to test lower velocities, especially for the yaw rate. The vessel response in calm seas is shown in Fig. 17. Note that the integral gain for the yaw rate controller unfortunately was set too high by accident, causing oscillation in the yaw rate.

From Fig. 17, we observe that the FB controller again suffers from poor tracking, and struggles with steady-state offset in yaw rate (despite the high integrator gain). The FF controller also struggles with steady-state offset, but has superior performance in the transients. The FF-FB controller combines the performance of the FB and FF controllers and provides good tracking and low steady-state offset.

From the performance metrics in Fig. 18, we can draw the same conclusions as for Test 1. However, concerning the IADC, the FF-FB controller has the most wear and tear, which is probably caused by the initial oscillatory behavior in yaw rate due to the high integrator gain resulting in high initial condition sensitivity.

5.5 Motion Control Experiments Summary

For controller evaluation, it is useful to compare the performance metrics. The final performance metric values for all the tests are presented in Table 4. We observe that:

- The FF and FF-FB controllers have the best performance:
  - The FF controller is best with respect to actuator wear and tear (IADC), also when scaled with the control precision (IAE-ADC).
  - The FF-FB controller is best with respect to control precision (IAE). In all tests except Test 2, it also has the best energy efficiency (IAEW). For Test 2, the FF and FF-FB controllers have near identical energy efficiency.

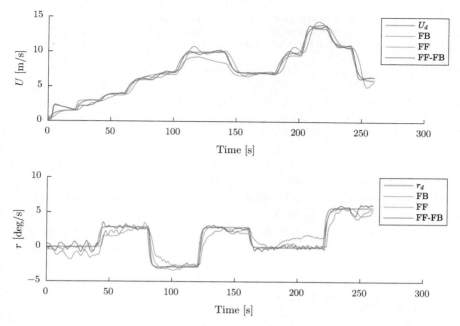

Fig. 17 Test 3 - Lower-speed trajectory tracking with steady states in calm seas. Observe the low amount of noise in the SOG-response compared to Fig. 13

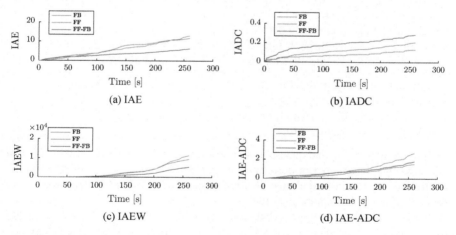

Fig. 18 Performance metrics for Test 3 in calm seas

- The FF-FB controller has the most consistent control precision performance (IAE) for varying environmental conditions.
- The FF and FF-FB controllers have quite similar consistency of energy efficiency (IAEW) for varying environmental conditions.
- The FB controller has the worst metrics in all the tests, except for IADC in Test 2.

Table 4 The performance metrics are normalized for each test, and the controller performing best for each metric in each test is highlighted in bold. C/S refer to calm (C) and slight (S) seas. *For Test 1 in slight seas, the FBL controller did not complete the entire test, hence the S metrics of FBL Test 1 are not comparable

Test case	Controller	IAE	IADC	IAE-ADC	IAEW
Test 1 C/S	FB	85.8 / 100.0	41.0 / 100.0	35.2 / 100.0	85.6 / 100.0
	FF	71.1 / 84.2	**21.0 / 22.1**	**14.9 / 18.6**	65.6 / 74.9
	FF-FB	**41.5 / 54.9**	55.1 / 55.4	22.9 / 30.5	**42.0 / 56.5**
	FBL*	84.9 / 78.3	45.7 / 33.9	38.8 / 26.5	87.2 / 71.5
Test 2 S	FB	100.0	96.9	100.0	100.0
	FF	60.4	**75.3**	**46.9**	**57.5**
	FF-FB	**53.4**	100.0	55.1	59.3
Test 3 C/S	FB	88.9 / 100.0	45.2 / 100.0	40.2 / 100.0	89.7 / 100.0
	FF	80.8 / 97.5	**28.7 / 26.4**	**23.2 / 25.8**	75.3 / 83.1
	FF-FB	**43.7 / 45.2**	62.7 / 70.2	27.4 / 31.7	**42.8 / 45.1**

6 Conclusion

In this paper, we have presented a powerful approach to modeling, identification and control of high-speed ASVs operating in the displacement, semi-displacement and planing regions. We have used this approach on a high-speed ASV to successfully identify a control-oriented model of the vessel covering all its operating regions. Furthermore, we have through full-scale motion control experiments compared the performance of four controllers all utilizing the identified model:

- A proportional-integral feedback (FB) controller with gain scheduling.
- A feedforward (FF) controller.
- A combined feedforward and feedback (FF-FB) controller.
- A feedback-linearizing (FBL) controller.

By both qualitative and quantitative comparisons, it is shown that the FF-FB and FF controllers have superior performance over the two others. The FF-FB and FBL controllers are formulated almost identically, but the FF-FB controller has superior robustness and performance over the FBL controller.

From the results, we observe that model-based feedforward control is a powerful tool, which when used correctly will result in outstanding performance. There are, however, pitfalls reducing the robustness with respect to time delays, sensor dropouts and noise. This is the case for the FBL controller, where using the measured vessel velocity in the damping feedforward term causes instability.

Possibilities for further work include:

- Use the SOG and ROT controllers for closed-loop pose control for, e.g., path following and target tracking scenarios.

- Use the SOG and ROT controllers and the identified model in combination with the dynamic window algorithm to achieve collision avoidance functionality, continuing the work in [5].
- Use the SOG and ROT controllers for manual velocity control through a joystick.
- Use the modeling approach for automatic and/or recursive model identification.
- Use the identified model to online modify the reference filter acceleration limits.

Acknowledgements This work was supported by the Research Council of Norway through project number 244116, and through the Centres of Excellence funding scheme with project number 223254. The authors would like to express great gratitude to Maritime Robotics for placing the Telemetron ASV at our disposal, and especially Thomas Ingebretsen for help with implementing the controllers and supporting the experiments. The authors would also like to thank Andreas L. Flåten for valuable discussions on the topic of linear regression and cross-validation.

References

1. Bertram, V.: Unmanned Surface Vehicles - A Survey. Skibsteknisk Selskab, Copenhagen, Denmark (2008)
2. Bishop, C.: Pattern Recognition and Machine Learning. Springer Science + Business Media, Berlin (2006)
3. Breivik, M.: Topics in Guided Motion Control of Marine Vehicles. Ph.D thesis, Norwegian University of Science and Technology, Trondheim, Norway (2010)
4. Breivik, M., Hovstein, V.E., Fossen, T.I.: Straight-line target tracking for unmanned surface vehicles. Model. Ident. Control **29**, 131–149 (2008)
5. Eriksen, B.-O.H., Breivik, M., Pettersen, K.Y., Wiig, M.S.: A modified dynamic window algorithm for horizontal collision avoidance for AUVs. In: Proceedings of IEEE CCA, Buenos Aires, Argentina (2016)
6. Faltinsen, O.M.: Hydrodynamics of High-Speed Marine Vehicles. Cambridge University Press, Cambridge (2005)
7. Fossen, T.I.: Handbook of Marine Craft Hydrodynamics and Motion Control. Wiley, New York (2011)
8. Liu, Z., Zhang, Y., Yu, X., Yuan, C.: Unmanned surface vehicles: an overview of developments and challenges. Ann. Rev. Control **41**, 71–93 (2016)
9. Prince, W.G., Bishop, R.E.D.: Probabilistic Theory of Ship Dynamics. Chapman and Hall, Boca Raton (1974)
10. Sonnenburg, C.R., Woolsey, C.A.: Modeling, identification, and control of an unmanned surface vehicle. J. Field Robot. **30**(3), 371–398 (2013)
11. Sonnenburg, C.R., Gadre, A., Horner, D., Kragelund, S., Marcus, A., Stilwell, D.J., Woolsey, C.A.: Control-oriented planar motion modeling of unmanned surface vehicles. In: Proceedings of OCEANS 2010, Seattle, USA (2010)
12. Sørensen, M.E.N., Breivik, M,: Comparing nonlinear adaptive motion controllers for marine surface vessels. In: Proceedings of 10th IFAC MCMC, Copenhagen, Denmark (2015)
13. Sørensen, M.E.N., Bjørne, E.S., Breivik, M.: Performance comparison of backstepping-based adaptive controllers for marine surface vessels. In: Proceedings of IEEE CCA, Buenos Aires, Argentina (2016)

Part VI
Coordinated and Cooperative Control of Multi-vehicle Systems

Part VI

From Cooperative to Autonomous Vehicles

Tom van der Sande and Henk Nijmeijer

Abstract What defines an autonomous vehicle? In this chapter the authors will try to answer this question and formulate the limitations of driver assistance systems as well as for—conditionally—autonomous vehicles. First a short summary of the levels of automation as provided by the society of automotive engineers (SAE) is given with a consideration on the individual levels and their implication. Following this, an overview of modern-day control techniques for advanced driver assistance systems such as cooperative adaptive cruise control is given and the requirements for automated driving are formulated. The chapter finishes with an outlook on automated driving and a discussion on the requirements and safety of automated driving.

1 Introduction

Driving a road vehicle requires mastering the control of both the longitudinal as well as the lateral degree of freedom. Besides this, careful path planning and interaction with other road users is of the utmost importance. Although we as humans have mastered these skills over the past 100 years, human error remains the primary cause of accidents [21]. Besides this, the growing number of vehicles on the road has increased congestion in some areas to unmanageable volumes [14]. One way of solving these problems is by helping a driver or removing the task of driving a vehicle from the human altogether, i.e., automated driving.

A way of helping the driver is by installing cruise control. It this case, the longitudinal degree of freedom of the vehicle is governed by a control system. With the addition of radar, the driver does not have to control the longitudinal velocity anymore in highway situations. Performance of such a system is, however, limited. By adding (wireless) communication with other road users, the performance can be

---

T. van der Sande (✉) · H. Nijmeijer
Eindhoven University of Technology, 5612 AZ Eindhoven, The Netherlands
e-mail: t.p.j.v.d.sande@tue.nl

H. Nijmeijer
e-mail: h.nijmeijer@tue.nl

© Springer International Publishing AG 2017
T.I. Fossen et al. (eds.), *Sensing and Control for Autonomous Vehicles*,
Lecture Notes in Control and Information Sciences 474,
DOI 10.1007/978-3-319-55372-6_20

significantly improved [16]. Note that the driver still has to perform a control task in such a vehicle, as he still has to take care of the lateral degree of freedom and that the systems only functions in limited ranges of operation, typically only highway use. Longitudinal driver support systems that are rapidly finding their way in even the lower class of automobiles include, for example, collision warning and city braking. In [2] it is argued that such systems reduce the amount of rear-end crashes with injuries by 42%.

Systems that help the driver with the lateral task, i.e., steering, are also commercially available. Examples of such systems are lane keep assist or automated parking. However, these systems only help in certain situations and are not capable following a preplanned path, let alone define it.

Similarly to road vehicles, research on automated systems is also done in the field of ships and airplanes. Both of these fields present their own set of challenges compared to road vehicles. For example in [11] a controller is developed that compensates for drift forces that are caused by disturbances such as ocean currents, wind, and waves. In the research a line of sight approach is used and the side slip angle of the vessel is estimated online compensated for in the controller. In path following control for road vehicles, the side slip angle is often ignored all together. Noteworthy is furthermore that, compared to road vehicles, the dynamics are an order of magnitude slower. Besides single vehicle automation, research has also been performed on cooperation [4] and coordinated control [23] of fleet ships and planes [12]. For ships the main objective is to make transferring goods between ships more easy, where for planes the focus lies on reducing fuel consumption. Another reason for research into synchronization and positioning is placing a ship close to, for example, an oil rig. In such a case disturbances still exist that force the ship out of position. A big challenge with ships exists, since they are typically under actuated [15]. The dynamics furthermore prevent the use of time-invariant feedback techniques. Furthermore important to note is that, since a typical ship is underactuated, constraints on the ships' motion make path planing import in the design of a feedback [23].

In the field of robotics much work has been performed on cooperation and synchronization [13, 20]. The primary objective for most of this research is that the group of robots tries to achieve a common goal together. Examples here are small robots lifting one larger object, or a formation of unmanned aerial vehicles that is used for surveillance. To work toward a common goal the first challenge is to achieve consensus, i.e., coordinate how the task is going to be performed. Difficulties with reaching consensus are on the one hand related to communication. The question that can be asked is which information is transmitted and what is the quality of this information? Furthermore, the interaction between robots changes continuously. The secondary objective lies with acting on the achieved consensus. The interaction between acting on the task, i.e., control and achieving consensus is still a topic of much research [20].

The aforementioned review shows that research on automation and cooperation is performed in numerous fields. This chapter will primarily focus on road transportation. An often made assumption within road transportation is that a vehicle equipped with one of the aforementioned systems is an autonomous vehicle. However, one should ask themselves, what defines an autonomous vehicle? In this chapter the authors will try to answer this question and formulate the limitations of driver assistance systems as well as for—conditionally—autonomous vehicles. The guide for this discussion will be one of the most advanced driver assistant systems, being cooperative adaptive cruise control, on the one hand and autonomous vehicles on the other hand.

The outline of this chapter is as follows. First the levels of automation as introduced by the Society of Automotive Engineers (SAE) are discussed. Following this discussion, a description of autonomous and cooperative systems is given, highlighting the requirements as well as discussing the features. An outlook of the future of mobility, including cooperative and autonomous driving is then presented. Finally, conclusions are presented.

2 Levels of Automation

Taking control of the driving task can be done a multitude of levels, as indicated by the Society of Automotive Engineers (SAE) in their J3016 standard [24]. A summary of these levels is given in Fig. 1. It indicates that with increasing level of automation, the car takes over more driving tasks. At level 0 no automation is installed in the vehicle and the driver is in full control. Installing either longitudinal or lateral automation increases the level of automation to level 1. Including both longitudinal as lateral automation raises it further to level 2 and adding monitoring creates a level 3 vehicle. A level 4 vehicle has the same capabilities as a level 3 vehicle, with the addition of fall-back performance in certain situations. Full autonomy is only reached in level 5, the vehicle is then capable of performing all driving tasks and does not rely on any human input during the execution of the driving task. Note that this implies that the vehicle automation should be unconditional. This does not mean that it should be able to solve all situations, the vehicle is still bound by the laws of physics.

Important to note in Fig. 1 is the column '*fall-back performance of dynamic driving task*', this indicates when the human should still be vigilant. It implies that only in level 5 full automation is reached and that in all the other levels at some point in time driver input is required. Especially level 3 might give a false sense of security since it still uses the driver as a fall-back device. For now, the authors therefore argue that at level 3 automation, leaving all driving tasks to the vehicle should only be done within a closed perimeter, design exclusively for autonomous driving.

Automation level 4 operation allows for full autonomy within certain bounds, for example, highway driving. Within its allowable range of operation the car is capable of performing all driving tasks, including the dynamic driving driving task fallback. Practically, this implies that in an emergency case, the vehicle is capable of safely taking the vehicle out of harms way. As the SAE rightfully indicates, the driver is a passenger when level 4 driver assist systems are enabled. Unconditional operation is only achieved in level 5.

The primary goal for automation within aviation is to reduce the amount of human work [7]. In larger commercial airliners, typically all three degrees of rotation are

Level	Name	Narrative definition	Execution of steering and acceleration/ deceleration	Monitoring of driving environment	Fallback performance of dynamic driving task	System capability (driving modes)
Human driver monitors the driving environment						
0	No Automation	the full-time performance by the *human driver* of all aspects of the *dynamic driving task*, even when enhanced by warning or intervention systems	Human driver	Human driver	Human driver	n/a
1	Driver Assistance	the *driving mode*-specific execution by a driver assistance system of either steering or acceleration/deceleration using information about the driving environment and with the expectation that the *human driver* perform all remaining aspects of the *dynamic driving task*	Human driver and system	Human driver	Human driver	Some driving modes
2	Partial Automation	the *driving mode*-specific execution by one or more driver assistance systems of both steering and acceleration/deceleration using information about the driving environment and with the expectation that the *human driver* perform all remaining aspects of the *dynamic driving task*	System	Human driver	Human driver	Some driving modes
Automated driving system ("system") monitors the driving environment						
3	Conditional Automation	the *driving mode*-specific performance by an *automated driving system* of all aspects of the *dynamic driving task* with the expectation that the *human driver* will respond appropriately to a *request to intervene*	System	System	Human driver	Some driving modes
4	High Automation	the *driving mode*-specific performance by an *automated driving system* of all aspects of the *dynamic driving task*, even if a *human driver* does not respond appropriately to a *request to intervene*	System	System	System	Some driving modes
5	Full Automation	the full-time performance by an *automated driving system* of all aspects of the *dynamic driving task* under all roadway and environmental conditions that can be managed by a *human driver*	System	System	System	All driving modes

Fig. 1 SAE levels of automation [24]

automated, with special stages identified for all the different stages of flight [10]. Special attention is given to automated landing, dividing it into three categories. Noteworthy here, is that instrumented landing strongly depends not only on the equipment of the aircraft, but also the runway.

The primary discussion with automation in aviation at this point is on the method of pilot fall-back [22]. Basically, a pilot is expected to intervene if an issue occurs, however, the pilots are also expected to fly using automation from a few minutes after takeoff until a few minutes before landing. This gives rise to the problem of pilots no longer knowing how to handle emergency situations in case the automation fails.

3 System Description

An autonomous vehicle is capable of (partially) performing the driving task normally performed by a driver. This is in sharp contrast with a vehicle equipped with cooperative adaptive cruise control, which is capable of only performing a part of the driving tasks. However, for a vehicle to be cooperative, it has to share its information with its surrounding vehicles and also receive information from neighboring vehicles. In this section, the requirements and specifications of cooperative and autonomous vehicles will be discussed in more detail.

3.1 Cooperative Adaptive Cruise Control

"*Cooperative driving may be described as influencing the individual vehicle behavior, either through advisory or automated actions, so as to optimize the collective behavior with respect to road throughput, fuel efficiency and/or safety...* " referring to the thesis of J. Ploeg [16]. The quote adequately describes the expected behavior of cooperation in a platoon of vehicles. To achieve cooperation, some form of communication is required. Typically the IEEE 802.11p [25] communication standard is used for this, conveying the desired acceleration of the preceding vehicle. Furthermore, as a feedback mechanism, the relative velocity and position is measured. Note that without communication CACC is no longer functional, since the vehicle then only relies on its onboard sensors. A suggested option is then to reduced to degraded CACC (dCACC) as proposed in [18]. With degraded CACC the follower vehicle estimates the acceleration of its predecessor, thereby being able to maintain a smaller headway time than ACC. Performance is worse than CACC however, since it now relies on the estimated actual acceleration in stead of the desired acceleration.

An overview of the basic setup of a CACC equipped vehicle is shown in Fig. 2. Additional to the required sensors, GPS can be used to determine the vehicle's absolute position. This addition benefits vehicle cooperation in two ways, first of all the vehicle knows who its closest neighbors are. Secondly, the absolute position

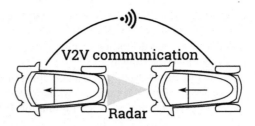

Fig. 2 Required sensors for cooperative adaptive cruise control

of two vehicles can be used to calculate a relative position and function as an additional safety check.

Besides sensing, for CACC to function from a control perspective, an actuator is required. This requires control over the drive line of the vehicle. Typically, a local controller is designed that compensates the vehicle dynamics through feedback linearization and reduces them to simpler dynamics, such that the vehicle behavior as incorporated in the platoon of vehicles is a point mass with drivetrain dynamics, typically of the form

$$\begin{bmatrix} \dot{d}_i \\ \dot{v}_i \\ \dot{a}_i \end{bmatrix} = \begin{bmatrix} v_{i-1} - v_i \\ a_i \\ -\frac{1}{\tau} a_i + \frac{1}{\tau} u_i \end{bmatrix}, \quad i \in S_m, \qquad (1)$$

with $d_i(t)$, $v_i(t)$, $a_i(t)$ the following distance, forward velocity and acceleration respectively. The input u_i can be seen as the desired acceleration and τ a time constant that represents the driveline dynamics. The set of all vehicles in the platoon is defined in $S_m = i \in \mathbb{N} 1 \leq i \leq m$ in a platoon of length m.

The main requirement for the design of a CACC controller is that the distance, d_i, between the vehicle i and its predecessor $i - 1$ equals a desired distance $d_{r,i}$. The desired distance is defined as a constance time gap spacing policy [17]

$$d_{r,i}(t) = r_i + h v_i(t), \quad i \in S_m. \qquad (2)$$

Here, h is the time gap, r_i the standstill distance and $v_i(t)$ the forward velocity. Note that for a homogeneous platoon the time gap h is chosen constant. The spacing error is now expressed as

$$e_i(t) = d_i(t) - d_{r,i}(t) = (q_{i-1}(t) - q_i(t) - L_i) - r_i + h v_i(t), \quad i \in S_m, \qquad (3)$$

where $q_i(s)$ is the rear bumper position of the vehicle and L_i the vehicle length. The first control objective is now defined as

$$\lim_{t \to \infty} e_i(t) = 0 \quad \forall \quad i \in S_m. \qquad (4)$$

Secondly, string stability of the platoon of vehicles is required, which is defined as

$$|\Gamma(s)|_{H_\infty} \leq 1. \tag{5}$$

with

$$\Gamma_i(s) = \frac{y_i(s)}{y_{i-1}(s)} \tag{6}$$

is required. A suitable controller can be designed by considering the error dynamics

$$\begin{bmatrix} e_{1,i} \\ e_{2,i} \\ e_{3,i} \end{bmatrix} = \begin{bmatrix} e_i \\ \dot{e}_i \\ \ddot{e}_i \end{bmatrix}, \quad i \in S_m \tag{7}$$

It is easy to see that the first and second term are simple integrators. By using (3) and (1), the time derivative of $e_{3,i}$ is defined as

$$\dot{e}_{3,i} = -\frac{1}{\tau}e_{3,i} - \frac{1}{\tau}\xi_i + \frac{1}{\tau}u_{i-1}, \quad i \in S_m \tag{8}$$

with

$$\xi_i := h\dot{u}_i + u_i. \tag{9}$$

To stabilize the error dynamics and compensate for the input u_{i-1}, ξ_i is chosen as

$$\xi_i = K \begin{bmatrix} e_{1,i} \\ e_{2,i} \\ e_{3,i} \end{bmatrix} + u_{i-1}, \quad i \in S_m \tag{10}$$

and $K := \begin{bmatrix} k_p & k_d & k_{dd} \end{bmatrix}$. With these extra controller dynamics the closed loop model becomes

$$\begin{bmatrix} \dot{e}_{1,i} \\ \dot{e}_{2,i} \\ \dot{e}_{3,i} \\ \dot{u}_i \end{bmatrix} = \begin{bmatrix} 0 & 1 & 0 & 0 \\ 0 & 0 & 1 & 0 \\ -\frac{k_p}{\tau} & -\frac{k_d}{\tau} & -\frac{1+k_{dd}}{\tau} & 0 \\ -\frac{k_p}{h} & -\frac{k_d}{h} & -\frac{k_{dd}}{h} & -\frac{1}{h} \end{bmatrix} \begin{bmatrix} e_{1,i} \\ e_{2,i} \\ e_{3,i} \\ u_i \end{bmatrix} + \begin{bmatrix} 0 \\ 0 \\ 0 \\ \frac{1}{h} \end{bmatrix} u_{i-1}. \tag{11}$$

Using the Routh–Hurwitz stability criterion it is shown that the equilibrium is asymptotically stable for any time gap $h > 0$ and for the choice $k_p, k_d > 0, k_{dd} + 1 > 0$. Hereby the vehicle following objective (4) is met [16].

The second requirement for the platoon of vehicle is string stability (5) and (6). For this a frequency domain approach is preferred, with the plant model defined as

$$G(s) = \frac{q_i}{u_i} = \frac{1}{s^2} \frac{1}{\tau s + 1}. \tag{12}$$

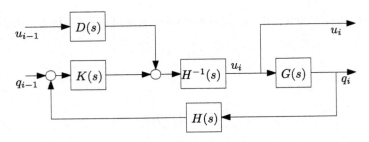

Fig. 3 Block scheme of the individual vehicle within a platoon

Furthermore the spacing policy of (2) in frequency domain can be written as

$$H(s) = hs + 1 \qquad (13)$$

and the feedback controller as

$$K(s) = k_p + k_d s + k_{dd} s^2. \qquad (14)$$

For practical reasons, the second derivative term k_{dd} is often set to zero. A CACC controlled vehicle is then represented by the block scheme as shown in Fig. 3. Here $D(s)$ represents a communication delay between vehicle i and $i-1$ and can be modeled as $D(s) = e^{-\theta s}$ or for controller synthesis a pade approximation can be made [26].

Recall that a concern with CACC is to find a controller such that the H_∞-norm of the string stability transfer function

$$\Gamma(s) = \frac{1}{H(s)} \frac{G(s)K(s) + D(s)}{1 + G(s)K(s)} \qquad (15)$$

is smaller than 1 [16], i.e.,

$$|\Gamma(s)|_{H_\infty} \leq 1. \qquad (16)$$

Without the influence of a communication delay θ it can be shown that the system is stable for any choice of controller gains and time gap. With increasing communication delay the headway time h will have to be increased to guarantee $|\Gamma(s)|_{H_\infty} \leq 1$. A simulation result of such a suitable choice is shown in Fig. 4 where a platoon of five vehicles is simulated. In this simulation a PD controller is designed (so $k_{dd} = 0$), such that with the choice of the headway time and communication delay a string-stable platoon is created. This can be verified from the velocity response, which shows no undershoot for increasing vehicle number. The parameters for this simulation are shown in Table 1. Note that in this table a vehicle internal delay is also mentioned which is not included in the analysis above. Measurements have however shown that there exists a delay between desired and actual vehicle acceleration.

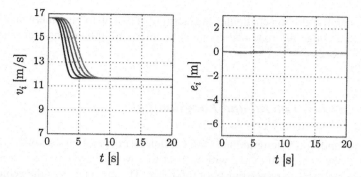

Fig. 4 Simulated response of a five vehicle platoon with *left* the velocity and *right* the tracking error [18]. Here the communication delay θ is set to 0.2 s and the vehicle time constant τ to 0.1 s

Table 1 Vehicle and controller parameters

Symbol	Value	Description
θ	0.02 s	Communication delay
τ	0.1 s	Vehicle time constant
ϕ	0.2 s	Vehicle internal delay
k_p	0.2	Proportional controller gain
k_d	0.7	Differential controller gain

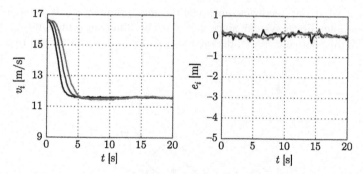

Fig. 5 Measured response of a three vehicle platoon with *left* the measured velocity and *right* the tracking error [18]

Practical tests have shown that with the control topology as introduced above, string stability can be guaranteed up to a headway time of 0.6 s [18]. Results of one of these tests is shown in Fig. 5, again, the controller parameters of Table 1 are used. In this test a three vehicle platoon shows string-stable behavior, as no undershoot of the velocity exists. If improvement in the wireless communication can be made, the headway time can even be reduced to 0.3 s. Another, more advanced test, the GCDC, was performed between Eindhoven and Helmond in May of 2016 [9]. In the grand cooperative driving challenge, the merging of two platoons, emergency vehicle passage on a congested road and cooperative intersection handling were displayed with great success.

3.2 Combined Lateral and Longitudinal

An obvious addition to longitudinal CACC is lateral cooperation. For this, define the location of the vehicle in (x, y) with respect to a world frame and ψ_i the rotation with respect to the x-axis, see Fig. 6 for reference. The kinematics of the vehicle then become

$$\begin{bmatrix} \dot{x}_i \\ \dot{y}_i \\ \dot{\psi}_i \end{bmatrix} = \begin{bmatrix} \cos(\psi_i) & 0 \\ \sin(\psi_i) & 0 \\ 0 & 1 \end{bmatrix} \begin{bmatrix} v_i \\ \omega_i \end{bmatrix}, \quad i \in S_m. \tag{17}$$

Here v_i and ω_i are the forward and rotation velocity of the vehicle respectively. Note that the elegant linear model as presented in the case of longitudinal CACC quickly disappears when introducing a second degree of freedom [1]. The output of the vehicle is now chosen as $\eta_i = [x, y]^T$ with its time derivative defined as

$$\dot{\eta}_i = \begin{bmatrix} \dot{x}_i \\ \dot{y}_i \end{bmatrix} = \begin{bmatrix} \cos(\psi_i) & 0 \\ \sin(\psi_i) & 0 \end{bmatrix} \begin{bmatrix} v_i \\ \omega_i \end{bmatrix}, \quad i \in S_m. \tag{18}$$

By using dynamic feedback linearization, the system can now be transformed into a decoupled linear system, since the output does not show the input ω_i directly. To achieve this, define $a_i := \dot{v}_i$, the second derivative of η with respect to time now

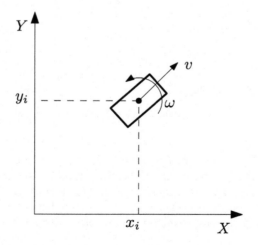

Fig. 6 Unicycle model

becomes

$$\ddot{\eta}_i = \begin{bmatrix} \ddot{x}_i \\ \ddot{y}_i \end{bmatrix} = \begin{bmatrix} \cos(\psi_i) & -v_i \sin(\psi_i) \\ \sin(\psi_i) & v_i \cos(\psi_i) \end{bmatrix} \begin{bmatrix} a_i \\ \omega_i \end{bmatrix} := H_i \begin{bmatrix} a_i \\ \omega_i \end{bmatrix}, \quad i \in S_m. \quad (19)$$

Define the new input $u_i = [u_{1,i}, u_{2,i}]^T$ as $u_i := H_i [a_i, \omega_i]^T$ such that the original input can be determined as

$$\begin{bmatrix} a_i \\ \omega_i \end{bmatrix} = H_i^{-1} u_i = \begin{bmatrix} \cos(\psi_i) & \sin(\psi_i) \\ -\frac{1}{v_i} \sin(\psi_i) & \frac{1}{v_i} \cos(\psi_i) \end{bmatrix} u_i. \quad (20)$$

Note that as long as $v_i \neq 0$, H_i is nonsingular. The full system now becomes

$$\begin{aligned}
\dot{x}_i &= v_i \cos(\psi_i) \\
\dot{y}_1 &= v_i \sin(\psi_i) \\
\dot{\psi}_i &= \frac{-u_{1,i} \sin(\psi_i) + u_{2,i} \cos(\psi_i)}{v_i} \\
\dot{v}_i &= u_{i,1} \cos(\psi_i) + u_{2,i} \sin(\psi_i)
\end{aligned} \quad (21)$$

such that the decoupled system becomes

$$\begin{aligned}
\ddot{x}_i &= u_{1,i} \\
\ddot{y}_i &= u_{2,i}
\end{aligned} \quad (22)$$

when the output η_i is considered. Note that the system is now basically reduced to two double integrators.

Considering constant time gap following policy as in (2) in two dimensions and similarly the error (3) the error dynamics of the vehicle become [1]

$$e_{x,i} = (x_{i-1} - x_i) - r \cos \psi_i + h \dot{x}_i \quad (23)$$
$$e_{y,i} = (y_{i-1} - y_i) - r \sin \psi_i + h \dot{y}_i. \quad (24)$$

Now define the state vector as $e_i = [e_{x,i}, e_{y,i}, \dot{e}_{x,i}, \dot{e}_{y,i}]$, the time derivative of e_i is then defined as

$$\dot{e}_{x,i} = (\dot{x}_{i-1} - \dot{x}_i) + r\dot{\psi}_i \sin \psi_i - h\ddot{x}_i \quad (25)$$

$$\dot{e}_{y,i} = (\dot{y}_{i-1} - \dot{y}_i) - r\dot{\psi}_i \cos \psi_i - h\ddot{y}_i \quad (26)$$

$$\ddot{e}_{x,i} = u_{1,i-1} - u_{1,i} + r\dot{\psi}_i^2 \cos(\psi_i) - h\dot{u}_{i-1} - \frac{r \sin(\psi_i)}{v_i} \left(\dot{u}_{1,i} \sin(\psi_i) - \dot{u}_{2,i} \cos(\psi_i) + 2\dot{\psi}_i \dot{v}_i \right) \quad (27)$$

$$\ddot{e}_{y,i} = u_{2,i-1} - u_{2,i} + r\dot{\psi}_i^2 \sin(\psi_i) - h\dot{u}_{2-1} - + \frac{r \cos(\psi_i)}{v_i} \left(\dot{u}_{1,i} \sin(\psi_i) - \dot{u}_{2,i} \cos(\psi_i) + 2\dot{\psi}_i \dot{v}_i \right) \quad (28)$$

With feedback linearization and a PD controller it is now possible to show that the error dynamics (23) can be stabilized as long as the feedback gains are chosen positive.

Now consider the case where vehicles are making a steady state turn, i.e., $\omega = c \in \mathbb{R}^+, \dot{v} = 0$. Following from (25), (26) and considering steady state, i.e., $\dot{e}_{x,i} = \dot{e}_{y,i} = 0$, it can now be written that

$$\dot{x}_{i-1} + \dot{y}_{i-1} = \left(\dot{x}_i - r\dot{\psi}_i \sin \psi_i + h\ddot{x}_i\right) + \left(\dot{y}_i + r\dot{\psi}_i \cos \psi_i + h\ddot{y}_i\right). \tag{29}$$

Squaring the equation and substituting the variables using (18), (19) and (21) gives

$$v_{i-1}^2 = v_i^2 + r^2\omega_i^2 + h^2(\ddot{x}^2\ddot{y}^2) + 2rhv_i\omega_i^2. \tag{30}$$

Remember that $r, h, v_i > 0$ gives that $v_{i-1}^2 \geq v_i^2$. Secondly, consider the orientation error of the follower vehicle and its time derivative

$$\begin{aligned} e_{\psi_i,i} &= \psi_{i-1} - \psi_i \\ \dot{e}_{\psi_i,i} &= \dot{\psi}_{i-1} - \dot{\psi}_i. \end{aligned} \tag{31}$$

Suppose that the lead vehicle is on a steady state curve with $\dot{\psi}_i = \dot{\psi}_{i,ss}$. Given that the follower vehicle is controlled, it follows that $\lim_{t \to \infty}(\dot{\psi}_{i-1,ss} - \dot{\psi}_{i,ss}) = 0$, implying that $\omega_{i-1} = \omega_i$. In steady state cornering the radius of the curve is defined as

$$R_i = \frac{v_i}{\omega_i}. \tag{32}$$

Referring back to (30), it follows that $R_{i-1} \geq R_i$, since $v_{i-1}^2 \geq v_i^2$ or in practical terms, the follower vehicle cuts the corner. Simulation results of the a controlled string of vehicle are shown in Fig. 7. From this figure it becomes apparent that with increasing vehicle number the platoon shows corner cutting behavior. Note that from vehicle $i - 1$ to i this behavior might be small, but with large platoons the corner radius is significantly smaller.

3.3 Autonomous Vehicles

For a vehicle to function autonomously, it should be fully aware of its surroundings, own location and state. An example of the required sensors to facilitate autonomous driving is

- Long range forward facing radar to detect traffic at high velocities.
- Camera information to detect road features such as lines, traffic signs, and other infrastructure. Furthermore, a camera can be used to identify other road users.
- Short range LIDAR sensor that detects objects close to the vehicle.

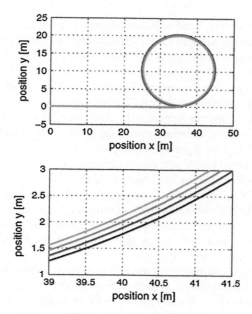

Fig. 7 Plot of a circular path, clearly showing corner cutting behavior. The vehicle number increases with a lighter *colored line*

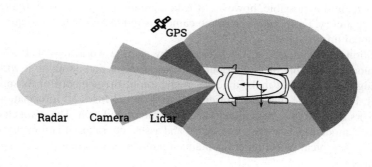

Fig. 8 Required sensors for autonomous driving

- Side and rear viewing sensors.
- Inertial measurement unit, measuring the vehicles planar motion.
- Forward velocity.
- GPS to determine the vehicle's absolute position in the world.

This is also captured in Fig. 8, which indicates the required sensors and their respective range. Being aware of its internals state and its surroundings is one thing, acting on it a different challenge altogether. Obviously, the actuators of the vehicle, i.e., throttle, brake and steering, need to be automated. The biggest challenge, however, is determining the required input for these actuators, i.e., path planning.

In the simplest example, driving from A to B reduces the control of an autonomous vehicle to a tracking problem, i.e., keeping within the road boundaries and taking the correct turn at the correct location. Typical driving, however, is not as straight forward as tracking a predefined trajectory, since the dynamics of modern day traffic complicate the task of driving significantly. Accounting for the movement of other vehicles makes the task of autonomous driving much more reactive than proactive. This calls for the design of algorithms that predict the motion of other traffic, such as predicting whether a vehicle in a another lane will cut in [5].

Another challenge is creating one coherent image of the world based on the set of sensors the vehicle is equipped with. For example, the GPS, inertial measurement unit and forward velocity in combination with steering angle can all be used to determine the traveled distance and part of the state of the vehicle. The research question here is to create an observer that fuses all relevant signals but also functions fault tolerant, indicating which sensors can be trusted.

Another example is combining all three front-facing sensors. Radar, LIDAR, and camera can all be used to detect objects in front of the vehicle. However, all have their merits and limitations. Radar systems work in almost all conditions, however it does not provide color or contrast and has limited resolution, making it difficult to determine what type of object is in front of the vehicle. LIDAR has better resolution, but fails in bad weather conditions and is typically large in volume. With a camera object detection is possible, however, it is computationally expensive, limiting the update frequency. Furthermore, a stereo camera is required to reliable acquire three dimensional information.

Additional performance can be unlocked by using a world model. In short the world model is capable of finding out common patterns and combine it with prior knowledge and with that make a better prediction of the behavior of its surroundings [3]. For example the object-oriented world model fuses all information from the sensors into a single dynamic model of the monitored area. The main challenge with world modeling, however, is to detect and assess complex situations that evolve over time. A typical world model contains a number of different processing steps, for example,

- Sensors, measuring the vehicle behavior and its surroundings.
- Prior knowledge, this contains previously encountered situations and ways of handling them, but also methods and algorithms on how to deal with sensor data.
- Learning, by recognizing patterns, the knowledge is updated.
- The world model.
- The acting part, including plan making and actuation.

A way of representing the one point in time of the world model is by considering a scene, which contains all observed and inferred object information. Here an object can have static information such as the width of car, but also dynamic properties such as its relative velocity [3]. Furthermore, a degree of belief is attributed to all objects, indicating how much its reading can be trusted. Important to note is that the scenes evolve over time. This also has a consequence for observed objects, which might not produce a new measurement value in a new scene. To facilitate removal of this object from the world model an aging mechanism is typically implemented which interacts closely with the degree of belief of the object. Once a lower threshold is surpassed, the object will be removed from the model.

From a practical point of view, Google produces reports on their autonomous driving projects each month. In September 2016 issue [8] the case for world modeling is made. Google indicates that their controllers rely heavily on prior knowledge. In September issue of their report the statement is made that the first 90% of driving is easy, it is the last 10% that requires the most effort. Examples such as horses on the road or cars driving in the wrong direction require experience on how to deal with these situations. The question on whether to store all this data and how to access it with over a million vehicles instead of the 58 vehicles Google is now running.

Concluding from the analysis stated above, numerous techniques exist to determine the state of the vehicle and its location relative to other road users. With increasing confidence in the world model, certain assumptions can be made with a higher degree of confidence. However, the main requirement is to guarantee safety at all times. A trade-off needs to be found between safety, i.e., maintaining a safe distance to other road users, and performance, i.e., minimizing road usage. Being too cautious might even cause accidents, as Google's report indicate [8]. Most of the collisions reported involve rear-end collisions, where the nonautonomous follower vehicle rear-ended one of Google's autonomous vehicles.

3.4 Outlook

Up to now, the discussion has been mostly held on vehicle level. Incorporating a multitude of sensors to determine the vehicle state and the state of its surroundings. Adding communication can improve throughput by reducing the gap at which vehicles follow each other. A promising addition to autonomous driving is vehicle to infrastructure (V2I) and the inverse (I2V) communication [6]. As was discussed before, an autonomous vehicle needs to maintain a certain safety bubble in order to guarantee safe operation. A way of helping an autonomous vehicle to reduce its safety bubble is adding V2I and I2V communication. Consider for example an obscured intersection with two vehicle approaching. An autonomous vehicle will know from map data that there is an intersection coming up and as a result reduce its velocity such that it can stop in time if another road user appears. Now consider the same situation, but with the addition of communication. The intersection (I2V) can now relay information about an approaching vehicle to the autonomous vehicle such that it can make a decision on whether it should stop or it can continue driving at its current velocity.

A unique quality of humans is that they can interact with other road users on a level that vehicles cannot. Quoting D. Dolgov from Google; *"Over the last year, we've learned that being a good driver is more than just knowing how to safely navigating around people, but also knowing how to interact with them.*" [8]. Subtle ques such as nudging to another driver are unknown to autonomous vehicles, but are required to interact successfully in mixed traffic. A large research piece of research is still required to successfully perform this unique human quality.

With the addition of more and more smart features and communication in vehicles, the largest pitfall is too much data. Not only is there a challenge with conflicting data, i.e., deciding which sensor gives the correct reading and thereby avoiding false positives on the one hand, but more importantly false negatives. A second challenge is communication of the data in a timely fashion, Ploeg et al. [19] already showed that with increasing time delay using cooperation is not beneficial anymore.

4 Discussion

Referring back to the introduction, the question one should ask themselves is what defines an autonomous vehicle? Referring to J3016 [24], autonomy refers to *"...the capacity for self-governance..."*. Which makes it easy to disprove CACC as a system that makes a vehicle autonomous, since it relies heavily on communication with other road users. This does not imply that CACC does not have a place in modern day traffic, as it can be very effective in minimizing road usage. It is furthermore not designed with a viable fall-back mechanism, making it effectively a level 2 system.

The analysis on autonomous vehicles clearly shows that the primary task for a control system is reacting to its surroundings, making it a reactive system, rather than proactive. The main concern lies with determining what to react to. Techniques such as sensor fusion and world modeling come to play here, determining state of the autonomous vehicle and its surroundings. Disconcerting the intention of all road users is key here. Performance can be improved by developing more detailed models, however, safety will always remain key.

In the following years many of these questions will be answered, promoting the usage of automated features in everyday traffic. The main suggestion in designing automated systems is to design for level 4 operation whilst maintaining vigilance of level 3 or lower. The primary reason for this statement is that the range of automation in which a level 4 vehicle is guaranteed to function is ever changing in everyday traffic. Can we truly guarantee that the fall-back mechanism functions properly? Furthermore, experience from automation in aviation teaches us that if the driver should still function as a fall-back, he should still be capable of driving and handling the situation. The question is how to maintain the driver at a level that he is still capable of performing this task. A second concern refers back to Fig. 1 and a horizontal transition within the driving level. For example consider level 2, the system executes the steering and acceleration task, however, the driver is still monitoring the driving environment. At the moment the driver takes over there is no smooth transition from one state to the other. More attention should be focussed on this transition and guaranteeing that it is safe.

As a final note it is worth asking the question on whether we can truly design a successful automated vehicle without communication? Obviously full autonomy must be a fall-back mechanism of automation in combination with communication, however to be successful as an individual vehicle, do we need to know more of our environment than our sensors tell us? Classical linear control theory teaches us that there is a large benefit in using feedforward.

Acknowledgements This work is part of the research program i-CAVE with project number 14893, which is partly financed by the Netherlands Organisation for Scientific Research (NWO).

References

1. Bayuwindra, A., Aakre, Ø., Ploeg, J., Nijmeijer, H.: Combined lateral and longitudinal cacc for a unicycle-type platoon. In: 2016 IEEE Intelligent Vehicles Symposium (IV), pp. 527–532 (2016)
2. Beene, R.: Automotive news - automatic braking reduces rear-end crashes, iihs study finds. Online, 1 (2016). Accessed 11 Dec 2016
3. Belkin, A., Kuwertz, A., Fischer, Y., Beyerer, J.: World Modeling for Autonomous Systems, Innovative Information Systems Modelling Techniques. InTech (2012)
4. Bondhus, A.K., Pettersen, K.Y.: Control of ship replenishment by output feedback synchronization. In: Proceedings of OCEANS 2005 MTS/IEEE, vol. 2, pp. 1610–1617 (2005)
5. Carvalho, A., Williams, A., Lefevre, A., Borreli, F.:. Autonomous cruise control with cut-in target vehicle detection. In: Proceedings of AVEC16 (2016)
6. Chen, L., Englund, C.: Cooperative intersection management: a survey. IEEE Trans. Intell. Transp. Syst. **17**(2), 570–586 (2016)
7. Chialastri, A.: Automation in Aviation. InTech (2012)
8. Dolgov, D.: Google self-driving car project monthly report - september 2016. Technical report, Google (2016)
9. Englund, C., Chen, L., Ploeg, J., Semsar-Kazerooni, E., Voronov, A., Bengtsson, H.H., Didoff, J.: The grand cooperative driving challenge 2016: boosting the introduction of cooperative automated vehicles. IEEE Wirel. Commun. **23**(4), 146–152 (2016)
10. FAA. Aeronautical information manual. Technical report, U.S. department of transportation (2015)
11. Fossen, T., Pettersen, K., Galeazzi, R.: Line-of-sight path following for dubins paths with adaptive sideslip compensation for drift forces. IEEE Trans. Control Syst. Technol. **23**(2), 820–827 (2015)
12. Hanson, C.: Automated cooperative trajectories for a more efficient and responsive air transportation system. Presentation, October 2015. NASA
13. Nijmeijer, H., Rodriguez-Angeles, A.: Synchronization of Mechanical Systems, vol. 46. World Scientific, Singapore (2003)
14. Nilsson, J., Brännström, M., Coelingh, E., Fredriksson, J.: Lane change maneuvers for automated vehicles. IEEE Trans. Intell. Transp. Syst. **PP**(99), 1–10 (2016)
15. Pettersen, K., Fossen, T.: Underactuated dynamic positioning of a ship-experimental results. IEEE Trans. Control Syst. Technol. **8**(5), 856–863 (2000)
16. Ploeg, J.: Analysis and design of controllers for cooperative and automated driving. Ph.D. thesis, Eindhoven University of Technology (2014)
17. Ploeg, J., van de Wouw, N., Nijmeijer, H.: Lp string stability of cascaded systems: Application to vehicle platooning. IEEE Trans. Control Syst. Technol. **22**(2), 786–793 (2014)
18. Ploeg, J., Semsar-Kazerooni, E., Lijster, G., van de Wouw, N., Nijmeijer, H.: Graceful degradation of cooperative adaptive cruise control. IEEE Trans. Intell. Transp. Syst. **16**(1), 488–497 (2015)

19. Ploeg, J., van de Wouw, N., Nijmeijer, H.: Fault tolerance of cooperative vehicle platoons subject to communication delay. IFAC-PapersOnLine **48**(12), 352–357 (2015)
20. Ren, W., Beard, R.: Distributed Consensus in Multi-vehicle Cooperative Control: Theory and Applications, 1st edn. Springer Publishing Company, Incorporated (2007)
21. Richards, D., Stedmon, A.: To delegate or not to delegate: a review of control frameworks for autonomous cars. Applied Ergonomics, vol. 53, Part B:383 – 388. Transport in the 21st Century: The Application of Human Factors to Future User Needs (2016)
22. Rosenkrans, W.: Top-level automation. Online - Flight safety foundation, September 2012. https://flightsafety.org/asw-article/top-level-automation/ - Consulted 11-2016
23. Shiriaev, A., Robertsson, A., Freidovich, L., Johansson, R.: Coordinating control for a fleet of underactuated ships. In: Group Coordination and Cooperative Control, pp. 233–250. Springer (2006)
24. Society of Automotive Engineers. Taxonomy and definitions for terms related to driving automation systems for on-road motor vehicles J3016, 09 (2016)
25. Strom, E.G.: On medium access and physical layer standards for cooperative intelligent transport systems in Europe. Proc. IEEE **99**(7), 1183–1188 (2011)
26. Xing, H., Ploeg, J., Nijmeijer, H.: Pade approximation of delays in cacc-controlled string-stable platoons. In: *Proceedings of AVEC16* (2016)

Coordination of Multi-agent Systems with Intermittent Access to a Cloud Repository

Antonio Adaldo, Davide Liuzza, Dimos V. Dimarogonas
and Karl H. Johansson

Abstract A cloud-supported multi-agent system is composed of autonomous agents required to achieve a common coordination objective by exchanging data over a shared cloud repository. The repository is accessed asynchronously by different agents, and direct inter-agent commuication is not possible. This model is motivated by the problem of coordinating a fleet of autonomous underwater vehicles, with the aim to avoid the use of expensive and power-hungry modems for underwater communication. For the case of agents with integrator dynamics, a control law and a rule for scheduling the cloud access are formally defined and proven to achieve the desired coordination. A numerical simulation corroborate the theoretical results.

1 Introduction

Networks of autonomous mobile agents are the subject of a vast body of research [3, 9, 10], since they have broad and heterogeneous application domains. For most realistic applications, employing a team of agents rather than a single agent has many advantages, such as robustness to the failure of one agent, or simply completing the task faster thanks to the agents' cooperation. However, the use of a team also brings about the challenge of coordination: namely, the agents need to adjust their behavior

---

A. Adaldo (✉) · D.V. Dimarogonas · K.H. Johansson
Automatic Control Department and ACCESS Linnaeus Center,
School of Electrical Engineering, KTH Royal Institute of Technology,
Osquldas väg 10, 10044 Stockholm, Sweden
e-mail: adaldo@kth.se

D.V. Dimarogonas
e-mail: dimos@kth.se

K.H. Johansson
e-mail: kallej@kth.se

D. Liuzza
Department of Engineering, University of Sannio in Benevento,
Piazza Roma 21, 82100 Benevento, Italy
e-mail: davide@liuzza.unisannio.it

© Springer International Publishing AG 2017
T.I. Fossen et al. (eds.), *Sensing and Control for Autonomous Vehicles*,
Lecture Notes in Control and Information Sciences 474,
DOI 10.1007/978-3-319-55372-6_21

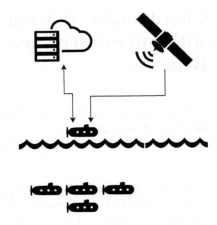

Fig. 1 Cloud-supported coordination of a team of AUVs. A vehicle is isolated when it is underwater, but when on the water surface, it receives GPS measurements and can access the cloud

according to what the other agents in team are doing. Coordination is particularly challenging in the case of autonomous underwater vehicles (AUVs), because they have limited sensing, localization and communication capabilities [6, 12, 13]. Underwater communication can be implemented by means of acoustic modems, but these are typically expensive, short-ranged and power-hungry.

To deal with communication constraints, event-triggered and self-triggered control [8] has been applied to multi-agent systems [4]. However, with these designs, the agents still need to exchange information directly (albeit intermittently), which could be problematic in underwater settings. In this chapter, we present a setup where the agents do not need to communicate directly; instead, they intermittently and asynchronously access a common cloud repository to upload and download information. Cloud accesses are scheduled independently by each agent with a self-triggered, recursive logic. In a team of AUVs, this setup is realized by letting the vehicles access the cloud via radio communication when they come to the water surface to get GPS measurements. Figure 1 illustrates a possible configuration. The surfaced vehicle receives a GPS position measurement and can exchange data with the cloud, while the vehicles that are under water are isolated. Later, a different vehicle can surface and exchange data with the cloud.

With the cloud's support, the agents can exchange information without opening a direct communication channel with each other. Cloud-supported coordination algorithms have been formally analyzed in the case of formation control for agents with integrator dynamics [1, 2]. In this chapter, we introduce the general framework of a cloud-supported multi-agent system, of which the setup in [1, 2] is a particular case where analytical convergence results are available. Cloud-supported control designs for multi-agent coordination have also been used in [5, 7, 11].

The rest of this chapter is organized as follows. In Sect. 2, we introduce some notation and standard results in multi-agent control theory. In Sect. 3, we describe the elements of a CSMAS and formally define the coordination objective. In Sect. 4, we find a sufficient condition, related to the scheduling of the cloud accesses, for achieving the desired coordination. In Sect. 5 we particularize the analysis to formation control for integrator agents, and we derive an explicit scheduling rule that achieves the desired coordination objective. In Sect. 6 we present simulation results. In Sect. 7, we summarize the results and present future developments of the work.

2 Preliminaries

A directed graph is a tuple $\mathcal{G} = (\mathcal{V}, \mathcal{E})$, with $\mathcal{V} = \{1, \ldots, N\}$ and $\mathcal{E} \subset \mathcal{V} \times \mathcal{V}$. Each element of $\mathcal{V}$ is called a *vertex* of the graph, while $(j, i) \in \mathcal{E}$ is called an *edge* from vertex j to vertex i. For each edge $e \in \mathcal{E}$, we denote as $\text{tail}(e)$ and $\text{head}(e)$, respectively, the first and the second vertex of the edge. Denoting the edges as $e_1, \ldots, e_M$, with $M = |\mathcal{E}|$, the *incidence matrix* of the graph is defined as $B \in \mathbb{R}^{NM}$, with

$$\{B\}_{ip} = \begin{cases} 1 & \text{if head}(e_p) = i, \\ -1 & \text{if tail}(e_p) = i, \\ 0 & \text{otherwise.} \end{cases} \quad (1)$$

Moreover, we define the *in-incidence matrix* as $C \in \mathbb{R}^{NM}$, with

$$\{C\}_{ip} = \begin{cases} 1 & \text{if head}(e_p) = i, \\ 0 & \text{otherwise.} \end{cases}$$

A *path* from vertex j to vertex i is a sequence of vertices $i_1, \ldots, i_P$ such that $i_1 = j$, $(i_k, i_k + 1) \in \mathcal{E}$ for all $k \in \{1, \ldots, P-1\}$, and $i_P = i$. A vertex $r \in \mathcal{V}$ is said to be a *root* of the graph if there exists a path from r to any other vertex in the graph. A subset $\mathcal{T}$ of the edges is said to be a *spanning tree* if it contains exactly $N - 1$ edges and $(\mathcal{V}, \mathcal{T})$ contains a root. If the graph admits a spanning tree, then let (without loss of generality) $\mathcal{T} = \{e_1, \ldots, e_{N-1}\}$ and $\mathcal{C} = \mathcal{E} \setminus \mathcal{T}$; then, partition the matrices B and C accordingly as $B = [B_\mathcal{T} \ B_\mathcal{C}]$ and $C = [C_\mathcal{T} \ C_\mathcal{C}]$. It can be verified [9] that $B_\mathcal{T}$ is full rank $N - 1$, and there always exists $T \in \mathbb{R}^{(N-1) \times (M-(N-1))}$ such that $B_\mathcal{C} = B_\mathcal{T} T$. The matrix

$$R = B_\mathcal{T}^\top (C_\mathcal{T} + C_\mathcal{C} T^\top) \in \mathbb{R}^{N-1 \times N-1}$$

is called the *reduced edge Laplacian* of the graph. Note that the reduced edge Laplacian is only defined if the graph admits a spanning tree, and it satisfies $RB_\mathcal{T} = B_\mathcal{T}^\top CB$. A notable result in graph theory states that $(-R)$ is Hurwitz [14]. If $(j, i) \in \mathcal{E}$, we say that vertex j is a *parent* of vertex i. The set of the parents of vertex i is denoted as $\mathcal{V}_i$; namely, $\mathcal{V}_i = \{j \in \mathcal{V} : (j, i) \in \mathcal{E}\}$.

3 System Model and Control Objective

In this section, we describe a cloud-supported multi-agent system. Such system is composed of $N \geq 1$ autonomous agents indexed from 1 to N, and a remote information repository (the cloud) which can be accessed asynchronously by the agents. The set of the agent indexes is denoted as $\mathcal{V} = \{1, \ldots, N\}$. In our motivating application, each agent is an AUV, and the repository is hosted on a base station, which can be accessed from an AUV when it is on the water surface.

The open-loop dynamics of each agent i is given by the differential equation

$$\dot{x}_i(t) = f(x_i(t), u_i(t), d_i(t)), \quad i \in \{1, \ldots, N\}, \tag{2}$$

where $x_i(t) \in \mathbb{R}^{n_x}$ is the state of the agent, $u_i(t) \in \mathbb{R}^{n_u}$ the control input applied to the agent, and $d_i(t) \in \mathbb{R}^{n_d}$ a disturbance input. We make the following assumptions.

Assumption 1 The function f is globally Lipschitz with respect to u_i and d_i for all $x_i \in \mathbb{R}^{n_x}$, i.e., there exists a *lipschitz constant* $L_f > 0$ such that

$$\|f(x, u_1, d_1) - f(x, u_2, d_2)\| \leq L_f(\|u_1 - u_2\| + \|d_1 - d_2\|),$$

for any $x \in \mathbb{R}^{n_x}$, $u_1, u_2 \in \mathbb{R}^{n_u}$, and $d_1, d_2 \in \mathbb{R}^{n_d}$, where $\|\cdot\|$ denotes the Euclidean norm.

Assumption 2 The disturbance signals are bounded by

$$\|d_i(t)\| \leq \delta(t) := (\delta_0 - \delta_\infty)e^{-\lambda_\delta t} + \delta_\infty,$$

with $0 \leq \delta_\infty < \delta_0$ and $\lambda_\delta \geq 0$.

The states, controls and disturbances of the agents are grouped into the stack vectors $x(t)$, $u(t)$ and $d(t)$, so that $x(t) = [x_1(t)^\top, \ldots, x_N(t)^\top]^\top$ etc. Similarly, we can stack the right-hand side of (2) for $i \in \mathcal{V}$ as $F(x, u, d) = [f(x_1, u_1, d_1)^\top, \ldots, f(x_N, u_N, d_N)^\top]^\top$, so that (2) is written compactly as

$$\dot{x}(t) = F(x(t), u(t), d(t)).$$

To achieve the desired coordination, the agents need to exchange information about their states, controls, and disturbances. To capture the topology of the information exchange among the agents, we use a directed graph $\mathcal{G} = (\mathcal{V}, \mathcal{E})$, where each vertex corresponds to an agent, and an edge $(j, i) \in \mathcal{E}$ means that agent i acquires information from agent j. Hence, the set $\mathcal{V}_i$ of the parents of vertex i corresponds to the agents whose information is received by agent i. The graph $\mathcal{G}$ is called the *communication graph*.

Assumption 3 The communication graph admits a spanning tree $\mathcal{T}$.

Assumption 3 is a connectivity requirement: it means that there must exist at least one agent whose information can reach all the other agents.

In most practical scenarios it is unrealistic to assume that the agents can communicate directly. For example, in our motivating application, having the AUVs exchange information while navigating underwater is problematic: underwater communication can be achieved with acoustic modems, but these are often expensive, short-ranged and power-hungry [6]. To address these concerns, we do not let the agents exchange information directly, but we let them intermittently exchange information with the cloud. Hence, the graph $\mathcal{G}$ does not represent a topology of direct communication links among the agents, but it captures the topology of the information exchanges through the cloud. In our case, a cloud access corresponds to a vehicle surfacing and connecting to a base station and to a GPS.

The desired coordination objective is captured by a function of the agents' states, which we call *objective function*. Namely, the objective function is a function $V : \mathbb{R}^{Nn} \to \mathbb{R}_{\geq 0}$, and the control objective is

$$\limsup_{t \to \infty} V(x(t)) \leq \varepsilon, \qquad (3)$$

where $\varepsilon \geq 0$ is an assigned tolerance on the desired coordination. We need the following technical assumption on $V(\cdot)$.

Assumption 4 The objective function $V(\cdot)$ is continuously differentiable, and

$$\left\| \frac{\partial V(x)}{\partial x} \right\| \leq \beta \sqrt{V(x)}$$

for some $\beta > 0$ and all $x \in \mathbb{R}^{Nn_x}$. Moreover, there exists a distributed feedback law that, in absence of disturbances, is able to drive the agents to the desired coordination; i.e., there exist functions $v_i : \mathbb{R}^{Nn_x} \to \mathbb{R}^{n_u}$, with $v_i(x)$ depending explicitly on x_j only if $j \in \mathcal{V}_i$, such that, for some $\gamma > 0$, we have

$$\frac{\partial V(x)}{\partial x}^\top F(x, v(x), 0_{Nn_d}) \leq -\gamma V(x),$$

where $v(x) = [v_1(x)^\top, \ldots, v_N(t)^\top]^\top$.

Remark 1 Assumption 4 corresponds to the requirement that the coordination problem under consideration can be solved by a continuous and distributed feedback law.

In what follows, we shall refer to $v_i(x)$ as the *virtual feedback law*.

Let $t_{i,1}, t_{i,2}, \ldots, t_{i,k}, \ldots$ be the sequence of time instants when agent i accesses the cloud. Conventionally, we let $t_{i,0} := 0$ for all the agents. For convenience, we introduce the *last-access function* $l_i(t)$, which returns the index of the latest cloud access of agent i before time t:

$$l_i(t) = \max\{k \in \mathbb{N} : t_{i,k} \leq t\}. \tag{4}$$

At each cloud access $t_{i,k}$, agent i receives a measurement of its current state $x_i(t_{i,k})$, downloads a packet $\mathcal{D}_{i,k}$, and uploads a packet $\mathcal{U}_{i,k}$. The packet $\mathcal{D}_{i,k}$ contains the information that agent i needs to acquire about other agents, while the packet $\mathcal{U}_{i,k}$ contains the information that agent i needs to share with other agents (the content of these packets is given below). When an agent uploads packet $\mathcal{U}_{i,k+1}$, this packet overwrites the previously uploaded packet $\mathcal{U}_{i,k}$. In this way, the cloud only contains the latest packet uploaded by each agent, and the size of the repository does not grow over time. Namely, the content of the cloud at a generic time instant $t \geq 0$ is

$$\mathcal{C}(t) = \mathcal{U}_{1,l_1(t)} \cup \cdots \cup \mathcal{U}_{N,l_N(t)},$$

where $l_i(t)$ is the last-access function defined by (4).

The agents apply piecewise constant control signals, updated at each cloud access:

$$u_i(t) = u_{i,k} \quad \forall t \in [t_{i,k}, t_{i,k+1}). \tag{5}$$

Algorithm 1 summarizes the operations executed by each agent when accessing the cloud.

Algorithm 1 Operations executed by agent i when accessing the cloud at time $t_{i,k}$

download packet $\mathcal{D}_{i,k}$
compute control value $u_{i,k}$
compute next access time $t_{i,k+1}$
upload packet $\mathcal{U}_{i,k}$ (overwriting $\mathcal{U}_{i,k-1}$)

Packet $\mathcal{U}_{i,k}$ is composed of the current time $t_{i,k}$, the current state $x_i(t_{i,k})$ of the agent, the newly computed control value $u_{i,k}$ and scheduled access $t_{i,k+1}$:

$$\mathcal{U}_{i,k} = (t_{i,k}, x_i(t_{i,k}), u_{i,k}, t_{i,k+1}). \tag{6}$$

Packet $\mathcal{D}_{i,k}$ must be a subset of $\mathcal{C}(t_{i,k})$ (i.e., the information contained in the cloud at time $t_{i,k}$). In particular, we let

$$\mathcal{D}_{i,k} = \bigcup_{j \in \mathcal{V}_i} \mathcal{U}_{j,l_j(t_{i,k})}, \tag{7}$$

where $\mathcal{V}_i$ is the parent set of vertex i in the communication graph. Evaluating (6) for agent j and access time $t_{j,l_j(t_{i,k})}$, we have

$$\mathcal{U}_{j,l_j(t_{i,k})} = (t_{j,l_j(t_{i,k})}, x_i(t_{j,l_j(t_{i,k})}), u_{j,l_j(t_{i,k})}, t_{j,l_j(t_{i,k})+1}),$$

which, using also (7), yields

$$\mathcal{D}_{i,k} = \bigcup_{j \in \mathcal{V}_i} (t_{j,l_j(t_{i,k})}, x_i(t_{j,l_j(t_{i,k})}), u_{j,l_j(t_{i,k})}, t_{j,l_j(t_{i,k})+1}). \tag{8}$$

Hence, the information exchanged among the agents (through the cloud) is defined by (6) and (8).

Each agent contains three modules: a *predictor*, a *controller*, and a *scheduler*.

The predictor module of agent i produces an estimate $\hat{x}_i^i(t)$ of the state of the agent, as well as an estimate $\hat{x}_j^i(t)$ of the state of the other agents $j \in \mathcal{V}_i$. These estimates are obtained by using the state measurements acquired from the cloud, and integrating the agents' dynamics under the assumption that no disturbances are acting on the system. Namely, for $\hat{x}_i^i(t)$, we have

$$\begin{aligned}\dot{\hat{x}}_i^i(t) &= f(\hat{x}_i^i(t), u_{i,k}, 0_{n_d}) \quad \forall t \in [t_{i,k}, t_{i,k+1}),\\ \hat{x}_i^i(t_{i,k}) &= x_i(t_{i,k}),\end{aligned} \tag{9}$$

while for $\hat{x}_j^i(t)$, we have

$$\begin{aligned}\dot{\hat{x}}_j^i(t) &= f(\hat{x}_j^i(t), u_{j,l_j(t_{i,k})}, 0_{n_d}) \quad \forall t \in [t_{j,l_j(t_{i,k})}, t_{j,l_j(t_{i,k})+1}),\\ \hat{x}_j^i(t_{j,l_j(t_{i,k})}) &= x_j(t_{j,l_j(t_{i,k})}).\end{aligned} \tag{10}$$

Note that $t_{j,l_j(t_{i,k})}$, $x(t_{j,l_j(t_{i,k})})$, $u_{j,l_j(t_{i,k})}$, and $t_{j,l_j(t_{i,k})+1}$ are available, since they have been downloaded from the cloud by agent i. The stack vector of the estimated states is denoted as $x^i(t)$, namely $\hat{x}^i(t) = [\hat{x}_1^i(t)^\top, \ldots, \hat{x}_N^i(t)^\top]^\top$.

The controller module has the task to compute the control values $u_{i,k}$. These control values are obtained evaluating the feedback law $v_i(\cdot)$ on the estimated states at time $t_{i,k}$; namely,

$$u_{i,k} = v_i(\hat{x}^i(t_{i,k})). \tag{11}$$

Finally, the scheduler module has the job to establish the time of the agent's next cloud access.

4 Convergence Analysis

In this section, we study the convergence properties of the described cloud-supported multi-agent system, and we derive a sufficient condition on the access scheduling to achieve coordination. Denote as $\tilde{u}_i(t)$ the mismatch between the actual control input $u_i(t)$ and the virtual feedback signal $v_i(x(t))$. Namely, denote $\tilde{u}_i(t) = u_i(t) -$

$v_i(x(t))$ and let $\tilde{u}(t)$ be the corresponding stack vector. Since the virtual feedback law $v_i(x(t))$ is known to drive the system to coordination in absence of disturbances, intuition suggests that coordination would be preserved if $\tilde{u}(t)$ and $d(t)$ are kept small. This intuition is rigorously formalized and demonstrated in the following Lemma.

Lemma 1 *Let Assumptions 1, 2, and 4 hold. Consider a positive bounded scalar signal $\varsigma(t)$. If $\|\tilde{u}(t)\| \leq \varsigma(t)$ for all $t \geq 0$, then $V(t) \leq W(t)$, where $W(t)$ is the solution to*

$$\begin{cases} \dot{W}(t) = -\gamma W(t) + \beta\sqrt{W(t)}L_f(\varsigma(t) + \delta(t)), \\ W(0) \geq V(0). \end{cases} \quad (12)$$

Moreover, system state converges to the set

$$X_\infty = \left\{ x : \sqrt{V(x)} \leq \frac{\sqrt{N}\beta L_f(\varsigma_\infty + \delta_\infty)}{\gamma} \right\},$$

where $\varsigma_\infty = \limsup_{t\to\infty} \varsigma(t)$.

Proof Using the agents' dynamics (2), we can write

$$\begin{aligned}\dot{V}(x(t)) &= \frac{\partial V(x(t))}{\partial x}^\top F(x(t), u(t), d(t)) \\ &= \frac{\partial V(x(t))}{\partial x}^\top F(x(t), v(x(t)), 0_{Nn_d}) \\ &\quad + \frac{\partial V(x(t))}{\partial x}^\top (F(x(t), u(t), d(t)) - F(x(t), v(x(t)), 0_{Nn_d})).\end{aligned} \quad (13)$$

Using Assumptions 1 and 4, we upper bound the right-hand side of (13) as

$$\dot{V}(x(t)) \leq -\gamma V(x(t)) + \beta\sqrt{V(x(t))}L_f(\|\tilde{u}(t)\| + \|d(t)\|), \quad (14)$$

where $L_f > 0$ is the Lipschitz constant of f. Using the Comparison Lemma, we know from (14) that $V(x(t))$ is upper-bounded by the solution $W(t)$ of (12). In particular, taking $t \to \infty$, and using Assumption 2, we have

$$\limsup_{t\to\infty} V(x(t)) \leq \lim_{t\to\infty} W(t) = \left(\frac{\beta L_f(\varsigma_\infty + \delta_\infty)}{\gamma}\right)^2. \quad \square$$

Thanks to Lemma 1, we know that, if we can guarantee that $\|\tilde{u}(t)\|$ is bounded, then the control objective (3) is satisfied with $\varepsilon = \left(\frac{\beta L_f(\varsigma_\infty + \delta_\infty)}{\gamma}\right)^2$. As a particular case, if disturbances asymptotically vanish ($\delta_\infty = 0$) and we manage to make $\tilde{u}(t)$ vanish, then we have $\varepsilon = 0$, which means perfect coordination. Boundedness (and possibly, asymptotic disappearance) of the control error $\tilde{u}(t)$ should be achieved by an opportune scheduling of the cloud accesses, as exemplified in the next section.

5 Cloud Access Scheduling for Formation Control

In this section, address the problem of formation control. To do this, we need to specify the general model dynamics (2) and the general objective function (3).

Specifically, the state, control and disturbance of each agent are related to the motion of the AUV, and their physical meaning depends on the mathematical model $f(\cdot)$ adopted to represent the vehicle motion. We use a planar integrator model, where $x_i(t)$ represents the horizontal position of the vehicle, while $u_i(t)$ and $d_i(t)$ represent the contributions to the horizontal velocity of the vehicle given respectively by the control input and by external disturbances. In this way, we have $n_x = n_u = n_d = 2$, and

$$f(x_i, u_i, d_i) = u_i + d_i, \tag{15}$$

which leads to the vehicle dynamics

$$\dot{x}_i(t) = u_i(t) + d_i(t).$$

Naturally, (15) satisfies Assumption 1 with $L_f = 1$.

The coordination objective is that the vehicles reach a specified formation, and in particular, each vehicle is required to occupy the position $b_i \in \mathbb{R}^2$ with respect to a virtual center of the formation. Interestingly, this control objective is captured by an objective function inferenced by the communication graph:

$$V(x) = \frac{1}{2}(x-b)^\top (B_T P B_T^\top \otimes I_2)(x-b), \tag{16}$$

where $\otimes$ denotes the Kronecker product, $b = [b_1^\top, \ldots, b_N^\top]^\top$, $B = [B_T, B_C]$ is the incidence matrix of the communication graph, and $P \in \mathbb{R}^{N \times N}$ is a positive definite matrix such that $PR + R^\top P = 2Q$ is symmetric and positive definite. (Such a P exists because $(-R)$ is Hurwitz.) From the definition of the incidence matrix (1), and from the properties of the Kronecker product, one can easily verify that $V(x) = 0$ if and only if $x_i = \bar{x} + b_i$ for some $\bar{x} \in \mathbb{R}^2$, which corresponds to the desired coordination; otherwise, $V(x) > 0$. Moreover, one can easily verify that (15) and (16) satisfy Assumption 4 with $\beta = \|B_T \sqrt{P}\|$, $\gamma = \frac{2 \min \text{eig}(Q)}{\max \text{eig}(P)}$, and the virtual feedback law

$$v_i(x) = \sum_{j \in \mathcal{V}_i} (x_j(t) - b_j - x_i(t) + b_i). \tag{17}$$

See the Appendix for a proof.

Note that Assumptions 1, 3, and 4 hold for this choice of agent dynamics and objective function; therefore, they are considered standing assumptions in this section.

From Lemma 1, we know that coordination is related to boundedness of $\|\tilde{u}(t)\|$; therefore, our first step in designing a scheduler is to compute an upper bound for the control errors $\tilde{u}_i(t)$ related to agent i when it accesses the cloud at time $t_{i,k}$. To this aim, consider the time interval $[t_{i,k}, t_{i,k+1})$ between the cloud access k and $k+1$ of vehicle i. From (5), (11), and (17), we have

$$\tilde{u}_i(t) = u_{i,k} - v_i(x(t)) = v(\hat{x}^i(t_{i,k})) - v_i(x(t))$$
$$= \sum_{j \in \mathcal{V}_i} (\hat{x}^i_j(t_{i,k}) - x_j(t) - \hat{x}^i_i(t_{i,k}) + x_i(t)). \tag{18}$$

The addends $x_j(t)$ and $x_i(t)$ in (18) can be written by integrating the system dynamics (15) in opportune time intervals. First, $x_i(t)$ can be computed by integrating (15) in the interval $[t_{i,k}, t)$, which gives

$$x_i(t) = x_i(t_{i,k}) + (t - t_{i,k})u_{i,k} + \int_{t_{i,k}}^t d_i(\tau)\mathrm{d}\tau. \tag{19}$$

Second, $x_j(t)$ can be obtained by integrating (15) in the interval $[t_{j,l_j(t_{i,k})}, t)$. However, for $t > t_{j,l_j(t_{i,k})+1}$, the control input $u_j(t)$ is not known by agent i, because the latest control input computed by agent j and uploaded to the cloud is $u_{j,l_j(t_{i,k})}$, which is applied only until $t_{j,l_j(t_{i,k})+1}$. Hence, we need to distinguish two cases: when $t \leq t_{j,l_j(t_{i,k})+1}$, we have

$$x_j(t) = x_j(t_{j,l_j(t_{i,k})}) + (t - t_{j,l_j(t_{i,k})})u_{j,l_j(t_{i,k})} + \int_{t_{j,l_j(t_{i,k})}}^t d_j(\tau)\mathrm{d}\tau; \tag{20}$$

conversely, when $t > t_{j,l_j(t_{i,k})+1}$, we have

$$x_j(t) = x_j(t_{j,l_j(t_{i,k})}) + (t_{j,l_j(t_{i,k})+1} - t_{j,l_j(t_{i,k})})u_{j,l_j(t_{i,k})}$$
$$+ \int_{t_{j,l_j(t_{i,k})+1}}^t u_j(\tau)\mathrm{d}\tau + \int_{t_{j,l_j(t_{i,k})}}^t d_j(\tau)\mathrm{d}\tau. \tag{21}$$

Similarly, the addends $\hat{x}^i_j(t_{i,k})$ and $\hat{x}^i_i(t_{i,k})$ can be evaluated by integrating the dynamics (9)–(10) of the predictor. For $\hat{x}^i_i(t_{i,k})$, we have simply $\hat{x}^i_i(t_{i,k}) = x_i(t_{i,k})$, which is the GPS measurement that the AUV obtains when it surfaces; for $\hat{x}^i_j(t_{i,k})$, we have

$$\hat{x}^i_j(t_{i,k}) = x_j(t_{j,l_j(t_{i,k})}) + (t_{i,k} - t_{j,l_j(t_{i,k})})u_{j,l_j(t_{i,k})}. \tag{22}$$

Note that we do not need to distinguish two cases as done for $x_j(t)$, since $t_{i,k} \leq t_{j,l_j(t_{i,k})+1}$ by definition. Using (19)–(22), and observing that $u_j(\tau) = v_j(x(\tau)) + \tilde{u}_j(\tau)$, we can rewrite (18) as

$$\tilde{u}_i(t) = |\mathcal{V}_i| \left((t - t_{i,k}) u_{i,k} + \int_{t_{i,k}}^{t} d_i(\tau) d\tau \right) - \sum_{j \in \mathcal{V}_i} \int_{t_{j,l_j(t_{i,k})}}^{t} d_j(\tau) d\tau$$

$$- \sum_{\substack{j \in \mathcal{V}_i \\ j: t \leq t_{j,l_j(t_{i,k})+1}}} (t - t_{i,k}) u_{j,l_j(t_{i,k})}$$

$$- \sum_{\substack{j \in \mathcal{V}_i \\ j: t > t_{j,l_j(t_{i,k})+1}}} \left((t_{j,l_j(t_{i,k})+1} - t_{i,k}) u_{j,l_j(t_{i,k})} + \int_{t_{j,l_j(t_{i,k})+1}}^{t} (v_j(x(\tau)) + \tilde{u}_j(\tau)) d\tau \right).$$

(23)

The term $v_j(x(\tau))$ in the right-hand side of (23) is not known by agent i, but it can be related to the value of $V(x)$. Namely, from (17), we have $v_j(x) = (C^j B \otimes I_2)(x - b)$, where C^j denotes the j-th row of the in-incidence matrix of the communication graph. Since the communication graph has a spanning tree, we can decompose $C^j = [C_T^j \ C_C^j]$ and $B = [B_T \ B_C]$, with $B_C = B_T T$, which leads to

$$v_j(x) = ((C_T^j + C_C^j T^\top) B_T \otimes I_2)(x - b).$$

Taking norms of both sides, and using the Cauchy–Schwartz inequality, we have

$$\|v_j(x)\| \leq \|C_T^j + C_C^j T^\top\| \|(B_T \otimes I_2)(x - b)\|. \quad (24)$$

On the other hand, using positive-definiteness of P, we have from (16)

$$\frac{1}{2} \|(B_T \otimes I_2)(x - b)\| \lambda_1^P \leq V(x), \quad (25)$$

where λ_1^P denotes the smallest eigenvalue of P. Using (25) in (24) we have finally

$$\|v_j(x)\| \leq \mu_j V(x), \quad (26)$$

where $\mu_j = \frac{2\|C_T^j + C_C^j T^\top\|}{\lambda_1^P}$. Taking norms of both sides in (23), and using the triangular inequality and (26), gives

$$\|\tilde{u}_i(t)\| \leq \left\| |\mathcal{V}_i| (t - t_{i,k}) u_{i,k} - \sum_{\substack{j \in \mathcal{V}_i \\ j: t \leq t_{j,l_j(t_{i,k})+1}}} (t - t_{i,k}) u_{j,l_j(t_{i,k})} \right.$$

$$\left. - \sum_{\substack{j \in \mathcal{V}_i \\ j: t > t_{j,l_j(t_{i,k})+1}}} (t_{j,l_j(t_{i,k})+1} - t_{i,k}) u_{j,l_j(t_{i,k})} \right\|$$

$$+ |\mathcal{V}_i| \int_{t_{i,k}}^{t} \delta(\tau) d\tau + \sum_{j \in \mathcal{V}_i} \int_{t_{j,l_j(t_{i,k})}}^{t} \delta(\tau) d\tau$$

$$+ \sum_{\substack{j \in \mathcal{V}_i \\ j: t > t_{j,l_j(t_{i,k})+1}}} \int_{t_{j,l_j(t_{i,k})+1}}^{t} (\mu_j V(x) + \|\tilde{u}_j(\tau)\|) d\tau,$$

(27)

where we have used Assumption 2 to bound the disturbance terms. As a particular case, for $t = t_{i,k}$, (27) reads

$$\|\tilde{u}_i(t_{i,k})\| \leq \sum_{j \in \mathcal{V}_i} \int_{t_{j,l_j(t_{i,k})}}^{t_{i,k}} \delta(\tau) d\tau. \tag{28}$$

The inequalities (27) and (28) suggest a recursive rule for the scheduling of the cloud accesses, which is formalized in the following lemma.

Lemma 2 *Let Assumption 2 hold. Consider a positive bounded scalar signal $\varsigma(t)$, and a scalar $\alpha \in (0, 1)$. Let each agent schedule its cloud access recursively as*

$$t_{i,k+1} = \inf\{t \geq t_{i,k} : \Delta_{i,k}(t) \geq \alpha\varsigma(t) \vee \sigma_{i,k}(t) \geq \varsigma(t)\}, \tag{29}$$

where

$$\Delta_{i,k}(t) = \max_{q:i \in \mathcal{V}_q} |\mathcal{V}_q| \int_{t_{i,k}}^{t} \delta(\tau) d\tau \tag{30}$$

and

$$\sigma_{i,k}(t) = \Big\| |\mathcal{V}_i|(t - t_{i,k})u_{i,k} - \sum_{\substack{j \in \mathcal{V}_i \\ j:t \leq t_{j,l_j(t_{i,k})+1}}} (t - t_{i,k})u_{j,l_j(t_{i,k})}$$

$$- \sum_{\substack{j \in \mathcal{V}_i \\ j:t > t_{j,l_j(t_{i,k})+1}}} (t_{j,l_j(t_{i,k})+1} - t_{i,k})u_{j,l_j(t_{i,k})} \Big\|$$

$$+ |\mathcal{V}_i| \int_{t_{i,k}}^{t} \delta(\tau) d\tau + \sum_{j \in \mathcal{V}_i} \int_{t_{j,l_j(t_{i,k})}}^{t} \delta(\tau) d\tau \tag{31}$$

$$+ \sum_{\substack{j \in \mathcal{V}_i \\ j:t > t_{j,l_j(t_{i,k})+1}}} \int_{t_{j,l_j(t_{i,k})+1}}^{t} (\mu_j W(x(\tau)) + \varsigma(\tau)) d\tau.$$

Then, we have $\|\tilde{u}_i(t)\| \leq \varsigma(t)$ for all $t \geq 0$ and all $i \in \mathcal{V}$.

We call $\varsigma(t)$ the *threshold function*, because it is used as a comparison threshold for $\Delta_{i,k}(t)$ and $\sigma_{i,k}(t)$ in triggering the cloud accesses.

Proof At each access time $t_{i,k}$, (28) holds. Using (30), we have

$$\|\tilde{u}_i(t_{i,k})\| \leq \sum_{j \in \mathcal{V}_i} \frac{\Delta_{j,l_j(t_{i,k})}(t_{i,k})}{\max_{q:j \in \mathcal{V}_q} |\mathcal{V}_q|} \leq \sum_{j \in \mathcal{V}_i} \frac{\Delta_{j,l_j(t_{i,k})}(t_{i,k})}{|\mathcal{V}_i|}. \tag{32}$$

Since all the agents apply scheduling rule (29), we have $\Delta_{j,l_j(t_{i,k})}(t_{i,k}) \leq \alpha\varsigma(t_{i,k})$, and we can further bound (32) as

$$\|\tilde{u}_i(t_{i,k})\| \leq \sum_{j \in \mathcal{V}_i} \frac{\alpha\varsigma(t_{i,k})}{|\mathcal{V}_i|} \leq \alpha\varsigma(t_{i,k}) = \varsigma(t_{i,k}) - (1-\alpha)\varsigma(t_{i,k}).$$

Hence, at the access times $t_{i,k}$, the inequality $\|\tilde{u}_i(t)\| \leq \varsigma(t)$ is satisfied with a margin of at least $(1-\alpha)\varsigma(t_{i,k})$. Suppose that there exist an agent i and a time $\bar{t}$ such that $\|\tilde{u}_i(\bar{t})\| = \varsigma(\bar{t})$, while $\|\tilde{u}_j(\tau)\| < \varsigma(\tau)$ for all times $\tau \leq \bar{t}$ and for all agents. Let $k = l_i(\bar{t})$. Comparing the right-hand sides of (27) and (31), and recalling from Lemma 1 that $V(t) \leq W(t)$ at least until time $\bar{t}$, we conclude that $\|\tilde{u}_i(\bar{t})\| \leq \sigma_{i,k}(\bar{t})$, which implies $\sigma_{i,k}(\bar{t}) \geq \varsigma(t)$. This inequality in turn implies (by the scheduling rule (29)) that there must have been another cloud access between $t_{i,k}$ and $\bar{t}$. However, since we have set $k = l_i(t)$, we conclude that $\bar{t}$ is itself an access time. Therefore, from (32), we know that, immediately after the access,

$$\|\tilde{u}_i(\bar{t})\| \leq \varsigma(\bar{t}) - (1-\alpha)\varsigma(\bar{t}).$$

Hence, $\|\tilde{u}_i(t)\| \leq \varsigma(t)$ can never be violated, since, whenever $\|\tilde{u}_i(t)\|$ should reach $\varsigma(t)$ from below, it would immediately trigger a cloud access that resets $\|\tilde{u}_i(t)\|$ to a lower value. □

When using event-triggered controllers, it is always necessary to demonstrate that the control law does not induce an accumulation of control updates, otherwise it would not be possible to implement the controller. The accumulation of control updates is known as Zeno behavior [8], and it is always an undesired phenomenon in event-triggered control. For our system, Zeno behavior corresponds to an accumulation of the cloud accesses, In the following lemma, we show that the scheduling law (29) does not induce Zeno behavior in the cloud accesses, as long as the threshold function $\varsigma(t)$ satisfies some conditions.

Lemma 3 *Let Assumption 2 hold. Let the threshold function be*

$$\varsigma(t) = \varsigma_0 e^{-\lambda_\varsigma t} + \varsigma_\infty, \tag{33}$$

with $0 \leq \varsigma_\infty \leq \varsigma_0$, and $\lambda_\varsigma > 0$. If either of the following conditions is satisfied:

(C1) $\varsigma_\infty > 0$;
(C2) $\varsigma_\infty = \delta_\infty = 0$ and $\lambda_\varsigma < \min\{\lambda_\delta, \frac{\gamma}{2}\}$;

then the scheduling rule (29) *does not induce Zeno behavior in the cloud accesses. In particular, there exists a finite lower bound for the inter-access times: $t_{i,k+1} - t_{i,k} \geq \Delta T$ for some $\Delta T > 0$.*

Proof We only give the proof for the case that condition C2 is satisfied. When C1 is satisfied, the proof is similar. If C2 holds, then $\varsigma(t) = \varsigma_0 e^{-\lambda_\varsigma t}$ and $\delta(t) = \delta_0 e^{-\lambda_\delta t} \leq \delta_0 e^{-\lambda_\varsigma t}$. Then, by the Comparison Lemma, we have that $W(t) \leq Z(t)$, where $Z(t)$ is the solution of

$$\begin{cases} \dot{Z}(t) = -\gamma Z(t) + \beta \sqrt{Z(t)} L_f (\varsigma_0 + \delta_0) e^{-\lambda_\varsigma t}, \\ Z(0) = V(0). \end{cases} \tag{34}$$

Solving (34), we have

$$\begin{aligned} W(t) \leq Z(t) &= e^{-\frac{\gamma}{2}t} V(0) + e^{-\frac{\gamma}{2}t} \frac{\beta L_f (\varsigma_0 + \delta_0)}{\gamma - 2\lambda_\varsigma} (e^{(\frac{\gamma}{2} - \lambda_\varsigma)t} - 1) \\ &\leq \left(V(0) + \frac{\beta L_f (\varsigma_0 + \delta_0)}{\gamma - 2\lambda_\varsigma} \right) e^{-\lambda_\varsigma t}, \end{aligned} \tag{35}$$

where, for the last inequality, we have used $\lambda_\varsigma < 2\gamma$. Denote $\bar{W} = \left(V(0) + \frac{\beta L_f (\varsigma_0 + \delta_0)}{\gamma - 2\lambda_\varsigma} \right)$, so that (35) reads

$$W(t) \leq \bar{W} e^{-\lambda_\varsigma t}. \tag{36}$$

Recalling that $u_i(t) = v_i(t) + \tilde{u}_i(t)$ with $\|\tilde{u}_i(t)\| \leq \varsigma(t)$, and using (26), (33) and (36), we can write

$$\|u_i(t)\| \leq \|v_i(t)\| + \|\tilde{u}_i(t)\| \leq (\mu_i \bar{W} + \varsigma_0) e^{-\lambda_\varsigma t}. \tag{37}$$

Using (37) as an upper bound for all the control signals in the right-hand side of (31), and recalling that $\delta(t) \leq \delta_0^{-\lambda_\varsigma t}$, we have

$$\begin{aligned} \sigma_{i,k}(t) \leq &\left(|\mathcal{V}_i| (\mu_i \bar{W} + \varsigma_0) + \sum_{j \in \mathcal{V}_i} (\mu_j \bar{W} + \varsigma_0) \right) \int_{t_{i,k}}^{t} e^{-\lambda_\varsigma \tau} d\tau \\ &+ 2|\mathcal{V}_i| \int_{t_{i,k}}^{t} \delta_0 e^{-\lambda_\varsigma \tau} d\tau + \sum_{j \in \mathcal{V}_i} \int_{t_{j,l_j(t_{i,k})}}^{t_{i,k}} \delta(\tau) d\tau. \end{aligned}$$

Reasoning as in Lemma 2, the last addend in the previous inequality can be bounded by $\alpha \varsigma(t_{i,k})$, which yields

$$\sigma_{i,k}(t) \leq \Xi_i \int_{t_{i,k}}^{t} e^{-\lambda_\varsigma \tau} d\tau + \alpha \varsigma(t_{i,k}), \tag{38}$$

where $\Xi_i = |\mathcal{V}_i|(\mu_i \bar{W} + \varsigma_0) + \sum_{j \in \mathcal{V}_i}(\mu_j \bar{W} + \varsigma_0) + 2|\mathcal{V}_i|\delta_0$. Consider now the two conditions that can trigger a cloud access. One condition is $\Delta_{i,k}(t) \geq \alpha\varsigma(t)$, which can be written as

$$\frac{\max_{q:i\in\mathcal{V}_q}|\mathcal{V}_q|}{\lambda_\delta}e^{-\lambda_\delta t_{i,k}}(1 - e^{\lambda_\delta(t-t_{i,k})}) \geq \alpha\frac{\varsigma_0}{\lambda_\varsigma}e^{-\lambda_\varsigma t_{i,k}}e^{-\lambda_\varsigma(t-t_{i,k})}. \quad (39)$$

Recalling that $\lambda_\varsigma < \lambda_\delta$, (39) implies

$$\frac{\max_{q:i\in\mathcal{V}_q}|\mathcal{V}_q|}{\lambda_\delta}(1 - e^{\lambda_\delta(t-t_{i,k})}) \geq \alpha\frac{\varsigma_0}{\lambda_\varsigma}e^{-\lambda_\varsigma(t-t_{i,k})},$$

meaning that $t - t_{i,k}$ is lower-bounded by the solution ΔT_1 of the equation

$$\frac{\max_{q:i\in\mathcal{V}_q}|\mathcal{V}_q|}{\lambda_\delta}(1 - e^{\lambda_\delta \Delta T_1}) \geq \alpha\frac{\varsigma_0}{\lambda_\varsigma}e^{-\lambda_\varsigma \Delta T_1},$$

which is a positive constant. Similarly, the second condition that can trigger a cloud access is $\sigma_{i,k}(t) \geq \varsigma(t)$. From (38), we have that this condition implies

$$\frac{\Xi_i}{\lambda_\varsigma}e^{-\lambda_\varsigma t_{i,k}}(1 - e^{-\lambda_\varsigma(t-t_{i,k})}) + \alpha\varsigma_0 e^{-\lambda_\varsigma t_{i,k}} \geq \varsigma_0 e^{-\lambda_\varsigma t_{i,k}}e^{-\lambda_\varsigma(t-t_{i,k})}.$$

Dividing both sides by $e^{-\lambda_\varsigma t_{i,k}}$, we can see that to satisfy the triggering condition $\sigma_{i,k}(t) \geq \varsigma(t)$, $t - t_{i,k}$ must be larger than the solution of the equation

$$\frac{\Xi_i}{\lambda_\varsigma}(1 - e^{-\lambda_\varsigma \Delta T_2}) + \alpha\varsigma_0 \geq \varsigma_0 e^{-\lambda_\varsigma \Delta T_2},$$

which is a positive constant. We conclude that

$$t - t_{i,k} \geq \min\{\Delta T_1, \Delta T_2\}$$

for the trigger of a new cloud access at t. □

The formation control for a multi-agent system with cloud-supported coordination is formalized in the following theorem. The proof is immediate from Lemmas 1–3.

Theorem 1 *Consider a CSMAS where the agents' dynamics is given by (15) and the objective function is given by (16). Let Assumptions 2 hold. Let the control signals be given by (11), (9), and (10), where $v_i(x)$ is defined by (17). Let the cloud accesses be scheduled recursively by (29), with threshold function (33). Then, if either conditions C1 or C2 in Lemma 3 is satisfied, the cloud accesses do not exhibit Zeno behavior, and the agents accomplish the coordination objective (3) with $\varepsilon = \left(\frac{\beta L_f(\varsigma_\infty + \delta_\infty)}{\gamma}\right)^2$. In particular, if C2 is satisfied, the agents accomplish (3) with $\varepsilon = 0$.*

6 Simulation

In this section, we simulate a formation control of $N = 5$ agents executing the cloud-supported control scheme described above.

The agents exchange information through the cloud according to the graph $\mathcal{G} = (\mathcal{V}, \mathcal{E})$, with $\mathcal{V} = \{1, \ldots, 5\}$ and $\mathcal{E} = \{(1, 2), (1, 3), (1, 4), (4, 5), (3, 4), (4, 3), (3, 1)\}$. The edges are indexed according to the order that they appear in $\mathcal{E}$, so that the first four edges constitute a spanning tree. The matrices $B = [B_T \ B_C]$, $C = [C_T \ C_C]$, T and R are computed accordingly. The agents dynamics are given by (15) and the objective function is given by (16), with P computed in such a way that $R^\top P + PR = 2I_4$. With these choices, we get $\gamma \simeq 1.21$ and $\beta \simeq 2.53$. The agents are subject to disturbances $d_i(t) = \delta(t)[\cos(2\pi t + \phi_i), \sin(2\pi t + \phi_i)]^\top$, where $\delta(t)$ is given in Assumption 2, and $\phi_i = i/5$. In this way, we always have $\|d_i(t)\| \leq \delta(t)$, so that Assumption 2 is satisfied. In particular, we choose $\delta_0 = 1, \delta_\infty = 0$, and $\lambda_\delta = 0.7$. Note that, since the disturbances vanish asymptotically, the algorithm accomplishes perfect coordination; i.e., (3) is attained with $\varepsilon = 0$. For the threshold function, we choose $\varsigma_0 = 15$, $\varsigma_\infty = 0$, and $\lambda_\varsigma = 0.42 < \min\{\lambda_\delta, \frac{\gamma}{2}\}$, so that Condition C2 in Theorem 1 is satisfied, allowing perfect coordination. The desired formation is described by the vectors $b_i = [\cos(2\pi \frac{i}{N}), \sin(2\pi \frac{i}{N})]$. The simulation is set in the time interval $[0, 10]$. We do not choose a specific time unit, since it depends on the practical application and on the type of vehicle. At time $t = 0$ all the agents are placed in the origin.

The results of the simulation are illustrated in Figs. 2, 3, 4 and 5. Figure 2 shows the position of the agents along the first coordination axis. From this figure, we can see that the agents converge after approximately $t = 4.0$. Figure 3 shows the paths of the agent in the plane. From this figure, we can see how the agents reach the desired formation. The baricenter of the formation is not in the origin, which is due to the disturbances and is coherent with the fact that the objective function is only concerned with the relative positions. Figure 4 shows the evolution of the objective function $V(x(t))$ and its upper bound $W(x(t))$ used in the access scheduling. We can see that the upper bound is conservative, in the sense that it has slower convergence

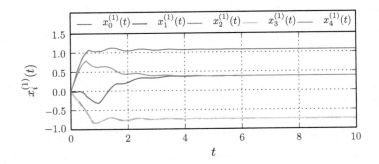

Fig. 2 Positions of the agents along the first coordination axis

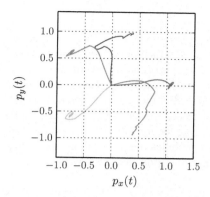

Fig. 3 Paths of the agents in the Euclidean plane. Color legend as in Fig. 2

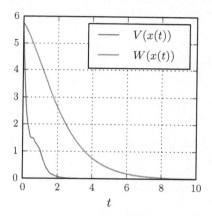

Fig. 4 Objective function $V(x(t))$ and its upper bound $W(x(t))$ used in the access scheduling

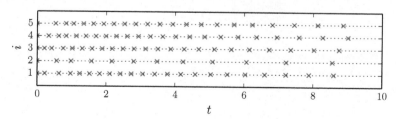

Fig. 5 Cloud accesses of each agent

than the actual objective function $V(x(t))$. Figure 5 shows the cloud accesses of each agent. Note that cloud access is more frequent in the first phase of the coordination, when the control signals have larger magnitude as the agents are far from their assigned formation.

7 Conclusion

In this chapter, we studied cloud-supported multi-agent systems, where coordination is achieved by having the agents exchange information over a shared repository. This framework is particularly useful when inter-agent communication is interdicted, such as in coordination of AUVs. For the case of formation control of integrator agents, a scheduling rule that guarantees convergence of the agents to the desired coordination objective was analyzed. The theoretical results were corroborated with a numerical simulation. Future work includes the use of more accurate AUV models as well as experimental validation of the proposed framework.

Acknowledgements This work has received funding from the European Union Horizon 2020 Research and Innovation Programme under the Grant Agreement No. 644128, AEROWORKS, from the Swedish Foundation for Strategic Research, from the Swedish Research Council, and from the Knut och Alice Wallenberg foundation.

Appendix

Lemma 4 *The agents' dynamics* (15) *and the objective function* (16) *satisfy Assumption 4 with* $\beta = \|B_{\mathcal{T}}\sqrt{P}\|$, $\gamma = \frac{2\min \text{eig}(Q)}{\max \text{eig}(P)}$, *and* $v_i(x)$ *given by* (17).

Proof Taking the derivative of (16), we have

$$\frac{\partial V(x)}{\partial x} = (B_{\mathcal{T}} P B_{\mathcal{T}}^\top \otimes I_2)(x - b),$$

which, taking norms of both sides and applying the triangular inequality, gives

$$\left\| \frac{\partial V(x)}{\partial x} \right\| \leq \|B_{\mathcal{T}}\sqrt{P}\|\sqrt{V(x)}.$$

Moreover, if we take the feedback law $v_i(x) = \sum_{j \in \mathcal{V}_i}(x_j(t) - b_j - x_i(t) + b_i)$, then we have $v(x) = -(CB^\top \otimes I_2)(x - b)$, and

$$\begin{aligned}\frac{\partial V(x)}{\partial x}^\top F(x, v(x), 0_{n_d}) &= -(x-b)^\top (B_{\mathcal{T}} P B_{\mathcal{T}}^\top C B^\top \otimes I_2)(x-b) \\ &= -(x-b)^\top (B_{\mathcal{T}} P R B_{\mathcal{T}}^\top \otimes I_2)(x-b) \\ &= -(x-b)^\top (B_{\mathcal{T}} Q B_{\mathcal{T}}^\top \otimes I_2)(x-b) \\ &\leq -\min \text{eig}(Q)\|(B_{\mathcal{T}} \otimes I_2)(x-b)\|^2.\end{aligned} \qquad (40)$$

Noting that $V(x) \leq \frac{1}{2} \max \text{eig}(P) \|(B_\mathcal{T} \otimes I_2)(x - b)\|^2$, we can further bound (40) as

$$\frac{\partial V(x)}{\partial x}^\top F(x, v(x), 0_{n_d}) \leq -\frac{2 \min \text{eig}(Q)}{\max \text{eig}(P)} V(x).$$

which completes the proof. □

References

1. Adaldo, A., Liuzza, D., Dimarogonas, D.V., Johansson, K.H.: Control of multi-agent systems with event-triggered cloud access. In: European Control Conference, Linz, Austria (2015)
2. Adaldo, A., Liuzza, D., Dimarogonas, D.V., Johansson, K.H.: Multi-agent trajectory tracking with event-triggered cloud access. In: IEEE Conference on Decision and Control (2016)
3. Bullo, F., Cortes, J., Martinez, S.: Distributed Control of Robotic Networks. Princeton University Press, Princeton (2009)
4. Dimarogonas, D.V., Johansson, K.H., Event-triggered control for multi-agent systems. In: IEEE Conference on Decision and Control (2009)
5. Durham, J.W., Carli, R., Frasca, P., Bullo, F.: Dynamic partitioning and coverage control with asynchronous one-to-base-station communication. IEEE Trans. Control Netw. Syst. **3**(1), 24–33 (2016)
6. Fiorelli, E., Leonard, N.E., Bhatta, P., Paley, D.A., Bachmayer, R., Fratantoni, D.M.: Multi-AUV control and adaptive sampling in Monterey Bay. IEEE J. Ocean. Eng. **31**(4), 935–948 (2006)
7. Hale, M.T., Egerstedt, M.: Differentially private cloud-based multi-agent optimization with constraints. In: Proceedings of the American Control Conference, Chicago, IL, USA (2015)
8. Heemels, W.P.M.H., Johansson, K.H., Tabuada, P.: An introduction to event-triggered and self-triggered control. In: IEEE Conference on Decision and Control (2012)
9. Mesbahi, M., Egerstedt, M.: Graph Theoretic Methods in Multiagent Networks. Princeton Univerisity Press, Princeton (2010)
10. Newman, M.E.J.: Networks: An Introduction. Oxford University Press, Oxford (2010)
11. Nowzari, C., Pappas, G.J.: Multi-agent coordination with asynchronous cloud access. In: American Control Conference (2016)
12. Paull, L., Saeedi, S., Seto, M., Li, H.: AUV navigation and localization: a review. IEEE J. Ocean. Eng. **39**(1), 131–149 (2014)
13. Teixeira, P.V., Dimarogonas, D.V., Johansson, K.H.: Multi-agent coordination with event-based communication. In: American Control Conference, Marriot Waterfront, Baltimore, MD, USA, (2011)
14. Zeng, Z., Wang, X., Zheng, Z.: Convergence analysis using the edge Laplacian: robust consensus of nonlinear multi-agent systems via ISS method. Int. J. Robust Nonlinear Control **26**, 1051–1072 (2015)

A Sampled-Data Model Predictive Framework for Cooperative Path Following of Multiple Robotic Vehicles

A. Pedro Aguiar, Alessandro Rucco and Andrea Alessandretti

Abstract We present a sampled-data model predictive control (MPC) framework for cooperative path following (CPF) of multiple, possibly heterogeneous, autonomous robotic vehicles. Under this framework, input and output constraints as well as meaningful optimization-based performance trade-offs can be conveniently addressed. Conditions under which the MPC-CPF problem can be solved with convergence guarantees are provided. An example illustrates the proposed approach.

1 Introduction

This chapter presents a sampled-data model predictive framework for cooperative path following (CPF) of multiple, possibly heterogeneous, autonomous robotic vehicles. In simple terms, the CPF problem, which is at the root of many practical applications, consists of given N vehicles and conveniently defined desired paths, designing the motion control algorithms for each vehicle to drive and maintain them in their respective desired paths at a common speed profile and holding a specified formation pattern. Different solutions to the CPF and similar problems can be found in the literature, see e.g., [1, 6, 8, 9, 12, 15, 17, 19, 20], and the references there in. An interesting strategy consists of decoupling the CPF problem into (i) a path following problem, where the goal is to derive control laws to drive each vehicle to its path at the reference speed profile, and (ii) a multiple vehicle coordination problem, where the objective is to adjust the speed of each vehicle so as to achieve the desired formation pattern. The papers [1, 12] offer a theoretical overview of the subject and introduce techniques to solve the CPF problem.

A.P. Aguiar (✉) · A. Rucco · A. Alessandretti
University of Porto, Porto, Portugal
e-mail: pedro.aguiar@fe.up.pt

A. Rucco
e-mail: alessandrorucco@fe.up.pt

A. Alessandretti
e-mail: andrea.alessandretti@fe.up.pt

Despite the active research in the field, most of the contributions rely on control algorithms that are derived using standard linear or nonlinear techniques and rarely explicitly consider the case of constrained input signals. This results in control laws that are only applicable with stability guarantees in a limited region where the control action, designed for the unconstrained vehicle, does not violate the system constraints.

Model predictive control (MPC) [7, 11, 16], given its ability to explicitly handle constraints, represents a natural direction to pursue. Moreover, the optimization-based nature of MPC schemes leads to high performance, in terms of convergence rates. Further, its ability to generate future state and input predictions of the system can be particularly suited for multi-vehicle control because the predicted input trajectories can be applied to the system in the case of temporarily unavailable communication link among agents.

It is important, however, to stress that MPC approaches are in general more complex to implement, and their complexity can even increase given the intrinsic time-varying nature of the problem. Consider for instance the trajectory tracking case, where the position of the vehicle is required to track a reference trajectory. This problem can be formulated as a stabilization problem in a conveniently defined error space. However, even in this basic single-agent problem, the error dynamic will be time-varying. In an MPC framework, this results in a difficult design of suitable terminal sets and terminal costs, which is usually computed locally around the origin of the error space. We refer to [10] for an overview on MPC for motion control.

In [4], the trajectory tracking problem and the path following problem for underactuated vehicles are solved using a nonlinear auxiliary control law. In this case, the resulting terminal set is only limited by the size of the system constraints, leading to global solutions for the case of unconstrained systems.

In this chapter, we argue that combining recent results of sampled-data model predictive control (that use nonlinear auxiliary control laws to help on the design of terminal costs and terminal sets) with CPF seems a direct and appealing technique. In particular, we propose a framework that exploits the potential of optimization-based control strategies with significant advantages on explicitly addressing input and output constraints and with the ability to allow the minimization of meaningful cost functions. Conditions under which the CPF problem can be solved with convergence guarantees are provided. Further, the result in [4] is extended using a design technique presented in [5] for the computation of the terminal cost, leading to a global region of attraction even in the case of constrained input signals. An example consisting of three autonomous underwater vehicles (AUVs) and one autonomous surface vehicle (ASV) that are required to carry out a survey mission, illustrates the proposed framework.

The remainder of the chapter is organized as follows. Section 2 introduces the coordinated output regulation problem. Section 3 presents the CPF problem and introduces the overall control architecture, which builds on the interaction of a low-frequency long horizon trajectory planning, Sect. 3.1, with a higher frequency MPC control for cooperative path following, Sect. 3.2. A numerical simulation of the overall architecture is presented in Sect. 4. Section 5 closes the chapter with some conclusions.

Notation: The terms $\mathscr{C}(a,b)$ and $\mathscr{PC}(a,b)$ denote the space of continuous and piecewise continuous trajectories, respectively, defined over $[a,b]$ or $[a,+\infty)$ for the case where $b = +\infty$. For a given $n \in \mathbb{N}$, $SE(n)$ denotes the Cartesian product of $\mathbb{R}^n$ with the group $SO(n)$ of $n \times n$ rotation matrices and $se(n)$ denotes the Cartesian product of $\mathbb{R}^n$ with the space $so(n)$ of $n \times n$ skew-symmetric matrices. A continuous function $\sigma : [0,a) \to [0,\infty)$ is said to belong to class $\mathscr{K}$, or to be a class $\mathscr{K}$ function, if it is strictly increasing and $\sigma(0) = 0$. It is said to belong to class $\mathscr{K}_\infty$, or to be a class $\mathscr{K}_\infty$ function, if $a = \infty$ and $\sigma(r) \to \infty$ and $r \to \infty$. A continuous function $\beta : [0,a) \times [0,\infty) \to [0,\infty)$ is said to belong to class $\mathscr{KL}$, or to be a class $\mathscr{KL}$ function, if for each fixed scalar s, the mapping $\beta(t,s)$ belongs to class $\mathscr{K}$ with respect to r and, for each fixed scalar r, the mapping $\beta(t,s)$ is decreasing with respect to s and $\beta(r,s) \to 0$ as $s \to \infty$. For a given set $\mathscr{A} \subseteq \mathbb{Z}$, we denote by $|\mathscr{A}|$ the cardinality of $\mathscr{A}$, i.e., the number of integers in $\mathscr{A}$. The term $\mathscr{B}(r) := \{x : \|x\| \leq r\}$ denotes the closed ball set with radius $r \geq 0$. For a given square matrix A, the notation $A \succ 0$ or $A \succeq 0$ denotes A to be a positive definite or positive semi-definite matrix, respectively.

2 Sampled-Data MPC for Coordinated Output Regulation

This section formulates the coordinated output regulation problem and provides the main results to solve it using a MPC approach. In particular, key conditions under which the proposed sampled-data MPC solves the problem with convergence guarantees are described.

2.1 Problem Formulation

Consider a set of $n_\mathscr{I}$ continuous-time dynamical systems (denoted as agents) where the generic i-th agent, $i \in \mathscr{I} := \{1, 2, \ldots, n_\mathscr{I}\}$ is described as

$$\dot{x}^{[i]}(t) = f^{[i]}(t, x^{[i]}(t), u^{[i]}(t)), \qquad x^{[i]}(t_0) = x_0^{[i]}, \qquad t \geq t_0 \qquad (1a)$$

with $x^{[i]}(t) \in \mathbb{R}^{n^{[i]}}$ and $u^{[i]}(t) \in \mathscr{U}(t) \subseteq \mathbb{R}^{m^{[i]}}$ denoting the state and input vectors at time $t \geq t_0$. The scalar $t_0 \in \mathbb{R}$ and the vector $x_0^{[i]}$ represent the initial time and state of the system, respectively, and the input $u^{[i]}(t)$ is constrained within the input constraint set $\mathscr{U}^{[i]} : \mathbb{R}_{\geq t_0} \rightrightarrows \mathbb{R}^{m^{[i]}}$.

Suppose that each system includes a special internal state denoted as coordination vector $\gamma^{[i]}(t) \in \mathbb{R}^{n_c}$ that evolves with time according to

$$\dot{\gamma}^{[i]}(t) = v_d + u_\gamma^{[i]}(t), \qquad \gamma^{[i]}(t_0) = \gamma_0^{[i]}, \qquad t \geq t_0 \qquad (1b)$$

for a constant predefined vector $v_d \in \mathbb{R}^{n_c}$ and an input signal $u_\gamma^{[i]}(t) \in \mathbb{R}^{n_c}$ that will be a function of the agreement vectors of the other agents. More precisely, the agents communicate among each others according to a communication graph $\mathscr{G} := (\mathscr{V}, \mathscr{E})$, where the vertex set $\mathscr{V}$ collects all the indexes of the systems, that is, $\mathscr{V} = \mathscr{I}$, and the edge set $\mathscr{E} \subseteq \mathscr{V} \times \mathscr{V}$ is such that $(i, j) \in \mathscr{E}$ if and only if the system i can access $\gamma^{[j]}(t)$. Therefore, system i can access from their neighborhoods $j \in \mathscr{N}^{[i]} := \{j : (i, j) \in \mathscr{E}\}$ at time t the agreement states $\gamma_{\mathscr{N}^{[i]}}(t) := \{\gamma^{[j]}(t) : j \in \mathscr{N}^{[i]}\}$.

The output of each agent is defined as

$$y^{[i]}(t) = h^{[i]}(t, x^{[i]}(t), \gamma^{[i]}(t)) \in \mathscr{Y}(t) \tag{1c}$$

that is constrained within the output constrain set $\mathscr{Y}^{[i]} : \mathbb{R}_{\geq t_0} \rightrightarrows \mathbb{R}^{p^{[i]}}$.

Given the above setup, we can now formulate the following problem:

Problem 1 (*Coordinated output regulation*) Design a control law for the input signals $u^{[i]} \in \mathscr{PC}(t_0, \infty)$ and $u_\gamma^{[i]} \in \mathscr{PC}(t_0, \infty)$ such that for every $i \in \mathscr{I}$ the state vector $x^{[i]}(t)$ is bounded for all $t \geq t_0$, and as time approaches infinity the following holds:

1. The output vector $y^{[i]}(t) \in \mathbb{R}^{p^{[i]}}$ converges to the origin;
2. The network disagreement function

$$\phi(t) := \sum_{(i,j) \in \mathscr{E}} (\gamma^{[i]}(t) - \gamma^{[j]}(t))^2, \tag{1d}$$

   converges to the origin;
3. The vector $\dot{\gamma}^{[i]}(t)$ converges to the predefined value $v_d \in \mathbb{R}^{n_c}$ for all $i \in \mathscr{I}$. □

2.2 Sampled-Data MPC

This section introduces a sampled-data MPC controller that solves Problem 1. The key idea behind the proposed scheme is to combine the performance index from output tracking MPC with another term that rewards the consensus among the coordination vectors of the systems in the network. To this end, we rely on existing results on consensus control laws for discrete-time systems, and in particular we assume the following:

Assumption 1 (*Consensus law*) Consider a set of discrete-time autonomous systems that communicate according to the same communication graph $\mathscr{G} = (\mathscr{V}, \mathscr{E})$ introduced in Sect. 2.1, and satisfies

$$\xi^{[i]}(k+1) = \xi^{[i]}(k) + k_{con}(\xi^{[i]}(k), \xi_{\mathscr{N}^{[i]}}(k)) + \eta^{[i]}(k) \tag{2}$$

with $i \in \mathscr{I}$, where $\xi^{[i]}(k) \in \mathbb{R}^{n_c}$, $\eta^{[i]}(k) \in \mathbb{R}^{n_c}$, and $\xi_{\mathscr{N}^{[i]}}(k)$ denote the i-th coordination vector, an external vector, and the coordination vectors from the neighborhood $\mathscr{N}^{[i]}$, respectively, at step $k \in \mathbb{Z}_{\geq k_0}$, with $\xi^{[i]}(k_0) = \xi_0^{[i]} \in \mathbb{R}^{n_c}$ denoting the initial condition of the system i at the initial time step $k_0 \in \mathbb{Z}$. Suppose further that $\|\eta^{[i]}(k)\| \leq a_\eta^{[i]} e^{-\lambda_\eta^{[i]}(k-k_0)}$ for some constants $\lambda_\eta > 0$, $a_\eta \geq 0$, and the following holds:

1. As $k \to \infty$, the disagreement function $\phi(k) = \sum_{(i,j) \in \mathscr{E}} (\xi^{[i]}(k) - \xi^{[j]}(k))^2$ converges asymptotically to zero;
2. The output of the consensus control law $k_{con} : \mathbb{R} \times \mathbb{R}^{|\mathscr{N}^{[i]}|} \to \mathscr{U}_c$, is bounded, i.e., $\mathscr{U}_c \subset \mathbb{R}^{n_c}$ is bounded;
3. There exists an integrable class-$\mathscr{KL}$ function $\beta : \mathbb{R}_{\geq 0} \times \mathbb{R}_{\geq 0} \to \mathbb{R}_{\geq 0}$ such that $\|k_{con}(\xi^{[i]}(k), \xi_{\mathscr{N}^{[i]}}(k))\| \leq \beta(\sum_{i \in \mathscr{I}} \|\xi_0^{[i]}\|, k - k_0)$. $\square$

To solve Problem 1, we propose a sampled-data MPC scheme where the control input is computed at the time instants $\mathscr{T} := \{t_0, t_1, \ldots\}$ with $t_{k+1} > t_k$ for all $k \geq 0$ and with $t_k \to \infty$ as $k \to \infty$. Within the generic intervals $[t_k, t_{k+1})$, an open-loop signal is applied that results from an optimization process. Toward the formulation of the open-loop MPC optimization problem, we first use the discrete-time control law from Assumption 1 to build a piecewise linear continuous-time control signal that addresses the points (2) and (3) of Problem 1 by only using the values of $(\gamma^{[i]}(t), \gamma_{\mathscr{N}^{[i]}}(t))$ evaluated at the time instants $t \in \mathscr{T}$.

To this end, for a generic time $t_k \in \mathscr{T}$ and $\gamma^{[i]}(t_k) \in \mathbb{R}^{n_c}$, an auxiliary signal $\bar{u}_{\gamma,aux_{t_k}}^{[i]} \in \mathscr{C}(t_k, +\infty)$ is defined as

$$\bar{u}_{\gamma,aux_{t_k}}^{[i]}(\tau) = \begin{cases} \frac{1}{t_{k+1}-t_k} k_{con}(\gamma^{[i]}(t), \gamma_{\mathscr{N}^{[i]}}(t)), & \tau \in [t_k, t_{k+1}] \\ 0, & \tau > t_{k+1} \end{cases}. \quad (3)$$

Let $\lfloor t \rfloor$ be the maximum sampling instant $t_k \in \mathscr{T}$ smaller than or equal to t, i.e., $\lfloor t \rfloor = \max_{k \in \mathbb{N}_{\geq 0}} \{t_k \in \mathscr{T} : t_k \leq t\}$. Then, by Assumption 1 the control signal $u_\gamma^{[i]}(t) = \bar{u}_{\gamma,aux_{\lfloor t \rfloor}}^{[i]}(t)$ solves points (2) and (3) of Problem 1.

At this stage, we are ready to state the MPC optimization problem associated with the proposed control scheme.

Definition 1 (*Open-loop MPC problem*) Given the tuple of parameters $p = (t, x^{[i]}, \gamma^{[i]}, \gamma_{\mathscr{N}^{[i]}}, \eta^{[i]}) \in \mathbb{R}_{\geq t_0} \times \mathbb{R}^{n^{[i]}} \times \mathbb{R}^{n_c} \times \mathbb{R}^{|\mathscr{N}^{[i]}|} \times \mathbb{R}^{n_c}$, an horizon length $T \in \mathbb{R}_{>0}$, and the auxiliary signal $\bar{u}_{\gamma,aux_t}^{[i]} \in \mathscr{C}(t, +\infty)$ from (3), the open-loop MPC optimization problem $\mathscr{P}(p)$ consists in finding the optimal control signals $\bar{u}^{\star[i]} \in \mathscr{PC}(t, t+T)$ and $\bar{v}_\gamma^{\star[i]} \in \mathscr{PC}(t, t+T)$ that solves

$$J_T^{\star[i]}(p) = \min_{\substack{\bar{u}^{[i]} \in \mathscr{PC}(t,t+T) \\ \bar{v}_\gamma^{[i]} \in \mathscr{PC}(t,t+T)}} J_T(p, \bar{u}^{[i]}, \bar{v}_\gamma^{[i]})$$

s.t. $\dot{\bar{x}}^{[i]}(\tau) = f^{[i]}(\tau, \bar{x}^{[i]}(\tau), \bar{u}^{[i]}(\tau)), \quad \bar{x}^{[i]}(t) = x^{[i]},$

$\dot{\bar{\gamma}}^{[i]}(\tau) = v_d + \bar{u}_\gamma^{[i]}(\tau), \quad \bar{\gamma}^{[i]}(t) = \gamma^{[i]},$

$\dot{\bar{\eta}}^{[i]}(\tau) = \bar{v}_\gamma^{[i]}(\tau), \quad \bar{\eta}^{[i]}(t) = \eta^{[i]},$

$\bar{y}^{[i]}(\tau) = h^{[i]}(\tau, \bar{x}^{[i]}(\tau), \bar{\gamma}^{[i]}(\tau)),$

$\bar{u}_\gamma^{[i]}(\tau) = \bar{u}_{\gamma,aux_t}^{[i]}(\tau) + \bar{\eta}^{[i]}(\tau),$ (4a)

$(\bar{y}^{[i]}(\tau), \bar{u}^{[i]}(\tau), \bar{v}_\gamma^{[i]}(\tau)) \in \mathscr{Y}^{[i]}(\tau) \times \mathscr{U}^{[i]}(\tau) \times \mathscr{V}_\gamma^{[i]}(\tau),$

$(\bar{y}^{[i]}(t+T), \bar{\eta}^{[i]}(t+T)) \in \mathscr{Y}_{aux}^{[i]}(t+T) \times \mathscr{B}(r_\eta^{[i]})$

$|\bar{\eta}^{[i]}(\tau)| \le a_\eta^{[i]} e^{-\lambda_\eta^{[i]}(t-t_0)}$ (4b)

for all $\tau \in [t, t+T]$ and with

$$J_T^{[i]}(p, \bar{u}^{[i]}, \bar{v}_\gamma^{[i]}) := \int_t^{t+T} l^{[i]}(\tau, \bar{x}^{[i]}(\tau), \bar{u}^{[i]}(\tau), \bar{\gamma}^{[i]}(\tau), \bar{u}_\gamma^{[i]}(\tau)) + l_c^{[i]}(\bar{\eta}^{[i]}(\tau), \bar{v}_\gamma^{[i]}(\tau)) d\tau$$
$$+ m^{[i]}(t+T, \bar{x}^{[i]}(t+T), \bar{\gamma}^{[i]}(t+T)) + \frac{1}{2} m_\eta^{[i]} (\bar{\eta}^{[i]}(t+T))^2, \quad (4c)$$

for four constants $\lambda_\eta^{[i]} > 0$, $a_\eta^{[i]} \ge 0$, $m_\eta^{[i]} \ge 0$, and $r_\eta^{[i]} \ge 0$, and where the constraint (4b) can be omitted in the case of $\mathscr{N}^{[i]} = \emptyset$. The *finite horizon cost* $J_T^{[i]}(\cdot)$, which corresponds to the *performance index* of the MPC controller, is composed of two stage costs, i.e., the *tracking stage cost* $l^{[i]} : \mathbb{R}_{\ge t_0} \times \mathbb{R}^{n^{[i]}} \times \mathbb{R}^{m^{[i]}} \times \mathbb{R}^{n_c} \times \mathbb{R}^{n_c} \to \mathbb{R}_{\ge 0}$ and the *consensus stage cost* $l_c^{[i]} : \mathbb{R}^{n_c} \times \mathbb{R}^{n_c} \to \mathbb{R}_{\ge 0}$, and two terminal costs, i.e., the *tracking terminal cost* $m^{[i]} : \mathbb{R}_{\ge t_0} \times \mathbb{R}^{n_c} \times \mathbb{R}^{n_c} \to \mathbb{R}_{\ge 0}$, and the *consensus terminal cost* $\frac{1}{2} m_\eta^{[i]} (\bar{\eta}^{[i]}(t+T))^2$. The tracking terminal cost is defined over the set of $(t, x^{[i]}, \gamma^{[i]})$ such that the associated $y^{[i]}(t) \in \mathscr{Y}_{aux}^{[i]}(t)$ belongs to the *tracking terminal set* $\mathscr{Y}_{aux}^{[i]} : \mathbb{R}_{\ge t_0} \rightrightarrows \mathbb{R}^{p^{[i]}}$. Similarly, the consensus terminal cost is evaluated with $\bar{\eta}^{[i]}(t+T)$ constrained in the *consensus terminal set* $\mathscr{B}(r_\eta^{[i]})$. $\square$

In the sequel, for the sake of clarity, we omit the explicit superscript label $[i]$ to the agent i whenever it is clear from the context. We will also use the simplified notation of omitting the time dependence and in particular, for the dependent signals $\bar{x}, \bar{u}, \bar{\gamma}, \bar{v}$, a function $l(\cdot)$ evaluated as $l(\tau, \bar{x}(\tau), \bar{u}(\tau), \bar{\gamma}(\tau), \bar{v}(\tau))$ is denoted by $l(\tau, \bar{x}, \bar{u}, \bar{\gamma}, \bar{v})$ or $\bar{l}(\tau)$. Moreover, for a generic time $t \ge t_0$, the superscript $^{\star t}$ is used to denote all the trajectories of a given signal associated with the optimal predictions of $\mathscr{P}(t)$.

The proposed sampled-data MPC approach is obtained by solving the optimization problem in Definition 1 at every time sample $t_k \in \mathscr{T}$ and applying the associated optimal input trajectories within the generic interval $[t_k, t_{k+1})$, with $k \in \mathbb{Z}_{\ge 0}$, as follows

$$u(t) = \bar{u}^{\star \lfloor t \rfloor}(t), \qquad u_\gamma(t) = \bar{u}_\gamma^{\star \lfloor t \rfloor}(t), \qquad \dot{\eta}(t) = \bar{v}_\gamma^{\star \lfloor t \rfloor}(t) \qquad (5a)$$

where $\eta \in \mathbb{R}^{n_c}$ denotes the state of the controller with initial condition $\eta(t_0) = \eta_0 \in \mathbb{R}^{n_c}$.

At this point, we are ready to state the sufficient conditions for the MPC controller (5) to solve Problem 1.

The first set of conditions are obtained by adapting the standard MPC sufficient conditions for state convergence to the origin in the state space, to convergence of the output signal to the origin in the output space.

Assumption 2 The input constraint set $\mathscr{U}(t)$ is uniformly bounded over time. Moreover, the function $f(\cdot)$ in (1a) is locally Lipschitz in x, piecewise continuous in t and u, and bounded for bounded x in the region of interest, i.e., the set $\{\|f(t, x, u)\| : t \geq t_0, x \in \bar{\mathscr{X}}, u \in \mathscr{U}(t)\}$ is bounded for any bounded $\bar{\mathscr{X}} \subset \mathbb{R}^n$. □

Assumption 3 Consider the output defined in (1c).

1. The system (1) satisfies for all $t \geq t_0$ the input–output-to-state stability condition

$$\|x(t)\| \leq \beta_x(\|x_0\|, t - t_0) + \sigma_u(\|u\|_{[t_0,t)}) + \sigma_y(\|y\|_{[t_0,t)}) + \sigma_x \quad (6)$$

for a class-$\mathscr{K}\mathscr{L}$ function $\beta_x : \mathbb{R}_{\geq 0} \times \mathbb{R}_{\geq 0} \to \mathbb{R}_{\geq 0}$, classes-$\mathscr{K}$ functions $\sigma_u, \sigma_y : \mathbb{R}_{\geq 0} \to \mathbb{R}_{\geq 0}$, and a constant scalar $\sigma_x \geq 0$;
2. The gradient of the right-hand side of the output equation (1c) $\nabla h(t, x(t), \gamma(t))$ is uniformly bounded over time for bounded values of x, i.e., for all x with $\|x\| \leq B$, there exists a scalar $b_B > 0$, possibly dependent on $B \geq 0$ such that $\|\nabla h(t, x, \gamma)\| \leq b_B$ for every time $t \geq t_0$ and $\gamma \in \mathbb{R}^{n_c}$. □

It is worth noticing that when condition (6) holds with $\sigma_x = 0$ and if $u(t) \to 0$ as $t \to \infty$, a controller that solves Problem 1 would also drive the state of the system to the origin. Notice, however, that this could be restrictive, as in the case of the considered example, and therefore, a positive term σ_x is introduced to guarantee only boundedness of the state trajectory.

The point (2) of the latter assumption is rather general, and we refer to the illustrative example for more insight on when such condition holds.

Assumption 4 For any given signals $\gamma \in \mathscr{C}(t_0, +\infty)$ and $u_\gamma \in \mathscr{PC}(t_0, +\infty)$ the following holds:

1. The output constraint set $\mathscr{Y}(t)$ and the terminal set $0 \in \mathscr{Y}_{aux}(t) \subseteq \mathscr{Y}(t)$ are closed, connected, and contain the origin for all $t \geq t_0$. The input constraint set $\mathscr{U}(t)$ is closed for all $t \geq t_0$;
2. The constraint set $\mathscr{V}_\gamma(t)$ is compact, uniformly bounded over time and such that $\mathscr{B}(r_\eta \lambda_\eta) \subseteq \mathscr{V}_\gamma(t)$, for all $t \geq t_0$;
3. The tracking stage cost $l(\cdot)$ is zero with $y(t) = 0$ and there is a class-$\mathscr{K}_\infty$ function $\alpha_s : \mathbb{R}_{\geq 0} \to \mathbb{R}_{\geq 0}$ such that $l(\tau, \bar{x}, \bar{u}, \bar{\gamma}, \bar{u}_\gamma) \geq \alpha_s(\|y\|)$ for all $(\tau, \bar{x}, \bar{u}, \bar{\gamma}, \bar{u}_\gamma) \in \mathbb{R}_{\geq t_0} \times \mathbb{R}^n \times \mathscr{U}(t) \times \mathbb{R}^{n_c} \times \mathbb{R}^{n_c}$;
4. The consensus cost function $l_c(\cdot)$ is zero with $\eta = 0$ and the function $\alpha_c : \mathbb{R}_{\geq 0} \to \mathbb{R}_{\geq 0}$ such that $l_c(\eta, v_\gamma) \geq \alpha_c(\|\eta\|) := m_\eta \lambda_\eta \eta^2$ for all $(\eta, v_\gamma) \in \mathbb{R}^{n_c} \times \mathbb{R}^{n_c}$;

5. For any given values of $(x, u, \gamma, u_\gamma) \in \mathbb{R}^n \times \mathbb{R}^m \times \mathbb{R}^{n_c} \times \mathbb{R}^{n_c}$ the functions $l(t, x, u, \gamma, u_\gamma)$ and $m(t, x, \gamma)$ are uniformly bounded over time $t \geq t_0$;
6. There exists a feasible auxiliary control law $k_{aux} : \mathbb{R}_{\geq t_0} \times \mathbb{R}^n \times \mathbb{R}^{n_c} \times \mathbb{R}^{n_c} \to \mathbb{R}^m$ such that, for the associated closed-loop system (1a) with $u(t) = k_{aux}(t, x, \gamma, u_\gamma)$ with initial time and states $(\hat{t}, \hat{x}, \hat{\gamma}) \in \mathbb{R}_{\geq t_0 + T} \times \mathbb{R}^n \times \mathbb{R}^{n_c}$ such that $y(\hat{t}) \in \mathcal{Y}_{aux}(\hat{t})$, and for all $\bar{u}_\gamma \in \mathscr{C}(t_0 + T, \infty)$ with $u_\gamma(\tau) \in \mathscr{B}(v_d + r_\eta)$, the input and output vector satisfy $u(t) \in \mathcal{U}(t)$ and $y(t) \in \mathcal{Y}_{aux}(t)$, the associated state $x(t)$ exists and is unique, and the condition

$$m(\hat{t} + \delta, x(\hat{t} + \delta), \gamma(\hat{t} + \delta)) - m(\hat{t}, x(\hat{t}), \gamma(\hat{t})) \leq -\int_{\hat{t}}^{\hat{t}+\delta} l(t) dt$$

holds for any $\delta > 0$. □

Assumption 5 Consider the open-loop MPC optimization problem from Definition 1. The tracking stage cost $l(\cdot)$ and the tracking terminal cost $m(\cdot)$ are Lipschitz continuous on (γ, u_γ), i.e., there exists a pair of constants $C_l \geq 0$ and $C_m \geq 0$ such that

$$|l(t, x, u, \gamma_1, u_{\gamma,1}) - l(t, x, u, \gamma_2, u_{\gamma,2})| \leq C_l \left\| \begin{bmatrix} \gamma_1 - \gamma_2 \\ u_{\gamma,1} - u_{\gamma,2} \end{bmatrix} \right\|$$

$$|m(t, x, \gamma_1) - m(t, x, \gamma_2)| \leq C_m \left\| \begin{bmatrix} \gamma_1 - \gamma_2 \\ u_{\gamma,1} - u_{\gamma,2} \end{bmatrix} \right\|$$

holds for any given $(t, x, u) \in \mathbb{R}_{\geq t_0} \times \mathbb{R}^n \times \mathbb{R}^m$, $(\gamma_1, u_{\gamma,1}) \in \mathbb{R}^{n_c} \times \mathbb{R}^{n_c}$, and $(\gamma_2, u_{\gamma,2}) \in \mathbb{R}^{n_c} \times \mathbb{R}^{n_c}$. □

In Assumption 5, the functions $m(\cdot)$ and $l(\cdot)$ are required to be Lipschitz only on the variables (γ, u_γ), which makes the assumption rather general. Moreover, since the Lipschitz constants are not used in the design phase, only their existence is required, and not their computation.

At this point, we are ready to state the main result of this section.

Theorem 1 *Consider a set of constrained systems (1) that communicates according to a communication network $\mathscr{G} = (\mathscr{V}, \mathscr{E})$ as described in Sect. 2.1. If Assumptions 1–5 hold, then the proposed sampled-data MPC control law (5) solves Problem 1. The region of attraction of the proposed controller corresponds to the set of initial conditions of the system such that the open-loop MPC problem of Definition 1 is feasible.* □

Proof A sketch of the proof follows. The convergence analysis is carried out using the value function defined as $V(t) := J_T^*(p(t))$ that is obtained by solving the problem $\mathscr{P}(p(t))$ with $p(t) = (t, x(t), \gamma(t), \gamma_\mathscr{N}(t), \eta(t))$ where, in order for $V(t)$ to be well-defined for all $t \in [t_k, t_{k+1})$ and $\gamma^{[i]}(t) \in \mathbb{R}^{n_c}$, we redefine the auxiliary trajectory (3) as $\bar{u}_{\gamma,aux_t}^{[i]} \in \mathscr{C}(t, +\infty)$ with

$$\bar{u}^{[i]}_{\gamma,aux_t}(\tau) = \begin{cases} \frac{1}{\delta_k} k_{con}(\gamma^{[i]}(t_k), \gamma_{\mathcal{N}^{[i]}}(t_k)), & \tau \in [t, t+\delta_k] \\ 0, & \tau > t+\delta_k, \end{cases}$$

where $\delta_k := t_{k+1} - t_k$ for all $k \in \mathbb{Z}_{\geq 0}$. Notice that, although the implementation of (5) only requires to solve $\mathscr{P}(p(t))$ at the time instants $t \in \mathscr{T}$, the value function considers its optimal value at every time $t \geq t_0$, and therefore is well defined for all time instants $t \geq t_0$. Under the assumptions of Theorem 1, it is possible to show that for any $\delta > 0$ the evolution of the value function satisfies

$$V(t_k + \delta) \leq V(t_k) - \int_{t_k}^{t_k+\delta} \alpha(\|z(\tau)\|) d\tau + \bar{\bar{\beta}} \tag{8}$$

with $z(t) := [y(t)', \eta(t)']'$, for a class-$\mathscr{K}_\infty$ function $\alpha : \mathbb{R}_{\geq 0} \to \mathbb{R}_{\geq 0}$, and where the constant $\bar{\bar{\beta}}$ captures the effect of the disagreement on the values of $\gamma^{[i]}$ in the network (resulting in a $\bar{u}^{[i]}_{\gamma,aux_t}$ different from zero). At this point, Barbalat's lemma can be used to show that $\alpha(\|z(t)\|) \to 0$ as $t \to +\infty$ and, by the continuity and positive definitiveness of $\alpha(\cdot)$, the vector $z(t)$, and therefore the vectors $y(t)$ and $\eta(t)$, converge to the origin with $t \to +\infty$. The proof is concluded by noticing that, by Assumption 1, as $\eta(t) \to 0$ the disagreement function is driven to the origin. ∎

3 MPC-CPF Framework

This section proposes a sampled-data model predictive control framework for cooperative path following (MPC-CPF) for a group of N autonomous robotic vehicles modeled by general continuous-time systems of the form

$$\dot{x}^{[i]}(t) = f^{[i]}(x^{[i]}(t), u^{[i]}(t)), \quad x^{[i]}(0) = x_0^{[i]}, \ u^{[i]}(t) \in \mathscr{U}^{[i]}(t), \ i \in \mathscr{I}, \ t \geq 0, \tag{9}$$

where for each vehicle $i \in \mathscr{I} := \{1, 2, \ldots N\}$, the vector $x^{[i]}(t) \in \mathbb{R}^{n^{[i]}}$ denotes the state and $u^{[i]}(t) \in \mathbb{R}^{m^{[i]}}$ the input vector, which can be constrained by the input constraint set $\mathscr{U}^{[i]} : \mathbb{R} \rightrightarrows \mathbb{R}^{m^{[i]}}$.

Figure 1 illustrates a motion control system architecture, which consists of several interconnected subsystems: the trajectory generator, the high-level motion controller that can be a trajectory tracking controller, a path following controller or a cooperative path following (CPF) controller, and a low-level controller all supported by the navigation and the communication systems. Next, we briefly describe the role of each block, which will be in the sequel formally defined in a sampled-data model predictive control context.

Trajectory Generator—a sampled (and in many cases event) based process whose inputs are the vehicle dynamical models (9), the boundary conditions (e.g., the initial and final poses and velocities of the vehicles), optionally nominal trajectories (usually prescribed as spacial curves not necessarily feasible in terms of the

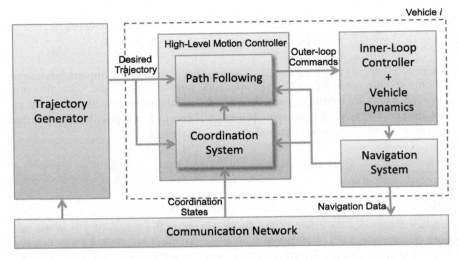

Fig. 1 An example of a simplified high-level cooperative path following scheme

vehicle dynamics), safety constraints (minimal allowed distances to obstacles and other vehicles as well as the map of the location of the obstacles). The outputs of this process are the (locally) optimal (in some sense) collision-free obstacle-avoiding desired trajectories for each vehicle. Depending on the complexity of the application scenario, the trajectory generator usually runs in a base station computer with relatively significant computational power.

High-level motion controller—a system that usually runs onboard of each vehicle and whose input is the desired trajectory provided by the trajectory generator. This system can be either a single vehicle motion controller (e.g., a trajectory tracking or a path following controller) or a multiple motion controller like a cooperative path following system as it is depicted in Fig. 1. In the latter case, there exist a subsystem, the coordination system, that is in charge of providing the adequate correction command signals to synchronize the coordination states of each vehicle.

Inner loop controller—a dynamical system (sometimes denoted as an autopilot) that is responsible for sending the low-level command signals to make the vehicle track a given desired signal (e.g., a desired heading) that is provided by the high-level motion controller (that usually operates at the kinematic level).

Navigation system—a filter-like process that uses local navigation data provided by the onboard sensors (e.g., an inertial measurement unit) as well as (when possible) information that is obtained over the communication network to compute the vehicle's state (e.g., linear and rotational positions and velocities of the vehicle).

Communication system—a system that is responsible for supervising and managing the flow of information (in and out).

We now describe in more detail the above-mentioned systems using optimal sampled based concepts. The last three mentioned systems are not discussed because they are out of the scope of this chapter.

3.1 Trajectory Generator

Let $\mathcal{T}_{TG} := \{t_0, t_1, \ldots\}$ be an infinite and increasing sequence of discrete sample times in $[t_0, \infty)$. Given a pair $(t, \{z^{[i]}\}_{i \in \mathcal{I}}) \in \mathcal{T}_{TG} \times (\mathbb{R}^{n^{[i]}})^N$ and a horizon length $T_{TG} > 0$, the trajectory generator outputs the (local) optimal pairs $\{x^{\star[i]}([t, t + T_{TG}]), u^{\star[i]}([t, t + T_{TG}])\}_{i \in \mathcal{I}}$ that solves

$$\min_{\{x^{[i]}(\cdot), u^{[i]}(\cdot)\}_{i \in \mathcal{I}}} \int_t^{t+T_{TG}} \sum_{i \in \mathcal{I}} \ell^{[i]}(x^{[i]}(\tau), u^{[i]}(\tau)) \, d\tau + \sum_{i \in \mathcal{I}} m(x^{[i]}(T)), \quad t \in \mathcal{T}_{TG}$$

$$\text{subject to} \quad \dot{x}^{[i]}(\tau) = f^{[i]}(x^{[i]}(\tau), u^{[i]}(\tau)), \quad \forall \tau \in [t, t + T_{TG}], \forall i \in \mathcal{I},$$
$$(x^{[i]}(\tau), u^{[i]}(\tau)) \in \mathcal{X}(\tau) \times \mathcal{U}(\tau) \quad \forall \tau \in [t, t + T_{TG}], \forall i \in \mathcal{I},$$
$$x^{[i]}(t) = z^{[i]}, \quad \forall i \in \mathcal{I}, \tag{10}$$

where $\ell^{[i]} : \mathbb{R}^{n^{[i]}} \times \mathbb{R}^{m^{[i]}} \to \mathbb{R}$ and $m^{[i]} : \mathbb{R}^{n^{[i]}} \to \mathbb{R}$ are the incremental and terminal costs. It is worth noting that (10) is formulated as a sampled-data MPC approach in which the vehicle model, the state and control constraints, and the cost function are described in continuous time. In the illustrative example, we solve the optimal control problem (10) using the PRojection Operator based Newton method for Trajectory Optimization (PRONTO), [13], combined with the barrier function relaxation developed in [14]. This method is a direct method based for solving continuous-time optimal control problems. We refer the reader to the accompanying book chapter [18] for a detailed discussion of the aforementioned optimization technique and its application to motion planning.

Following a sampled-data MPC approach, at the current discrete-time sample $t \in \mathcal{T}_{TG}$, PRONTO computes the (local) optimal trajectory $x^{\star[i]}([t, t + T_{TG}])$, $u^{\star[i]}([t, t + T_{TG}])$. The computed optimal trajectory is set as reference signal (desired reference trajectory) to the high-level motion controller.

3.2 High-Level Motion Controller

Using the framework presented in Sect. 2, we now describe how one can develop for the general case a high-level motion controller for trajectory tracking and path following in case of a single vehicle, and for cooperative path following in case of multiple vehicles. Particular applications are described in Sect. 4.

Trajectory Tracking

The trajectory tracking motion control problem is concerned with the design of control laws that force a given vehicle to track a desired temporal/spatial trajectory. In this chapter, we adopt the following definition.

Definition 2 (*Trajectory tracking problem*) Consider a single vehicle described by (9) and let $y_d(t) : \mathbb{R}_{\geq 0} \to \mathbb{R}^p$ be a given desired reference trajectory. Design a feedback control law such that all the relevant closed-loop signals are bounded and the tracking error defined as a function of the state $x(t)$ and $y_d(t)$ converges to zero as $t \to \infty$.

To solve the trajectory tracking problem using the framework presented in Sect. 2, we need to set in (1b) $v_d = 1$, $u_\gamma(t) = 0$, $t \geq t_0$, and $\gamma_0 = 0$, thus yielding $\gamma(t) = t$. Thus, if in addition we associate the output in (1c) as the tracking error, and in the open-loop MPC problem, we set the cost $l_c(\cdot) = 0$ and $m_\eta = 0$, it follows that Theorem 1 applies and the tracking problem is solved.

Path Following

In contrast with trajectory tracking, path following is less restrictive in the sense that the objective is to steer the vehicle toward a path and make it follow the path with an assigned speed profile.

Definition 3 (*Path following problem*) Consider a single vehicle described by (9) and let $y_d(\gamma) \in \mathbb{R}^p$ be a given desired path parameterized by $\gamma \in \mathbb{R}$, and $v_d \in \mathbb{R}$ a desired speed assignment. Design a feedback control law such that all the relevant closed-loop signals are bounded and the path following error defined as a function of the state $x(t)$ and $y_d(\gamma(t))$ converges to zero as $t \to \infty$, and the parameter γ satisfies the speed assignment $\dot{\gamma} \to v_d$ as $t \to \infty$.

Notice from the definition, that there are no explicit temporal specifications, that is, the vehicle is not required to be at a certain point at a desired time. This fact has some considerable implications in terms of performance if the problem setup is a priori a path following problem. In that case, it is not recommended to implement the naive approach of converting the path following as a tracking problem at the level of the desired reference to track. See e.g., [2, 3] for a comparison and discussion between the trajectory tracking and path following approaches.

To implement a path following algorithm using the framework presented in Sect. 2, we only have to associate the output in (1c) as the path following error, and in the open-loop MPC problem, by setting the cost $l_c(\cdot) = 0$ since there are no requirements of coordination with other vehicles.

Cooperative Path Following

Cooperative path following deals with the case where a group of vehicles is required to follow predefined spatial paths, while holding a desired formation pattern at a desired formation speed. To this end, the vehicles need to be supported by an intervehicle communication network, as it is described in Sect. 2.1.

Definition 4 (*Cooperative path following problem*) Consider a group of vehicles described by (9) and let $y_d^{[i]}(\gamma^{[i]}) \in \mathbb{R}^{p^{[i]}}$ be given desired paths parameterized by $\gamma^{[i]} \in \mathbb{R}$, $i \in \mathscr{I}$, and $v_d \in \mathbb{R}$ a desired common speed assignment. Design feedback control laws such that all the relevant closed-loop signals are bounded, and as t approaches to infinity,

1. The path following errors converge to zero;
2. The network disagreement function $\phi(t)$ defined in (1d) converges to the origin;
3. The signals $\dot{\gamma}^{[i]}(t)$ converge to the desired common speed assignment v_d.

The implementation of a CPF using the MPC coordinated output regulation framework calls for the execution of a consensus algorithm (see Assumption 1) that is responsible to change the nominal speeds of the vehicles so as to achieve the desired temporal synchronism. The framework in Sect. 2 can then be applied directly by associating the outputs in (1c) as the path following errors.

4 Illustrative Example

In this section, as an illustrative example of the proposed approach, we address the CPF problem for marine vehicles that are tasked to do a seabed mapping of a given area. The involved vehicles are three autonomous underwater vehicles (AUVs) and one autonomous surface vehicle (ASV). The three AUVs have to survey the mission area. One of them acts as formation leader in terms of communication for the remaining ones, which are assumed to be equipped with high-resolution scanning equipment for bottom surveys. The ASV is employed to help the navigation system of the AUVs and to work as a relay of communication for a base station. An a priori desired lawnmower path is assigned offline to the vehicles and is replanned online based on potential obstacles. In this scenario, we assume that the replanning path only happens for the ASV when it may encounter obstacles (like other surface vehicles). The proposed sampled-data MPC-CPF framework is an appealing approach to tackle this scenario. Specifically, we deal with (i) the computation of (local) optimal trajectories, (ii) the design of MPC controllers for formation keeping of the vehicles along the reference paths.

4.1 Trajectory Generator

The goal is to compute collision-free obstacle-avoiding trajectories. To formulate the optimal control problem described in Sect. 3.1 we need to introduce the vehicle models, define the cost function to be optimized and specify the state/input constraints and boundary conditions. Due to space limitations, in the sequel, we will focus only on the ASV case because it is the one that requires replanning.

Vehicle Model

In this example, for easiness of presentation, we use simplified planar dynamic models for the vehicles. Let I be an inertial coordinate frame and $\{B^{[i]}\}$ be a body coordinate frame attached to the generic vehicle i. The pair $(p^{[i]}(t), R^{[i]}(t)) \in SE(2)$ denote the configuration of the vehicle, position and orientation, where $R^{[i]}(t)$ is the rotation matrix from body to inertial coordinates. Now, let $(v^{[i]}(t), \Omega(\omega^{[i]}(t))) \in SE(2)$ be the twist that defines the velocity of the vehicle, linear, and angular, where the matrix $\Omega(\omega^{[i]}(t))$ is the skew-symmetric matrix associated to the angular velocity $\omega^{[i]}(t) \in \mathbb{R}$, defined as

$$\Omega(\omega) := \begin{pmatrix} 0 & -\omega \\ \omega & 0 \end{pmatrix}.$$

We consider the following dynamical model:

$$\dot{p}^{[i]}(t) = R^{[i]}(\theta) v^{[i]}(t) + v_c^{[i]}(t),$$
$$\dot{\theta}^{[i]} = \omega^{[i]},$$
$$M^{[i]} \dot{v}^{[i]}(t) = -\left(\Omega(\omega^{[i]}) M^{[i]} + D_v^{[i]}\right) v^{[i]}(t) + \begin{pmatrix} 1 \\ 0 \end{pmatrix} u_v^{[i]},$$
$$J^{[i]} \dot{\omega}^{[i]} = -d_\omega^{[i]} \omega^{[i]} + u_\omega^{[i]}$$

where $\theta^{[i]} \in \mathbb{R}$ is the heading angle that parametrizes $R^{[i]} \in SO(2)$, $v_c^{[i]} \in \mathbb{R}^2$ is the ocean current disturbance velocity vector, $M^{[i]} \in \mathbb{R}^{2 \times 2}$ is the mass matrix, $J^{[i]} \in \mathbb{R}$ is the moment of inertia, $D_v^{[i]} \in \mathbb{R}^{2 \times 2}$ and $d_\omega^{[i]} \in \mathbb{R}$ captures the damping effects, and the control input $u^{[i]}(t) = [u_v^{[i]}(t), u_\omega^{[i]}(t)]'$ is composed by the external force in forward direction and the external torque in the vertical axis.

Problem Formulation: Desired Curve, Cost Function, and Constraints

The lawnmower maneuver is specified with a concatenation of straight lines (i.e., zero curvature) of 50 m and circular arcs (i.e., constant curvature) with radius 15 m and a constant speed of 2 m/s assigned on it. The desired curvature σ_d is approximated analytically by a suitable combination of hyperbolic tangent functions and the longitudinal and lateral coordinates and the tangent angle are given by integration of the unicycle model, $\dot{x}_d = v_d \cos \chi_d$, $\dot{y}_d = v_d \sin \chi_d$, $\dot{\chi}_d = v_d \sigma_d$. Given the desired lawnmower maneuver (i.e., path and speed profile), we are interested to compute the ASV trajectory that is close in the L_2 sense to the desired maneuver. We propose to penalize deviations of the ASV trajectory from the desired maneuver without forcing it to exactly match the desired maneuver (it can be unfeasible for the ASV, both in terms of dynamics constraints and collision-free obstacle avoidance path). To this end, the incremental and terminal costs in (10) are set as

$$\ell^{[i]}(x^{[i]}(\tau), u^{[i]}(\tau)) = \frac{1}{2}\left(\|x^{[i]}(\tau) - x^d(\tau)\|_Q^2 + \|u(\tau)^{[i]} - u^d(\tau)\|_R^2\right)$$

$$m(x^{[i]}(T)) = \frac{1}{2}\|x^{[i]}(T) - x^d(T)\|_{P_f}^2,$$

where $(x^d(\cdot), u^d(\cdot))$ is the desired lawnmower maneuver, Q, R, and P_f are suitable positive definite weighting matrices.

The obstacle avoidance constraint can be formulated as follows. Let an obstacle be defined through the coordinates of its center x_{obs} and y_{obs} and its radius r_{obs}. Then, in order to achieve collision-free obstacle trajectories, the trajectory generator must ensure that the constraint

$$c_{obs}(x^{[i]}(\tau)) = \frac{(p_x^{[i]}(\tau) - x_{obs})^2}{r_{obs}^2} + \frac{(p_y^{[i]}(\tau) - y_{obs})^2}{r_{obs}^2} - 1 \geq 0,$$

is satisfied for all $\tau \in [t, t + T_{TG}]$. As mentioned in Sect. 3.1, the input constraints and the collision avoidance constraint are taken into account by using a barrier function relaxation proposed in [14].

Numerical Results

Figure 2 shows numerical computations for the first two straight lines of a lawnmower maneuver in the presence of ocean currents. We model the space-varying current as follows (see blue arrows):

$$v_{cx}(x, y) = 0.5 \sin\left[\left(\frac{x}{70}\right)^2 + \left(\frac{y}{70}\right)^2\right], \quad v_{cy}(x, y) = 0.5 \cos\left(\frac{xy}{70^2}\right).$$

We also consider that there is one obstacle centered in $x_{obs} = 40\,\text{m}$ and $y_{obs} = 35\,\text{m}$ and with radius $r_{obs} = 10\,\text{m}$ (see solid circle in Fig. 2). The (local) optimal trajectory is shown in Figs. 2 and 3. Initially, the trajectory generator system does not take into account the obstacle since it is not detected by the ASV, see Fig. 2a. Due to the current disturbance, the ASV is "drifting" laterally, i.e., the body frame of the ASV is not aligned with the desired straight line. A nonzero heading error angle is observed, yet the ASV is able to follow the desired lawnmower path with high accuracy. Once the obstacle is detected, the ASV trajectory is replanned since the desired path crosses the obstacle field (see dash-dot line in Fig. 2b). The result of the replanning is shown in Fig. 2b: the ASV is able to avoid the obstacle (even though the current is pushing the ASV against the obstacle) and, at the same time, to stay as close as possible to the desired path. In order to ensure the trajectory feasibility (in terms of the obstacle collision avoidance constraint), we observe a deviation of the ASV trajectory from the desired maneuver, see Fig. 3.

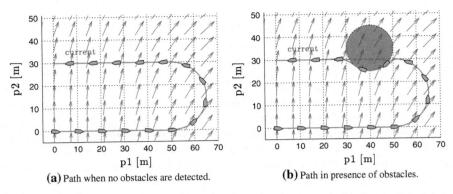

(a) Path when no obstacles are detected.

(b) Path in presence of obstacles.

Fig. 2 Lawnmower maneuver. The desired path (*dash-dot line*) and the local optimal collision-free obstacle path (*solid line*) are shown. The obstacle is depicted as *solid circle*. The *blue arrows* represent the current field

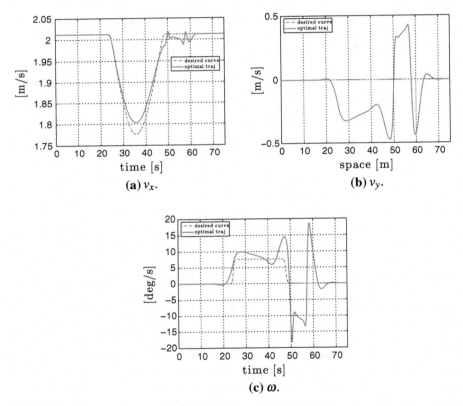

(a) v_x.

(b) v_y.

(c) ω.

Fig. 3 ASV trajectory for the lawnmower maneuver: the desired (*dash-dot line*) and the local optimal trajectory (*solid line*) are shown. **a** Longitudinal velocity. **b** Lateral velocity. **c** Yaw rate

4.2 MPC for Cooperative Path Following

In this example, to develop the CPF high-level motion controller, for simplicity we consider the underactuated kinematic model that is described in Sect. 4.1 but with

$$v^{[i]}(t) = \left[v_1^{[i]}(t), 0\right]', \quad u^{[i]}(t) := \left[v_1^{[i]}(t), \omega^{[i]}(t)\right]' \in \mathscr{U}^{[i]}, \quad v_c(t) = \left[0, 0\right]'. \tag{11}$$

Let $c^{[i]}(t) \in \mathbb{R}^2$ be a constant point in the body frame placed at a constant distance $\varepsilon_{B^{[i]}}^{[i]} \in \mathbb{R}^2$ from the center of rotation of the body frame $\{B^{[i]}\}$ of the vehicle such that

$$c^{[i]}(t) := p^{[i]}(t) + R^{[i]}(t)\varepsilon_{B^{[i]}}^{[i]}.$$

Each vehicle is associated with a differentiable desired path $c_d^{[i]} : \mathbb{R} \to \mathbb{R}^2$ parametrized by $\gamma^{[i]}(t) \in \mathbb{R}$, which is assumed together with $\|\frac{\partial}{\partial \gamma} c_d^{[i]}(\gamma)\|$ to be uniformly bounded over $\gamma^{[i]}$.

The cooperative path following problem Definition 4 can be recovered from Problem 1 by choosing as states the position $p^{[i]}(t)$ and the rotation matrix $R^{[i]}(t)$, and as output

$$y^{[i]}(t) = R^{[i]}(t)'(p^{[i]}(t) - c_d^{[i]}(\gamma^{[i]}(t))) + \varepsilon_{B^{[i]}}^{[i]} \tag{12}$$

with $t \geq t_0$. Indeed, $y^{[i]}(t) = 0$ if and only if $c_d^{[i]}(\gamma^{[i]}(t)) = c^{[i]}(\gamma^{[i]}(t))$.

In what follows, we proceed with the design of the MPC controller to satisfy the conditions of Sect. 2.2.

Consensus Law

Let $A := [a_{ij}]$ be the adjacency matrix of the communication graph $\mathscr{G}$, such that $a_{ij} > 0$ for all $j \in \mathscr{N}^{[i]}$ and $a_{ij} = 0$ otherwise. Moreover, consider the maximum degree of the network $\Delta_d := \max i(\sum_{j \neq i} a_{ij})$. Then, the consensus control law

$$k_{con}(\gamma_i, \gamma_{\mathscr{N}^{[i]}}) = \varepsilon \sum_{j \in \mathscr{N}^{[i]}} a_{ij}(\gamma_i - \gamma_j),$$

with $0 < \varepsilon < \frac{1}{\Delta_d}$ satisfies Assumption 1.

Auxiliary Control Law

First, we proceed with the design of an auxiliary control law required by Assumption 4 point (6). Consider the tracking output (12) with time derivative

$$\dot{y} = -\Omega y - R'\frac{\partial}{\partial \gamma}c_d(\gamma)\dot{\gamma} + \Delta u, \quad \Delta := \begin{pmatrix} 1 & -\varepsilon_2 \\ 0 & \varepsilon_1 \end{pmatrix}$$

and the Lyapunov-like function $W = \|y\|$. Then, for any $\varepsilon_2 \neq 0$, choosing

$$u = k_{aux}(t, x, \gamma, u_\gamma) = \begin{cases} \Delta^{-1}\left(R'\frac{\partial}{\partial \gamma}c_d(\gamma)(v_d + u_\gamma) - K\frac{y}{\|y\|}\right), & \|y\| \neq 0 \\ \Delta^{-1}R'\frac{\partial}{\partial \gamma}c_d(\gamma)(v_d + u_\gamma) & \|y\| = 0 \end{cases} \quad (13)$$

results in

$$\dot{y} = \begin{cases} -\Omega y - K\frac{y}{\|y\|}, & \|y\| \neq 0 \\ 0 & \|y\| = 0 \end{cases} \implies \dot{W} = \begin{cases} \frac{y'\dot{y}}{\|y\|} = \frac{-y'Ky}{\|y\|^2} \leq -\lambda_{min}(K), & \|y\| \neq 0 \\ 0 & \|y\| = 0, \end{cases}$$

where we used the fact that Ω is skew-symmetric and therefore $y'\Omega y = 0$ for all $y \in \mathbb{R}^2$. As a consequence, the vector y converges in finite time to the origin as follows

$$\|y(\tau)\| \leq \begin{cases} \|y(t)\| - \lambda_{min}(K)(\tau - t), & \tau \in [t, t + \frac{\|y(t)\|}{\lambda_{min}(K)}] \\ 0, & \tau > t + \frac{\|y(t)\|}{\lambda_{min}(K)} \end{cases}. \quad (14)$$

Tracking Stage Cost, Terminal Cost, and Terminal Set

Tracking stage cost. The tracking stage cost

$$l(t, x, u, \gamma, u_\gamma) = y'Qy, \quad Q \succ 0, \quad (15)$$

satisfies Assumption 4 points (3) and (5).

Tracking terminal cost. At this point let (y_{aux}, u_{aux}) be the pair of output and input trajectories of the system in closed-loop with the auxiliary control law (13), starting at the time and output pair $(\hat{t}, \hat{y})$. Then, combining (14) with (15), it follows that the associated tracking stage cost is upper bounded as

$$l(\tau, y_{aux}, u_{aux}) \leq \hat{l}(\tau; \hat{t}, \hat{y}) = \begin{cases} \lambda_{max}(Q)(\|\hat{y}\| - \lambda_{min}(K)(\tau - t))^2, & \tau \leq \hat{t} + \frac{\hat{y}}{\lambda_{min}(K)} \\ 0, & \tau > \hat{t} + \frac{\|\hat{y}\|}{\lambda_{min}(K)} \end{cases}$$

where $\hat{l}(\cdot)$ satisfies

$$\hat{l}(\tau; \hat{t} + \delta, y_{aux}(\hat{t} + \delta)) \leq \hat{l}(\tau; \hat{t}, \hat{y})$$

and $\lim_{\tau \to \infty} \hat{l}(\tau; \hat{t}, \hat{y}) = 0$ for all $\delta \geq 0$. We can now conclude by Lemma 24 of [5] that the tracking terminal cost

$$m(\hat{t}, \hat{y}) = \int_{\hat{t}}^{+\infty} \hat{l}(\tau; \hat{t}, \hat{y}) d\tau = \lambda_{max}(Q) \left[-\frac{1}{3\lambda_{min}(K)} (\|\hat{y}\| - \lambda_{min}(K)\tau)^3 \right]_0^{\frac{\|\hat{y}\|}{\lambda_{min}(K)}}$$
$$= \frac{\lambda_{max}(Q)}{3\lambda_{min}(K)} \|\hat{y}\|^3 \tag{16}$$

satisfies the terminal cost decrease of Assumption 4 - item 6.

Tracking terminal set. Next we design a terminal set to satisfy Assumption 4 - item 6. In particular, we show that for a specific selection of the input constraint set, the auxiliary law is always feasible, and therefore we can omit the tracking terminal set. From (13) we have

$$\|[u_{aux}(\tau)]_1\| \leq \|[\Delta^{-1}]_1\| \left\| \frac{\partial}{\partial \gamma} c_d(\gamma) \right\| (\|v_d\| + r_\eta) + \|[\Delta^{-1}K]_1\| =: v_{max} \tag{17a}$$

$$\|[u_{aux}(\tau)]_2\| \leq \|[\Delta^{-1}]_2\| \left\| \frac{\partial}{\partial \gamma} c_d(\gamma) \right\| (\|v_d\| + r_\eta) + \|[\Delta^{-1}K]_2\| =: \omega_{max}, \tag{17b}$$

where for a generic matrix A, the term $[A]_i$ denotes the i-*th* row of A. Moreover, the input constraint set $\mathscr{U}(t)$ is uniformly bounded over time by making

$$\{v, \omega : |v| \leq v_{max}, |\omega| \leq \omega_{max}\} \subseteq \mathscr{U}(t) \tag{18}$$

for all $t \geq t_0$ with v_{max} and ω_{max} from (17).

Thus, combining (18) with (17), the auxiliary controller is always feasible and therefore the tracking terminal set can be omitted, i.e., choosing $\mathscr{Y}_{aux}(t) = \mathbb{R}^2$ for all $t \geq t_0$. The chosen tracking terminal set, tracking terminal cost, and constraint sets satisfy Assumption 4 item 1 and 6.

Consensus stage cost. Lastly, choosing the consensus stage cost as

$$l_c(\eta, v_\eta) = \eta' Q_c \eta + v_\eta' O_c v_\eta, \quad Q_c \succ 0, \quad \lambda_{min}(Q_c) \geq 2m_\eta \lambda_\eta, \quad O_c \succeq 0 \tag{19}$$

Assumption 4 -item 4 is satisfied.

Theorem 2 *Consider a set of constrained vehicles described by* (11) *communicating according to a communication network* $\mathscr{G} = (\mathscr{V}, \mathscr{E})$ *as presented in Sect. 2.1 and the open-loop MPC problem from Definition 1 with output* (12), *tracking stage cost* (15), *consensus stage cost* (19), *auxiliary control law* (13), *tracking terminal cost* (16), *and tracking terminal set* $\mathscr{Y}_{aux}(t) = \mathbb{R}^2$. *Then, the proposed sampled-data MPC control law* (5) *solves the cooperative path following problem.* □

Proof The proof follows from Theorem 1, where the assumptions of the latter are verified throughout Sect. 4.2. ∎

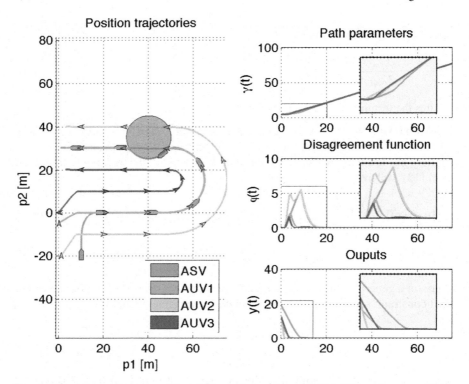

Fig. 4 Closed-loop trajectories of the vehicles with the proposed MPC-CPF motion control laws

Numerical Results

This section presents a simulated scenario where the four vehicles perform CPF and communicate in a sparse fashion with the following communication structure: all the communications between the AUVs are bidirectional, but AUV 1 and 3 only communicate with AUV 2, but not among each other. The ASV, which is in charge of bridging the communication within the underwater vehicles and a ground station, communicates bidirectionally with AUV 2.

The controllers for the vehicles are designed using Theorem 2 with $\lambda_\eta = r_\eta = m_\eta = 0.1$, $O_c = 0.01$, $\varepsilon_B = 0.5$, $K_e = 0.1 I_{2\times 2}$, $O = I_{2\times 2}$, $Q = 100 I_{2\times 2}$, where $I_{2\times 2}$ is the two by two identity matrix. Constraint (4b) is never active. Figure 4 displays the trajectories of the vehicles. It can be seen that the vehicles converge to the desired formation, with the ASV navigating over the AUV 2 and avoiding detected surface obstacles, and the AUVs moving in the desired parallel formation performing seafloor mapping.

Figure 4, right column, provides further insights on the decision-making process of the MPC controllers, which are all initialized with $\gamma(t_0) = 0$ and $\eta(t_0) = 0$. In an initial phase, the MPC controllers increase the disagreement among the path parameters to reduce the outputs y, which are associated with a higher cost in the performance index, and therefore favorite a rapid convergence of the vehicles to the paths. Then, as the vehicles approach the desired paths, the disagreement is driven to zero, and therefore the vehicles achieve the desired formation pattern.

5 Conclusion

This chapter described a sampled-data model predictive control framework for cooperative path following. Under this framework, input and output constraints as well as meaningful optimization-based performance trade-offs can be conveniently addressed. Conditions under which the MPC-CPF problem can be solved with convergence guarantees were provided. An illustrative example was presented using simplified dynamic models, but the proposed approach can be applied to more complex dynamic models subject to disturbances and communication constraints, which is the subject of ongoing research.

References

1. Aguiar, A.P., Pascoal, A.M.: Coordinated path-following control for nonlinear systems with logic-based communication. In: IEEE Conference on Decision and Control, pp. 1473–1479 (2007)
2. Aguiar, A.P., Hespanha, J.P.: Trajectory-tracking and path-following of underactuated autonomous vehicles with parametric modeling uncertainty. IEEE Trans. Autom. Control **52**, 1362–1379 (2007)
3. Aguiar, A.P., Hespanha, J.P., Kokotovic, P.V.: Performance limitations in reference tracking and path following for nonlinear systems. Automatica **44**(3), 598–610 (2008)
4. Alessandretti, A., Aguiar, A.P., Jones, C.N.: Trajectory-tracking and path-following controllers for constrained underactuated vehicles using Model Predictive Control. In: 2013 European Control Conference (ECC), pp. 1371–1376 (2013)
5. Alessandretti, A., Aguiar, P.A., Jones, C.N.: On convergence and performance certification of a continuous-time economic model predictive control scheme with time-varying performance index. Automatica **68**, 305–313 (2016)
6. Børhaug, E., Pavlov, A., Panteley, E., Pettersen, K.Y.: Straight line path following for formations of underactuated marine surface vessels. IEEE Trans. Control Syst. Technol. **19**(3), 493–506 (2011)
7. Chen, H., Allgöwer, F.: A quasi-infinite horizon nonlinear model predictive control scheme with guaranteed stability. Automatica **34**(10), 1205–1217 (1998)
8. Cichella, V., Kaminer, I., Dobrokhodov, V., Xargay, E., Choe, R., Hovakimyan, N., Aguiar, A.P., Pascoal, A.M.: Cooperative path following of multiple multirotors over time-varying networks. IEEE Trans. Autom. Sci. Eng. **12**(3), 945–957 (2015)
9. Egerstedt, M.B., Hu, X.: Formation constrained multi-agent control. IEEE Trans. Robot. Autom. **17**(6), 947–951 (2001)
10. Faulwasser, T.: Optimization-based Solutions to Constrained Trajectory-tracking and Path-following Problems. Ph.D. Thesis. Otto-von-Guericke-Universität Magdeburg (2012)
11. Fontes, F.A.C.C.: A general framework to design stabilizing nonlinear model predictive controllers. Syst. Control Lett. **42**(2), 127–143 (2001)

12. Ghabcheloo, R., Aguiar, A.P., Pascoal, A.M., Silvestre, C., Kaminer, I., Hespanha, J.P.: Coordinated path-following in the presence of communication losses and time delays. SIAM J. Control Optim. **48**(1), 234–265 (2009)
13. Hauser, J.: A projection operator approach to the optimization of trajectory functionals. In: 15th IFAC World Congress, pp. 377–382 (2002)
14. Hauser, J., Saccon, A.: A barrier function method for the optimization of trajectory functionals with constraints. In: IEEE Conference on Decision and Control, pp. 864–869 (2006)
15. Ihle, I.A.F., Arcak, M., Fossen, T.I.: Passivity-based designs for synchronized path-following. Automatica **43**(9), 1508–1518 (2007)
16. Rawlings, J.B., Mayne, D.Q.: Model Predictive Control: Theory and Design. Nob. Hill Publishing, Madison (2009)
17. Rucco, A., Aguiar, A.P., Fontes, F., Pereira, F.L., Sousa, J.: A Model Predictive Control-Based Architecture for Cooperative Path-Following of Multiple Unmanned Aerial Vehicles. Developments in Model-Based Optimization and Control, pp. 141–160. Springer, Berlin (2015)
18. Saccon, A., Aguiar, A.P., Bayer, F A., Hausler, A.J., Notarstefano, G., Pascoal, A.M., Rucco, A., Hauser, J.: Constrained Optimal Motion Planning for Autonomous Vehicles using PRONTO In Sensing and Control for Autonomous Vehicles: Applications to Land, Water and Air Vehicles. In: Fossen, T.I., Pettersen, K.Y., Nijmejier, H. (eds.) (2017)
19. Skjetne, R., Moi, S., Fossen, T.I.: Nonlinear formation control of marine craft. In: IEEE Conference on Decision and Control, vol. 2, pp. 1699–1704, Las Vegas, NV (2002)
20. Xargay, E., Kaminer, I., Pascoal, A., Hovakimyan, N., Dobrokhodov, V., Cichella, V., Aguiar, A.P., Ghabcheloo, R.: Time-critical cooperative path following of multiple UAVs over time-varying networks. AIAA J. Guid. Control Dyn. **36**(2), 499–516 (2013)

Coordinated Control of Mobile Robots with Delay Compensation Based on Synchronization

Yiran Cao and Toshiki Oguchi

Abstract This chapter considers a consensus control problem for two-wheel mobile robots with input time delay. To solve the problem, an anticipating synchronization-based state predictor is applied to each robot. First, we propose a consensus controller with an angle predictor to compensate the effect of time delay in order to apply feedback linearisation. Then, a consensus condition for this controller is derived and investigated. Extending this idea, a controller with a full state predictor is given and a sufficient condition for consensus is provided. Finally, an example of formation control using the proposed controller for a group of two-wheel mobile robots is given to illustrate the usefulness of the proposed control scheme and the validity of the derived condition.

1 Introduction

In many applications, such as environmental monitoring, security and surveillance, scientific exploration, and intelligent transportation, to organise multiple agents to accomplish multiple tasks through cooperation is an attracting research field for researchers in recent years. The growing demand for autonomous multi-agent networks has stimulated a broad interest in formation control [1–5], flocking control [6–8] and tracking control [9–11]. These control algorithms are based on local information exchange generally performed using communication channels between agents according to their interconnection graph topology or using sensors to measure the information of other agents. Since there exist communication, measurement and computation, the existence of time lags is inevitable in real applications. Because the existence of time delay might degrade the system performance or even destroy the

Y. Cao (✉) · T. Oguchi
Department of Mechanical Engineering, Graduate School of Science and Engineering, Tokyo Metropolitan University, 1-1, Minami-osawa, Hachioji-shi, Tokyo 192-0397, Japan
e-mail: cao-yiran@ed.tmu.ac.jp

T. Oguchi
e-mail: t.oguchi@tmu.ac.jp

stability, it is significant to design a suitable control strategy for this kind of systems to decrease the effect of time delay.

Several studies [12–18] have focused on the consensus problem on multi-agent systems (MASs) with time delay. In [12, 13], an upper bound of the allowable input time delay is given under which consensus can be achieved for networks of dynamic agents with a fixed, undirected and connected graph topology. The paper [14] derives a consensus condition for linearly coupled networks with time-varying delays, and [15] deals with the average-consensus problem for MAS with non-uniform and asymmetric time delays. The article [16] investigates the rightmost eigenvalue behaviour associated with the network topology and the coupling strength for MAS with time delay and show the simulation and experimental results by using two-wheel mobile robot in [17], and [18] investigates the communication delays that can satisfy the consensus properties of MAS endowed with nonlinear dynamics. These studies fundamentally focused on analyses of the effect of communication delays for consensus control.

A different direction to solve the consensus problem with communication delay is to compensate the effect of time delay caused in communication by using a delay compensator at each agent or a central controller. The paper [19] proposes a controller design method for nonlinear systems with time delays by a state feedback and a state predictor based on synchronization of coupled systems. By using the control scheme with synchronization-based state predictor, [20, 21] solve the tracking control problem of a two-wheel mobile robot with communication time delay. In our previous works [22, 23], we proposed a consensus controller combined with a state predictor proposed in [19] based on anticipating synchronization for integrator systems. The usefulness of the proposed scheme was examined through numerical simulations and experiments of two-wheel mobile robots. Since the two-wheel mobile robot is subject to a nonholonomic constraint, to reduce the dynamics of each two-wheel mobile robot to an integrator system, each system had to have a local controller to realize input–output linearization besides the consensus controller. In this chapter, we consider a consensus problem for two-wheel mobile robots with time delays without using such a local controller. To compensate the effect of time delay in communication, we introduce a full state synchronization-based predictor and apply a consensus controller combined with the predictor to multiple two-wheel mobile robots. We derive a sufficient condition for a global consensus of the MASs in directed graph networks.

The rest of this chapter is organized as follows. In Sect. 2, we briefly introduce the coordinated consensus problem of multi-robot systems and the basic notations used in this chapter. Then, we propose a steering angle predictor-based controller and derive a global consensus condition for the system in Sect. 3. After that, we propose a controller with a full state predictor and derive a corresponding globally sufficient consensus condition. A numerical simulation of formation control for a group of two-wheel mobile robots with the full state predictor-based controller is given in Sect. 4, which satisfies the consensus condition and the formation is finally achieved. At the end, conclusions are given in Sect. 5.

2 Problem Formulation

Considering the two-wheel mobile robot shown in Fig. 1, the kinematic model of the point O^i on mobile robot i is expressed as

$$\begin{bmatrix} \dot{x}_i(t) \\ \dot{y}_i(t) \\ \dot{\theta}_i(t) \end{bmatrix} = \begin{bmatrix} \cos\theta_i(t) & -R\sin\theta_i(t) \\ \sin\theta_i(t) & R\cos\theta_i(t) \\ 0 & 1 \end{bmatrix} \begin{bmatrix} v_i(t) \\ \omega_i(t) \end{bmatrix}, \qquad (1)$$

$$\boldsymbol{B}(\theta_i(t)) := \begin{bmatrix} \cos\theta_i(t) & -R\sin\theta_i(t) \\ \sin\theta_i(t) & R\cos\theta_i(t) \end{bmatrix}$$

for $i = 1, \ldots, N$. Vector $\boldsymbol{\xi}_i(t) := [x_i(t), y_i(t)]^T \in \mathbb{R}^2$ denotes the coordinates of the point O^i on the edge of the mobile robot. $\theta_i(t) \in \mathbb{R}$ is the angle between the direction of the velocity $v_i(t)$ and the x-axis at t. $\omega_i(t)$ is the angle velocity and $R \in \mathbb{R}^+$ is the radius of the mobile robot.

Then we consider a group of robots forming a communication network as a graph $\mathcal{G}$ and each mobile robot as an agent in this graph. $L(\mathcal{G})$ is the graph Laplacian of graph $\mathcal{G}$. If the information communication between agents i and j is bidirectional, the graph $\mathcal{G}$ is undirected; If the communication is unidirectional, the graph is called a directed graph. The Laplacian $L(\mathcal{G})$ has the following entries

$$\ell_{ij} = \begin{cases} -1 & \text{if } j \in \mathcal{N}_i \\ 0 & \text{if } j \notin \mathcal{N}_i \text{ and } j \neq i \\ |\mathcal{N}_i| & \text{if } j = i \end{cases}, \qquad (2)$$

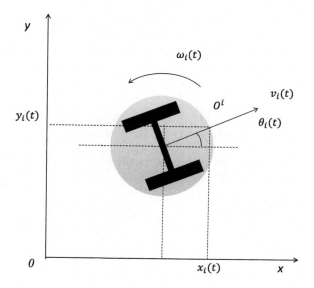

Fig. 1 The kinematic model of mobile robot i. The coordinates (x_i, y_i) and angle θ_i of point O^i are the states

where $\mathcal{N}_i$ denotes the set of agents that have a directed connection to agent i in graph $\mathcal{G}$ and $|\mathcal{N}_i|$ denotes the cardinality of set $\mathcal{N}_i$.

Assumption 1 Let the communication network of mobile robots with dynamics (1) be represented as a directed graph which contains a directed spanning tree.

Remark 1 A directed graph is said to have a spanning tree if there exists an agent such that it has a directed path to all of the other agents. If a graph contains a directed spanning tree, the corresponding graph Laplacian $L(\mathcal{G})$ always has one zero eigenvalue and $N-1$ eigenvalues have positive real parts cf. [1]. Especially, if a graph is an undirected and connected graph, the corresponding graph Laplacian $L(\mathcal{G})$ always has one zero eigenvalue and $N-1$ positive real eigenvalues.

Remark 2 For a graph $\mathcal{G}$ satisfying Assumption 1, matrix Q is given as

$$M_0 L(\mathcal{G}) M_0^{-1} = \begin{bmatrix} 0 & \mathbf{1}_{N-1}^T \\ \hline 0 & ML(\mathcal{G})M^+ \end{bmatrix} = \begin{bmatrix} 0 & \mathbf{1}_{N-1}^T \\ \hline 0 & Q \end{bmatrix}$$

where

$$M_0 = \begin{bmatrix} 1 & 1 & \cdots & 1 \\ 1 & -1 & & 0 \\ \vdots & & \ddots & \\ 1 & 0 & & -1 \end{bmatrix} = \begin{bmatrix} \mathbf{1}_N^T \\ \hline M \end{bmatrix},$$

$$M_0^{-1} = \begin{bmatrix} \frac{1}{N} & \frac{1}{N} & \frac{1}{N} & \cdots & \frac{1}{N} \\ \frac{1}{N} & -\frac{N-1}{N} & \frac{1}{N} & \cdots & \frac{1}{N} \\ \vdots & \frac{1}{N} & -\frac{N-1}{N} & & \vdots \\ & & & \ddots & \frac{1}{N} \\ \frac{1}{N} & \frac{1}{N} & & \cdots & -\frac{N-1}{N} \end{bmatrix} = \begin{bmatrix} \frac{1}{N} \\ \vdots & M^+ \\ \frac{1}{N} \end{bmatrix}.$$

Therefore, $Q = ML(\mathcal{G})M^+$ has $N-1$ eigenvalues with positive real parts and it equivalently means $-Q$ is a Hurwitz matrix.

For this multi-robot system, the consensus problem to be considered here is formulated as follows.

Definition 1 (*Coordinate Consensus Problem*)
For a multi-robot system (1), find a control protocol which makes $\xi_i(t) - \xi_j(t) = 0$ for all $i, j \in \{1, \ldots, N\}$ as $t \to \infty$. This problem is called the coordinate consensus problem.

The coordinate consensus problem is called rendezvous in some studies [1, 24]. The relative-position-based formation control problem can be transformed into a coordinate consensus problem which is given as an example in Sect. 4.

The following notations are used throughout this chapter. $\|\cdot\|$ denotes the Euclidean vector norm in $\mathbb{R}^m$ and its induced norm on matrices which is defined $\|A\| = \sqrt{\lambda_{max}(A^T A)}$ as the norm of matrix $A \in \mathbb{R}^{m \times n}$ where $\lambda_{max}(\cdot)$ denotes the maximum eigenvalue of the corresponding matrix. $\lambda_{min}(\cdot)$ denotes the minimum eigenvalue of the corresponding matrix. The induced norm satisfies the inequality $\|AB\| \le \|A\| \|B\|$ for matrices $A^{m \times n}$ and $B^{n \times s}$. $\|\phi\|_c$ is the continuous norm defined by $\max_{a \le s \le b} \|\phi(s)\|$ for $\phi \in \mathscr{C}([a,b], \mathbb{R}^N)$. $|\cdot|$ denotes the absolute value and $\otimes$ denotes the Kronecker product.

3 Predictor-Based Controller

In this section, we propose the control scheme for the multi-robot system with input time delay. In real applications, since the mobile robots cooperate with each other through network, there exist time delays in information communication or states acquisition. Based on this background, we assume there exists a constant time delay $\ell \in \mathbb{R}^+$ for each input.

3.1 Predictor for Steering Angle

We assume that each robot given by (1) has input time delay ℓ. The dynamics (1) with a delayed input vector can be written as follows.

$$\begin{bmatrix} \dot{\xi}_i(t) \\ \dot{\theta}_i(t) \end{bmatrix} = \begin{bmatrix} B(\theta_i(t)) \\ \hdashline 0 \quad 1 \end{bmatrix} \begin{bmatrix} v_i(t-\ell) \\ \omega_i(t-\ell) \end{bmatrix} \quad (3)$$

Defining $u_i(t) = [v_i(t), \omega_i(t)]^T$, the dynamics (3) can be written as follows.

$$\begin{cases} \dot{\xi}_i(t) = B(\theta_i(t)) u_i(t-\ell) \\ \dot{\theta}_i(t) = \omega_i(t-\ell) \end{cases}, \quad (4)$$

where ξ_i is considered as the output vector. Since $\det(B(\theta_i(t))) = R$, $B(\theta_i(t))$ is invertible for any $\theta_i(t)$. Then each system achieves input–output linearisation by applying $u_i(t) = B^{-1}(\theta_i(t)) w_i(t)$, if there is no delay at the input, and the consensus problem can be reduced to the one for integrator systems. However, due to the existence of the input time delay, we need the future value of $\theta_i(t)$ to achieve linearisation. Therefore, to estimate the future value of $\theta_i(t)$, the predictor for angle $\theta_i(t)$ is stated as

$$\dot{\hat{\theta}}_i(t) = \omega_i(t) - k_p(\hat{\theta}_i(t-\ell) - \theta_i(t)), \tag{5}$$

where $\hat{\theta}_i(t)$ is the predicted angle and $k_p \in \mathbb{R}^+$ is the prediction gain. This predictor is based on anticipating synchronization [19], which is a kind of master–slave synchronization. The total dynamics is constituted by the nominal systems and the correction term of the difference of the system output and the delayed predictor output. Here, if $\hat{\theta}_i(t-\ell)$ asymptotically converges to $\theta_i(t)$, it means that $\hat{\theta}_i(t)$ converges to $\theta_i(t+\ell)$. Thus the predictor (5) can estimate the future value of angle $\theta_i(t)$. By combining the feedback linearisation technique with the predictor for angle, the control algorithm can be given as

$$\boldsymbol{u}_i(t) = -k\boldsymbol{B}^{-1}(\hat{\theta}_i(t))\Sigma_{j\in\mathcal{N}_i}(\boldsymbol{\xi}_i(t) - \boldsymbol{\xi}_j(t)). \tag{6}$$

Then, the total system for N mobile robots can be written as

$$\begin{cases} \dot{\boldsymbol{\xi}}(t) = \bar{\boldsymbol{B}}(\boldsymbol{\theta}(t))\boldsymbol{u}(t-\ell) \\ \dot{\boldsymbol{\theta}}(t) = \boldsymbol{\omega}(t-\ell) \\ \boldsymbol{u}(t) = -k\bar{\boldsymbol{B}}^{-1}(\hat{\boldsymbol{\theta}}(t))(L(\mathcal{G}) \otimes \boldsymbol{I}_2)\boldsymbol{\xi}(t) \\ \dot{\hat{\boldsymbol{\theta}}}(t) = \boldsymbol{\omega}(t) - k_p(\hat{\boldsymbol{\theta}}(t-\ell) - \boldsymbol{\theta}(t)), \end{cases} \tag{7}$$

where

$$\bar{\boldsymbol{B}}(\boldsymbol{\theta}(t)) = \begin{bmatrix} \boldsymbol{B}(\theta_1(t)) & \cdots & 0 \\ & \ddots & \\ 0 & \cdots & \boldsymbol{B}(\theta_N(t)) \end{bmatrix},$$

$\boldsymbol{\theta}(t) = [\theta_1(t), \ldots, \theta_N(t)]^T$ and $\hat{\boldsymbol{\theta}}(t) = [\hat{\theta}_1(t), \ldots, \hat{\theta}_N(t)]^T$.

The prediction error is defined as

$$\hat{\boldsymbol{e}}_\theta(t) = \hat{\boldsymbol{\theta}}(t-\ell) - \boldsymbol{\theta}(t) \tag{8}$$

where $\hat{\boldsymbol{e}}_\theta(t) = [\hat{e}_{\theta 1}(t), \ldots, \hat{e}_{\theta N}(t)]^T$. The prediction error dynamics for $\hat{\boldsymbol{e}}_\theta(t)$ is given as

$$\dot{\hat{\boldsymbol{e}}}_\theta(t) = -k_p\hat{\boldsymbol{e}}_\theta(t-\ell), \tag{9}$$

while the output synchronization error of each robot is summarized as

$$\boldsymbol{e}(t) = \begin{bmatrix} \boldsymbol{\xi}_1(t) - \boldsymbol{\xi}_2(t) \\ \vdots \\ \boldsymbol{\xi}_1(t) - \boldsymbol{\xi}_N(t) \end{bmatrix} = (\boldsymbol{M} \otimes \boldsymbol{I}_2)\boldsymbol{\xi}(t). \tag{10}$$

The dynamics is derived as

$$\begin{aligned}
\dot{e}(t) &= (M \otimes I_2)\dot{\xi}(t) \\
&= (M \otimes I_2)\bar{B}(\theta(t))u(t-\ell) \\
&= -k(M \otimes I_2)\bar{B}(\theta(t))\bar{B}^{-1}(\hat{\theta}(t-\ell))(L(\mathcal{G}) \otimes I_2)\xi(t-\ell) \\
&= -k(M \otimes I_2)\bar{A}(\hat{e}_\theta(t))(L(\mathcal{G})M^+ \otimes I_2)e(t-\ell),
\end{aligned} \quad (11)$$

where $M^+ \in \mathbb{R}^{N \times (N-1)}$ denotes the Moore–Penrose pseudo-inverse matrix of M,

$$\bar{A}(\hat{e}_\theta(t)) = \bar{B}(\theta(t))\bar{B}^{-1}(\hat{\theta}(t-\ell)) = \begin{bmatrix} A(\hat{e}_{\theta 1}(t)) & \cdots & 0 \\ & \ddots & \\ 0 & \cdots & A(\hat{e}_{\theta N}(t)) \end{bmatrix}$$

and

$$A(\hat{e}_{\theta i}(t)) = \begin{bmatrix} \cos \hat{e}_{\theta i}(t) & \sin \hat{e}_{\theta i}(t) \\ -\sin \hat{e}_{\theta i}(t) & \cos \hat{e}_{\theta i}(t) \end{bmatrix}.$$

The dynamics (11) can be equivalently written as

$$\dot{e}(t) = -k\bar{Q}e(t-\ell) + g(t, e, \hat{e}_\theta)$$

where $g(t, e, \hat{e}_\theta)$ is the difference of the right term of (11) and $-k\bar{Q}e(t-\ell)$. It is defined as $g(t, e, \hat{e}_\theta) := -k((M \otimes I_2)(\bar{A}(\hat{e}_\theta(t)) - I_{2N})(L(\mathcal{G})M^+ \otimes I_2))e(t-\ell)$ and $\bar{Q}$ is defined as $\bar{Q} = (ML(\mathcal{G})M^+) \otimes I_2$. Note that if the angle prediction error $\hat{e}_\theta(t)$ satisfies $\hat{e}_\theta(t) = 0$, matrix $\bar{A}(0) = I_{2N}$ and $g(t, e, 0) = 0$.

Reducing the consensus problem to a stability problem of both the prediction error and the synchronization error dynamics (9) and (11), we consider to derive a consensus condition.

The prediction error dynamics (9) and synchronization error dynamics (11) are summarised as

$$\dot{\hat{e}}_\theta(t) = -k_p \hat{e}_\theta(t-\ell) \quad (12a)$$

$$\dot{e}(t) = -k\bar{Q}e(t-\ell) + g(t, e, \hat{e}_\theta). \quad (12b)$$

Considering the stability of (12), we obtain the following sufficient condition for the consensus problem.

Theorem 1 *Assume that the multi-robot system (7) satisfies Assumption 1. If, for a given time delay $\ell > 0$, there exit $k_p > 0$ satisfying*

$$0 < k_p \ell < \frac{\pi}{2}, \quad (13)$$

$k > 0$ and symmetric matrices $P > 0$, $W_1 > 0$ satisfying the following inequalities

$$P\bar{Q} + \bar{Q}^T P - W_1 \geq 0 \qquad (14)$$

$$2k\ell\sqrt{\frac{\lambda_{max}(P)}{\lambda_{min}(P)}}\|\bar{Q}\|^2\|P\| < \lambda_{min}(W_1), \qquad (15)$$

the origin of equation (12) is globally asymptotically stable and then the system (7) achieves consensus globally.

Following the below procedures, we can obtain suitable values for k, k_p, P and W_1 for a given time delay ℓ:

Algorithm 1

Step 1. For a given time delay ℓ, determine a prediction gain k_p satisfying the inequality $0 < k_p < \frac{\pi}{2\ell}$.

Step 2. Determine a positive number p and solve the LMI $P\bar{Q} + \bar{Q}^T P - pI_{2N-2} \geq 0$ with respect to P. Then, set $W_1 = P\bar{Q} + \bar{Q}^T P$.

Step 3. Determine the value of k satisfying the inequality (15).

3.2 A Consensus Controller with Full State Predictor

In the previous subsection, we introduced a predictor to estimate the future value of angle θ and applied the feedback linearisation technique to remove the nonlinearity in the input–output relationship. In this subsection, we propose a full state predictor based on synchronization in order to compensate for the effect of time delay. The controller is given as

$$u(t) = -k\bar{B}^{-1}(\hat{\theta}(t))(L(\mathcal{G}) \otimes I_2)\hat{\xi}(t) \qquad (16)$$

for $i = 1, \ldots, N$. Here $\hat{\xi}(t) = [[\hat{x}_1(t), \hat{y}_1(t)]^T, \ldots, [\hat{x}_N(t), \hat{y}_N(t)]^T]^T$ denotes the predicted coordinates of the system. The full state predictor is designed as follows.

$$\begin{cases} \dot{\hat{\xi}}(t) = \bar{B}(\hat{\theta}(t))u(t) - k_{p1}(\hat{\xi}(t-\ell) - \xi(t)) \\ \dot{\hat{\theta}}(t) = \omega(t) - k_{p2}(\hat{\theta}(t-\ell) - \theta(t)). \end{cases} \qquad (17)$$

k_{p1} and $k_{p2} \in \mathbb{R}^+$ are prediction gains for $\xi(t)$ and $\theta(t)$, respectively. The initial condition of the predictor is given as $[\hat{x}(t), \hat{y}(t)]^T = [\hat{\psi}_x(t), \hat{\psi}_y(t)]^T$ ($-\ell \leq t < 0$)

where $\hat{\psi}_x(t), \hat{\psi}_y(t) \in \mathscr{C}([-\ell, 0], \mathbb{R}^N)$ and $\hat{\boldsymbol{\theta}}(t) = \hat{\boldsymbol{\phi}}(t)$ ($-\ell \leq t < 0$) where $\hat{\boldsymbol{\phi}}(t) \in \mathscr{C}([-\ell, 0], \mathbb{R}^N)$.

Thus the dynamics of the total system can be summarized as

$$\begin{cases} \dot{\boldsymbol{\xi}}(t) = \bar{\boldsymbol{B}}(\boldsymbol{\theta}(t))\boldsymbol{u}(t-\ell) \\ \dot{\boldsymbol{\theta}}(t) = \boldsymbol{\omega}(t-\ell) \\ \boldsymbol{u}(t) = -k\bar{\boldsymbol{B}}^{-1}(\hat{\boldsymbol{\theta}}(t))(L(\mathscr{G}) \otimes \boldsymbol{I}_2)\hat{\boldsymbol{\xi}}(t) \\ \dot{\hat{\boldsymbol{\xi}}}(t) = \bar{\boldsymbol{B}}(\hat{\boldsymbol{\theta}}(t))\boldsymbol{u}(t) - k_{p1}(\hat{\boldsymbol{\xi}}(t-\ell) - \boldsymbol{\xi}(t)) \\ \dot{\hat{\boldsymbol{\theta}}}(t) = \boldsymbol{\omega}(t) - k_{p2}(\hat{\boldsymbol{\theta}}(t-\ell) - \boldsymbol{\theta}(t)). \end{cases} \quad (18)$$

Next we derive a consensus condition for the above system. The prediction errors of all states should converge to 0. The prediction error of the predicted state $\hat{\boldsymbol{\xi}}$ is defined as

$$\hat{\boldsymbol{e}}(t) = \hat{\boldsymbol{\xi}}(t-\ell) - \boldsymbol{\xi}(t). \quad (19)$$

The synchronization error of this system is given by Eq. (10). The synchronization error dynamics for system (18) is given as

$$\begin{aligned} \dot{\boldsymbol{e}}(t) &= (\boldsymbol{M} \otimes \boldsymbol{I}_2)\dot{\boldsymbol{\xi}}(t) \\ &= -k(\boldsymbol{M} \otimes \boldsymbol{I}_2)\bar{\boldsymbol{B}}(\boldsymbol{\theta}(t))\bar{\boldsymbol{B}}^{-1}(\hat{\boldsymbol{\theta}}(t-\ell))(L(\mathscr{G}) \otimes \boldsymbol{I}_2)\hat{\boldsymbol{\xi}}(t-\ell) \\ &= -k(\boldsymbol{M} \otimes \boldsymbol{I}_2)\bar{\boldsymbol{A}}(\hat{\boldsymbol{e}}_\theta(t))(L(\mathscr{G}) \otimes \boldsymbol{I}_2)(\hat{\boldsymbol{e}}(t) + (\boldsymbol{M}^+ \otimes \boldsymbol{I}_2)\boldsymbol{e}(t)). \end{aligned} \quad (20)$$

The prediction error dynamics for $\hat{\boldsymbol{e}}(t)$ can be derived as

$$\begin{aligned} \dot{\hat{\boldsymbol{e}}}(t) &= \dot{\hat{\boldsymbol{\xi}}}(t-\ell) - \dot{\boldsymbol{\xi}}(t) \\ &= (\bar{\boldsymbol{B}}(\hat{\boldsymbol{\theta}}(t-\ell)) - \bar{\boldsymbol{B}}(\boldsymbol{\theta}(t)))\boldsymbol{u}(t-\ell) - k_{p1}(\hat{\boldsymbol{\xi}}(t-2\ell) - \boldsymbol{\xi}(t-\ell)) \\ &= -k(\boldsymbol{I}_{2N} - \bar{\boldsymbol{A}}(\hat{\boldsymbol{e}}_\theta(t)))(L(\mathscr{G}) \otimes \boldsymbol{I}_2)\hat{\boldsymbol{\xi}}(t-\ell) - k_{p1}\hat{\boldsymbol{e}}(t-\ell) \\ &= -k(\boldsymbol{I}_{2N} - \bar{\boldsymbol{A}}(\hat{\boldsymbol{e}}_\theta(t)))(L(\mathscr{G}) \otimes \boldsymbol{I}_2)(\hat{\boldsymbol{e}}(t) + (\boldsymbol{M}^+ \otimes \boldsymbol{I}_2)\boldsymbol{e}(t)) - k_{p1}\hat{\boldsymbol{e}}(t-\ell). \end{aligned} \quad (21)$$

The prediction error for predicted $\hat{\theta}$ is given in Eq. (8) and the corresponding error dynamics for $\hat{\boldsymbol{e}}_\theta(t)$ of system (18) can be derived as

$$\dot{\hat{\boldsymbol{e}}}_\theta(t) = -k_{p2}\hat{\boldsymbol{e}}_\theta(t-\ell). \quad (22)$$

The error dynamics (20)–(22) can be equivalently rewritten as

$$\dot{\boldsymbol{e}}(t) = -k(\boldsymbol{M}L(\mathscr{G}) \otimes \boldsymbol{I}_2)(\hat{\boldsymbol{e}}(t) + (\boldsymbol{M}^+ \otimes \boldsymbol{I}_2)\boldsymbol{e}(t)) + g_1(t, \hat{\boldsymbol{e}}_\theta, \hat{\boldsymbol{e}}, \boldsymbol{e}) \quad (23a)$$
$$\dot{\hat{\boldsymbol{e}}}(t) = -k_{p1}\hat{\boldsymbol{e}}(t-\ell) + g_2(t, \hat{\boldsymbol{e}}_\theta, \hat{\boldsymbol{e}}, \boldsymbol{e}) \quad (23b)$$
$$\dot{\hat{\boldsymbol{e}}}_\theta(t) = -k_{p2}\hat{\boldsymbol{e}}_\theta(t-\ell) \quad (23c)$$

where we define $g_1(t, \hat{e}_\theta, \hat{e}, e) := k(M \otimes I_2)(I_{2N} - \bar{A}(\hat{e}_\theta(t)))(L(\mathscr{G}) \otimes I_2)(\hat{e}(t) + (M^+ \otimes I_2)e(t))$ and $g_2(t, \hat{e}_\theta, \hat{e}, e) := -k(I_{2N} - \bar{A}(\hat{e}_\theta(t)))(L(\mathscr{G}) \otimes I_2)(\hat{e}(t) + (M^+ \otimes I_2)e(t))$.

For dynamics (23a) and (23b), we can rewrite them in matrix form as follows.

$$\begin{bmatrix} \dot{e}(t) \\ \dot{\hat{e}}(t) \end{bmatrix} = \begin{bmatrix} -k\bar{Q} & -k(ML(\mathscr{G}) \otimes I_2) \\ 0 & 0 \end{bmatrix} \begin{bmatrix} e(t) \\ \hat{e}(t) \end{bmatrix} + \begin{bmatrix} 0 & 0 \\ 0 & -k_{p1}I_{2N} \end{bmatrix} \begin{bmatrix} e(t-\ell) \\ \hat{e}(t-\ell) \end{bmatrix} \\ + \begin{bmatrix} g_1(t, \hat{e}_\theta, \hat{e}, e) \\ g_2(t, \hat{e}_\theta, \hat{e}, e) \end{bmatrix}. \tag{24}$$

For simplicity, we rewrite Eq. (24) without the last term as

$$\dot{E}(t) = -C_1 E(t) - C_2 E(t - \ell), \tag{25}$$

where $E(t) = [e(t), \hat{e}(t)]^T$,

$$C_1 = \begin{bmatrix} k\bar{Q} & k(ML(\mathscr{G}) \otimes I_2) \\ 0 & 0 \end{bmatrix}$$

and

$$C_2 = \begin{bmatrix} 0 & 0 \\ 0 & k_{p1}I_{2N} \end{bmatrix}.$$

In a similar way to the proof of Theorem 1, we can derive a global stability condition for the error dynamics (20)–(22).

Theorem 2 *Assume that the multi-robot system (18) satisfies Assumption 1. If, for a given time delay $\ell > 0$, there exist $k_{p2} > 0$ satisfying*

$$0 < k_{p2}\ell < \frac{\pi}{2}, \tag{26}$$

$k, k_{p1} > 0$ *and symmetric matrices $P_1 > 0$, $W_3 > 0$ satisfying the following inequalities*

$$P_1(C_1 + C_2) + (C_1 + C_2)^T P_1 - W_3 \geq 0 \tag{27}$$

$$2k_{p1}\ell(k\|\bar{Q}\| + k_{p1})\sqrt{\frac{\lambda_{max}(P_1)}{\lambda_{min}(P_1)}}\|P_1\| < \lambda_{min}(W_3), \tag{28}$$

the origins of Eqs. (20)–(22) are globally asymptotically stable. Then the system(18) achieves consensus.

Following the below procedures, we can obtain k, k_{p1}, k_{p2}, P_1 and W_3 satisfying inequalities (26)–(28) for a given time delay ℓ.

Algorithm 2

Step 1. For a given time delay ℓ, determine the prediction gain k_{p2} satisfying the inequality $0 < k_{p2} < \frac{\pi}{2\ell}$.
Step 2. Choose candidates for the parameters k and k_{p1}.
Step 3. Determine a positive number p_1 and solve the LMI $P_1(C_1 + C_2) + (C_1 + C_2)^T P_1 - p_1 I_{4N-2} \geq 0$ with respect to P_1. Then, set $W_3 = P_1(C_1 + C_2) + (C_1 + C_2)^T P_1$.
Step 4. Evaluate the inequality (28). If the inequality (28) holds, this system with these parameter settings can achieve consensus. If not, return to **Step 2**.

4 Example

In this section, we attempt to apply the proposed controllers to the formation control of two-wheeled robots. As mentioned in [13], the relative-position-based formation control is a sort of consensus control problem. The control protocol is given as

$$u_i(t) = -k\bar{B}^{-1}(\hat{\theta}_i(t)) \sum_{j \in \mathcal{N}_i} (\hat{\xi}_i(t) - \hat{\xi}_j(t) - r_{ij}) \tag{29}$$

where $r_{ij} = [r_{xij}, r_{yij}]^T \in \mathbb{R}^2$ is the desired relative constant horizontal and vertical distances between robots i and j, i.e. $r_{ij} = r_i - r_j$. By applying the coordinate transformation $\hat{z}_i = \hat{\xi}_i - r_i$ and $z_i = \xi_i - r_i$ for $i = 1, \ldots, N$, the dynamics of the total system can be summarized as

$$\begin{cases} \dot{z}(t) = \bar{B}(\theta(t))u(t - \ell) \\ \dot{\theta}(t) = \omega(t - \ell) \\ u(t) = -k\bar{B}^{-1}(\hat{\theta}(t))(L(\mathcal{G}) \otimes I_2)\hat{z}(t) \\ \dot{\hat{z}}(t) = \bar{B}(\hat{\theta}(t))u(t) - k_{p1}(\hat{z}(t - \ell) - z(t)) \\ \dot{\hat{\theta}}(t) = \omega(t) - k_{p2}(\hat{\theta}(t - \ell) - \theta(t)). \end{cases} \tag{30}$$

From (30), we know that r_{ij} is independent of the stability of this system and Theorem 2 holds for this formation problem.

Figure 2 shows the formation pattern includes four robots ($R = 50$ mm). In this figure, the weight on each edge means the desired relative distance where $r = 200$ mm, and each desired relative positions is defined as $r_{12} = [r, 0]^T$, $r_{13} = [r, r]^T$, $r_{14} = [0, r]^T$, respectively. The graph Laplacian $L(\mathcal{G})$ for four robots is

$$L(\mathscr{G}) = \begin{bmatrix} 0 & 0 & 0 & 0 \\ -1 & 1 & 0 & 0 \\ -1 & 0 & 1 & 0 \\ -1 & 0 & 0 & 1 \end{bmatrix} \tag{31}$$

and the corresponding eigenvalues are $\lambda_0 = 0, \lambda_1 = \lambda_2 = \lambda_3 = 1$. The input time delay is set as $\ell = 1\,[\mathrm{s}]$, predictor gain $k_{p1} = 0.05$, $k_{p2} = 0.2$ and the coupling strength $k = 0.05$. From Theorem 2, we obtain matrices $\boldsymbol{P}_1$ and $\boldsymbol{W}_3$ satisfying the condition (27) and (28) as

$$\boldsymbol{P}_1 = \begin{bmatrix} 47.1639 & -6.4077 & -6.4077 & 3.0709 & -3.4606 & 0.1949 & 0.1949 \\ -6.4077 & 47.1639 & -6.4077 & 3.0709 & 0.1949 & -3.4606 & 0.1949 \\ -6.4077 & -6.4077 & 47.1639 & 3.0709 & 0.1949 & 0.1949 & -3.4606 \\ 3.0709 & 3.0709 & 3.0709 & 62.7615 & 0.7996 & 0.7996 & 0.7996 \\ -3.4606 & 0.1949 & 0.1949 & 0.7996 & 64.0377 & 0.1615 & 0.1615 \\ 0.1949 & -3.4606 & 0.1949 & 0.7996 & 0.1615 & 64.0377 & 0.1615 \\ 0.1949 & 0.1949 & -3.4606 & 0.7996 & 0.1615 & 0.1615 & 64.0377 \end{bmatrix} \otimes \boldsymbol{I}_2,$$

and

$$\boldsymbol{W}_3 = \begin{bmatrix} 4.7164 & -0.6408 & -0.6408 & 2.0245 & -2.7043 & 0.3399 & 0.3399 \\ -0.6408 & 4.7164 & -0.6408 & 2.0245 & 0.3399 & -2.7043 & 0.3399 \\ -0.6408 & -0.6408 & 4.7164 & 2.0245 & 0.3399 & 0.3399 & -2.7043 \\ 2.0245 & 2.0245 & 2.0245 & 7.1974 & -0.2271 & -0.2271 & -0.2271 \\ -2.7043 & 0.3399 & 0.3399 & -0.2271 & 6.7498 & -0.0033 & -0.0033 \\ 0.3399 & -2.7043 & 0.3399 & -0.2271 & -0.0033 & 6.7498 & -0.0033 \\ 0.3399 & 0.3399 & -2.7043 & -0.2271 & -0.0033 & -0.0033 & 6.7498 \end{bmatrix} \otimes \boldsymbol{I}_2.$$

The initial state of each robot $(\boldsymbol{x}_i(0), \boldsymbol{y}_i(0), \boldsymbol{\theta}_i(0))$ is set as $(0.2, 0.5, 0)$, $(0.15, 0.2, 0.2)$, $(0.5, 0.1, 0.4)$, $(0.8, 0.7, 0.1)$ for $i = 1, \ldots, 4$, respectively. The initial states of predictors $(\hat{\boldsymbol{x}}_i(0), \hat{\boldsymbol{y}}_i(0), \hat{\boldsymbol{\theta}}_i(0))$ are set as $(0, 0, 0)$ for $t \in [-\ell, 0]$ and $i = 1, \ldots, 4$.

Figure 3 shows the trajectories of four mobile robots, and the start positions of the robots are marked by asterisks. From this figure, we know that robot 1 keeps still and the other three robots reach the delayed relative positions. The square plotted by deep green dashed lines denotes the formation pattern that the robots finally achieved

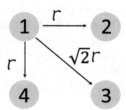

Fig. 2 Network topology $\mathscr{G}$ and formation pattern for four mobile robots

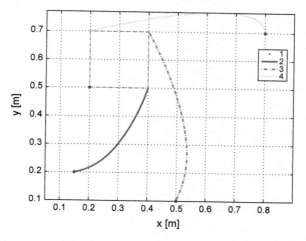

Fig. 3 Trajectories of robots 1, 2, 3 *and* 4

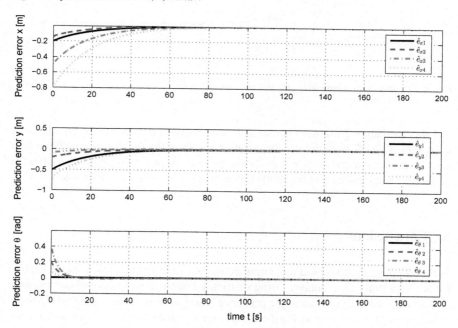

Fig. 4 Prediction errors $\hat{e}_{\theta i}(t)$ and $\hat{e}(t)$ for each robot

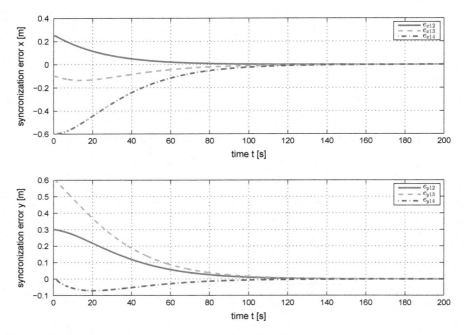

Fig. 5 Synchronization error $e(t)$ of states x, y for each robot

and this formation pattern coincides with the desired one. As can be directly seen from Figs. 4 and 5, both the synchronization errors of the outputs x, y of the robots and the prediction errors of x, y, θ converge to 0. As a result, time delay effect is perfectly compensated and the multi-robot system can achieve consensus.

5 Conclusions

In this chapter, we discussed the output consensus problem for two-wheeled mobile robots with input time delays. Using distributed control, we proposed a consensus controller combined with an angle predictor based on synchronization and the feedback linearisation technique. Furthermore, we extended this idea to a consensus controller with a full state predictor based on synchronization to perfectly compensate the effect of time delays. For these two controllers, we derived consensus conditions by reducing the problem to the stability problem of the error dynamics. Under the derived conditions, it is guaranteed that both the proposed controllers achieve consensus globally. The usefulness of the proposed controllers and the validity of stability analysis were shown by numerical simulations for a formation problem of two-wheeled mobile robots.

Appendix

Proof of Theorem 1

Considering the following nonlinear retarded cascaded system:

$$\dot{x}(t) = f_x(t, x_t) + g_{xy}(t, x_t, y_t) \tag{32a}$$

$$\dot{y}(t) = f_y(t, y_t) \tag{32b}$$

where $x \in \mathbb{R}^m$, $y \in \mathbb{R}^n$, $x_t \in \mathscr{C}([-\tau, 0], \mathbb{R}^m)$, and $y_t \in \mathscr{C}([-\tau, 0], \mathbb{R}^n)$. Assume that $f_x(t, 0) = f_y(t, 0) = g_{xy}(t, \varphi_x, 0) = 0$ for all $t \in \mathbb{R}^+$ and $\varphi_x \in \mathscr{C}([-\tau, 0], \mathbb{R}^m)$. In the absence of the coupling term $g_{xy}(t, x_t, y_t)$, system of x has the following form:

$$\dot{x}(t) = f_x(t, x_t). \tag{33}$$

The following theorem in [25] gives sufficient conditions for the global uniform asymptotic stability of the system (32).

Theorem 3 ([25], Theorem 4) *Assume that for system (33) there exists a function $V(t, x)$ of the Lyapunov–Razumikhin type which satisfies the following assumptions:*

(a) $V(t, x)$ is continuously differentiable, positive definite, and has the infinitesimal upper limit with $\|x\| \to 0$ and the infinitely great lower limit with $\|x\| \to \infty$;

(b) the time derivative of the function V, given by the functional $\dot{V}(t, \varphi_x) = \frac{\partial V}{\partial t}(t, \varphi_x(0)) + \frac{\partial V}{\partial x}(t, \varphi_x(0)) f_x(t, \varphi_x)$, satisfies the estimate $\dot{V}(t, \varphi_x) \le 0$ for all $\varphi_x \in \Omega_t(V) = \{\varphi \in \mathscr{C}(n) : \max_{-\tau \le s \le 0} V(t+s, \varphi_1(s)) \le V(t, \varphi_1(0))\}$;

(c) $|\dot{V}(t, \varphi_x)| \ge U(t, \varphi_x)$ for all $(t, \varphi_x) \in \mathbb{R}^+ \times \mathscr{C}(n)$, where the functional $U(t, \varphi_x)$ is uniformly continuous and bounded in each set of the form $\mathbb{R}^+ \times \varkappa$ with a compact set $\varkappa \subset \mathscr{C}$;

(d) the intersection of the sets $V_{\max}^{-1}(\infty, c) := \{\varphi_x \in \mathscr{C}(n) | \exists \varphi_n \to \varphi_x, t_n \to +\infty : \lim_{n \to \infty} \max_{-\tau \le s \le 0} V(t_n + s, \varphi_n(s)) = \lim_{n \to \infty} V(t_n, \varphi_n(0)) = c\}$ and $U^{-1}(\infty, 0)$ is empty with $c \ne 0$;

(e) for all $x \in \mathbb{R}^n$ such that $\|x\| > \eta$, the inequality $\|\frac{\partial V}{\partial x}\| \cdot \|x\| \le c_1 V(t, x)$ holds, and, for all $x \in \mathbb{R}^n$ such that $\|x\| \le \eta$, the estimate $\|\frac{\partial V}{\partial x}\| \le c$ is valid with certain constants $\eta, c_1, c > 0$;

(f) for $\varphi_y \in \mathscr{C}(n)$ and some continuous functions $\alpha_1, \alpha_2 : \mathbb{R} \to \mathbb{R}$, the functional g_{xy} admits the following estimate: $\|g_{xy}(t, \varphi_x, \varphi_y)\| \le (\alpha_1(\|\varphi_y\|_c) + \alpha_2(\|\varphi_y\|_c)) \|\varphi_x(0)\| \|\varphi_y\|_c$;

(g) solutions of system (32b) admit the estimate $\|y(t; t_0, \varphi_y)\| \le k_1 \|\varphi_y\|_c e^{-k_2 t}$ with certain constants $k_1, k_2 > 0$;

Then, $[x^T, y^T]^T = 0$ is a globally uniformly asymptotically stable equilibrium point of system (32).

Remark 3 From [25], we have that if the function $V(t, x)$ is quadratic in x, the bounds on its growth posed in the (e) condition are automatically satisfied and $\|\varphi_x(0)\|$ can be replaced by $\|\varphi_x\|_c$ in the estimation of the functional g_{xy} in the (f) assumption.

Proof of Theorem 1

Proof The proof is given by using Theorem 3. First, we show that there exists a function V satisfying Lyapunov–Razumikhin Theorem [26, 27] for equation (12b) without the second term,

$$\dot{e}(t) = -k\bar{Q}e(t - \ell), \tag{34}$$

satisfying assumptions (a–e) of Theorem 3.

Define a quadratic Lyapunov–Razumikhin function candidate given as $V(e(t)) = e^T(t)Pe(t)$ where P is a positive definite matrix. Using the Leibniz–Newton formula, the time derivative of $V(e(t))$ is as follows.

$$\begin{aligned}
\dot{V}(e(t)) &= 2e^T(t)P\dot{e}(t) \\
&= -2ke^T(t)P\bar{Q}e(t - \ell) \\
&= -2ke^T(t)P\bar{Q}\left(e(t) - \int_{t-\ell}^{t} \dot{e}(\theta)d\theta\right) \\
&= -2ke^T(t)P\bar{Q}\left(e(t) + \int_{t-2\ell}^{t-\ell} k\bar{Q}e(\theta)d\theta\right) \\
&= -ke^T(t)(P\bar{Q} + \bar{Q}^T P)e(t) - 2\int_{t-2\ell}^{t-\ell} k^2 e^T(t)P\bar{Q}\bar{Q}e(\theta)d\theta.
\end{aligned} \tag{35}$$

Under the Razumikhin condition $V(e(s)) < q^2 V(e(t))$, where $t - 2\ell \leq s \leq t$ for some $q > 1$, we obtain

$$\dot{V}(e(t)) < -ke^T(t)W_1 e(t) + 2k^2\ell q\sqrt{\frac{\lambda_{max}(P)}{\lambda_{min}(P)}}\|\bar{Q}\|^2\|P\|\|e(t)\|^2, \tag{36}$$

where W_1 is a positive definite matrix satisfying $P\bar{Q} + \bar{Q}^T P - W_1 \geq 0$. From Remark 2, since $-\bar{Q}$ is a Hurwitz matrix, there exists a unique symmetric matrix $P > 0$ satisfying $P\bar{Q} + \bar{Q}^T P = W_1$ for a given matrix $W_1 > 0$. Therefore, since a positive definite matrix W_1 satisfies the relation

$$\lambda_{min}(W_1)\|e(t)\|^2 \leq e^T(t)W_1 e(t) \leq \lambda_{max}(W_1)\|e(t)\|^2,$$

the inequality (36) can be rewritten as

$$\dot{V}(e(t)) \leq \left(-k\lambda_{min}(W_1) + 2k^2\ell q \sqrt{\frac{\lambda_{max}(P)}{\lambda_{min}(P)}} \|\bar{Q}\|^2 \|P\|\right) \|e(t)\|^2.$$

Consequently, if $2k\ell\sqrt{\frac{\lambda_{max}(P)}{\lambda_{min}(P)}}\|\bar{Q}\|^2\|P\| < \lambda_{min}(W_1)$ holds, we can choose $1 < q < \frac{\lambda_{min}(W_1)}{2k\|\bar{Q}\|^2\|P\|\ell}\sqrt{\frac{\lambda_{min}(P)}{\lambda_{max}(P)}}$ and the zero solution of dynamics (12a) without the second term is globally asymptotically stable. Thus $V(e(t)) = e^T(t)Pe(t)$ is a quadratic Lyapunov–Razumikhin function that satisfies (a–e) of Theorem 3 in Appendix for dynamics (12a) without the second term.

Moreover, we show the second term in dynamics (12b) satisfies the assumption (f) of Theorem 3 and Remark 3. The norm of $g(t, e, \hat{e}_\theta)$ satisfies the following inequality

$$\|g(t, e, \hat{e}_\theta)\| \leq \|M \otimes I_2\| \|\bar{A}(\hat{e}_\theta(t)) - I_{2N}\| \|L(\mathscr{G})\| \|M^+ \otimes I_2\| \|e(t-\ell)\|.$$

Since

$$\|\bar{A}(\hat{e}_\theta(t)) - I_{2N}\| = \sqrt{\lambda_{max}((\bar{A}(\hat{e}_\theta(t)) - I_{2N})^T (\bar{A}(\hat{e}_\theta(t)) - I_{2N}))}$$

and

$$\bar{A}(\hat{e}_\theta(t)) - I_{2N})^T (\bar{A}(\hat{e}_\theta(t)) - I_{2N})$$
$$= diag((A(\hat{e}_{\theta 1}(t)) - I_2)^T (A(\hat{e}_{\theta 1}(t)) - I_2), \ldots, (A(\hat{e}_{\theta N}(t)) - I_2)^T (A(\hat{e}_{\theta N}(t)) - I_2)),$$

the diagonal blocks are rewritten as

$$(A(\hat{e}_{\theta i}(t)) - I_2)^T (A(\hat{e}_{\theta i}(t)) - I_2) = (2 - 2\cos(\hat{e}_{\theta i}(t)))I_2 = 4\sin^2\left(\frac{\hat{e}_{\theta i}(t)}{2}\right)I_2.$$

Therefore,

$$\|\bar{A}(\hat{e}_\theta(t)) - I_{2N}\|^2 \leq \max\left\{4\sin^2\left(\frac{\hat{e}_{\theta i}(t)}{2}\right)\right\}$$

and we get

$$\|\bar{A}(\hat{e}_\theta(t)) - I_{2N}\| \leq \max\left\{2\left|\sin\left(\frac{\hat{e}_{\theta i}(t)}{2}\right)\right|\right\}$$

for $i = 1, \ldots, N$. Thus the inequality

$$\|g(t, e, \hat{e}_\theta)\| \leq k\|M \otimes I_2\| \|M^+ \otimes I_2\| \|L(\mathscr{G})\| \max\left\{2\left|\sin\left(\frac{\hat{e}_{\theta i}(t)}{2}\right)\right|\right\} \|e\|_c$$
$$\leq k\|M \otimes I_2\| \|M^+ \otimes I_2\| \|L(\mathscr{G})\| \max\{|\hat{e}_{\theta i}(t)|\} \|e\|_c$$

holds, since $|\sin(\frac{\hat{e}_{\theta i}(t)}{2})| \leq |\frac{\hat{e}_{\theta i}(t)}{2}|$ holds for all $\hat{e}_{\theta i}$. Thus $g(t, e, \hat{e}_\theta)$ satisfies the assumption (f) in Theorem 3.

Finally, it is obvious that dynamics (12a) is globally exponentially stable under the condition $0 < k\ell < \frac{\pi}{2}$. The solution of the dynamics is bounded by $\|\hat{e}_\theta(t)\| \leq L e^{\alpha t} \|\varphi_{\hat{e}_\theta}\|_c$ where $\alpha < 0$, L is a certain constant, and $\varphi_{\hat{e}_\theta}$ is the initial function for dynamics (12a). As a result, (g) of Theorem 3 in Appendix is fulfilled.

Therefore, we conclude that the solution of dynamics (11) is bounded and then the zero solution of dynamics (11) is globally uniformly asymptotically stable under the conditions (13) and (15) in Theorem 1.

Proof of Theorem 2

Proof At first, we consider the stability of Eq. (25). Define a quadratic Lyapunov–Razumikhin function candidate $V(E(t)) = E^T(t) P_1 E(t)$ where P_1 is a positive definite matrix. By using the Leibniz–Newton formula, we derive the time derivative of $V(E(t))$ along the trajectory of (25) as follows.

$$\begin{aligned}
\dot{V}(E(t)) &= 2E^T(t) P_1 \dot{E}(t) \\
&= -2E^T(t) P_1 (C_1 E(t) + C_2 E(t-\ell)) \\
&= -E^T(t) \left(P_1(C_1 + C_2) + (C_1 + C_2)^T P_1 \right) E(t) \\
&\quad - 2 \left(\int_{t-\ell}^{t} E^T(t) P_1 C_2 C_1 E(\theta) d\theta + \int_{t-2\ell}^{t-\ell} E^T(t) P_1 C_2 C_2 E(\theta) d\theta \right)
\end{aligned} \quad (37)$$

If $V(E(s)) < q_1^2 V(E(t))$ where $t - 2\ell \leq s \leq t$ for some $q_1 > 1$, we get

$$\dot{V}(E(t)) \leq -E^T(t) W_3 E(t) + 2q_1 \ell \sqrt{\frac{\lambda_{max}(P_1)}{\lambda_{min}(P_1)}} (\|C_1\| + \|C_2\|) \|C_2\| \|P_1\| \|E(t)\|^2 \quad (38)$$

where W_3 is a positive definite matrix satisfying $P_1(C_1 + C_2) + (C_1 + C_2)^T P_1 - W_3 \geq 0$. From Remark 2, since $-(C_1 + C_2)$ is a Hurwitz matrix, there exists a unique symmetric matrix $P_1 > 0$ satisfying $P_1(C_1 + C_2) + (C_1 + C_2)^T P_1 = W_3$ for a given symmetric matrix $W_3 > 0$. Since the positive definite matrix W_3 satisfies the following relation

$$\lambda_{min}(W_3) \|E(t)\|^2 \leq E^T(t) W_3 E(t) \leq \lambda_{max}(W_3) \|E(t)\|^2,$$

the inequality (38) can be rewritten as

$$\dot{V}(E(t)) \leq -\lambda_{min}(W_3) \|E(t)\|^2 + 2k_{p1} q_1 \ell (k\|\bar{Q}\| + k_{p1}) \sqrt{\frac{\lambda_{max}(P_1)}{\lambda_{min}(P_1)}} \|P_1\| \|E(t)\|^2. \quad (39)$$

Therefore, if $2k_{p1}\ell\sqrt{\frac{\lambda_{max}(P_1)}{\lambda_{min}(P_1)}}(k\|\bar{Q}\| + k_{p1})\|P_1\| < \lambda_{min}(W_3)$ holds, we can choose $1 < q_1 < \frac{\lambda_{min}(W_3)}{2k_{p1}\|P_1\|(k\|\bar{Q}\|+k_{p1})\ell}\sqrt{\frac{\lambda_{min}(P_1)}{\lambda_{max}(P_1)}}$ implies that the zero solution of dynamics (25) is globally asymptotically stable. Thus $V(E(t)) = E^T(t)P_1E(t)$ is a quadratic Lyapunov–Razumikhin function that satisfies Theorem 3 for dynamics (25).

Next in a similar way as the foregoing subsection, we show the last term $[g_1(t, \hat{e}_\theta, \hat{e}, e), g_2(t, \hat{e}_\theta, \hat{e}, e)]^T$ satisfies the assumption (f) of Theorem 3 and Remark 3 in Appendix. g_1 and g_2 satisfy the following inequalities.

$$\|g_1(t, \hat{e}_\theta, \hat{e}, e)\| \leq k\|M \otimes I_2\|\|L(\mathscr{G})\| \max\left\{2\left|\sin\left(\frac{\hat{e}_{\theta i}(t)}{2}\right)\right|\right\}(\|\hat{e}\|_c$$
$$+ \|M^+ \otimes I_2\|\|e\|_c)$$
$$\leq k\|M \otimes I_2\|\|L(\mathscr{G})\| \max\{|\hat{e}_{\theta i}(t)|\}(\|\hat{e}\|_c + \|M^+ \otimes I_2\|\|e\|_c)$$

$$\|g_2(t, \hat{e}_\theta, \hat{e}, e)\| \leq k\|L(\mathscr{G})\| \max\left\{2\left|\sin\left(\frac{\hat{e}_{\theta i}(t)}{2}\right)\right|\right\}(\|\hat{e}\|_c + \|M^+ \otimes I_2\|\|e\|_c)$$
$$\leq k\|L(\mathscr{G})\| \max\{|\hat{e}_{\theta i}(t)|\}(\|\hat{e}\|_c + \|M^+ \otimes I_2\|\|e\|_c).$$

Thus $g_1(t, \hat{e}_\theta, \hat{e}, e)$ and $g_2(t, \hat{e}_\theta, \hat{e}, e)$ satisfy assumption (f) in Theorem 3.

Finally, the steering angle prediction error dynamics (23c) is globally exponentially stable under the condition $0 < k_{p2}\ell < \frac{\pi}{2}$ and the solution of the dynamics is bounded by $\|\hat{e}_\theta(t)\| \leq Le^{\alpha t}\|\varphi_{\hat{e}_\theta}\|_c$ where $\alpha < 0$, L is certain constant, and $\varphi_{\hat{e}_\theta}$ is the initial function for dynamics (23c). Thus, assumption (g) in Theorem 3 is satisfied.

Therefore, the solution of dynamics (24) is bounded and the zero solution of dynamics (23) is globally uniformly asymptotically stable under conditions (26) and (28) in Theorem 2.

References

1. Ren, W., Beard, R.W: Distributed consensus in multi-vehicle cooperative control. Springer, Heidelberg (2008)
2. Listmann, K., Masalawala, M., Adamy, J.: Consensus for formation control of nonholonomic mobile robots. In: IEEE International Conference on Robotics and Automation (2009)
3. Guillet, A., Lenain, R., Thuilot, B., Martinet, P.: Adaptable robot formation control: adaptive and predictive formation control of autonomous vehicles. IEEE Robot. Autom. Mag. **21**(1), 28–39 (2014)
4. Nascimento, T.P., Moreira, A.P., Conceição, A.G.S.: Multi-robot nonlinear model predictive formation control: moving target and target absence. Robot. Auton. Syst. **61**(12), 1502–1515 (2013)
5. Wachter, L.M., Ray, L.E.: Stability of potential function formation control with communication and processing delay. In: American Control Conference, pp. 2997–3004. IEEE (2009)

6. Ferrari, S., Foderaro, G., Zhu, P., Wettergren, T.A.: Distributed optimal control of multiscale dynamical systems: a tutorial. IEEE Control Syst. **36**(2), 102–116 (2016)
7. Antonelli, G., Arrichiello, F., Chiaverini, S.: Flocking for multi-robot systems via the null-space-based behavioral control. Swarm Intell. **4**(1), 37–56 (2010)
8. Moshtagh, N., Jadbabaie, A., Daniilidis, K.: Vision-based control laws for distributed flocking of nonholonomic agents. In: IEEE International Conference on Robotics and Automation, pp. 2769–2774 (2006)
9. Dai, G.B., Liu, Y.C.: Leaderless and leader-following consensus for networked mobile manipulators with communication delays. In: IEEE Conference on Control Applications (CCA), pp. 1656–1661 (2015)
10. Cai, Y., Zhan, Q., Xi, X.: Path tracking control of a spherical mobile robot. Mech. Mach. Theory **51**, 58–73 (2012)
11. Lan, Y., Yan, G., Lin, Z.: Synthesis of distributed control of coordinated path following based on hybrid approach. IEEE Trans. Autom. Control **56**(5), 1170–1175 (2011)
12. Olfati-Saber, R., Murray, R.M.: Consensus problems in networks of agents with switching topology and time-delays. IEEE Trans. Autom. Control **49**(9), 1520–1533 (2004)
13. Olfati-Saber, R., Fax, J.A., Richard, M.: Murray: consensus and cooperation in multi-agent networked systems. Proc. IEEE **95**(1), 215–233 (2007)
14. Liu, X., Lu, W., Chen, T.: Consensus of multi-agent systems with unbounded time-varying delays. IEEE Trans. Autom. Control **55**(10), 2396–2401 (2010)
15. Sakurama, K., Nakano, K.: Average-consensus problem for networked multi-agent systems with heterogeneous time-delays. IFAC Proc. Vol. **44**(1), 2368–2375 (2011)
16. Qiao, W., Sipahi, R.: A linear time-invariant consensus dynamics with homogeneous delays: analytical study and synthesis of rightmost eigenvalues. SIAM J. Control Optim. **51**(5), 3971–3992 (2013)
17. Qiao, W., Sipahi, R.: Consensus control under communication delay in a three-robot system: design and experiments. IEEE Trans. Control Syst. Technol. **42**(2), 687–694 (2016)
18. Papachristodoulou, A., Jadbabaie, A., Munz, U.: Effects of delay in multi-agent consensus and oscillator synchronization. IEEE Trans. Autom. Control **55**(6), 1471–1477 (2010)
19. Oguchi, T., Nijmeijer, H.: Control of nonlinear systems with time-delay using state predictor based on synchronization. In: The Euromech Nonlinear Dynamics Conference (ENOC), pp. 1150–1156 (2005)
20. Kojima, K., Oguchi, T., Alvarez-Aguirre, A., Nijmeijer, H.: Predictor-based tracking control of a mobile robot with time-delays. In: IFAC Symposium on Nonlinear Control Systems, pp. 167–172 (2010)
21. Alvarez-Aguirre, A., van de Wouw, N., Oguchi, T., Nijmeijer, H.: Predictor-based remote tracking control of a mobile robot. IEEE Trans. Control Syst. Technol. **22**(6), 2087–2102 (2014)
22. Cao, Y., Oguchi, T., Verhoeckx, P.B., Nijmeijer, H.: Predictor-based consensus control of a multi-agent system with time-delays. In: IEEE Conference on Control Applications (CCA), NSW, pp. 113–118 (2015)
23. Cao, Y., Oguchi, T., Verhoeckx, P.B., Nijmeijer, H.: Consensus control for a multiagent system with time delays. Mathematical Problems in Engineering (2017). doi:10.1155/2017/4063184
24. Dimarogonas, D.V., Kyriakopoulos, K.J.: On the rendezvous problem for multiple nonholonomic agents. IEEE Trans. Autom. control **52**(5), 916–922 (2007)
25. Sedova, N.O.: The global asymptotic stability and stabilization in nonlinear cascade systems with delay. Russ. Math. **52**(11), 60–69 (2008)
26. Gu, K., Chen, J., Kharitonov, V.L.: Stability of Time-Delay Systems. Springer Science and Business Media, New York (2003)
27. Hale, J.K.: Introduction to Functional Differential Equations, vol. 99. Springer Science and Business Media, New York (1993)

Index

A

Acoustic navigation, 156, 188
Acoustic sensors, 74, 229
Actuator dynamics, 410, 417
Adaptive training, 387, 388, 397, 398, 400, 401, 403
Aerial robots, 123
Aerial vehicles, 18, 89, 90, 208, 229, 289
Anticipating synchronization, 495, 496, 500
Artificial potential field, 198
ASV control architecture, 66
Attitude observer, 6, 32
Automatic Identification System (AIS), 277
Autonomous agents, 453, 456
Autonomous driving, 437, 447–449
Autonomous ground vehicles, 183
Autonomous networks, 453
Autonomous robotic systems, 473
Autonomous Surface Vehicles (ASV), 111, 269, 407, 474
Autonomous surface vessels, 187
Autonomous Underwater Vehicle (AUV), 52, 89, 183, 187, 208, 322, 453, 454, 474, 485
Autonomous vehicles, 101, 114, 183, 192, 208, 229, 230, 407, 435, 439, 446, 449, 450
Autonomy, 51–53, 66, 67, 99, 101, 111, 113, 148, 188, 203, 208, 365, 366, 382, 437, 450, 451
Azimuth angle, 271

B

Basis functions, 416, 417
Bearing-only SLAM, 122, 127, 129, 133, 137
Bearing-only tracking, 289, 301, 302
Biologically inspired robots, 347

C

Camera, 3, 4, 6, 16, 17, 25–30, 38, 39, 47, 48, 52, 54, 60, 61, 73, 82, 84, 91, 92, 106, 107, 111, 144, 145, 148–150, 170, 175, 225, 229, 249–252, 262, 263, 265, 280, 289, 290, 292, 294, 298, 300–302, 305, 307, 309, 311, 313, 350, 351, 357, 358, 368, 371–373, 378, 446, 448
Cell decomposition, 197
Cloud access, 453, 454, 457–460, 462, 464, 465, 467, 469
Cloud-supported systems, 454
Clustering, 102, 104, 172, 243, 273, 280, 281, 286
Cluster(s), 102
COLAV, 269, 270, 286, 287
Collective firing rate, 389, 395, 397–399, 401, 403
Collective response, 388, 389, 395
Collision constraints, 212, 482, 486, 487
Combined Feedforward and Feedback (FF-FB) controller, 420, 421, 430
Communication delay, 442, 443, 496
Computer vision, 4, 73, 113

Consensus control, 476, 477, 489, 495, 505
Consensus controller, 496
Constrained systems, 480
Control of marine robots, 344
Control-oriented modeling approach, 408
Control-Oriented Vessel Model, 409
Cooperative driving, 439
Cooperative localization, 156, 157
Cooperative Path Following (CPF), 473, 474, 483–485, 489, 493
Coordinated control, 436
Coupling, 217, 230, 321, 322, 379, 389, 396, 399, 402, 403, 496, 506, 509
Cross-Validation (CV), 319, 321, 322, 332, 334, 336, 339, 416

D

Data association, 250, 251, 269, 276, 286, 287, 289, 291–293, 299, 303, 306, 310, 314
Directed graph, 199, 397, 455, 456, 497, 498
Displacement region, 409, 410
Distance keeping, 51, 53, 66
Distributed control, 508
2-D mapping, 121
3-D mapping, 121
3-D modelling, 62
Doppler Velocity Logger (DVL), 51
3D perception, 73
Driver assistance systems, 435
3D triangulation, 79
Dual, 408
Dubins path, 194, 195
Dynamic inversion, 216

E

Edge detection, 293, 296
Electronic brain, 388–391
Elevation angle, 307
Estimation, 123, 128, 129, 132, 135–138, 157, 166, 251, 264, 281, 301, 510
Event-triggered control, 465
Exogenous Kalman filter, 25, 26, 34, 48
Extended Kalman filter, 25, 48, 122, 144, 300, 301
Extended target, 287

F

Factor graph, 107, 145–147, 153, 154, 157, 158
Feedback-linearizing controller, 421

Feedforward controller, 408
Filter consistency, 283, 284
Flight test, 35
FMCW, 270
Formation control, 454, 461, 467, 468, 470, 495, 496, 499, 505
Frenet–Serret frame, 189, 190, 240
Froude number, 409

G

Geolocation, 249, 250, 252, 253
Georeferencing, 52, 290, 293, 298, 300, 303–306, 309, 311, 313
Global exponential stability, 12–14, 38, 48
Graph Laplacian, 497, 498, 505
Graph search, 191, 200, 273
Ground vehicles, 187, 188, 191

H

Hamilton–Jacobi equation, 196
Hindmarsh–Rose (HR) model neurons, 391
Hyperparameters, 416–418

I

Inequality constraints, 212–214, 216, 217, 220, 225
Inertial Measurement Unit (IMU), 148, 447, 482
Inertial navigation, 26, 320
Inertia matrix, 409, 410
Intermittent control, 93

L

Land masking, 272
Laser model, 81
Laser scanning, underwater, 74
Lasso, 416
Lateral control, 435
Level sets, 196
Levels of automation, 435, 437, 438
LiDAR, 121, 133, 269, 446, 448
Line-of-sight, 173, 252, 253, 343, 344, 436
Line-of-sight guidance, 343, 344
Linearized Kalman filter, 123
Linear regression, 415, 416
Localization, 106, 113, 124, 143, 152, 157, 168, 171, 178, 290, 454
Longitudinal control, 322
Loss function, 416
Lyapunov-Razumikhin theorem, 510

Index

M

Machine vision, 28, 35, 44, 47, 293, 299
Mahalanobis distance, 292, 304
Maneuverability, 283, 284
Marine robotics, 110–112, 114, 143, 158, 344
Marine vessels, 289
Measurement extraction, 412, 414
Minimum energy, 222, 224
Mobile robot, 186, 187, 191, 201, 387–392, 394, 496–500, 506, 508
Model Predictive Control (MPC), 238, 473, 474, 481, 493
Monocular camera, 123
Motion, 207
Motion planning, 201, 202, 207–209, 212, 215, 221, 223–225, 344, 483
Motion primitives, 198–200
Moving obstacles, 191
Multi-agent system, 454
Multi-agent coordination, 454
Multi-agent system, 470
Multi vehicle coordination, 143
Multiple target tracking, 250, 251, 253, 290–292, 299, 300, 308
Multiple vehicles, 90, 99, 101, 186, 207–209, 215, 221, 223, 224, 483

N

Navigation, 25, 26, 31, 35, 54, 62, 93, 103, 104, 106, 122, 134, 143, 148, 151, 152, 156, 157, 161, 168, 183, 188, 223, 269–271, 279, 290, 300, 312, 370, 383, 426, 481
Neural networks (NN), 387, 388
Nonidentical neurons, 388
Nonlinear control, 6, 209, 210
Nonlinear observer, 4, 25, 26, 31, 32, 48
Nonlinear systems, 6, 25, 128, 202, 216, 229–231, 234, 236, 243, 496

O

Object detection, 289–291, 293, 310–314, 448
Obstacle avoidance, 214, 486, 487
Optical flow, 25–27, 29, 47, 137, 291
Optimal control, 183, 192, 193, 198, 207–209, 211–213, 218, 219, 221, 225, 477, 483, 485
Optimization, 66, 107, 109, 123, 132, 143, 145, 147, 150, 158, 191–193, 198, 208–211, 213–217, 219–221, 223, 225, 229, 231, 236, 238, 239, 242, 243, 256, 371–373, 378, 388, 473, 474, 477, 478, 480, 483, 493
Over behavior, 450

P

Parameter mismatch, 393
Parameter vectors, 416–418
Path planning, 183–189, 191, 192, 199, 202, 203, 388, 435, 448
Path smoothing, 183, 197, 198, 201
Path-following control, 344–347, 349, 350, 353, 355, 356
Path-following controller, 348, 353, 481, 482
PDAF, 269, 270, 273–275, 286
Performance metrics, 407, 408, 423, 427, 428
Photogrammetry, 51, 52, 62, 66
Pinhole camera model, 29, 30, 301
Planning, 90, 151, 183–185, 188, 191, 199, 201–203, 207, 212, 219, 223, 344, 435, 474
Planning region, 191
Pose graph, 106–110, 143, 147, 150–152, 154, 155
Practical synchronization, 396, 399, 403
Probabilistic data association, 269
Process noise covariance, 283, 311
PRONTO, 207–209, 211, 213–216, 218, 221, 223–225, 483
Proportional-integral feedback (FB) controller, 420, 430

R

Radar, 113, 269–271
Range Data, 60
Range-and-bearing SLAM, 121, 122, 126, 127, 129, 130, 133
Range-only SLAM, 122, 126, 129, 136
Rapidly exploring random tree, 200
Rate of turn, ROT, 410
Ray-tracing, 73, 83, 85
Reachability front, 196
Recursive RANSAC, 249, 292
Reference filter, 422, 423, 431
Region matching, 28
Regularization, 367, 375, 376, 416–418
Regularization weight, 416
Remotely Operated Vehicles (ROV), 365
RGB-D camera, 121, 123, 133
Robotics, 343, 344, 436

S

Safety requirements, 435
Sampling-based motion planning, 201
Seafloor mapping, 492
Semi-displacement region, 409
Sensor-based simultaneous localization and mapping, 124
Sensor-based filter, 124, 132, 134, 136
Sensor fusion, 67, 161, 163, 164, 166
Ship hull inspection, 143, 144, 150–152, 158
SIFT, 28, 29, 39, 148, 291
Simultaneous Localization And Mapping (SLAM), 74, 122, 139, 144
Speed Over Ground (SOG), 277, 409, 410
Spiking Neural Networks (SNN), 387, 388
State estimation, 34, 37, 143, 278, 287, 292, 301
State predictor, 495, 496, 502, 508
State-lattice graph, 199
Structured light, 74
Synchronization, 107, 300, 398–400, 404, 436, 496, 500–503, 508

T

Target tracking, 250, 253, 269, 273, 284, 290, 292, 304, 308, 310, 408
Telemetron, 270, 280, 283–285, 408, 411, 423, 431
Terrain-based navigation, 223, 224
Terrain-following, 188
Thermal images, 289, 293, 299, 310, 314
Thresholding, 116, 150
Time delay, 176, 254, 422, 426, 430, 450, 495, 496, 499, 501, 502, 504, 505, 508

Track continuity, 286, 287
Track Fusion, 251, 261, 264–266
Track management, 270
Track-to-track association, 249–251, 256, 257, 263, 264, 266
Translational motion observer, 33
TRIAD, 33
Two-wheel mobile robots, 495, 496

U

Underwater mapping, 74
Underwater snake robots, 343–345, 353, 362
Unknown disturbances, 26
Unmanned Aerial Vehicles (UAVs), 114, 208, 224, 314, 436
Unmanned air vehicle, 249
Unmanned underwater vehicles, 387
Unmanned aircraft systems, 188

V

Vehicle dynamic model, 184, 191, 198, 203
Vehicle environment, 114
Vehicle state estimation, 185
Vessel detection, 293
Visual mapping, 51, 52
Visual tracking, 108

W

World modeling, 448, 450

Z

Zeno behavior, 465, 467

CPSIA information can be obtained
at www.ICGtesting.com
Printed in the USA
LVOW02*0705020617
536712LV00002B/5/P